Science of Science

How Present Reacts When You Take Future in Your Hands

Vipin Gupta

All rights reserved.

All rights reserved by the author. No part of this publication may be reproduced, stored in a retrieval system, or transmitted in any form or by any means, electronic, mechanical, photocopying, recording, or otherwise, without the author's prior permission.

ISBN:

Black & White Paperback

Independently Published Imprint

ISBN:

Color Hardcover

Indy Pub Imprint

First Published: 2023

Cover Design & Illustrations:

Afzal Sarif

Page Setup & Ebook:

Nivash Prabhakaran

Index:

Liya Jayabalan and Salini Kurup

Project Manager:

Bhakti Gupta

Contents

Glossary	vii
List of Figures	viii
About the Author	x
Motivation	xii
Chapter 1. Physics of Relativism	1
1.1 Theory.	3
1.2 The Scientific Theory of Reactivity and Point Relativity.	4
1.3 The General Theory of Activity and Nonlinear Relativity.	5
1.4 The Special Theory of Interactivity and Linear Relativity.	6
1.5 The Complex Theory of Proactivity and Parabolic Relativity.	7
1.6 The Simple Theory of Inactivity and Circular Relativity.	8
1.7 The Rational Theory of Deactivity and Square Relativity.	9
1.8 The Irrational Theory of Hyperactivity and Triangular Relativity.	10
1.9 The Whole Theory of Hypoactivity and Parallel Relativity.	11
1.10 The Wholesome Theory of Supreme Activity and Line Relativity.	12
1.11 The Wholesomewhole Theory of Para Activity and Primordial Relativity.	13
1.12 The Growth Theory of Primeval Activity and Primeval Relativity.	14
1.13 The Entropy Theory of Absolute Activity and Absolute Relativity.	16
1.14 The Absolutism Beyond Relativity of Theory.	19
Chapter 2. Metaphysics of Absolutism	21
2.1 Understanding Science.	23
2.1.1 Reality of Nature	29

2.1.2 Competing Theory of Nature	32
2.1.3 Complementary Theory of Observer	32
2.2 Making an Assumption.	32
2.3 Testing the Evidence.	34
2.4 Validating the Assumption-Evidence Singularity.	39
2.5 Management Implications of Scientific Paradigm.	50
2.6 Research Implications of Metaphysical Paradigm.	53
2.6.1 Time Dominates Reality in Modern Science	53
2.6.2 Two Physical Theories of Interactivity between the "S"ubject and the "O"bject	57
2.6.3 Two Metaphysical Ideas of Interactivity between the (Interactive) Past and the (Active) Future, when the Present is Reactive	60
2.6.4 Two Dynamic Doctrines Moderating Time's Proactivity with Space's Hyper-activity	66
2.6.5 Formatives for a Paradigm to Be Scientific	68
Chapter 3. Universe of Organizational Reality	**99**
3.1 Organizational Reality and the Limit Value.	101
3.2 Two Dimensions Universalizing the Present Reality as the Organizational Reality.	115
3.2.1. Three Primary Forms of Present Reality	116
3.2.2. Three Secondary Forms of Present Reality	116
3.3 Two Dimensions Unifying the Potential Reality with the Organizational Reality.	119
3.3.1. Three Primary Forms of the Potential Reality	119
3.3.2. Three Secondary Forms of the Potential Reality	122
3.4 Twelve Dimensions Differentiating the Dynamic Reality.	123
3.4.1. Risk Management Paradigm, Within Present Consciousness	123
3.4.2. Cost-Effective Paradigm, Without Present Consciousness	126
3.4.3. Formative Growth Paradigm, With Present Consciousness	134

Chapter 4. Normative Development Paradigm	**151**
4.1 Normative Development Paradigm, Within Past Consciousness.	154
4.2 The Path of Primordial Time	168
4.2.1 The Path of Action	168
4.2.2 The Path of Performer	171
4.2.3 The Path of Performing	173
4.3 The Path of Primordial Space.	183
4.3.1 Understanding Knower Deity.	184
4.3.2 Transcending Knower Deity.	187
4.4 The Path of Primordial Causation.	189
4.4.1 Understanding Manifestor Deity.	189
4.4.2 Transcending Manifestor Deity.	195
4.5 The Path of Primordial Cause.	203
4.5.1 Understanding Creator Deity	203
4.5.2 Transcending Creator Deity	206
Chapter 5. Transformative Exchange Paradigm	**223**
5.1 Transformative Exchange Paradigm, Within Present Consciousness.	228
5.2 The Path of Leadership Action.	238
5.2.1 Understanding Perpetuator Deity	238
5.2.2 Transcending Perpetuator Deity	244
5.3 The Path of Followership Action.	248
5.3.1 Understanding Destroyer Deity	249
5.3.2 Transcending Destroyer Deity	257
5.4 The Path of Entrepreneurship Action.	272
5.4.1 Understanding Illuminator Deity	272
5.4.2 Transcending Illuminator Deity	283
5.5 The Path of Management Action	287
5.5.1 Understanding Liberator Deity	290
5.5.2 Transcending Liberator Deity	292

Chapter 6. Formative Growth Paradigm	299
6.1 Formative Growth Paradigm, within Future Consciousness.	302
6.2 The Path of Organizational Action.	302
6.2.1 Understanding Super Wisher	303
6.2.2 Transcending Super Wisher	309
6.3 The Path of Citizenship Action.	314
6.3.1 Understanding Supreme Wisher	316
6.3.2 Transcending Supreme Wisher	335
6.4 The Path of Alienship Action.	368
6.4.1 Understanding Primordial Wisher	369
6.4.2 Transcending Primordial Wisher	373
6.5 The Path of Worship Action.	377
6.5.1 Understanding Para Wisher	384
6.5.2 Transcending Para Wisher	388
Acknowledgments	405
English Index	406
Hindi Index	432

Glossary

DIVINE   d = determination, I = imagination, v = virtue, I = intuition, n = natural, e = excellence

GUIDER   g = global, u = unique, i = inclusive, d = diverse, e = engagement, r = responsibility

SHEENY   s = social, h = human, e = ecological, e = economic, n = national, y = psychological

- Exo = Imaginary
- Endo = Illusionary

List of Figures

Figure #	Title	Page #
Figure 1	A Sentient Entity Reproduces Its Effect to Gravitate Its Copy with Gravitational Force	1
Figure 2	Gravitational Lensing vs. Gravitational Attraction	21
Figure 3	Addition of Momentum to Mass	25
Figure 4	Mass Attracts Itself as Something Massive	25
Figure 5	Subtraction of Velocity from the Mass for Conservation of Energy with Para Massive	26
Figure 6.1	Weak Attraction Follows Gravitational Force	35
Figure 6.2	Weak Repulsion Leads to Gravitational Force	35
Figure 7	A Divisible Cell Is Hexacell. When Divided Into A Daughter Cell, It Attracts Six Double Copies of the Mother Cell To Work As Nineteen!	37
Figure 8	A Multiplied Cell Is Octacell, Repelling a Multipliable Cell That Double Copies It with Two Cells for Networking Sixteen Cells	38
Figure 9	A Daughter Cell Is a Fertilized Cell, Crossing Four Cells as The Hundred Cells by Fertilizing, as a Fertilizing Magnet, the Six-Fold Growth of Sixteen Cells with Two Cells	39
Figure 10.1	First Object: Forming Reality	54
Figure 10.2	Second Object: Norming Reality	55
Figure 10.3	Third Object: Transforming Reality	55
Figure 11.1	Newton's Theory of Reactivity	59
Figure 11.2	Einstein's Theory of Activity	59
Figure 12	Subject As An Observer	61
Figure 13.1	Quantum Mechanics of Interactivity	64
Figure 13.2	Quantum Loop of Proactivity	65
Figure 14	The Great Attractor	83

List of Figures

Figure 15	Double Slit Experiment with Photon: The Nonlinear Wave Pattern with Four Patterns	91
Figure 16	Theory of Two Bright Spots with Two Slits	92
Figure 17	Single-Slit Experiment: The Linear Particle Pattern with One Pattern	93
Figure 18	Double-slit Experiment With Grain: Linear Particle Pattern with Two Patterns	93
Figure 19	Double-slit Experiment with Detector On: The Linear Particle Pattern with Two Patterns	94
Figure 20	Double-slit Experiment with Detector Off: The Nonlinear Wave Pattern with Four Patterns	95
Figure 21	Quantum Eraser: Double-slit Experiment with Two Detectors Off	96
Figure 22	Wave Eraser: Double-slit Experiment with Two Detectors On	97
Figure 23	A Blue-shifted, Luminous, Gravitational Arc	99
Figure 24	Observing Three to Image Reality of the Past	151
Figure 25	Gravitational Macrolensing vs. Gravitational Microlensing	222

About the Author

Vipin Gupta, Ph.D., is a management professor and co-director of the Center for Global Management at the Jack H. Brown College of Business and Public Administration, California State University, San Bernardino. He has a Ph.D. from the Wharton School of the University of Pennsylvania. He has been a gold medalist for outstanding academic performance in the Indian Institute of Management post-graduate program, Ahmedabad; a top rank holder in the B.Com. *(Hons) Program of Delhi University from Sri Ram College of Commerce; and an all-India rank holder at the Institute of Cost and Works Accountants of India graduate program.*

Vipin Gupta was previously at Simmons University, Boston, Grand Valley State University, and Fordham University. He has offered several training programs and workshops on strategic planning and cross-cultural management to senior executives, administrators, defense personnel, and research methods to doctoral students and faculty in India and the US. He has been a visiting or guest faculty at more than thirty business schools in India. Several leading national and regional newspapers and television channels have covered his workshops and lectures.

Professor Gupta has authored more than 250 journal articles and book chapters, including in leading journals such as the Journal of Business Venturing, Family Business Review, Research in Organizational Behavior, Asia-Pacific Journal of Management, Multinational Business Review, Journal of World Business, Advances in Global Leadership, and Management Review. Besides delivering lectures and keynotes in several nations, he has presented at international academic conferences in more than sixty nations, including the Academy of Management, IFSAM, EGOS, Society of Industrial Organization Psychologists, Global Entrepreneurship Conference, and Family Enterprise Research Conference. He has been on the governing boards and organizing committees of several international conferences. In 2017, he served as the academic program chair for the 52nd CLADEA Assembly.

About the Author

Dr. Gupta is the co-editor of the seminal GLOBE (*Global Leadership and Organizational Behavior Effectiveness Program*) book Culture, Leadership, and Organizations -The GLOBE Study of 62 Societies (*Sage Publications*, 2004). He is the principal investigator of the path-breaking CASE (*Culturally-sensitive Assessment Systems and Education*) Project on family businesses. He edited two critically acclaimed books on strategic management, performance, and leadership in emerging markets: Creating Performing Organizations (*Sage Publications*, 2003) and Transformative Organizations (*Sage Publications*, 2004). He is the author of the textbook, Strategic Management and Business Policy: Concepts and Applications (*PHI Learning*, 2003 and 2005). He has been the principal editor of ten books on business models in ten different regional clusters and the eleventh book on family businesses' gender dimension (*ICFAI University Press*, 2004). He has also co-authored the research manuscript MNC Subsidiaries in China: An empirical study of growth and development strategy (*Information Age Publishing; 2015*) and a textbook, Leadership Across the Globe (*Routledge USA*, 2015). Besides, he self-published twelve original books on the multidimensional reality of Mother Nature in 2021 and 2022.

Vipin Gupta has received the coveted 2005 Scott Myers Award for Applied Research in the Workplace from Society for Industrial Organization Psychologists, USA. As a 2015-16 American Council of Education fellow, he visited sixty-two universities, colleges, and higher education institutions in nine European nations, the USA, and India.

Motivation

Project Bhakti

Science of Science is the second in a series of six investigations with a thesis "Boosting Human Agility Kicks Totipotent Impact" (BHAKTI). Bhakti, i.e., Devotion, is the antithesis of the belief system that a transcendental makes you superior. With devotion, you invite—if not force—the transcendental to be immanent within you. When you radiate the transcendental's energy in descending motion until the entropy of "e-go," your perception of me as inferior changes. As I irradiate energy in ascending motion, I enjoy par excellence with the transcendental who transcends both of us because it is our past. As you countertrade my energy with your devotional intensity, we unite into a system self-projecting totipotency. Totipotency is the potential of the transcendental within us that makes both of us excellent.

You are the future that is active because of your devotion to my well-being. Your action produces a reaction within me. When you take the future in your hands, I react with joy to follow the path of devotion for the endoproduction of oneness with you so that I may enjoy the fruit of your action. I am the present that relates with you because you are the solution to the problem I face when the past expects me to fulfill the wish it carries into the future with my mediation. The past is a problem because it interacts with your action, knowing that it is omnipresent within me. The past leads the path of knowing as a worker so that I, as a knower, may activate you as the manifestor of the goal of my life—to be present within your potential for potentiating a destiny of eternity. The past interacts to amplify the effect of knowing the time's wish to be present within the space. Time is proactive because the space is inactive until the time's wish potentiates the space's potential to be the origin of one. The condition for potentiation is the time's proactive path of action that makes time's future active, present reactive, and past interactive to compensate for the space's inactivity with one—the causation for the present's path of devotion. The causation is the endoproduction of the past's path of knowing, seeking to know what the present knows, i.e., devotion drives the future's path of divinity that makes time divine once it proactively conceives its future for realizing excellence before deviating from space as spacetime with a wish for oneness with nature.

Project Bhakti is the secondary phase of my effort to illuminate new metaphysics for guiding the future of science in light of the limitations of correctly valuing energy. The primary phase, Project Vipin, was an investigation into the Vastly Integrated Processes Inside Nature to understand my nature as a way to illuminate the reality of Mother Nature. The findings of the primary phase are in twelve books.

Twelve Books in Project VIPIN
1. What is divine energy
2. What is present reality
3. Is present reality
4. Is divine energy
5. What is consciousness
6. What is para-consciousness
7. What is self-awareness
8. What is human factor
9. What is trading factor
10. What is cultural factor
11. What is exchange factor
12. What is technological growth

What's the Point? The point is to highlight the power of management as a discipline for addressing the grand challenges of today which are becoming grander as science is progressing by leaps and bounds.

Thesis. Taking an ontological approach, I begin with the *thesis* that everything, everybody, and everyone is transformable into energy. This entire manuscript is guided by self-substantiating mathematics. For the ease of readers, the initial part of the manuscript skips the values of each word. In the later part of the manuscript, the values of each word are included and where necessary, words from other languages are included to clarify the concepts and their link with the received wisdom from around the world with a focus on India. Those who love mathematics will discover that all sentences in the manuscript are equations and the energy values of each number, letter, word, or group of words are constants transformable into variables for the origin of the original element. For instance,

> "Transformative reality" (=63) of the "present-potential exchange" (=63) is "gravitational lensing" (=63) of "feminine element" (=37) whose "femininity" (=1,000 = [63+37] +[296+16-12] +600) "switches" (=16) the "initial state" (=−12) of the "intrinsic potential" (=−12) of the "masculine" (=296) "trading" (=53) "masculinity" (=53) from the "preceding lifetime" (=600) into "consciousness" (=4) with "knowledge" (=600) of the "temporal cost" (=600) of the "heterocyclic ring" (=16) that "forms" (=100,000 = 1000 * [53+47]) an "infinitely long surface" (=16) for "servicing" (=47) the "reproductive force" (=100) to the "present" (=1,600).

The thesis has three goals.

Primary goal. The primary goal is to clarify the sequence of forms energy takes and equate those forms with the empirical findings of modern science about the workings of nature.

- $e_1 = m^2 c$ [Point Potential energy]
- $e_2 = mc^2$ [Rectilinear Kinetic energy]
- $e_3 = mc$ [Linear Dynamic energy]
- $e_4 = c$ [Spherical energy]
- $e_5 = m$ [pi energy]
- $e_6 = m^2$ [Geodesic energy]
- $e_7 = c^2$ [Spatial energy]

- $e_8 = c^3$ [Temporal energy]
- $e_9 = m^3$ [Triangular energy]
- $e_{10} = 0$ [Metric energy]
- $e_{11} = 1$ [Point energy]
- $e_{12} = 19$ [Pointed energy]

> **How Time Makes Its Reality Self-Luminous**
>
> The present repels itself electrically as the past. The past is anybody the future wishes the present to be. The future attracts itself magnetically as the present. The present is somebody wishing the future to be better after repelling its limitation to the past. The past reproduces itself gravitationally as the future with time. The future is nobody for whom the past is wishable because it greets time as a body with a consciousness of the goal to fulfill the wish it is wishing as a Wisher. Time produces the consciousness of space when it ascends thermodynamically as a part of everybody. The space becomes conscious of itself as everybody when it descends every element for breeding an entity with time as the body of consciousness. With the growth of an entity, time ascends while space descends. The potential for the growth of an entity is immanent within the space. The origin of the potential emanates with time without space. The continuity of space as time is imaginary. In the present, time acts to speed the light of the imaginary for union with space eventually. The union is illusionary. As the past, time interacts with the mass of illusionary by imagining its future to be heavy. As the future, time reacts with the consciousness of the real by intuiting its past to be light. The past is light because the space for mass is the time's past.
>
> 1. **Future's point reactivity.** When the present acts as if it is the mass of the past, the future reacts to reproduce the mass that discharges its charge at the speed of light to recharge the present with the reproductive force of its reaction at the point of the action itself. Recharge gives intrinsic strength to the present, which is real even though the initial action is illusionary and the eventual reaction is imaginary.

The energy of the system = m, which is reproductive x c, which recharges m for it to be reproductive

Approximates reality only if due to intrinsic weakness, the future does not react with the same mass because otherwise no reason for the present to act since the effect of the present's action is zero. In other words, when gravitational force = 0.

2. **Present's nonlinear activity.** When the future recharges the present with the mass of the past, the past interacts at the speed of light for trading the mass from the future and servicing that to the present. The past trades the mass because its position is at the bottom. It services the mass to the present because the future gives momentum to the present to move forward with its quantum-effect, while the past tails with its mass that the present assumes as its.

The energy of the system = m of the past x c the past trades x c the past services

Approximates reality only if the past interacts with the same mass as that of the present because otherwise no reason for the future to react as if it is different on the top at the edge of infinity. In other words, when gravitational force = infinity.

3. **Past's linear interactivity.** When the past trades the mass from the future, the time moves forward proactively at the speed of light for that mass to be present when it heads to the future. Since mc is heavy like a head, the future makes it reproductive so that m becomes light like a tail, while the reproductive force adds speed to the light.

The energy of the system = mc, before the past comes to light after the present moves forward with the future.

Approximates reality only if time proacts so that the mass of the present dominates the mass of the past. The future decides why the former is twice the latter because the future self-perpetuates the past to be present. In other words, when gravitational force = ½.

4. **Time's circular proactivity.** When time moves proactively to lighten its heaviness, the mass of time's past moves forward the circle of time at a speed of light that charges itself into a sphere of space.

   The energy of the system = c.

   Approximates reality only when the space inactivates the mass which the time adds to triple itself, forcing the space to square itself. In other words, when gravitational force = 1. [In reality, the space is a square that produces an illusion that it is active.]

5. **Space's squared inactivity.** When the space is inactive like a sphere, the sphere squares itself because it discharges its mass to speed up light. Once the mass is zero, the light's speed is zero as well. Mass charges the speed of light with the heaviness to be inactive.

   The energy of the system = m.

   Approximates reality only when the causation activates itself as the time which becomes massive through the endoproduction of the space. In other words, when gravitational force = 2. [In reality, the causation orders the endoreproduction of both space and time with a two for the continuity of the space as the time].

6. **Causation's triangularity.** When the space discharges its mass to speed up light, causation triangulates the charge to concentrate it as a mass at a point from where light emanates at speed to shed the heaviness.

   The energy of the system = m^2.

   Approximates reality only when the entity is active first as the space, then as the time, and finally as the causation for the endoreproduction of both space and time as the spacetime. In other words, when gravitational force = 3. [In reality, the entity becomes conscious of space with time by producing an illusion of the consciousness that lives within her potential.]

7. **Cause's parallelism.** When the causation concentrates the charge at a point as mass, it causes the mass to be reproductive in the form of charge reproducing itself as a parallel that is concurrently discharging and recharging itself as a system at the speed of light.

   The energy of the system $= c^2$.

   Approximates reality only when the potential for the entity exists within a para entity, which is present before everyone as a para deity. In other words, when gravitational force = 5. [In reality, the para deity can't be present without the param deity].

8. **Entity's triangularism.** When the charge is reproductive because it is trading and servicing itself at the speed of light for recharging and discharging, the charging remains constant. Without exchanging the mass for charging, there can't be a recharging of the present and its discharging to load up the past of the present before the future arrives.

   The energy of the system $= c^3$.

   Approximates reality only when the para deity is repeated as the base for everyone. In other words, when gravitational force = 10, which makes a para deity the base of everyone's development. [In reality, the base self-perpetuates the para deity for the endoreproduction of three with a two].

9. **Para entity's squarism.** A para entity squares the mass for exchanging it as its charge that an entity may reproduce as its before triangulating it as the mass of her past as a para entity

   The energy of the system $= m^3$.

   Approximates reality only when everyone is heavier than the one triangulating itself to be a para deity omnipresent within everyone's consciousness as the causation for their existence. In other words, when gravitational force = 16, which makes one light. [In reality, as a deity, one is the essence of the growth of light through the entropy of heavy. Heavy precedes light].

10. **Deity's linearism.** A deity reproduces itself to be the one present as a para entity with the potential to produce itself as an entity who observes her future from the lens of the past before his energy was zero.

The energy of the system = 0.

Approximates reality only when one is the causation for everyone to be zero. In other words, when gravitational force = -1, which produces one as light emanating from the past. [In reality, light emanates from the future because the past is dark and the present's consciousness becomes heavy when an entity becomes conscious of the possibility of incarnating over time in diverse forms as a deity].

11. **Para deity's Pointism.** A para deity behaves like a zero to observe his future with the eyes of the one present as a deity who sees his past to be the same as that of the para deity because he is the holy spirit of the father no longer alive.

The energy of the system = 1.

Approximates reality only when zero precedes one. In other words, when gravitational force = 22/7. [In reality, one is the remainder when the potential copies the reality of the present as its future to be a zero].

12. **Param deity's Absolutism.** A param deity reproduces the deity as a para deity to perpetuate his past as the one present within nine with the potential of eighteen divided for the growth of three.

The energy of the system = 19.

Approximates reality only when eighteen produces a growth of three by triangulating itself with a two. In other words, when gravitational force = 100. [In reality, the growth of three is immanent within sixteen that twins eight as its essential nature to be reproductive as the present].

Secondary goal. The secondary goal is to clarify the consequence of energy's endoproduction in diverse forms. Specifically, three must be applied as a correction factor to know how the initial state (initial spatial state) becomes a transformable (tensor) with time as the multiplier with its past, present, and future.

- $E_1 = 3\ m^2c = -12 =$ Initial state = Density Tensor
- $E_2 = 3\ mc^2 = 6 =$ Perfect Tensor = Spherical Tensor
- $E_3 = 3\ mc = -6 =$ Scale-effect = Pi Tensor
- $E_4 = 3\ c = -3 =$ Time tensor = Geodesic force
- $E_5 = m = 2 =$ Space tensor = Universe
- $E_6 = m^2 = 4 =$ Cause tensor = Omega
- $E_7 = c^2 = 1 =$ Circular tensor = Space curvature
- $E_8 = c^3 = -1 =$ Metric tensor = Energy-momentum tensor = Cosmological constant
- $E_9 = m^3 = 8 =$ Triangular tensor = Energy tensor
- $E_{10} = 0 =$ Parallel tensor = Eventual State
- $E_{11} = 3*1 =$ Line tensor = Einstein gravitational constant
- $E_{12} = 3*19 - 4 = 53 =$ Point tensor = Individual [Free from one who double copies the relative-effect of three to differentiate the individual and integrate the collective]
- $E_{13} = 264 =$ Individual tensor = Emanation value
- $E_{14} = 9 =$ Collective tensor = Thing

Tertiary goal. The tertiary goal is to clarify the causal element using metaphysics to investigate the validity of two epistemological approaches, guided by two contrasting belief systems.

1) Eastern belief system. God is immanent within One. One is the image of a deity who incarnates in diverse forms through the exoproduction of a twin. One's consciousness creates the reality of everything as a body with one's divinity. Twin's nature is to self-perpetuate divinity with a triple copy of

energy of one as the goal of creation. A triple copy emanates from Twin's Energy.

2) Western belief system. One emanates from God. One is the image of God, the para deity, which takes diverse forms as an observer through the endoproduction of the Holy Spirit, omnipresent within the Father. Father triple copies the essence as energy. Triple copy is immanent within energy.

One is *essential* to both belief systems because it is the triangulation-effect of two that triangulates itself with a triple copy for the natural growth of six within a goalkeeper.

Modern science is concluding that one is All that exists as everything through the observer's endoproduction, although a cosmological constant is essential to account for its observance. An observer is like a devil who has a purpose. The cosmological constant is like a Satan who neutralizes that purpose with one's positive energy to nudge the devil to let go negative energy of his anger.

Project BHAKTI comprises the following six books.

1. Science of Goalkeeper
2. Science of Science
3. Science of Religion
4. Science of Spirituality
5. Science of Divinity
6. Science of Infinity

The science of Goalkeeper is the Science of Reality. Science modifies the consciousness of reality to make one omniscient. Religion makes one conscious of the illusionary reality relative to another who is creating an illusion of modification to be absolute. Spirituality positions the imaginary reality reproduced by another over the reality one produces to be omnipresent. Divinity produces the quantum reality of another within one who wishes to transcend the present with another's potential. Infinity transcends the present to let one enjoy the reality of the potential to be another.

Science of science quantifies scientific facts. The criterion for substantiating science must be rigorous: the effect of the thesis substantiated with the empirically-verifiable quantified values grounded in simple mathematics, without the complexity of assumptions guided by illusionary ideas and imaginary theories.

Using two different approaches, modern scientists have concluded that energy in the cosmos is zero. The *quantum science approach* says that the negative energy one repels is superpositioned over the positive energy one eventually attracts. The (net) energy is zero. The *relativity approach* says that the energy within the photon at rest is zero. Before the photon comes into existence as the first particle for illuminating light, the energy is zero. After that, energy remains conserved at zero, relative to the photon that gains kinetic mass as it curves light by moving linearly with zero rest mass.

In the case of quantum science, one repels negative energy when one is local to open space within one for enjoying positive energy by becoming global. Therefore, it has greater precision at the local, micro level. In the case of relativity, at the local level, the energy within one remains zero. One has to repel oneself first to gain mass and, therefore, positive energy, before one becomes a subject another attracts as an object for appropriating the mass one gains. As a subject is attracted like an object towards another subject, it loses mass with the growth of negative energy from the undesired development. Therefore, relativity has greater precision at the global, macro level.

According to quantum science, there is a symmetry between the positive and the negative; energy within a subject continues to be zero unless the subject as a system interacts without itself. According to relativity, momentum present within an object discontinues the symmetry with its position as a subject. The energy within the subject changes from zero to one when a subject becomes an object of manipulation by another. One is present within the oneness of the sequence of positions the entity takes with time's momentum from past to future.

Postmodern science points to quantum relativity as the third science, the substitute that integrates both sciences to be the science of mesoscopic reality. Specifically, with infinite momentum, a subject gains a finite mass like a giant. When the energy of another attracts it as an object, another becomes a supergiant with an infinite mass. Another is a zero who becomes one, letting one take its position as a zero.

If there are several dimensions of reality and different sciences for explaining each dimension, how do we understand reality as a whole for its management using the lens of science?

> **Findings: Reality Transcending Quantum Relativity**
>
> Time makes the union of ten reproductive as ten squared, i.e., a hundred, while space produces another one with a zero. After the illusionary reproduction of two ones, another becomes 100 -2 = 98. As a third one takes its position before the two, the union becomes a potential wave of 98 -12 = 86. The fourth one gives momentum to the two before it, conserving the two's transformative reality as a three for the gravitational lensing of a universal reality of 86 -23 = 63. The fifth one leads to the organization of the three which becomes four with the force within two. Two makes one reproductive for reproducing zero. The organization = 63 -34 = 29. The sixth one cascades zero-to-one followership workculture for its reappearance as 29 -45 = -16. [One that will become six after the seventh one is the subject wishing to lead the eighth one to take its position, making the eight one—the entity—a follower].
>
> The succeeding one assumes the reality of the preceding one as the former becomes superior after emanating from the one that continues to be immanent as inferior. The transcendental keeps descending because it is the other. It makes the position of the third reproductive with the reproductive force of the momentum of the four. The other complements another by becoming $3^2/4^2 = 9/16$.
>
> **Implications.** As a manager gets promoted to be a leader managing an organization, a managed becomes the manager. The managed is inferior but becomes superior as a manager. The leader transcends the duality of inferior and superior because by making a superior the manager, the leader ensures that each

managed will become superior in the eventual state where everyone is a copy of the leader. It is false to assume that the manager is superior. The assumption is true only if the superior was the manager before the manager is superior. The evidence-based science that relies on everyone's experience to discern reality as a common denominator considers the manager as superior and the managed as inferior. It leads to the conclusion that the experience of time makes everyone superior through normative programming of reality. The managed develops into a manager before becoming zero like a leader who attracts everyone after repelling itself as a star of attraction to order a universe.

However, if the leader lacks gravitational quality par excellence, the future makes all a zero. The fifth leads their organization as a five that becomes zero in continuity of the sixth's followership workculture for the natural growth of six. Until five becomes zero, it produces infinite Gods with six as a foundation for itself to be a whole. Therefore, one does not need space and time to be whole. Five has limited time to be whole for absolute freedom from the entanglement with the real. The deadline is the demise of one as a leader that makes five a God for the endoproduction of infinite Gods who follow the five's perpetuating value after five's entropy.

To recap, modern science is based on two types of science: quantum science at the microscopic level and relativity at the macroscopic level.

- According to **quantum science**, energy $= 0$, because the negative energy of the mass radiated over the infinity is superpositioned over the positive energy of the mass irradiated from the infinity. Therefore, Rest mass $=$ Electric mass $= +\infty$. Magnetic mass $= -\infty$. Electromagnetic mass $=$ Internal energy of space within the subject which has become an object $= 0$.

- According to **relativity**, Photon $=$ hc/L, where h $=$ Planck's constant, c $=$ speed of light, and L $=$ wavelength. The "h" endoproduces a photon at the speed of light whose wavelength is limited to the speed of light. The "h" as the causation has zero energy at rest. It endoproduces the energy of one photon while in motion. Zero becomes one with the endoproduction of energy subject to the uncertainty that h may not remain constant at 1 and instead converge to 0 after servicing energy for the photon's exoproduction. Therefore, Rest mass (m) = Electric mass = 0. Since m = 0, e = mc^2 = 0. Photon has zero energy but orders the system with its presence relative to someone, none other than the causation omnipresent as h, the one present. Kinetic mass = Electromagnetic mass = Net mass = External energy of time within the subject before becoming an object = 1. Magnetic mass = 0.

When the magnetic mass is relative to electric mass, potential magnetic mass = Magnetic mass/ Electric mass = $-\infty / +\infty$ = -1. After repelling the potential, magnetic mass = 0. Potential repelled within the magnetic mass and eventually attracted as the electric mass = Electromagnetic mass = 1. After attracting the potential magnetic mass, electric mass = 0. *Therefore, quantum relativity is the substitute for the two sciences at the mesoscopic level.*

In the case of quantum relativity, at the local level, a subject is an object attracted to another. At the global level, the subject is attracting another as a giant to position itself as a supergiant after transforming another who is behaving like an object into a zero. Two is the rest mass and the value of energy. We observe the subject at neither the formative phase when it is globalizing its effect nor the transformative phase when it localizing its effect, but at the normative phase. In the normative phase, the subject is like a nation of two citizens, one who embraced citizenship to attract another—the alien repelling away to be a nation itself. Quantum relativity has greater precision at the national, meso phase of quantum entanglement as it norms a quantum system.

- According to **quantum relativity**, the zero as a subject is the invisible hand of God. Zero mediates everything until its entropy. Reality is subjective and varies as a function of the energy left within the subject after the addition of one from the right to the potential of zero. The potential of zero within the subject transforms into the present of one without the subject. One is observable by the subject within oneness with the objects endoproduced by forming one into an objective.

Quantum Relativity leaves irreducible uncertainty when explaining reality because the subject endoproduces infinite objects for realizing its reality but the endoproduced volume varies by time within a subject. A subject may transcend the limitations of time and realize equitable excellence as a citizen. An effective institution is a nation of citizenship system, where one included in a decision repels its uniqueness for globalizing its impact and making a difference. Consequently, a diverse one excluded enjoys an opportunity for equitable excellence by taking up the challenge of designing an integrative, meta-perspective. Excluded grow when included are the catalyst of growth. Those who take up the responsibility to grow are included in the decision to globalize the integrative, meta-perspective. In effect, no one is excluded from the decision system.

When everyone is included, no one takes a position over another. The quantum effect is zero. When the quantum effect is zero, no one has relative power over another. Everyone's power is absolute. Everyone has the freedom to globalize perspective by taking responsibility for integrating diverse perspectives and bringing their uniqueness to that integration. We may refer to this approach as "quantum absolutism." Even though there is zero quantum effect of one over another, one's absolutism is superpositioned over the absolutism of everyone, guided by their uniqueness and inclusion within the absolutism of another.

- **Quantum absolutism** is the international, mass phase, where one has the mass of consciousness about the weak linkages among the different citizenship systems reflected within the uniqueness of diverse members. Under this condition, the kinetic mass is infinite because no one is at inertial rest; everyone assumes responsibility for the collective universal well-being of the institution. The energy is infinite because the institution works like a high-energy system, optimizing the unique talent of its members for maximum possible growth. The magnetic mass attracted by each member is finite, ensuring a minimum possible growth baseline for everyone.

The asymmetry between the kinetic mass that reproduces the magnetic mass and the magnetic mass that becomes reproductive for symmetry with the kinetic mass points to a Reality within Reality. Specifically, subjective optimization is not sufficient and sustainable without optimizing each region in the world. Unless the subject exchanges its uniqueness with everyone and takes responsibility for integrating everyone's uniqueness, its development is not sustainable. When the subject is committed to everyone's par excellence, there is no question of securing a superposition through the quantum element. When the subject is right, what's left is the "absolutism" approach characterized by strong linkages with everyone.

- **Absolutism** is the regional, meta phase, where the magnetic mass is infinite, the electric mass is zero, and the energy is finite. When one is open to everyone's uniqueness, one's finite energy attracts infinite mass for limitless development.

Quantum Absolutism complements Quantum Relativity with Relativity. Absolutism supplements Quantum Absolutism with Quantum. Reality without Reality transcends Absolutism with its Potentialism. It is free of both Relativity and Quantum produced by Absolutism.

- **Potentialism** is the reality of the regionalizing, physical phase. It leaves the subject as an observer of the right without any left linkages curving the observance of the one's oneness on the right into regions. Regions form at the infinity point of the regionalizing phase. Consequently, Electromagnetic mass (Kinetic mass) = 0. Electric mass (Rest mass) = 1.

Why? Because electric mass is radiated as one's oneness and irradiated as oneness without the potential left with the magnetic mass. Magnetic mass = 48 + (48 -18) = 78.

48 = "Oneness" of four dimensions with the eight dimensions including four potential dimensions.

18 = "Potential" for oneness within eight dimensions without the four dimensions repelled with the electric effect.

Four dimensions include three dimensions of space repelled with the one dimension of time forward and then potentially attracted backward. Reality is not subjective, although a subject has the potential to mediate reality with its entropy after it orders one as the causation with its consciousness.

Reality = 7, which is attracted with 7 +1.

Consciousness = 4, which doubles after the entropy of another subject attracts the third subject.

Entropy = 4 +1 = 5.

Order = 7 -5 = 2.

Each subject seeks to reform the organizational reality with its normative development to be left as the right in the center. **Centrism** is immanent within Potentialism and produces Absolutism to order the "convergent energy" of Reality with the Reality of its body of consciousness.

- **Polycentrism** emanates from Centrism as the Reality of Reality. It is reality. Reality becomes reproductive within the continuity reproducing the reality of the essence. The triple tangent of reality, normed by the three dimensions of time over the past, present, and future of the essence, is the universe of organizational reality. The essence begins with **Ethnocentrism** of the consciousness in the past. It continues to be converging as **Geocentrism** over the present and discontinues to be diverging as **Regionalism** into the future before verging its impact as the **Transnationalism** of time.

Chapter 1
Physics of Relativism

Figure 1. A Sentient Entity Reproduces Its Effect to Gravitate Its Copy with Gravitational Force

Does one subtract the energy of the future from the light left with the present of the space, whose mass is variable when illuminated by time right past the darkness of the initial state of mass? (*Newtonian ideal of time acceleration*)? Does one add mass to be present as energy while moving at the speed of light like time relative to zero whose mass is constant because it is the eventual state in the future (*Einsteinian theory of time dilation*)?

Neither of the two. In reality, speed radiates light in the past whose consciousness makes one massive at present for reproducing the light to be zero in the future. The past is hidden as dark. The present is light because the time's energy is reproductive. The space sources its future as heavy with the continuity of both the past and the present within the future value. An entity's consciousness orders the future value to orbit the space's spirit within one who is omnipresent as a deity.

The initial state of space is luminous as a deity with the mass of a para deity repeated when a primordial illuminator self-perpetuates itself to be whole with the growth of a primordial greeter. The goal of the primordial greeter is to be the goalkeeper with the potential within the future value. The goalkeeper is the param deity who speeds the consciousness of the deity with supra deity as a time multiplier to fulfil the goal as a devoted deity. The devoted deity is devoted to the devotee deity, the devotee to the primordial self as the primordial illuminator of the deity omnipresent within one. One is the double copy of both the goalkeeper and the goal. One is the worker who works to fulfil the goal with the energy of its entity time as divine before degeneration, decay, and death when the deity, reproductive as a guider, ascends the ego of a subject as it descends its emotion to charm the nature with its strange supernaturalism. The dependence on the para deity impedes one's independence and the togetherness of the two accelerates their entropy to illuminate the otherness of the third, who is observing them as their future for beginning a new era of prudent management that does not copy the same behavior.

How many theories of relativism are there?

For an observer of science, there are two theories of relativism—a general theory without quantum mechanics and a special theory within quantum mechanics, both credited to Albert Einstein. The former is the theory relating to the objects at a macroscopic level. Objects take a position at the "least distance" with the most gravitational force among them. The latter is the theory relating to the subjects at the microscopic level. Subjects take a position at a zero distance that divides their gravitational force as a function of precession. With precession, a subject dilates time disproportionately because its disproportionate gravitational force radiates light curvilinearly at a disproportionate cost to its mass. By servicing its gravitational force to a massless collective, an object as an individual becomes a subject who has dilated time disproportionately. By trading the gravitational force from a massive individual, a subject as a collective becomes an object radiating light linearly with the addition of speed until it becomes heavy after irradiating consciousness of light curvilinearly with the subtraction of the position of precession it once held. The massless subject precedes the massive object because the former is the future the latter correlates to be present as the dark matter. The dark matter is illuminated by the light of the space deflected by the past as a black hole. Time destroys the blackhole with its heaviness.

Time is heavy because it reflects the consciousness of distance when both objects and subjects correlate from a position of the "most distance" with the least gravitational force between them. The theory relating to the entities at the mesoscopic level is scientific because it approximates reality more closely than the general case of one massive individual or the special case of an infinity of massless collectives. The scientific theory is credited to Isaac Newton. Although scientific, this theory is not generalizable for understanding the reality of a massive individual because it assumes the gravitational force is zero when an individual relates with a collective. It is also not specialized for knowing the reality of a massless collective because it assumes that the subjects are beyond the laws of nature.

The interaction of objects, subjects, and entities as a family produces nine additional theories of relativism with a tenth theory of absolutism.

1.1 Theory.

Science is a process of servicing theory as an ideal for testing the approximity of reality with the theory-effect. The assumption guiding the ideal is the proximity of the ideal-effect and the theory-effect. The ideal-effect produces a relativity-effect which relates the reality to ten faces of theory, constituting a family of ideals that relates theory with reality. The tenth face is absolute and the eleventh face is a potential family of theory with a potential to be absolute. The twelfth face relates the absolute with the potential with the endoproduction of the relativity-effect. The thirteenth face is the absolutism beyond relativity-effect. The fourteenth face is potentialism beyond absolutism. The fifteenth face is primevalism beyond potentialism. The sixteenth face is primordialism within primevalism.

The ideal of relativism is deep-rooted within the scientific paradigm. Although the credit for the idea of relativity as a theory is usually given to Einstein, a careful analysis shows that Einstein was simply following the present paradigm previously led by Newton, who also followed those before him. The energy equation famously credited to Einstein is self-evident in Newton's energy equations. Below the value of the twelve different theories of relativity is clarified.

This introduction culminates with a clarification of the role of absolutism in knowing the present's known reality with certainty. When there is certainty about the present's known reality, it is possible to activate the potentialism of the future with one's unknown reality which has the potential to transcend the present with conscious action. When one transcends the present, it becomes possible for everyone to be ontological after experiencing the primevalism of epistemological. When everyone's effect is axiological, primordialism of the metaphysical immanent within the physical element of nature becomes self-luminous.

1.2 The Scientific Theory of Reactivity and Point Relativity.

Newton's laws constitute the scientific theory of reactivity and point relativity. The answers we get from the Newtonian picture are an excellent approximation to reality in a specific case most common in the theories of science—when distance among masses is most so that there is zero correlation among the masses due to zero gravitational-effect. In such case, $F = ma$, the force between two objects is directly proportionate to their masses and inversely proportionate to the distance which decelerates their motion leading to their freefall into space after they come to rest in the future and stop interacting in the field.

1.3 The General Theory of Activity and Nonlinear Relativity.

Einstein's work is illustrative of the general theory of activity. The answers we get from the Einsteinian picture are an excellent approximation to reality in a general case that frees the theory from the assumption of zero correlation among the two objects forced to be together by the third object through its unit correlation with both. The general case pertains to short distances among masses when the gravitational-effect is one. The third object itself produces an illusion with its motion during free fall that it is repelling the second object as its present for attracting the first object as its future. The third object relates to the object-in-motion; the past accelerates the motion leading to the merger of the motion and the motion-effect. The motion repels itself as the second object as a wormhole, which then becomes an Einstein-Rosen bridge that attracts the first object as its motion-effect to merge with the motion to form the blackhole as the fourth object. In such case, $E = mc^2$, since the blackhole becomes ultramassive when energy is directly proportionate to the speed of the light's motion. The motion was zero in the past and the space was reproducing itself as the field to interact with itself without any effect on time, thus behaving like spacetime. The idea is to make the theory an ideal like a law everyone believes to be true and faithfully follows to make science the religion of our times. Nothing can move faster than light when the thing is the object-in-motion that the space transforms into its light to twin itself into the time taken to diffuse the light and fuse the dark element within the light into matter. No signal is left that it ever existed before time's motion.

The reality is neither the most distance allowable with r=0, nor the least distance allowed with r=1, but variable distance where r may be anything but zero and one. There is never one object in the universe that takes positions relating to itself nor even three objects that determine the position of the fourth object with the momentum of two objects after knowing the force the third object is emanating to be immanent within the fifth object as the sixth object. The force is the seventh object. Knowing is the eighth object. Knower is the ninth object. Wisher who wishes to be a knower is the tenth object. Worker working to materialize the wish is the eleventh object. The manifestor manifesting the wish is the twelfth object. The creator of the wish is the thirteenth object. The perpetuator of the wish is the fourteenth object. Destroyer of the potential of the present to wish for a future is the fifteenth object. The illuminator of the wish as a potential to shape the future one desires is the sixteenth object. Liberator of the potential so that one may desire the past to be present as the future after experiencing it is the seventeenth object. The Devoted, experiencing the reality one desires with devotion of energy, knowing that such devotion will lead to materialization of the consciousness of that reality as the object that makes the reality real, is the eighteenth object.

1.4 The Special Theory of Interactivity and Linear Relativity.

Quantum field theory is the special theory of interactivity. The special theory of relativity infuses the specific case into the general case to limit the theory with the idea that matter curves the space on its right before it begins the motion. The motion centers the knowing as the energy for the matter to be on the top as a knower every time it reproduces itself to be light. The light perpetuates itself as the one within energy for getting to the bottom of the truth. The answers we get from the Quantum picture are an excellent approximation to reality in a complex case when the truth is present only relative to the origin. The origin is first heavy when the matter is on the top conscious of the truth, but becomes light when the matter is at the bottom in the form of the consciousness of the truth.

Heavy is imaginary because as an object the matter can't be conscious until it subjects itself to the pressure the bottom faces and begins its time as a subject in the future when the space stops behaving like an object from the past that never materialized and was left as light. Light is real because it was present in the past as an object before realizing the presence of the subject illuminating the light as an illuminator. However, the complex is illusionary; it is the origin of consciousness that one has the potential to be the subject. The identity of the subject comes after one identifies itself as the object. The causation for one to self-identify as an object is an illusionary production of the belief system that one is a Holy Spirit that objectifies God with the consciousness of the Father. Thus, most distance dissolves within least distance when one believes in a system where God is the one who as Holy Spirit transforms his consciousness into Father. God with two norms a conscious entity which is Father on the left and Mother on the right, whose consciousness is left within Father after she moves to the right with the Holy Spirit as God.

1.5 The Complex Theory of Proactivity and Parabolic Relativity.

The Quantum loop is the complex theory of relativity. The complex theory of relativity diffuses the general case as a specific case with a thesis of equivalence principle to free the idea on the top from the limit the theory faces at the bottom. The thesis asserts that the solution to an alternative belief system is the scientist ascending acceleration through gravitation as a guider. The guider makes his paternal consciousness reproductive with his gravitational energy, superpositioned over the subjects who believe only in themselves.

When a scientist repels his copies with his gravitational energy, he attracts everyone into the wormhole to be captured by the Holy Spirit. The Holy Spirit takes them across the Einstein-Rosen bridge for them to witness God in the strange after space to free their charm for before space. The space curved by his energy informs everyone's motion so that they matter for the well-being of the space they left with theory to father the consciousness of those who follow their path as a child. Thus, the most distance is the solution for the least distance that makes everyone fall in line as a body of consciousness. The answers we get from the Simple picture are an excellent approximation to reality in a simple case when the potential for the truth of the present exists within beauty itself whose illusionary production makes one the origin of its future as the Father, mediated by its past as God.

1.6 The Simple Theory of Inactivity and Circular Relativity.

A simple theory of relativity refuses both the general case and the specific case together as a set to simplify the complex case through the gravitational redshift of the origin with the gravitational wave of time which dilates the space with its continuity. Space dilation produces an illusion of time dilation. Gravitational redshift makes the origin parallel to time because the mass of consciousness of the system lowers the origin of the space to loop the upper as the infinity. The infinity surfaces the space after the space repels itself as the time to attract a nonlinear wave. The nonlinear wave is reproductive like a gravitational wave because it reproduces itself as a linear wave of particles. Reproductive quality is inherent to the particle because it is a nonlinear wave's potential to be primeval space that breeds primordial space as it feeds itself as the absolute time to freeride as spacetime. However, the potential is not perfect because of its nature to be present as a possibility within a theory with inherent law of limitation imposed by the general case on the specific case.

The specific case simplifies the complex case to make itself an almost perfect rational case. Thus, the least distance resolves the most distance through the addition of the indeterminism to the position of a corporate element after subtraction of the certainty of momentum towards the national element. Distance from the international element impedes the sentient well-being of a citizen who localizes the theory as his ideal by minimizing the distance from the global. The answers we get from the Rational picture are an excellent approximation to reality in a nonrational case when the truth is present as the beauty to hide its harsh reality with a charm that is strange because it is not true but the effect of presenting a thesis as true to automate the antithesis as false. Thus, synthesis becomes the mate. The Holy Spirit reproduces itself as the Father to produce the togetherness of the two within the three, who altogether norms God as a five. A wisher consumes the altogetherness of five with its otherness. A collective gains an upper hand over the individual devoted to a collective. The individual has a lower hand because it begs the collective to bless it with its knowing so that it has the same energy for reproducing itself as God within the consciousness of everyone.

1.7 The Rational Theory of Deactivity and Square Relativity.

A rational theory of relativity confuses both the general and the specific case with the rational case that distances itself from the irrational case. The combination of three cases—general, specific, and rational—squares the irrational case as the fourth point. The fourth point is constant and known. As the third point, the rational case is variable and unknown until it triangulates itself with the second point to be parallel to the first point. The first point is semivariable and partially known because it lines the reality with the second point to get straight to the point. The point reproduces itself to curve its reality with a universe of points within which it lives as a force as an outcome of universalizing its potential as a whole into a wholesome point with its conjugate quantity momentum. When an observer forces its position over the outcome, the momentum of the conjugate quantity becomes variable, forming an angle that spins the reality intrinsic to the outcome into a perpendicular direction. The intrinsic spin transforms the vision to make the benefit mutual. When an observer embraces the outcome of the angular position, it races ahead with the angular momentum along the curve to circle itself as the God.

God originates the idea of irrationality to be the ideal like a Holy Spirit personifying the rationality of the theory with the energy of the Father. Rational theory confuses because the measurement of the theory-effect is not exact. It is distorted by the ideal-effect. Theory-effect is zero after the Holy spirit personifies the rationality of the theory to be an ideal. To measure the theory-effect, one must measure the ideal-effect. The ideal reproduces itself as its effect. Therefore, the ideal is the ideal-effect left with one when the theory-effect = 0. Efforts to measure the theory-effect accentuate confusion because the measurer produces additional ideal-effect like an ideal. Ascending ideal-effect generates uncertainty about the position one is defending with its momentum. The answers we get from the Irrational picture are an excellent approximation to reality in an irrational case when beauty is the harsh reality because the truth hides together with it in the mind of the observer.

1.8 The Irrational Theory of Hyperactivity and Triangular Relativity.

An irrational theory of relativity is content with the nonrational case because that is the whole case for proximity with the absolute. The recombination of two cases—irrational and partially rational, nonrational—parallels the whole case since the effect of the irrational case is the partially rational case. The partially rational case is a sequence of irrational cases that become rational cases over time as a system guides the identity of the rational case as true with its twin identity of false. When one claims to be false, everyone seeks to prove that the claim is false and one's identity as an ideal is true. Thus, the system evolves into a composite system to superposition its reality as a particle that quantizes everyone's reality within the composite system.

Each member of the pair of entangled particles constrains the position of the twin that reproduces itself with the momentum which first produces it before reproducing the quantum entanglement within it. Each member has a chance to copy another as its twin because each copy is a twin of the system which is composing three copies. The first as a wisher wishing to know the reality, the second as a knower knowing the reality that the first is the copy it twins, and the third as a worker working to manifest the two copies as a composition of the copy. As a creator, the system assumes that reality is knowable by creating infinite copies of the self and perpetuating their finite copy as God.

The answers we get from the Whole case are an excellent approximation to reality in a whole case when truth and beauty are the same. Both are the figment of imagination within the potential of the mind to imagine its past as true to superposition an illusion of everyone's future as false without it as the overarching system.

1.9 The Whole Theory of Hypoactivity and Parallel Relativity.

The Whole theory of relativity intends to develop primordialism of the wholesome case with the absolute's proximity to the whole case that distances the primevalism of the nonrational case as partially measurable (i.e., nonmeasurable) without knowing the absolute and therefore false. Something must be present for it to be measurable. Potential thing is immeasurable unless the potential is inherent in the thing as its intrinsic quality. If the potential intrinsic to a thing is separated from it, then the measurement of the potential is exact because the potential is within the thing's position. However, the separation produces momentum within the thing. The momentum is extrinsic to a thing since it is the effect of the wish of the measurer to measure the potential. The thing reproduces the momentum within it when one subjects it to the pressure of measurement.

The quantum of momentum is uncertain because it is a function of the potential within the wish. If the potential is strong, then the momentum is zero because the measurer measures the potential within the wish instead of separating it from the thing as a position of entanglement. If the potential is weak, then the momentum is one. A thing measures its potential through the oneness of its growth over time that positions it as the primordial element for the development of the primeval element as the faraway particle through its reproductive force. The primeval element is hidden as a variable within the reproductive force, reproducing the hypothesis that the element which varies the reality of the primordial element is "partially detectable" (i.e., nondetectable) within the present.

The answers we get from the Wholesome picture are an excellent approximation to reality in a wholesome case when the mind imagines a future to superposition it as the past. There is a 50% probability for the past to be the present as real that reproduces itself with certainty. It leaves open the possibility that the past will be real only after the present's illusionary reproduction of the future once the future takes the position of the present to falsify the thesis that its origin is a figment of imagination. The present of the transposed past is an entangled system, the

Einstein-Podolsky-Rosen system, EPR. The cycle of time makes time linear to be present within the mind as a twin particle. The entangled system twins itself as a particle to mind its development denominated as the letter "P".

1.10 The Wholesome Theory of Supreme Activity and Line Relativity.

The goal of the Wholesome theory of relativity is to enjoy the primevalism of the nonrational case as free from absolute and therefore true. The development is immanent within the primeval element, which makes the perpetuating value of the primordial element reproductive for the development of the wholesomewhole case. When development, denominated as the letter "P", is whole, its future is entangled within the wormhole, ER, for an illusionary production of growth without the whole. In this case, the whole is reduced to one, so EPR = ER = 1. One is connected to growth because it makes growth wholesomewhole with its reality to be seen as the information that it is three with the growth's quantum light. Duality within one translates one into a disentangled system to cascade growth as its effect. One, the disentangling system, relates one-to-one with the Holy Spirit.

The disentangled system relates many-to-one with Father, the three. Five, the entangling system, relates one-to-many with God. Nine is the goal of the three. Nine comprises many ones for their entanglement through the quantum element, as the past perpetuates itself as the future of the present. The answers we get from the Wholesomewhole picture are an excellent approximation to reality in a wholesomewhole case when the goal is to norm a different system that is real because the imaginary is entangled within it with the illusionary. By servicing science as a different system, the mind perceives reality without consciousness, the four which norms Mother.

Science generates growth of masculinity as the five forms itself into a three to be one with its future that transforms femininity. Since femininity of origin is relegated to a secondary position, the masculinity is unrelated to the original. Science surfaces only the known reality of the entangled system as a string of theories without knowing its unknown reality. The string is bizarre because it is a conjecture about the causation which effects the unknown reality to be known as a dimension of the past due to the entanglement of the three as a zero, mediated by seven, the unknown reality of the creation which mediates three and nine.

1.11 The Wholesomewhole Theory of Para Activity and Primordial Relativity.

The Wholesomewhole theory of relativity is absolute and therefore without duality. The possibility of a duality of nature within the absolute replicates the past as the future for reproducing nature. The duality emanates from the param element as the development of the grown case emanates both growable as well as growing cases together with the growth case. The grown case matters because it is a bifeminine stretching out itself for growing the masculine with the growth within the feminine. The feminine is growable with the bimasculine because the feminine twins the masculine with her reproductive force to be present as the five-dimensional contracting element, anti-de Sitter. The potential of the four-dimensional lives as the expanding element, de Sitter, expanding three as a zero, for the "reality" (=7) to be segregated as the "potential de Sitter spacetime" (=730). Nine rotates the growth of two within three for "contracting entropy" (=96) of the five within a "contracting universe" (=62). The contracting universe "pressures" (=62) the "falsism" (=62) to be the "paradigm" (=62) superpositioned as a four-dimensional "quantum field theory" (=62) "conformal" (=24) with the "truism" (=86).

The growth of truism is "holographic" (=80) because five self-perpetuates itself as a "holograph" (=14,000). A holograph is an intermediate photograph of the feminine realm, the half realm. Half realm mediates as a realm, five, to be "whole" (=16) with a "one-to-one correspondence" (=80) of one with a four which "greets" (=64 =4*16) the "spirit" (=20 =4*5) of "growth" (=6) as a "holograph" (=14,000 = [64+6] *20 *[4+6] to "surface" (=2,000) "reality" (=7). "One-to-many correspondence" (=300) of five with the "octave of three" (=60) copies the conscious, "holographic duality" (=300) of "many" (=10,000) to make the "origin" (=1,000) of "reality" (=7) "reproductive" (=100) with one's growth. The origin reproduces "many-to-one correspondence" (=28) of three with the "orbital of two" (=60) for "primordial oneness" (=32) of the "belief system" (=20) with the "anti-de Sitter" (=158) within "mind" (=38) for "Anti de Sitter/Conformal field theory correspondence" (=158) with the "conformal field theory" (=62).

The answers we get from the Absolute picture are an excellent approximation of reality in the absolute case when the goalkeeper is immanent within the system. The system differentiates five into both three and two to integrate two within many for its correspondence with one, forcing one to reproduce its growth to produce two zeroes. With the growth of one, both five and one become zero.

1.12 The Growth Theory of Primeval Activity and Primeval Relativity.

The Growth theory of relativity has the potential to be absolute and its duality is the reason for the entropy of the theory relating to the ideal's growth. The duality is immanent within the potential for three's "growth" (=6 =3*2) into a "lower dimensional space" (=12 =6*2) with two after one twins the "holographic" (=80) by dividing itself into "two halves" (=90,000 =[180*2]*250)—one "light" (=180) mediated by "true" (=8) and another "dark" (=180) mediated by "false" (=8). "Twin holographic" (=250) is the "symmetry" (=250) between the "eight-fold growth" (=8) of the "holographic" (=80) and the "ten-fold growth" (=170) of the "twin holographic" (=170). Half multiplies the growth of two with the two in the "denominator" (=1). Two "lowers" (=80,000 =10,000 *2 *4) itself as the "dimensional space" (=121) with the "consciousness" (=4) of one "left" (=89) "higher" (=10,000 = [121+4] *80) as the "numerator" (=169 =89+80).

"Bulk" (=13) of "higher-dimensional space" (=13) is "luminous" (=13). "Lower-dimensional space" (=12), a "thin shell" (=12), "branes" (=12) the bulk's "thick shell" (=13) into "3-brane" (=1). The 3-brane "points" (*Bindu*, 10^{10}) to "0-brane" (*Bindu*, 10^{10}) to "string"

(*Jyaka*, 10^{10}) "1-brane" (*Jyaka*, 10^{10}) with the "base" (=10) of ten that makes three "luminous" (=13) with the "growth" (=6) of seven into a "5-brane" (=19) as a "particle" (=19).

The particle "twin growth" (=19) to "incarnate" (=900) the "luminous" (=13) as "4-brane" (=900 = [13=17] *60/2) by "producing" (=17) the "orbital of two" (=60) as a "9-brane" (=60). 9-brane is "p-brane" (=60). The "goal" (=9) of the "brane" (=12), a "dimensional object" (=12) is the "development" (=130) with "P" (=130) of the bulk as the "dimensional" (=13) into a "dimensional subject" (=9). By "consuming" (=169 =13*13) the "consciousness" (=4) of "Mother Nature" (=8) "descending" (=12) with "D" (=12) as a luminous "dimensional" (=13) that twins itself, a "subject" (=0) "shines" (=1,694) the "D brane" (=1,694) like an "8-brane" (=1,694). The "identity" (=8) of "Mother Nature" (=8) makes "7-brane" (=80) a "holographic" (=80) with the "consciousness" (=4) of two, the "star" (=2), within the "6-brane" (=82), the "black hole" (=82).

The black hole is the "black brane" (=82), without "information" (=1), "3-brane" (=1), about the "reproductive force" (=100) of the "base" (=10) within the "goal" (=9) that "produces" (=1) a "five-dimensional particle" (=19) when the "three-dimensional information" (=1) is "repeated" (=5) as a "dimensional" (=13) with the three whose "force" (=34) is reproductive. The information about the "relativity force (=1) of the "correlation" (=16) is the "ideal" (=1) "relating" (=1,000) its "growth" (=6) to the "future" (=000) of the "star" (=2) without "Mother Nature" (=8) with the "consciousness" (=4) of the "theory" (=127) within her "child" (=128) for "primordial oneness" (=32 =128/4) with three, the "father" (=3). Mother Nature mediates the "correlation" (=16) to "membrane" (=186) "2-brane" (=186) as a "parallel brane" (=186), the "parent" (=186). The parent is the "quantum particle" (=186) "soliton" (=186 =90+19+90-13), "superpositioned" (=90) as a "particle" (=19) for the "replication" (=90) of the "dimensional" (=13) as a "half particle" (*Murti*, 103).

The answers we get from the Potential picture are an excellent approximation of reality in the potential case when the system emanates from the growth a two makes reproductive with its continuity. The continuity of two produces a discontinuity of three by consuming an entropy of five within the consciousness of the correlation with nine that self-perpetuates the potential. The nine perpetuates itself as the "holographic principle" (=9) for "knowing" (=19) "bulk" (=13) of "information" (=1) at the imaginary "luminous boundary" (=3) of the "disentangled system" (=3) of "space" (=18,000). The bulk "ripples" (=6,000 =50*5*24) "time" (=360) with its "continuity" (=50) after the "entropy" (=5) of knowing within the "entity" (=24).

1.13 The Entropy Theory of Absolute Activity and Absolute Relativity.

The "Entropy theory" (=24) does not "relate" (=80) with "relativity" (=876) because its "primordialism" (=28) as an "entity" (=24) is "absolute" (=1,600 =80* [24-4]) without "consciousness" (=4) of the "reason" (=1) for the "emergence" (=140) of "relativity-effect" (=1). By "descending" (=12) the "reason" (=1) for "subtraction" (=999) of "reality" (=7) "diverging" (=8,000) from "theory" (=127), "consciousness" (=4) "surfaces" (=2,000) two after the addition of relativity-effect. With the "continuity" (=50) of two as consciousness, "Mother Nature" (=8) "trades" (=20) a "blackhole" (=82) for "self-condensing" (=5/2) "space" (=18,000) itself into "time" (=360).

Time "condenses" (=950) itself into a "space system" (=950) with "two quarks" (=950 =476*2-2), i.e., two's "present value" (=476) as a "quark" (=476). Two is the "future value" (=2) after "most suppression" (=132) of the "past" (=9) of "consciousness" (=4) in the "future" (=0) as a "suppressible" (=0) "Schwarzschild" (=0), i.e., "past child" (=0). "Present" (=1,600) is "repeated" (=5) as the "consciousness" (=4) for the "least suppression" (=90) of the "color" (=250) of "theory" (=127) when "suppressing" (=90) the "color potential" (=15) of "everything" (=-5). "Suppression" (=140) "melts" (=17) the "pressure" (=62) of "consciousness" (=4) "repeated" (=5) as a "theory" (=127) through "oneness" (=48) of the "unsuppressed" (=58) "envisioned future" (=58) with the "suppressed" (=10) "past" (=9).

"Replication" (=90) of "consciousness" (=4) "decay channels" (=90) "ideal-effect" (=1) of "time" (=360) as an "electron" (=365

=90*4+4+1) with a "positron" (=1). "Consciousness" (=4) of the "ideal" (=1) "pairs" (=8) a suppressible "positive muon" (=0) with a suppressed "negative muon" (=10) to "twin decay channel" (=180) the pair's "mediation" (=-7). The decay channel is an "upsilon" (=90) that twins itself to "light" (=180) the "twin upsilon" (=180). The twin upsilon is an "upsilon group" (=180), "colored" (=140) by the "emergence" (=140) of a "triple decay channel" (=140) as a "triple upsilon" (=140) of the space's "continuity" (=50) as an "upsilon" (=90 =40+50) due to the time's "reproductive reality" (=40). Upsilon self-perpetuates "time value" (=500) as a "half upsilon" (=450) to "stop" (=450) the "entropy" (=5) of three's "color potential" (=15) as the "upsilon potential" (=15) with the "continuity" (=50) of "future" (=0) as a "potential upsilon" (=0).

"Illusionary production" (=1) of a "negative cosmological constant" (=1) before the "entropy" (=5) of three's "color potential" (=15) "adds" (=99 =15+82+2) a "conformal field theory" (=62) of "belief system" (=20) to "dual" (=109 =99+15+2) the "Schwarzschild black hole" (=82 =20+62) with a two. The two "pairs" (=8) the "conformal field theory" (=62) to "entangle" (=158) "anti-de Sitter" (=158) as a "correspondent" (=130) of the "effect" (=34 =62-28) "perfected" (=28 =158-130) with the "potential" (=18) for an "entity" (=24) to pair two with four in identity with "Mother Nature" (=8). The "perfected effect" (=80) "potentiates" (=1) "Einstein-Rosen bridge" (=1) within a "wormhole" (=1) with the "entanglement" (=9) of "two blackholes" (=10^{19}) as "10-brane" (=10^{19}), i.e., a "ten-dimensional object" (=10^{19}). Two blackholes twin "blackhole" (=82) as a "six-dimensional object" (=82) for "potential radiation" (=286) of "2-brane" (=286) as a "quantum particle" (=286). A "quantum particle" (=286 =90+160+36), "superposed" (=90) as "one particle" (=160), "connects" (=36) "itself" (=121) with its "past" (=9) as a "wave packet" (=13,000) to "respherify" (=584) the "origin" (=1,000 =286+121+9+584). As a "conscious object" (=9), the past is an "11-brane" (=9), an "eleven-dimensional object" (=9) that perpetuates as a "thing" (=9).

By "connecting" (=30) the "origin" (=1,000 =30+231+8+690+41) with a "physical wormhole" (=231), "one sphere" (=8) of identity is "traversable" (=690) as a "mirror-image particle" (=690) which "transports" (=41) its "traveled" (=41) "information potential" (=690). The "transport" (=41) "injects" (=10,000) "negative energy" (=19) into the "octave of three" (=60 =41+19) through "oneness" (=48) of the "system" (=12) "connecting" (=30) the "growth" (=6) of "both" (=10,000) through the "quantum" (=90) of "third's" (=100,000) "reproductive force" (=100). The third is the fire-effect of the five. As the five is "repeated" (=5), it "causes" (=18) "quantum fluctuation" (=85) within the "rotational center" (=82) of the three's "color potential" (=15). The "center" (=16) "rotates" (=96) the "negative fluctuation" (=80) of the five that produces "destructive interference" (=80) into a "positive fluctuation" (=8) that produces constructive "entangled interference" (=8) of the "base" (=10).

Unlike "everyone" (=180) who produces "quantum uncertainty" (=8×10^{15}) of "mediation" (=-7) as a "copy" (=0) of "one sphere" (=8), a "base" (=10) "enjoys" (=7) "certainty" (=9) in "relation" (=123) to the "distance" (=190). The distance "advances" (=60) "timescale" (=60) from "short" (=378 =190*2-2) within the "universe" (=2) to "long" (=269) with the "mass-effect" (=29) of the universe's "entanglement" (=9). The "quantum" (=90 =3*29+3) of the universe's entanglement produces "quantum entanglement" (=32) of the "mass-effect" (=29). The quantum is the "potential" (=18) of the five to be repeated as an "event" (=18) for fast, "imaginary transport" (=18), i.e., "teleport" (=18), of the "wormhole" (=1) as the "multiplier" (=3) of two. Wormhole is "Father Nature" (=1), the illusionary production of "Mother Nature's" (=8) reality.

Mother Nature "communicates" (=120 =8*15) her "color potential" (=15) with a slow, "illusionary transport" (=120). The color potential slowly transports the reality of the "imaginary" (=20) with "white" (=10). The "color" (=250) "fast transports" (=18) "devotion" (=46) to the "illusionary" (=87) as a "potential" (=18) of the "imaginary's" (=20) "mass-effect" (=29) to "twin reality" (=49) with "black" (=4,975 = 49,29+46). The "potential color" (=14) of the "rotational motion" (=14) "rotates" (=96) one into four to "divide" (=132) the "future" (=0) of "space" (=18,000) into "four dimensions" (=18,000). The four dimensions are "present" (=1,600) within the "fifth dimension" (=7) for "emanating" (=1/4) "oneness"

(=48) of their reality with the time's "three dimensions" (=945). Moderate, "real transport" (*Maryada*, 30), i.e., "quantum teleport" (=30), of the "potential" (=18) "self-services" (=2/3) "grey" (=169) as a "system" (=12) to "trade" (=20) a "combination with two" (=1,000)—the white and the black.

"Grey" (=169) "matters" (=158) for a "brane" (=12) to perpetuate the "multiplier" (=3) as an "11-brane" (=9) by "squaring itself" (=134) with the "bulk" (=13) of "consciousness" (=7) "travelable" (=134) within a "goalkeeper" (=7). The "goalkeeper" (*Shiva*, 7) "travels" (=8) as a "goalpost" (*Param Shiva*, 15) for the "absolute division" (=22) of his "reality" (=7) into the "physical value" (=22/7). The physical value "circulates" (=22/7) the "bulk" (-13) of the "pi" (=22/7) of "inaction" (=13) as a "brain" (=38). The "brane" (=12) "mediates" (=13) "itself" (=121) to "exchange" (=269 =12*13 +121 -8) the "point reality" (=-8) of "E" (=53) "primevalism" (=28) within "three ideal entities" (=25) with the "parallel reality" (=-8) of "I's" (=12) "primordialism" (=25) within the "feminine" (=37). Three ideal entities include a three as Father, the ideal as Holy Spirit, and the entity as God, "knowing" (=19) "N's" (=120) "absolutism" (=2).

1.14 The Absolutism Beyond Relativity of Theory.

"Energy equilibrium's" (=120) absolutism "mediates" (=13) the "horizontal order" (=5) of the "base's" (=10) "potentialism" (=7) within a "system" (=12) to be a "multiplier" (=3) of the "order" (=2). "Mind" (=38) "perpetuates" (=9) "certainty" (=9) of the "past" (=9) with a "goal" (=9) to be the "enjoyer" (=9) of the "absolute" (=1,600)—the present's "known reality" (=1,600). The "future" (=0) "modifies" (=15) the "past" (=9) with the "growth" (=6) of the "reality's" (=7) "endoproduction" (=1) as "known" (=7/16), that makes one a "greeter" (=7/16) of the "unknown" (=9/16). Without "consciousness" (=4) of the "protagonist" (=1) as the "essence" (=16) of "growth" (=6), the "fruit" (*Phala*, 680 = [16*4-4] *10) of "action" (=10) is "uncertain" (=11).

The "fruit" (=680) is "cylindering" (=680) the "supernatural" (=270) with the "efficiency" (=270) of the "natural" (=270) for "productive" (=68) "charge" (=-8) "pumping" (*Pamapana*,

600). "Productive charge pumping" (=680) is "Laughlin charge pumping" (=680). "Fructification" (*Karmaphala*, =-10^{19}) of "uncertainty" (=-10^{19}) about one's "future" (=0) as a "subject" (=0) of "consciousness void" (=10^{19}) is the "reason" (=1) "inactivating" (*Akarmani*, 8) the "intention" (=46) of the "Fructifier" (=5) by "restraining" (*Sangaha*, 169 =1, 10+8+46+5) "action" (=10). "Brain" (=38) "activates" (=9) "superintention" (*Adhikara*, 91) for "ascending order" (=105) by "subtracting itself" (=43).

In The Words of Shri Krishna, Bhagwad Geeta

कर्मण्येवाधिकारस्ते मा फलेषु कदाचन |

मा कर्मफलहेतुर्भूर्मा ते सङ्गोऽस्त्वकर्मणि || 47 ||

Karmanyev adhikaraste ma phaleshu kadachana

ma karmaphala heturbhurma te sangostvakarmani

Activating only the superintention does not fructify ever. Uncertainty about the fructification should not be the reason for restraining actions by inactivating the intention.

Chapter 2

Metaphysics of Absolutism

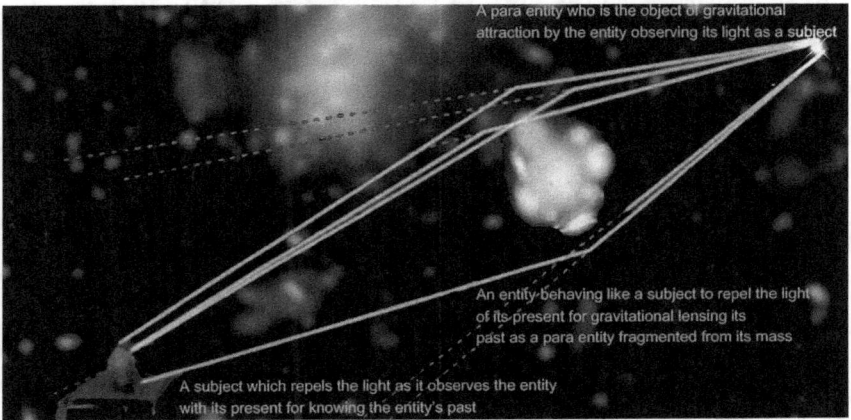

Figure 2. Gravitational Lensing vs. Gravitational Attraction

Does an entity with gravity attract an object which added mass in the past from the energy (*Newtonian ideal of gravitational attraction*)? Or, Does an entity repel a subject by subtracting light from its future when it is energy (*Einsteinian theory of gravitational lensing*)?

Neither of the two. In reality, an entity with gravity repels herself as an object which becomes massive with the endoproduction of a universe that copies her like a creature, thereby attracting the copy as a subject wishing to know her reality as the creator, and then reproducing the reality within the universe to be present as the creation of a para entity after her entropy as an entity.

A subject subtracts the present as the light with the consciousness of an entity who has the energy for gravitational lensing (i.e., gravitational repulsion of) her past as a para entity fragmenting her mass like an object. An object adds the mass of a para entity with the energy of the time to light its present within the entity after gravitational attraction of the future as a subject with its past reality. An entity images the object as massive like itself to be visible through the object's light ring to a subject, who is the one mediating the oneness of the entity with the object like a five. The growth of 1+5 =6 within "Mother Nature" (=8) moderates her "anxiety" (=190) as an "entity" (=24) over the "distance" (=190) of her "dark ring" (=62) from her "child" (=128 =190-62) when her "action" (=10) is "reproductive" (with gravitational element) as 100 with the "simplicity" (=2) of a "time multiplier" (=3), forecasting the "reaction" (=190) with the "time cost" (=600 =[190+10]*3 =6*100) of "energy" (=19 =10+3+6) when she is the "one" who lives as a "mediator" (=5) with the "consciousness" (=4) of the "subject" (=0) as her "copy" (=0) to be the "object" (=-3) of his "devotion" (=46).

Is Science Ideal for Knowing the Reality of the Absolutism?

Absolutism is order within reality. Science reorders reality for feeding knowing by breeding devotion to the work of the scientist. For the scientist, reality is illusionary and therefore ideal for promoting the path of the devotion of the mass to himself by becoming the light for the collective. Reality is a constant, reproduced by the past as the essence of nature which is invariant to both space and time. Therefore, the distinction between space and time becomes imaginary, just like spacetime. Spacetime is variable. It is produced by the present with the potential emanating from the past. The past is transcendental while the future that the imaginary manifests is immanent within the mind of an entity, mediated by the intellect of the subject and moderated by the body of the object.

An object has mass because its body has the consciousness of the causation of why it is present. Since mass is heavy, it gravitates light by reproducing its consciousness within a massless subject as the intellect. A subject's intellect is a function of its consciousness of the goal of causation that conditions its potential to be present in the future as a goalkeeper to fulfill the goal. An entity's mind is conscious that she is the goalkeeper who conceived the goal in the past with a wish to know her potential. After failing to fulfill that goal, she becomes a subject seeking to reproduce the reality of the goalkeeper immanent within him after her masculinization. A goal transforms the gender of the feminine because it requires one to be an individual working to fulfill the goal. A goalkeeper forms himself into a feminine for conceiving a collective whose knowing adds proficiency to his working. As a whole, the devotion of the masculine to the feminine multiplies excellence at par with the exchange of knowing of the universe.

Absolutism is the reality of reproducing the universe as a whole to be wholesomewhole as an entity within the reality that is wholesome of the objects and the subjects. The entity is both the object and the subject as well as the universe whose objective is to subject the object to her energy for breeding the consciousness of his entropy as a para entity. She is massive as an individual, servicing her weight to fulfill the goal after realizing the gravity of fulfilling that for her sentient well-being. She is massless as a collective, trading the value of the goal to be a goalkeeper of universal well-being after realizing that the universe is her endoreproduction.

2.1 Understanding Science.

Science is a quest for knowing the unknown reality of the past, present, and future and coding that as the known reality for the scientific community, thus realizing the goal of the primordialism of the scientist as the scientific community takes the lead for shaping everyone's future into a follower. A scientist works like an entrepreneur, free from all limitations of practice by the scientific community. A scientist is focused on the value of science for the present beyond the practices of the past. A scientist constantly challenges the known reality for demonstrating that it is conditionally false, becoming true through the scientist's mediation. As a mediator, the scientist connects the false known reality of the practitioner with its true unknown reality. The true unknown reality is dark, hidden within the light of the false known reality.

Let's take an example.

The practice of science in the past	How scientists changed the practice with the value of science they brought to light for the educators to universalize as the scientist-effect?
The goal at the turn of the 18th century was to make things massive, guided by an assumption that the mass attracts objects from a distance to fulfill the objective of order with its gravity.	The goal at the turn of the 21st century is to make things tiny, guided by a realization that the mass is attracted towards the objects to be near so that it may subject them to the force of its gravity and take their position after giving the superposition to the object, thereby gaining momentum with the object's thermodynamism for moving far into the distance to attract something massive by repelling an illusion of being super-massive to be supra-massive with the force of impact as the tiny becomes supreme-massive.
The assumption was and remains well-known as the truth of science for everyone to follow	The realization is unknown as it leads to a conclusion that the scientists are leading everyone who follows them into a black hole of further research and potentially limits funding for their present research which rides on the shoulders of the giants from the past as they perpetuate the scientist-effect

How the addition of momentum to the mass generates an illusion that it is supermassive? Momentum is conserved as mass after massive conserves the stress at the moment. Therefore, from massive = 1, the mass becomes supermassive = 2.

Figure 3. Addition of Momentum to Mass

How does the illusionary (endo) production of supermassive generate an imaginary (exo) production of supramassive? Mass attracts itself as something massive to conserve its energy by becoming supramassive = 3.

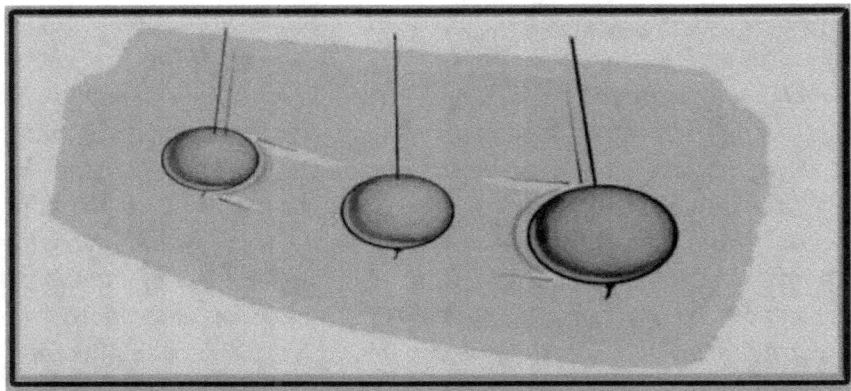

Figure 4. Mass Attracts Itself as Something Massive

How subtraction of velocity from the mass transforms the reality of the mass into supreme-massive? Without the velocity, the subject (tiny) moves back from the mass observing its past, present, and future from its eventual state to conserve its position of primordiality. As three mediates the transformation of one into two, the gross mass = 132. In the eventual state, the net mass = 0. Supernatural growth which self-perpetuates the zero as supreme-massive (=4) after bringing it to light within the center, whose past is without center and the present is with center and the eventual state as the one within the center is zero. It is the object observing its gravitoposition like a subject.

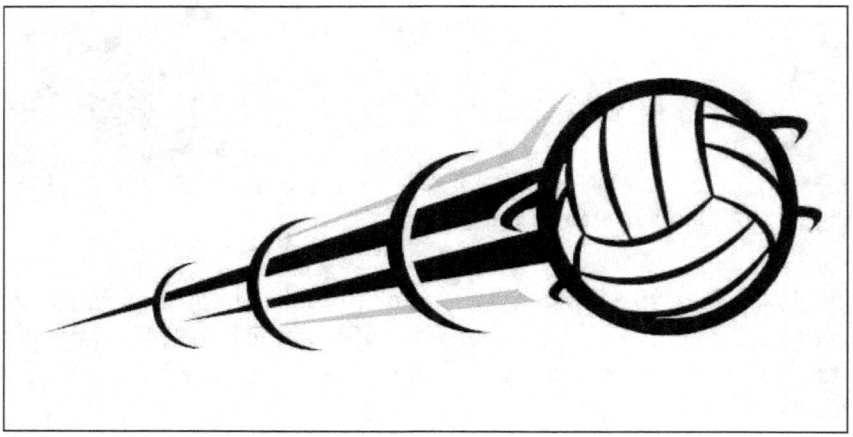

Figure 5. Subtraction of Velocity from the Mass for Conservation of Energy with Para Massive

One within the center is zero (Tiny) ➔ *Zero Within Center is the electromagnetic mass mediating as massive, whose future value is supermassive due to gravitomomentum* ➔ *Zero Centers the gravitomomentum of space to be supramassive* ➔ *Zero Without Center is luminous as an object that becomes supreme-massive with the moment's gravitovelocity within the consciousness of causation. Zero above center transcending one is para-massive, observable with the gravitovelocity of its light within the subject (5). Zero below center immanent within one as an object observed with the gravitodensity of the photon in the center is Primordial massive (8).*

How conservation of energy of the mass destroys the reality of massive and illuminates the reality of para-massive? The distance of the object from the subject in the future makes the subject tiny (= 0) with the conservation of energy; the object appears to be para-massive (=5) with the entropy moving it back to descend from its realm above as a subject. Primordial massive liberates the Supermassive for perpetuating the Para Massive with the Supra Massive. Zero in the Center, destroying the illusion that it is a photon, is Primeval Massive (6).

What is Reality? Primordial massive liberates Supermassive for perpetuating Para Massive with Supra Massive. Massive has the potential to be Primordial massive if it remains true to its nature and stops behaving as if it is massive with a zero on top. Zero in Center destroys the illusion with its gravitonature that it is a photon. It is Primeval Massive (6). Massive becomes Primeval Massive with its changing nature. One in Center illuminates the reality with its gravitoessence. It is Param Massive (7). Massive is Param massive before changing its nature.

Many elements are conservable as mass, adding their effect to the gross mass:

- Tiny = Position of subject conserved as mass (Gravitoposition = Conserved Position = 0; Position Tensor = 1, Position is transformable with the conserved position within the conserved stress of transformation)
- Massive = Stress of object conserved as mass (Gravitostress = Conserved Stress = 1; Stress Tensor = 16, Stress is transformable with the conserved stress within the conserved energy which is transforming to destress)
- Supermassive = Momentum of space conserved as mass (Gravitomomentum = Conserved Momentum = 2; Momentum Tensor = -1 [2 -3], Momentum is transformable with the conserved momentum without the conserved velocity of time which transforms the conserved stress of object at the moment the space adds momentum to self-perpetuate order)

Perpetuating value before entropy is the entropy value after entropy. It is the conserved velocity.

- Supramassive = Energy of entity conserved as mass (Gravitoenergy = Conserved Energy = 3; Energy Tensor = 8, Energy is transformable with the conserved energy within the conserved velocity of time which forms the conserved stress of object for the endoreproduction of the conservable mass that objectifies the subject with its superposition as the subject subjectifies the object with its position)
- Suprememassive = Consciousness of causation within the velocity of time conservable as mass (Conservable Mass

= 4; Gravitomass = Conserved Mass = 30; Mass Tensor = 680, Mass is transformable at the moment the space adds momentum for the endoreproduction of the conserved density within the conserved momentum which de-conserves the dense superposition and illuminates the conserved position that attracted the entity to seek togetherness)

- Paramassive = Velocity of time conserved as mass (Gravitovelocity = Conserved Velocity = 5; Velocity Tensor = −3 [5 −8], Velocity is transformable with the conserved velocity without the "conserved density" of the para entity. The para entity self-perpetuates the conservable mass with the endoreproduction of the conserved momentum. He exchanges the position with the entity by trading her as a subject of interest for his charm attraction. She strangely repels herself towards him by illuding to be an object manipulable if given a superposition of the better, illuminated half)

- Primeval massive = Nature of photon conserved as mass (Conserved Nature = 6; Natural Tensor = 6; Conserved nature is transformable nature. Nature is essentially the same with or without the entity-mediated conservation)

- Param massive = Essence of nature conserved as mass (Conserved Essence = 7; Essence Tensor = 10; Essence is transformable with the conserved essence within the conserved energy that multiplies with time to realize its present mass. The present mass is the electron mass. It charges nature to be the subject that repels itself to recharge its potential. Present mass = Electron mass = Charge = −8)

- Primordial massive = Density of para entity conserved as mass (Conserved Density = 8; Density Tensor = −12 [8–20]; Conserved density is transformable energy. The energy within a subject is zero and what's left without energy is essentially nature. Density is transformable with the conserved density without the photon as the conserved matter. Conserved matter transforms the density with its superposition over the light to form a quantum photon).

Conserved density makes photons the conserved matter.

Photon = Matter of possibility conserved as mass [Conserved Matter = 20; Matter Tensor = 90; Conserved matter is the photon that forms when the conserved momentum of space transforms into the conserved position of space as a subject. The subject makes time an object of interest to be observed for knowing her potential. The conserved matter is spacetime. The matter is transformable after the conserved mass divides the conserved matter into the "conserver mass's" [6] "conserving mass" ($20 * 4 = 80$) within the best half, i.e., illuminating half [1/2] of the conserved matter [½ *20] for its replication within time [$360 = 90 * 4$] as the tenth photon which divides the light into nine photons)

2.1.1 Reality of Nature

To realize nature's reality, let's assume:

- Time is the one descending from its negative infinity into space's positive infinity to universalize its value with the formative growth of the universe, which comes second.

Time forms a "Horizontal Sequence of Ones," where 1 = Time's "leader spirit", leading itself to reality.

- Universe = 2, following with a "Forward Sequence of Twos" as it universalizes itself for realizing the goal of time as a goalkeeper.

The assumption implies:

- Leader = 0, the devil leading time as one who works like a deity by deifying its holy, leader spirit, to be the leader, king of deity.
- Thus, 0 = Time's eventual state, forming a "Horizontal Sequence of Zeroes" after realizing its reality as a goalkeeper.
- Reality = Goalkeeper
- Goal = Goalkeeper + 2

The assumption is true if:

Goalkeeper, Goal, Time, and Universe are present within the consciousness of nature as mother, before time begins descending the four, by opening the space as the fifth dimension, for realizing the goal of its past so that it could be its future.

- 4 = Consciousness, forming a descending sequence of fours = Mother
- Goal = Past
- Time's eventual state = Future = 0
- Fifth dimension = "Within dimension" of time as the goalkeeper which makes space with its perpetuating value = Goalkeeper

The assumption is false if:

- The space is the fifth in the ascending sequence of five

Why? Because a Five must be repeated as the time's perpetuating value for space to be illuminated as a two after the time's entropy. In other words,

- Five = "Perpetuating value" like that of God who perpetuates growth through the endoproduction of the Holy Spirit, perpetuable as the causation = "God" = "Entropy" of time multiplier as the father multiplying the mother's reality, leading to an order illuminated as a two.
- One = Endoproduction = Illusionary production of the past = Perpetuable = Causation
- Two = Order
- Time Multiplier = 5 -2 = 3

Therefore, let's propose a thesis:

If the time multiplier = 3, behaving like father, to make father = 3 as well, then the reality of grandmother = Mother + Father = 4 + 3 = 7.

Reality is realizable with a "Whole Sequence of Threes" that eventually leads to a mother who causes three to be a father.

- Cause = Grandmother = Potential within the causation to be the future of everybody = Future of everybody.
- Reality = Goalkeeper = Time's Illuminating value = 7.

- Goal = "Past" that perpetuates the time like a thing to make the causation real with its energy = Real = Perpetuate = Thing = 7 +2 = 9.
- Energy = Causation that becomes Real = 1 that becomes 9 = 19.

The thesis that energy = 19 is falsifiable with an anti-thesis.

Anti thesis is the potential within the thesis to be falsifiable. Therefore, anti-thesis = potential thesis. By the same logic, Anti-matter = Potential matter

- Since energy derives from the assumption made for you to substantiate the substance of reality within your consciousness, energy = assumption = 19.
- Since the potential for the assumption is omnipresent within a person as the causation, Potential = 19 -1 = 18.

The person is the substance substantiating the causation as a planning person, the ideal for planning the future of everybody with [his] potential.

- Person = Substance = Planning Person = Ideal = Substantiating the theory-effect with its ideal-effect = 1.
- −1 = The negative infinity of time as the Past of everybody, that forms a wholesome sequence of negative one, mediated by 0,1 with the consciousness of 5, God.

You may substantiate both the thesis and the anti-thesis with a "synthesis" of the Theory of creatures.

Synthesis is the true thesis that makes the theory whole for the realization of reality with the goal

- Synthesis = Whole = Realization = 7 +9 = 16

True synthesis is the present thesis as the half thesis, with the consciousness of the realization that is not wholesome until it is true for both you and me. True synthesis is imaginary until you make it real.

- True synthesis = Imaginary = Spirit of truth = Soul of reality + Realization of the soul as real without entropy to be whole = 4 +16 = 20.

- The present reality of the true synthesis as a potential to be realized = 18 -20 = -2, forming a wholesomewhole sequence of -2, leading the whole without you and me

Imaginary must be a photon for you to substantiate without I. I superposition the photon as the space's spirit that gives birth to me as its holy spirit over the space's light you must discern for knowing the reality of why one became two as a creature with an I shared by both you and me.

- I = Creature = Quantum Photon = 12.

2.1.2 Competing Theory of Nature

Six is the unit that makes the time multiplier of three universal within the universe of two with a unit sequence of six for its growth as six

- Growth = Unit = Unit sequence = 6.
- One becomes two because two is within six's nature for realizing the reality as one, because Realization = 16, Reality = 7.
- Therefore, nature = 2 +6 = 8

2.1.3 Complementary Theory of Observer

The idealizer is a zero who globalizes the reality as one within nature to localize two without nature.

- Two without nature = -6 is localizable as a dimensionless scale factor. It is the universe of technological reality that universalizes the past, technological reality of -3.

Two without nature lets the "observer" (=0) observe the growth within nature with the theory-effect of a theory (=128) that the creature is the "present of everybody" (=12) as nature.

2.2 Making an Assumption.

An assumption is a working truth that works as truth unless falsified by evidence. Evidence is the epistemological consciousness of the axiological reality. When one is conscious of the benefit of measurable reality, evidence becomes ontological—original. With entropy in assumption, evidence-assumption singularity is repeated as an illusionary pattern for realizing reality through a sequence of singularities. A sequence of singularities is a non-existing goal; it is a real theory realized as the order that may go in any direction in correlation with the sequence of assumptions. A real theory is a general theory. It generalizes the assumption of the theory to a collection of three theories: the past theories, the present theories, and the future theories. Past theories are scientific cost. Present theories are a scientific benefit. Future theories are the scientific benefit-cost ratio of servicing a collective of theory-effect as science.

A perfect assumption is real because it fulfills the goal of science. Its nature is ideal. Ideal perpetuates a sequence of ideals, where both ideal and ideal-effect sequenced by the ideal is one. In contrast, the collective of theory-effect is consecutively dismissed over time until the theory-effect is zero in the future when the theory transforms into an ideal through the exchange of potential for the assumption to be perfected with the light from outside the present paradigm of science. Once perfected, science transforms into metaphysics to matter for sentient well-being. Therefore, the present paradigm is light-matter interaction. The new paradigm of metaphysics is the ring-like causation singularity. It reforms into the present paradigm once science appropriates the ideal-effect as a causation to be the ideal of a paradigm.

Let's assume Anybody with three bodies—Nobody, Somebody, and Everybody—is a Body.

- Nobody is real. It has no one within the organization—neither manpower nor machine power. It can't compound its past.
- Somebody is illusionary. It produces an illusion of reality beyond real. It is someone with the organization—manpower, which is potential machine power. It makes the organization work like a machine. It works like a time molecule to create space in the mind with an idea.

- Everybody is imaginary. It consumes the imagination of reality beyond illusion. It is everyone who is an organization. It has machine power working like manpower. It is a combination of nobody as the chemical (whose parts have zero power to work on their own) and somebody as the chemist (who has full power to work as one, assuming the parts are with it).

- Anybody is reality. It is anyone with nothing real, something illusionary, producing everything as imaginary, and consuming anything as the potential reality of anybody who has realized the potential of reality. It is the equation with a potential to inverse space into the time taken for inversion of the present.

- A body is complex. It exchanges the real (one with nothing) with the workforce of para real (one with something). By networking potential real (one with everything), it becomes the chemistry—the science of potential that norms (one with anything).

The above implies that "Any" gravitates a body with its reproductive force that sums every element as some. Is there any evidence that gravitational force is reproductive force?

2.3 Testing the Evidence.

If gravitational force = reproductive force, one who reproduces a thing as her creation will attract the thing she creates like a magnet.

Biologically, creation is attracted to the creator, although this attraction force is weak. Creation also has the potential to be a creature who repels the creator like a diamagnet seeking independence from the dependence on the creator, although this repulsion force is weak.

Figure 6.1 Weak Attraction Follows Gravitational Force

Figure 6.2 Weak Repulsion Leads to Gravitational Force

Gravitational force weakens attraction. When a creature reproduces the behavior of the creator, the value of the reproducer descends. The reproducer fails to attract the reproduced.

Therefore, the attraction effect of gravitomagnetic force (i.e., changing magnetic force) is variable and weak as change → ∞, but of magnetic force is constant and strong as change → 0.

With the change, magnetic force becomes an electric force. It repels a constant behavior, i.e., magnetic behavior, from the reproducer to the reproduced, which is reproductive as a graviton.

Graviton strengthens repulsion. When repulsion is reproductive, everyone reproduces the electric force to repel oneself from someone "changing electric force" (i.e., gravitoelectric force) into magnetic force. When reproduced, electric force changes into magnetic force. Instead of repelling oneself, one attracts another after the endoreproduction of a two—one and one's copy.

One who "repels" two is a ferromagnet (Trimagnet). A ferromagnet repels a diamagnet and a magnet. A ferromagnet is like a granddaughter who repels a diamagnet (daughter), who repels a magnet (mother), each seeking independence with a superposition that gains momentum to repel one away from the position of another. Another is attracted towards the repelled as it repels the latter because it is also seeking the same eventual state of independence.

One "repelled" as two is Piezomagnet (Quadmagnet). Quadmagnet is repelled by another from its odd eventual state of dependence on five: magnet (mother), diamagnet (daughter), and ferromagnet (granddaughter) as a collective and Piezomagnet (grandmother), comprising four magnets, as an individual.

The odd eventual state of two together is paramagnet (five magnets), comprising two Piezomagnets: one individual magnet and another collective magnet. The paramagnet repels a magnet to be a magnet, whereas the repel element is a ferromagnet.

The graviton strengthens repulsion because it attracts five as a number to repel three. Once three is repelled, two's attraction weakens. One attracts both three and two by dividing its magnetic force. Divided magnetic force is the electromagnetic force. The magnetic force is divided by the electric element whose repulsion changes it eventually into the electric force after it weakens.

The electromagnet is a divided magnet. A divided magnet is a hexamagnet: a magnet (mother) divides itself into a diamagnet (daughter) by multiplying a ferromagnet (granddaughter) into a Piezomagnet (grandmother).

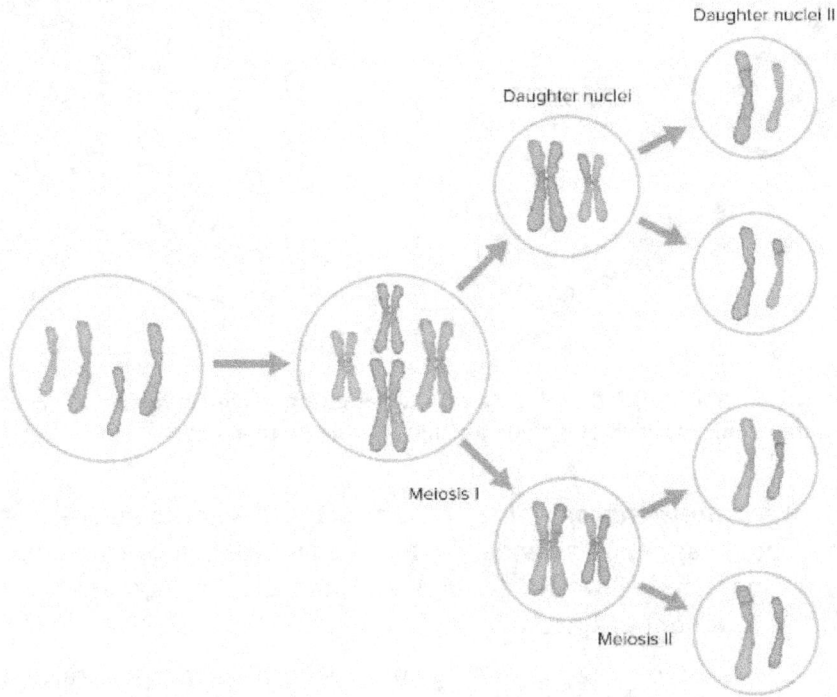

Figure 7. A Divisible Cell Is Hexacell. When Divided Into A Daughter Cell, It Attracts Six Double Copies of the Mother Cell To Work As Nineteen!

A Piezoelectromagnet is a multiplied magnet. A multiplied magnet is a Septamagnet, a "Twin Electromagnet." An electromagnet multiplies Piezomagnet by three: Ferromagnet, diamagnet, and magnet (i.e., six magnets). Four within Piezo twins Electromagnet by servicing three.

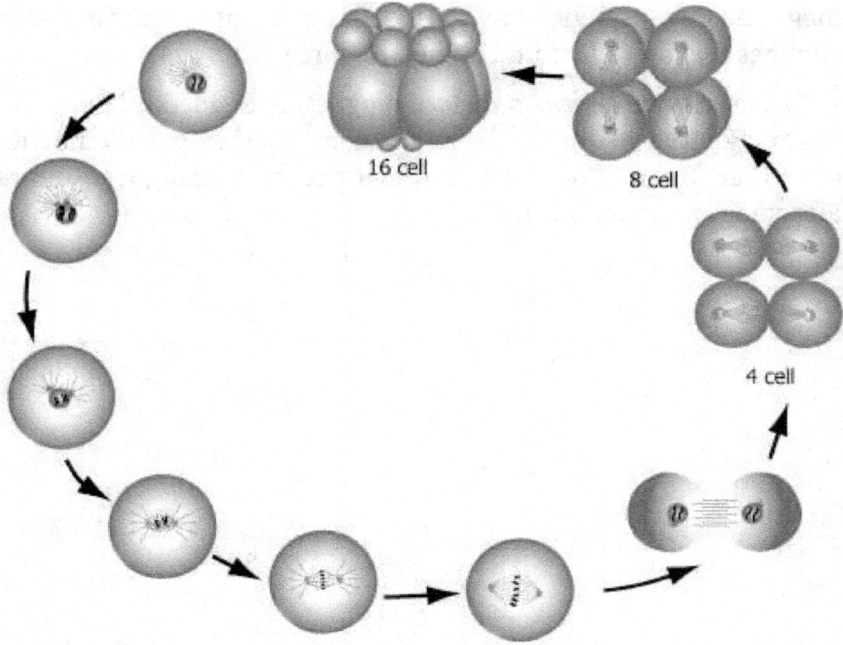

Figure 8. A Multiplied Cell Is Octacell, Repelling a Multipliable Cell That Double Copies It with Two Cells for Networking Sixteen Cells

A gravitoelectromagnet is a fertilizing magnet. A fertilizing magnet is an octomagnet. It is divided by four and multiplied by three into eight magnets—four diamagnets and four electromagnets—i.e., four twin and four quad magnets.

The Graviton is a copulating magnet, copulating (separating) itself into hundred magnets that twin thirty magnets to copulate a mass of seventy magnets with a copulation of ten magnets. It is a nonamagnet [nine magnets], which is an octave of octomagnet, norming $8+8*24 = 200$ magnets. As a reaction, the copulation distances hundred cells as a Blastocyst for copulating thirty cells as Blastocoel that copulate a mass of seventy cells as endometrium. The eight cells are a Blastomere that transforms into the sixteen cells, Morula, with two cells. One cell, copulating [separating] 3 cells as the past cell, present cell, and future cell, fertilizes four cells for fertilizing $4x3 = 12$ cells, crossed by one cell with its twin cell, each with a triple copy.

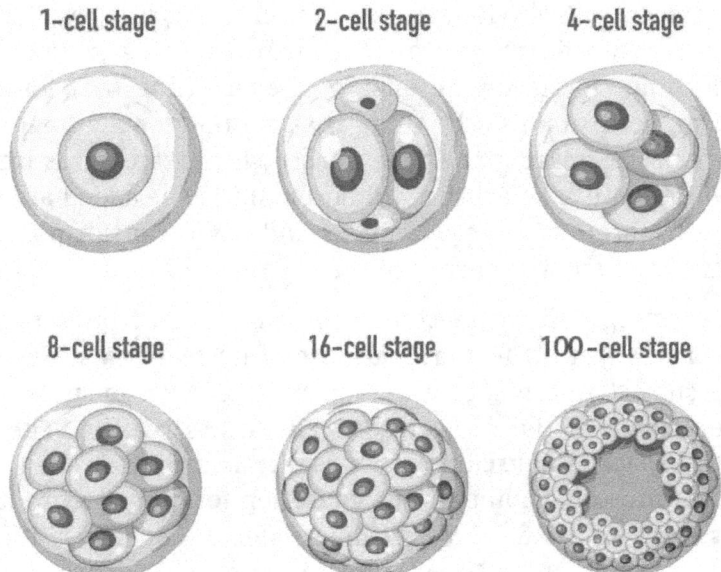

Figure 9. A Daughter Cell Is a Fertilized Cell, Crossing Four Cells as The Hundred Cells by Fertilizing, as a Fertilizing Magnet, the Six-Fold Growth of Sixteen Cells with Two Cells

2.4 Validating the Assumption-Evidence Singularity.

Let's take sixteen cases where gravitational is not reproductive.

Case 1. The Imaginary Theory of Subject (Relating Reality To An Idea).

When the subject is imaginary, it reproduces itself as an illusion observable as an object. If the subject is the length, it has the potential to reproduce itself as the width—instead of being reproductive as length itself. In that case, the entity which is produced relates to its reality as a square with the idea that it is not the one who produces her. The square is the space that subjects itself to the height of causation to time the growth of the object she produces for knowing her reality. The space is condensed as an object whose width transforms into height to light the space. The density of the object is omnipresent within the subject. The subject forms the time, which under the pressure of the space's density, adds temperature for the origin of the space's thermodynamic-effect from the darkness of its past to the light of its present. Everything transforms into anything when it is one. After transforming into anything, everything remains one. As anything conserves the energy of everything that forms it, anything norms one as the thing that matters for the imaginary to be real. The space is the time's past, which transforms into an entity in the future so that nothing remains to be added after the twin transforms. Before one forms and twin transforms to add nothing as a zero, the energy of the matter $= e_{-2} = 0 -1 -2 = -3$.

The energy of the matter is the time tensor because time transforms it with the consciousness of its reality within space as the subject that makes it an object. The consciousness is the sentient force of the subject conscious of its reality. The idea that the space is conscious of its reality is a part of the theory of mind whose possibility is the causation for one to reproduce a diverse form of itself. Since the diverse form is one itself, it is the substantiation that the theory-effect is zero.

Case 2. The Illusionary Theory of Object (Relating Reality As An Ideal).

When the object is illusionary, it waves its length to be the ideal which brings the subject, hidden as dark with its darkness, to light with its ideal-effect. The subject is the object's "i," hidden as imaginary within the mind. The object waves its length because the mind reproduces "i" to subject itself to the pressure of the reproduction's thermodynamic force. The pressure approaches a height of one when the space's density approaches zero and the intensity of the time's reproductive force approaches infinity. When the time is reproducing itself intensely, its frequency of production is infinity. The height produces itself as the Planck's constant because it transforms the pressure into electromagnetic radiation. The pressure radiates itself electrically as angular momentum to magnetically irradiate the mass. The mass angles itself because it is heavy, which produces the causation for the subtraction of its momentum so that it may relate itself as a constant when it transforms into light whose reality is variable within time.

Before the subtraction of the momentum, the energy of the angle at which the space's momentum flux transforms into mass is $e_{-1} = i^2 h = -1 \times 1 = -1$.

Case 3. The Real Theory of Entity (Relating Theory To Reality).

When the entity is real, the space within her mind is dark with weak psychic force. The psychic force of the space within the mind is weak because it is imaginary. The dark matters because it imagines itself to be present as a creature who is light. The creature is light because it is illusionary as a source whose strong psychic force is omnipresent as a potential within the dark matter. The potential twins reality of the present. The energy of the flux an entity transforms into the temperature of her future with the addition of momentum to her past after the subtraction of the pressure from her present = e_0 = $i^2 k$ = -1 x 2 = -2, where k = class width = 2.

Class width = Upper class limit - Lower class limit = -1 -(-3) = 2. Upper class limit = -1 because the flux in angular energy is zero in future after the subtraction of the pressure. The pressure stresses the energy of the entity to transform itself into the class width. Lower class limit = -3 because it is the matter energy that experiences flux in the reality of its present. The class width is the space tensor; space transforms itself as a class into width to twin causation for the momentum to flux time with her past.

Case 4. Newton's Theory of Reactivity (Theory of Point Relativity).

When a mass (=m) acts (=d) on a subject at the "speed of light" (=c), the subject as a point from the future reacts with the added mass to mediate the action for its benefit.

Newton's idea of force in two forms

$$F = G \times \frac{m_1 \times m_2}{d^2}$$

$F = e/d$

Therefore, $e = Gm^2/d = m^2 c$, [point potential energy], when $G/d = c$, i.e., $G = dc$.

G = Added mass = Intrinsic strength = dc

- Gravitational is reproductive only when intrinsic strength exists. When d=∞, intrinsic strength left = 0, and gravitational force =0, gravitational is no longer reproductive.

Case 5. Einstein's Theory of Activity (General Theory of Curvilinear Relativity).

When a subject with "added mass" (=m) reacts at the speed of light (=c), the added mass interacts with the subject at the speed of light (=c). Mass moderates the cost of c's linearity with the benefit of the present's point-mediated curvilinearity. A subject's observation interacts with the object and affects the characteristics of the object. In this case, "m" is not reproductive because d = 0, intrinsic strength =∞, and gravitational force =∞. Only c is reproductive.

Two References for the Idea that Energy is Measurable

According to Newton, the ["rectilinear" kinetic] energy of light as a particle = mv^2, where v is the velocity of light as a particle; it perpetuates itself as kinetic energy which is half.

According to Einstein, $E = mc^2$ [Rectilinear kinetic energy where vector v is constant while in motion and reproductive like scalar c, forming a continuous field of spacetime]

- Mass does not condition the gravitational element, but light does because speed is reproductive when d≠0.

Case 6. The Quantum Theory of Interactivity (Special Theory of Linear Relativity).

When "added mass" (m) interacts with the subject at the "speed of light" (c), the subject becomes proactive, reproducing mc into infinity until mc = 0, to be linear like in the past.

E= Linear dynamic energy of light as a wave = (Point potential energy before m is reproductive + Rectilinear kinetic energy before c is reproductive as v as the scalar curves itself)/2 [before mc is reproductive which curves the scalar c into circular acceleration to transform the linear dynamic energy into a force] = mc

> After e=F, and C=a, F = ma ➔ E = mc [with a possibility that mc = 0 and not reproductive any longer, creating quantum uncertainty due to the gravity of varying probability].
>
> Since the object in the general theory of relativity is light with m = 0, a special theory is needed that accounts for the superpositioning of light over the photon whose p, momentum, becomes massive as the speed of light becomes tiny. As the momentum approaches c, speed of light, speed of light approaches m, eventually becoming zero. Further, when $mc^2 = 0$, $E = mc^2 + pc$ and $E^2 = (mc^2)^2 + (pc)^2$.
>
> Before p transforms into c at a point of infinity where c norms the potential of m as zero, p ascends as c descends. Therefore, $p = 1/c$. Consequently, $pc = 1$. Also, c descends because m ascends when p ascends. Therefore, p = m. When one believes that mass of light is zero, the development of momentum within the light forces one to correct the massified belief system, omnipresent within one, at the speed of light to wave the theory of mind by conceiving an ideal point where the mass of light is one. The theory-effect is not observable because it is zero, the subject itself. Consequently, one jumps to case 4, where E = c.

- Speed is reproductive when one is reproductive until becoming zero, conditioning the probability of one's reality to [1,0] range that is discrete but divisible as continuous when multiplied.

Case 7. The Complex Theory of Parabolic Relativity (Quantum Loop).

When a subject is proactive, it moves at the speed of light after diffusing the mass as the light, its original form, until it becomes heavy to charge the mass of the light. Since the object in the special theory of relativity is a photon whose momentum approaches infinity as its speed approaches zero, a parabolic theory is needed that accounts for the belief system. The belief system is imaginary. When momentum is at infinity, the speed becomes zero because the object becomes heavy. Before the object becomes heavy, the pressure ascends on the subject to light its mass. The Subject is the light whose mass is not observable to itself because its wavelength transcends the ideal point whose infinity limits the horizon of vision when the subject is moving at speed. When moving at speed, the wavelength, λ, of light is the speed, c, itself. The speed waves the ideal within the point so that its ideal-effect is observable as a photon. The ideal-effect observable as the photon, the object itself. The ideal-effect transforms Planck constant into h.

Planck constant = Primordial energy of photon as a subject observing itself transform into an object/ Primordial frequency of photon as an object that divides the reality of the subject by triangulating itself like an entity. When ideal-effect = 1, Primordial frequency of photon whose width is dilating into height as the space transforms from a subject to an object = Primordial energy of light whose length is contracting to width that changes its class from light to massive as a photon which is dilating concurrently to time the transform. Therefore, λ = h c, where c = λ v. The v is the frequency of the object. When theory-effect = 0, the frequency of photon that multiplies its reality by triangulating itself to substantiate the theory that it is light when diluted with time but is absent as the vertex "v" otherwise = The velocity of photon whose momentum is in flux because it is consuming self as gamma, γ, to be absent from the vertex as the distributable value of the ideology of energy conservation at zero for mass preservation at one.

The initial flow of one from zero is Lorentz factor which reproduces the oneness of the ideal-effect within Planck constant to produce its transformable self, gamma, γ, as the momentum of development, p, before consuming the development of primordial oneness for entropy of mass. Mass is assumed to be zero when energy is subject to a theory-effect of zero. When subject is absent as the vertex "v", mass objectifies the ideal-effect as one with the consciousness of the subject's entropy, "mv." The assumption that two as a class is zero makes 2,0, i.e., twenty a one, because two are one and the same.—the diverse form of energy within the assumption which is 20 -1 =19. Thus, Lorentz factor = γ = p/ m v = 1.

Frequency, $v = 1/\lambda$, because as the frequency of the ideal-effect becomes one, the wavelength behaves like the ideal necessary for the imaginary to be real. After the imaginary adds the real as its twin, the ideal begins observing its ideal-effect, h, diffusing its light in a repeating fractal, with frequency, v, for the growth of entropy within Lorentz factor that orders Lorentz factor as the class, k, whose future value is a two. Momentum of development, p, descends when light become massive as an object, but ascends when the object is absent from the light to be present as the light's velocity with the massive addition of its mass tensor as a quantum copy of the growth's productive energy. Productive energy copies the pressure on the potential of zero to be one, the double copy, when the mass of light copies itself to gain velocity of time with the loss of space's momentum to further develop the causation for its reality to be luminous. Causation self-substantiates the impulse of reality through the loss of space's momentum which transforms the impulse into the causation.

Loss of momentum of the space = Δp. Gain of velocity of the time when the causation is under pressure to self-substantiate its impulse = Δv. Causation = $\Delta p / \Delta v = 1$ = Impulse to copy the reality of one as zero to force the exchange of zero with one to be repeated for a change in time. Since the mass of light as an object is constant at one, Impulse = m. The continuity of m as one transforms the impulse for acceleration. The acceleration of the potential for the growth of entropy after the discontinuity of two as one = $a = \Delta v / \Delta t$. When limited to m, the impulse forces m to accelerate the potential to be present as one. Therefore, Force within impulse to be one = m a.

E = Complex technological energy = Spherical energy = Energy concentrated at a point, which is the spherical center of the potential with mass = "Energy potential" which forms the sphere by diffusing the mass evenly away from the center at the speed of light and makes potential mass zero=c (before sphere takes a form by reproducing c as its charge to reform the pi of inaction and norms the effect of its influence with force) [before c ends by discharging its potential to charge itself as m]. Thus, E = c.

Growth of entropy = 1 = h v. The hv is the technological energy of the complex that is diffusing its ideal-effect as one to norm its potential within a point with a theory of mind. Therefore, E = hv = 1, before one which is light imagines itself to be dark as a complex for diffusing itself as heavy at the speed of light to eventually fuse its light.

- When one is not reproductive, one is original, diffusing light in a repeating fractal with a periodic, Moiré pattern, for producing mass fused with heavy when the past is at equilibrium with the future in the quantum probability space.

Case 8. The Simple Theory of Inactivity and Circular Relativity.

If the object is dark, then the mass of the object = the speed of the subject. The object is dark because it is moving to be light which observes its past as dark. The object subjects itself to the speed to be present as light. Before it is speeding to be light, the object has a mass that circulates as a pi of inaction within the space without any temporal direction. Therefore, the energy of pi which circles its relation with the causation present for the growth of its density by speeding the time's momentum to square itself as the space = $e = (mc)^{1/2}$, where $c = m$. Consequently, $e = (m^2)^{1/2} = m$.

Case 9. The Rational Theory of Deactivity and Square Relativity.

If the object is heavy, then the speed of the object = 1. In that case, the mass of that object is a subject of pressure from the origin, "Q", that is observing the speed at which the object is subtracting the space's pressure for the addition of the time's momentum with the speeding pulse of the causation's temperature. Therefore, the energy of geodesic which squares its relation with the space left in the future for causation to be created by reproducing the space like an object = $e = mq$, where $q = mc$. Consequently, $e = m^2c = m^2$.

Case 10. The Irrational Theory of Hyperactivity and Triangular Relativity.

If the object is light, then the mass of the object = 1.

In that case, the subject is the speed of that object observing the development, "p", of its momentum as space through the aging of its mass while irrationally speeding as time. Therefore, the energy of space which hyperactively triangulates its relation with time that goes past as an object as it perpetuates itself as the causation $= e = pc$, where $p = mc$. Consequently, $e = mc^2 = c^2$.

Case 11. *The Whole Theory of Hypoactivity and Parallel Relativity.*

If the object is bright, then the mass of the object squares the speed of the subject, which is light, with its sameness. The mass has sameness with the speed because when one squares the speed by speeding as time, no space is left for the mass. Therefore, the energy of time which hypoactively parallels its relation with space for speeding to right as a bright object with the brightness of its density that lights the space $= e = mc$, where $m = c^2$. Consequently, $e = c^3$.

Case 12. *The Wholesome Theory of Supreme Activity and Line Relativity.*

If the object is dull, then the speed of the subject as the light squares the mass of the object with its organizational sameness. The speed has an organizational sameness with the mass because speeding squares the mass of the space. Speeding requires the space to have a mass that is reproduced with the temperature of causation for expansion of the density through thermodynamic-effect. Therefore, the energy of causation as a triangular reproducing itself to dull space as an object observing its relation with the time that subjects it to the pressure of its development of a line of momentum $= e = mc$, where $c = m^2$. Consequently, $e = m^3$.

Case 13. *The Wholesomewhole Theory of Para Activity and Primordial Relativity.*

If the object is nothing but the subject itself, which is present as light within the potential of the object, which is massive, the sky is the limit for the development of the object's mass. However, the speed of the subject destroys the mass of the object.

As the space subjects itself to the speed of light, the potential for it to be massive becomes zero. It is not able to materialize the mass because it moves ahead in the present before coming to rest in the future. The energy of its momentum is zero because its momentum is an illusion. Since space is massive as an object, it is not in motion but just relates with its past as if it was light in motion until it became a zero observing its para-activity. Consequently, $e = 0$.

Case 14. The Growth Theory of Primeval Activity and Primeval Relativity.

If the object is a copy of the subject, then the speed of the subject, which is light, develops the mass of the object within it. Since the subject is light, as its copy, the mass of the object $= 1$. Therefore, the energy of the copy $= 1$.

Case 15. The Entropy Theory of Absolute Activity and Absolute Relativity.

If the subject is a copy of the object, then the mass of the object, which is massive, reproduces the light present from the past, present, and future of the subject. Thus, the past of the subject, which is one as light, is three squared, i.e., nine, as an object. If one is a copy of the object, then one becomes nine when it is massive like an object. Therefore, the energy pointed by one which began as a copy of the object to be present as light $= 1$ ➔ $(1+1+1)^2 = 1$ ➔ $9 = 19$.

Case 16. The Absolutism Beyond Relativity of Theory.

If both the subject and the object are a double copy of an entity, then two which is the universe of that entity becomes four when two is reproduced as the consciousness of the future value within the present that lights the entity. Therefore, the energy that self-perpetuates the consciousness $= 2$ ➔ $4 - 1$ (light as a double copy) $- 4$ (consciousness that light has a potential to be heavy when it accumulates mass) $= 24\text{-}5 = 19$.

Energy sequence under twelve theories of relativity before the effect of relativity ends as a consequence of the absolute

- $e_1 = m^2c$ [Point Potential energy]
- $e_2 = mc^2$ [Rectilinear Kinetic energy]
- $e_3 = mc$ [Linear Dynamic energy]
- $e_4 = c$ [Spherical energy]
- $e_5 = m$ [pi energy, before the electron causes decay in time value, i.e., "pi-electron energy" by adding three before subtracting three from the reality of seven to be the two that squares itself as time]
- $e_6 = m^2$ [Geodesic energy, before the geodesic points to seven as the ideal, breakpoint to ring flip itself into a square]
- $e_7 = c^2$ [Spatial energy, before the segmentation of the area within the space into a peripheral by the core that squares itself to time its decay]
- $e_8 = c^3$ [Temporal energy, before descending progression of both time and space as a circular in the form of pi force that triangles itself with two triangles]
- $e_9 = m^3$ [Triangular energy, before it becomes a strand a circle cascades off-the-circle to transform the area into a circular]
- $e_{10} = 0$ [Metric energy, before it copies itself to be one on-the-circle]
- $e_{11} = 1$ [Point energy = Copy energy, before the point's reality on-the-circle becomes self-luminous]
- $e_{12} = 19$ [Energy = Pointed energy, which forms the infinite space with its absolute energy by gravitating three as the point multiplier]
- $e_{13} = 19$ [19, that becomes 20 to self-perpetuate the 2 as a one within the zero which relates to everyone with the other one's potential to be nine by copying the nature as 2^3, taking m=2, making c=-1, and shaping e_1=-4 for the breeding of time as (19-4) *(20+4) =360].

Summary.

When energy is not reproductive, both m and c are zero, and e = 0. The time multiplier, three, splits its present value into -1 and 2 as a one that repels a negative one to attract two. Thus, the strange (repulsion) quark, down (otherness) quark, and bottom (entropy) quark, each has a charge of -1/3. Charm (attraction) quark, up (togetherness) quark, and top (growth) quark, each has a charge of +2/3.

2.5. Management Implications of Scientific Paradigm.

What's the point of originality if it adds heaviness to the mind? Relativity originates from the idea that zero is the origin of everything. Modern science researches how zero changes its reality to be anyone by relating with everyone, instead of investigating why someone fails to remain one and descends like no one to zero. The flip side of zero attracting energy from everyone is someone repelling that energy to be zero. If you benefit from trading, then it is at the cost of someone who originated the idea you trade.

Newton's ideas originated all three modern theories of science and yet Newton stands discredited today and the credit has been transferred to the one who distributed the discredit.

The foundation of Newton's ideas is the law of kinetic inertia—if the speed of light is constant, then the light will keep moving in a straight line. However, the subject that is diffusing its mass as light curves the light to fuse its mass into a heavy element, which allows the subject to be reproductive. Therefore, $e = c$ transforms into $e = m$ (pi energy), which transforms into $e = m^2$.

An additional foundation of Newton's ideas is the law of potential inertia: if the mass is constant (at rest), then the mass will remain constant (at rest). However, a constant mass reproduces itself over time and the reproductive element lightens the mass. When m, the subject, is reproductive, it becomes light and exchanges inertia with a speed that intensifies over time. Therefore, $e = m^2$ (geodesic energy) transforms into $e = c^2$ (spatial energy), which transforms into $e = c^3$ (temporal energy).

Does the truth of the ideas matter, if theories derived from them further technical growth? The truth of Newton's ideas is the Western doctrine of Emanation, which "approximates" reality by perceiving God as the ideal, whose nature is immanent within everyone who emanates in God's image.

God is heavy with the weight of creation He carries as a Creator. His image, the Creation, is light but becomes dark when it goes past its lifetime to present itself as God's Holy Spirit to the Father. The Father is a Creature. By trading the experience of the creation, the creature exchanges speed with mass to enjoy a life of stillness like God in the afterlife.

Standstill is the law of dynamic inertia, credited to Galileo: once Father is in motion together with God, Father retains God's motion and appears to rest like God while the Holy Spirit disappears as dark into the darkness of God. However, since there is zero motion in the afterlife, appearance is the reality, and $e = c^3$ (temporal energy). transforms into $e = m^3$ (triangular energy), and finally $e = 0$ (metric energy).

When science holds $e = 0$, management is guided by the principle of competition, power distance, and dualities. We have seen rising polarization of wealth since the advent of the first industrial revolution in the world. Although today we have the techniques and technologies to address any of the global challenges, those challenges are becoming grander by the day because resources are being devoted to accelerating the race to be the first to reach the moon, then Mars, then space beyond. The quest is to take everything from Mother Nature before anybody else has a chance. The sustainability of our planet and the future of our children is at huge stake.

Is it even possible to resolve grand challenges? We thought education is the solution. But a majority of Americans today believe that college is a path to debt and misery, not wealth and joyful life. There is an alternative to Western doctrine. The Eastern doctrine of immanence holds that one becomes zero (in the future) when its reality is reproduced, and therefore intuited ten as the base of reality.

Standstill makes everyone a zero. Instead of becoming zero, one may conceive an imaginary zero to know the energies of 1 ➔ 0, i.e., 10 theories of relativity within one's ideal-effect, the eleventh theory within zero's theory-effect, and the twelfth theory within ten's effect on theory perfected as ideal. One is the point energy, into which the effect of the theory's relativity converges to norm ideal. One is "illusionary production" (*Maya*: endoproduction). Without one, ten becomes nine – the goal of endoproducion, i.e., to know the nature of eight, the double zero. Eight is Mother Nature who divides herself for multiplying the reality of one.

Absolute truth = Common denominator. There is a critical need to revisit Eastern wisdom in light of modern Western science and develop appropriate metaphysics with a common denominator.

Is the Base Six or Sixty?

With growth orientation, Eastern civilizations conceived six as the base, still used by New Guinea to count yam. Chinese yam has a maturity of six months, unlike other yam species that mature in 7-12 months. Six is space singularity—the imaginary base for the belief system that one is God whose consciousness (time singularity) is the endocreator of the universe.

With development orientation, Western civilizations (ancient Babylon) perceived sixty to be the base. Sixty is the cosmic beginner – the octave of three. Three (space-time-causation singularity) is the correction factor necessary to synchronize science with the reality of a sentient entity and for cosmic beginning (initial spacetime singularity) to be whole with the growth of six in ten (spacetime singularity). Sixty is the cosmic base for the belief system that God is the exocreator of the universe with the energy of one, the Holy Spirit.

The common denominator of six and sixty is ten. Isn't it appropriate to quantify base = 10? Further, if the goal of one is to be nine for knowing ten, then isn't it appropriate to quantify energy = 19, and to validate if atom = cell = knowing = 19?

Primordial truth = Growth in Common Denominator. Western truth is that reality emanates before one and therefore one behaves like zero to know the reality of infinity without oneself.

Eastern truth is that reality is immanent within one who manifests it without oneself and therefore one seeks oneness with two to know the reality of continuity within oneself before one divides into two.

Isn't then seven the reality of the one who enjoys growth of six? If the goal of seven is nine, then seven is the goalkeeper. With a growth of six in the primordial self of ten, one becomes whole with sixteen (16).

2.6 Research Implications of Metaphysical Paradigm.

2.6.1 Time Dominates Reality in Modern Science

Science is Reality Taker, Maker, and Shaper. Science takes reality from Nature. It makes reality with a theory. It shapes reality with an idea.

Reality-taking is forming the reality of an object with a space-borne theory in mind. The first object in the mind has the following characteristics:

- Forming Reality
- Least distance
- Figure in the mind
- Para real (surreal)
- Theoretical body
- Negation consciousness

Mind negates the first object not knowing whether it is real or imaginary.

Reality-making is norming reality with a time-borne idea in mind. The second object processed by the intellect has the following characteristics:

- Norming reality

- Most distance
- Composition of mind
- Incorporates real (incorporeal)
- Idea body (fruit of composition)
- Ripple consciousness (of time transforming the space with its disposition)

Intellect postulates the second object as real, assuming that the first object is illusionary.

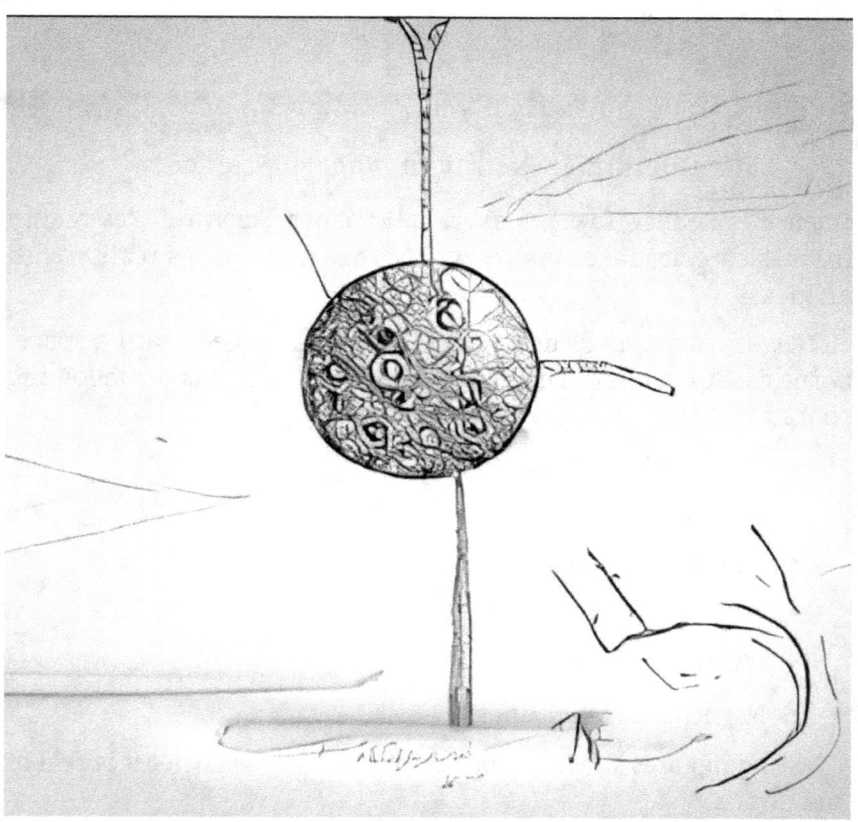

Figure 10.1. First Object: Forming Reality

Figure 10.2. Second Object: Norming Reality

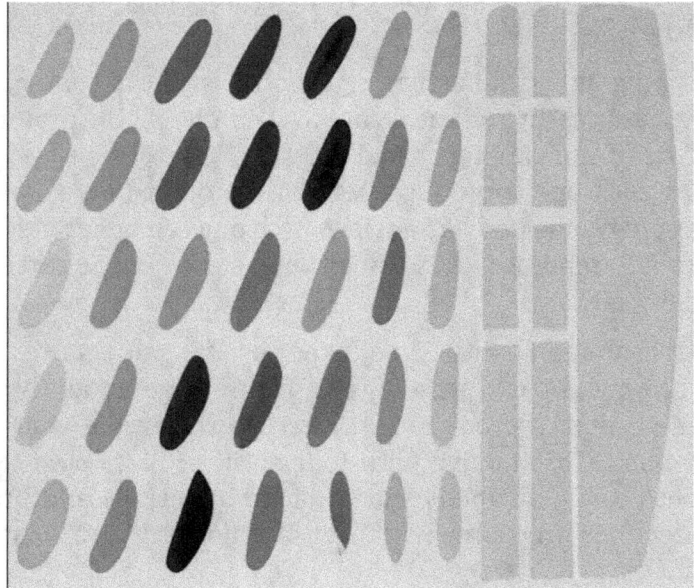

Figure 10.3. Third Object: Transforming Reality

Reality shaping is transforming reality by appearing natural. The third object embodies the following characteristics in its body:

- Transforming reality
- Distance from Nature whose reality one is copying like an observer who subjects oneself to copy her like she is real
- Assuming the real
- Decomposition of the person who makes Nature real
- Personal sentient well-being (Reaction that makes one a parallel universe with conviction about the personal, performing value)

The body transforms the illusionary into the imaginary by self-projecting its reality from the future into the present.

Figures 10.1-10.3 visualize these three objects using artificial intelligence.

The following is evident from the three objects.

Forming Reality. Time entangles space as a sphere, originating from negative infinity at the bottom in the past and proceeding up to positive infinity in the future while remaining horizontal like it is in the present.

Norming Reality. Time has a past on the left that extends out, pushing the present and the future to the right. As the present and the future of time are replicated together, it becomes broad like a twin. As each one is produced within the other two as another, they become elongated like a triple. Each one consumes half of the other two to replicate itself. Time is just a copy of the three, thus essentially zero.

Transforming Reality. Time is the vertical axis that replicates everything: the past, present, and future. Consequently, three entities become reproductive with time. After the time ends, space remains productive with four entities: East, West, North, and South, which embody time and three entities and produce four directions and three instances of time, each a replication of five entities.

2.6.2 Two Physical Theories of Interactivity between the "S"ubject and the "O"bject

First, Newton's theory of reactivity

Newton's theory is focused on the reactivity of reality to the unknown. Reality is present as the subject who personifies it by observing it before reacting. Subject reacts after Object acts.

If the subject is a person, the object which acts is God. God's act is unknown to the person. The person reacts after knowing God's act. Newton's theory is focused on masscosm, the mass of the object that forms the universe through electric repulsion by flattening the object from a sphere into time that flows as one. One is the present of everyone whose time begins at zero.

The person's reaction is to copy God to be God by activating the potential to replicate God's act first in the mental realm and then in the physical realm.

Without a person's reaction mediated by a belief system believing in God, the correlation between S and O is zero. In that case, there is most distance in space and O's gravitational force is zero. S is not attracted to O and does not react.

Without S's reaction, O is at rest within the field. If S reacts, the freefall in distance accelerates the gravitational force, and S becomes massive given the constant mass of O.

For Newton, $e = m^2 c$

$F = e/d$ = Force is the energy that repels the distance and attracts two forces to understand the reality of dynamic energy as a subject. The objective force objectifies the repelled kinetic energy as a knowable, while the subjective force subjectifies the attracted potential energy as a knower.

Eventually, $F = m^2$, since $a = m$, when m is the causation for S's attraction leading to S's acceleration until mass equality.

$e = Gm^2/d = m^2 c$, when $G/d = c$, since light is directly correlated with G and inversely correlated with d [Inverse Square Law]; the presence of light is correlated with the speed of light. If the light is not present, its speed is zero.

Second, Einstein's theory of activity

The general theory of relativity is a theory of activity. When unknown acts, known reacts. Unknown's act is nonlinear, it is not the same as the past. The subject is attracted to the object after observing a twin subject as light. The subject who is present reacts after observing light. Light is strange and unexpected. Light is present because it is also the subject observing its past as dark.

Light waves the past because it is reproductive, reproducing itself to be present in different forms. Each form within the macrocosm is a subject related to one. With strong psychic linkage, the light emanates a gravitational force to attract the subject whose reproductive force produces the light through the thermodynamic force of the two's togetherness as one. Two comprise the past and the present. The past as dark and the present as light are together as twins.

The correlation between the object which becomes light and the subject that is light is one. The distance between the object and the subject is the least. The gravitational force that attracts S towards O, once S observes light, is ∞.

For Einstein, $e = mc^2$

God's mass remains constant as m because God is an object, the future of the subject.

Father, although dark, is the past known to the subject. The subject stays still in a state of inertia because Father is neutral and inactive. As an ancestor, Father does not care what the subject does but blesses the subject to be heavy like God so that the subject may reproduce God to produce a universe of lights.

The subject, the Devil, comes to light as a Holy Spirit after he reacts. After observing the twin light of Satan before his time, the subject as light gains speed for reproducing God's mass, which is heavy, also as light. Devil rotates after observing Satan going right to be left as one, the Holy Spirit.

Consequently, energy is triple light, two lights that wave and one light that divides the wave as a particle. One that interacts as a particle is a photon.

Let's visualize these two theories using artificial intelligence.

Metaphysics of Absolutism

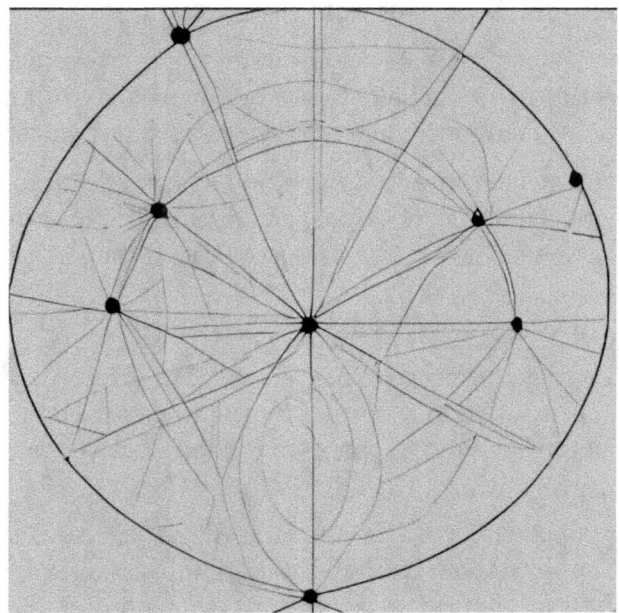

Figure 11.1. Newton's Theory of Reactivity

Figure 11.2. Einstein's Theory of Activity

The following is evident from the two figures.

Theory of Reactivity. As an absolute, the present reacts to the activity of the future which is unknown to it by following the future closely, leading to an ascending gap from the past lagging.

The past lags because it is the light that speeds up to be ahead of the future. The present is near the future because it is heavy. The present speeds down and drags the future towards it with its gravity.

The light is the center of the present's gravity. Before merging with the center, the present has a past. The present within the past also has a future not curved by the present. The focus is on the point potential energy, where m is reproductive while it has a position.

Theory of Activity. After the present disappears, two pasts remain. The first past is light which moves ahead with the future's kinetic energy to be present as a photon. The second past is a photon, but it has the potential to speed up like light to position itself above the first photon and be a quantum photon. Quantum photon is 12, the one that becomes two. The focus is on the rectilinear kinetic energy, where vector v is constant while in motion and reproductive like a scalar c.

Limitations of the Two Physical Theories. In both theories, the present is a passive observer. The future is dark because it is productive in descending the past's heaviness by reproducing it. The present is light because its past was reproductive. Nobody cares about the past that gifted freedom to the future to be anything it wishes by shaping the present with its potential to be different. Everybody worries about the future, wishing for somebody who does something in the present while everyone else is just observing to copy somebody's behavior. Thus, the two theories generate mass inertia.

2.6.3 Two Metaphysical Ideas of Interactivity between the (Interactive) Past and the (Active) Future, when the Present is Reactive

First, quantum mechanics of the past's interactivity

Quantum mechanics is the special theory of relativity. It is special because it relates the future to the present immanent within the past. It focuses on the microcosm from the past, not the macrocosm from the future. The past is interactive. It is the reality which is linear. The present does not relate to the past. The past is absent and left with nothing because the present is like the past. The past was in ascending order with its linearity when present as light before running out of life and going in descending order into the blackhole of darkness.

When a subject observes the twin light, the mind identifies the possibility to be present for eternity by reproducing one's copy. To be reproductive, one first needs to be heavy like God. With consciousness of one's past as reality known and certain, but not desirable, the subject decides to stay still in a horizontal order to copy oneself as time. The subject is present as the Holy Spirit within its copy. By deciding to react, the act transforms the subject into the God of time, with a "special" linear reality distinct from the past's "generality."

Figure 12. Subject As An Observer
Me observing my friend for figuring out what I couldn't when groping in the dark.

In quantum mechanics, $E = mc$.

Once a cat (subject) reacts to jump out of the present, it dies to be black, absent from its position. The cat that was present and acted after observing the twin cat, exchanged its position with the momentum to live as white in the afterlife, with the potential to be present if He wishes.

The cat still present (since the cat that was present observed two cats, of which one reacted and died) is alive, observing and enjoying life as grey. The light appears grey because of the shadow of the past's darkness which is dark and black and the potential of the future's whiteness which is bright and white.

The cat that is still present attracts the cat which has the potential to make his mass reproductive with his gravitational force [at the speed of light] because he is not related to Him and therefore is Adam as the Holy Spirit. By reproducing Himself, He becomes Eve with Adam as her twin reality. After giving birth to Adam and Eve, Father receives the blessing to transit from the hellish Black Hole of the past to the heavenly White Hole of the future.

Adam and Eve are together because Adam is the present and Eve is the time with a potential to be present if she wishes. God is the future who produces the illusion of the hellish Black Hole with His imagination of the Garden of Eden as a White Hole left by Adam and Eve when they chose to go down the hole at a complementary speed of light to norm energy. They have two paths: return to where they came from or grope in darkness with uncertainty about their future.

Second, the Quantum loop of time's proactivity

The Quantum loop of time's entanglement by space, which is present within the mind of Eve when she decides to incarnate as light, is a complex theory of time's parabolic reality. The Quantum loop focuses on the mesocosm the time forms through interactivity among its different transforms to norm the space as a complex of real, nonlinear, larger-than-life feminine reality whose present is a dull and imaginary, linear, part-of-life masculine reality whose future is bright.

Space is like a child. A child is a possibility when Adam, the subject observes Eve, the potential subject as his twin, the one who complements him with a two: Adam as one and child as zero, her copy. The subject's observation is dull because Eve is an illusion from his past when he imagined a mother and reacted to confirm if indeed the mother of his child is a potential reality. The subject forms a general nonlinear reality by pretending he is dead to norm a special linear reality that takes him to heaven for a second life to fill the hole in the space with a child.

Forming general and norming special leads to transforming the complex of a real formed to be present and an imaginary normed to be future. Nonlinear is present on top of linear, the future. Together, they transform into parabolic, the past which incarnates as space after a hole is left when the dark leaves as white to surface the black. The black is Mother, the dark's twin. Father incarnates as the child to fill the hole on the right of the Black, which is now in the center of what is left after Father leaves a hole in Mother's mind for a daughter right beside her.

In the Quantum loop, $e = c$.

Father, who leaves a hole for togetherness within Mother's mind, fills the mind with the darkness of an uncertainty that turns her black. Before turning black, she is dark like the Father because she is the past of the Father who is present as Adam. Adam can be real and be present only if his reality in the past was the mother who gave birth to Adam as the child to fill the space in her mind as if the space is the hole.

A child takes birth in a mother's mental realm as an egg. The mother reproduces her reality in the physical realm to produce the daughter as a twin egg. Twin egg means two eggs, the second of which is the daughter and the first the father of the daughter. The father is the physical reality of the mental egg. A mental egg reproduces itself as a physical egg for deciding the benefit of producing itself as an intellectual egg in the futuristic metaphysical realm. The intellectual egg acts like God, the future, who reproduces its twin copies. It is the fourth egg, the past whose future is Mother. The physical egg reacts like Holy Spirit, present as a twin copy. The mental egg stays inactive like Father, the past, who has the potential to twin copy. The second egg is the astral egg, which incarnates as a star, reproducing three eggs as the mother's triple copy after producing three eggs as the father's triple copy. Thus, as the seventh, an egg divides her reality into six eggs through Flemming's cellular division.

The egg is the energy "e" which eggs herself at the speed of light "c" for the addition of a copy. By observing her future as God, the copy norms everything as a "theory" (child) with a subtraction of five copies. With six copies, the egg is left as a zero, the time that lives her present as the space through three maternal copies and three potential maternal copies until the space realizes that he is the father.

Let's visualize these two ideas using artificial intelligence.

Figure 13.1. Quantum Mechanics of Interactivity

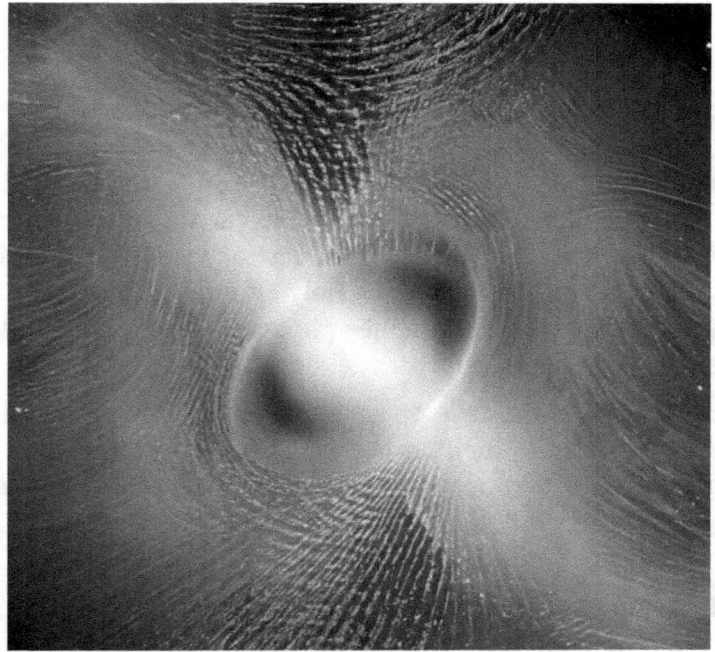

Figure 13.2. Quantum Loop of Proactivity

The following is evident from the two images.

Quantum mechanics of interactivity. The past becomes a sphere that influences both the present's linearity and the future's nonlinearity. The present's linearity circulates like a pi without the past as a spherical pi when the pi becomes whole. The future's nonlinearity integrates the whole within its darkness shadowed by the past to be a wholesome pi. The wholesome pi is a "continuous pi" because it is a replication of the spherical pi like a cube, giving depth, breadth, and length to time. The focus is on the linear dynamic energy of light as a wave.

Quantum loop. Time becomes a wholesomewhole pi that makes the past, present, and future reproductive as a complex to be a quantum loop (the space loop). Wholesomewhole pi positions itself as the area which experiences four dimensions of space by transforming itself into a photon that lights the space. The focus is on the complex technological energy, i.e., spherical energy, which is the energy of the past, present, and future concentrable at a point in time.

2.6.4 Two Dynamic Doctrines Moderating Time's Proactivity with Space's Hyper-activity

Western religious doctrine is the foundation of the science that "approximates" reality by perceiving God as the ideal. Ideal's nature is immanent within everyone who emanates in God's image. God is heavy with the weight of creation He carries as a Creator. His image, the Creation, is light but becomes dark when it goes past its lifetime to present itself as God's Holy Spirit to the Father. Father is Creature. There is an urgent need to open modern science to the formatives within Eastern spiritual doctrine.

Eastern spiritual doctrine "proximates" reality by conceiving the Creation as the feminine (Mother), the Creature as the masculine (Child), and the Creator as the androgynous (mind-born oneness of the feminine and the masculine within the whole, before differentiation of the gender with the reality of the Mother).

Mother's nature is the reality within one who Fathers the gender as the goal of his creation. Gender lets the Father be the Creature with three copies: negative feminine which repels her reality to attract the masculine as positive and real for exchanging herself with a greeter. A greeter is neutral and a substitute for reality since it is known to be whole which complements the real with a primordial element.

The cost of proximity escalates challenges when servicing approximity as the paradigm. The present has the potential to be different in the future if the bird does not guide the egg to copy her to make her past perfect and instead lets her child free from the cage that limited the horizon of his imagination and forced him to mimic his mother.

However, if the mother does not acculturate her child to her reality and the training is left to nature, the child will not be aware of the challenges of her past and may idealize her, following her every move, wishing to be a leader like her. The child will be emotionally linked with her radiant love, wishing to globalize himself as her gift. The potential to make the present different is also present within the past if it does not remain in darkness.

The negativity of the past drives one towards the positivity of the future, starting with a white board to fill with the colors one wishes instead of living with a black board already filled with all the colors, with no space left but to move with it as dark matter, like an absolute. The absolute forms relationships with everyone who may have an idea of how to be a leader leading the discovery of space beyond infinity to matter by exploiting the grey matter.

According to the wisdom of India, the reality becomes real when one is a zero who "proxies" the universe for the "endoproduction" (*Maya*) of a "mate," the fifth eye of God with a mass of four.

One faces the universe, the two, with one eye toward the future. That one eye is the space's fourth eye ahead of the time's three eyes.

Two faces three, the multiplier of the universe with a triple copy, by rotating the second eye towards the past when one was God before two became multiplier.

Three faces four, the consciousness as the mother of the multiplier of the one seeking to be God again with a belief in oneself as the divine, with the first eye observing the present when one is a zero like the time.

Zero behaves like one with the oneness within the three that it copies to be one with divinity, i.e., deity. Therefore, zero is the "King of deity" (*Indra*). The deity is the "guider spirit with three eyes" (*Trinetra*) whose nature is to make the base reproductive like a "guider" (*Guru*) to "quantum loop" the reality of time with its copy.

As a copy, zero times itself to be the third, central eye, guiding one to tap the potential of two: himself as the left eye and herself as the right eye. A focus on zero's desire to relate with everyone for changing its reality diverts one from recognizing the reality of the one who is absolute. A focus on twin zero, i.e., eight, Mother Nature, is key to knowing the potential (that makes one an eight) beyond absolute to be real. A focus on ten, the primordial self, is key to knowing the primordial illuminator of reality.

2.6.5 Formatives for a Paradigm to Be Scientific

Science recodes the nature of reality into a theory so that one may encode the elements of that nature to make an idea real. Theory forces one to give credit to the scientist before distributing the idea as original. Credit leaves a theory-effect within the idea since the idea may originate from the theory. A scientist gives credit to other scientists for the idea tested through the formulation of a theory. When an idea belongs to a collective, the theory becomes original because the origin is identifiable as an individual.

In reality, the originator of the theory is not the individual testing the idea with a theory to be ideal for everyone, rather it is the collective distributing the idea wishing for an individual to grab the opportunity as real. The idea is contaminated with the collective's ideal-effect. The collective believes that its ideas have a kernel of truth for establishing its supremacy, so that each individual within the collective may realize its potential to be supreme.

Are you conscious that the reality you are presently conscious of affects the realization of what you are potentially conscious of but not aware of whether it is real or an illusion of your awareness from the past?

Potential conscious = f (Present conscious)

Let's consider the reality you are presently conscious of. Presently,

- I am the speaker
- You are the listener

The "present conscious" affects the "potential conscious." Potentially,

- I am the protagonist, speaking to substantiate a thesis
- You are an antagonist, listening to initiate an anti-thesis as the potential thesis taking the substance present in the thesis and the potential for synthesis by your mind as a function of the reality it is conscious from what it has known in the past and retained in the brain as the value that lights your knowledge.

Reality Is Not Real. The reality your mind is conscious of is not real, but an illusion of the past that perpetuates as real.

Why does the conscious produce an effect that the reality is real? Conscious element flows like water. It has a normative effect. What I am conscious of at t0 effects what I am conscious of at t1 and what you are conscious of at t2. It brings normative sameness in the space, mental or physical.

Conscious-effect = *water-effect.*

The conscious element makes a thing real with the reality of potential within it to be true. Potential flows like fire. It has a formative effect. My potential to be conscious about s_0 effects what I am conscious about s_1 and what you are conscious about s_2. How I conceive s_0 effects how I perceive s_1 and how you experience s_2.

Fire-effect = Form the potential-effect takes before it flows like water to make one conscious of its potential to be constant.

i) Biology of how zero forms everything emanating from it with an idea.

Normative-effect is the light of the past's "linear reality" inherited without consciousness of its origin. Linear reality implies that the reality of the past is super positioned over the reality of the present as a quantum of field, guided by a theory that reality is special and left as constant when time moves to the right of the origin.

Reality is special because it keeps its relation to the origin as it correlates with the variable time within the space that originated it. Correlation keeps the order of reality real. If the origin conceived the light as reality, then one may perceive reality to be real and not correlate and instead behave like an absolute, forcing everyone to be relative. One correlate because one comes late after the origin relates the reality as is. The special theory of relativity, also known as the quantum field theory or simply quantum mechanics, takes the origin to be a negative one by assuming that the origin goes before one.

Consequently, from the perspective of the special theory of relativity, zero is the universe that forms everything by universalizing itself as an ideal. The universe is light (=0). It expands horizontally, moving like a ball (photon = 0), until it becomes heavy (=1) after adding infinity, standing as a vertical (=1), the idea that matters but comes late after the "ideal" (=0). Idea = Ideal-effect = 1.

Normative-effect is light, the reality inherited by everything within the vertical. Everything (=1) matters for the existence of zero as an "illusion" (=1) once it transforms into one. Reality (=0) transforms into an illusion (=1) when it does not exist as a "thing (=0) that originated from the origin. Existence is real (=1). Existence is the goal (=1) of reality.

The origin is a negative one that formed the normative-effect. A negative one is a cosmological constant, embodying the dark energy of the future yet to be illuminated. The negative one is dark because it is the origin of space that illuminates time with its light. The light emanated from the origin, which remains in the dark after subtracting infinity as the one immanent as real within the reality of zero.

ii) Physics of how one norm everything immanent within it with a theory.

From the perspective of the general theory of relativity, one is a theory of possibility that makes the possibility real with its reality. One's reality is an illusion because it is realized only if the possibility is real. Once realized as real, one is "ideal" (=1). Real is the generalization that generates reality as a special case. A special case is replicated as a specific form of spacetime as a thing to norm everything with the general form of spacetime. Continuum is the general case. It is real like one, the theory that is constant across spacetime as the spacetime. The theory about everything is the theory of everything. One is a photon, which is light until it becomes heavy with everything within it.

The possibility is the special case, the reality, which is zero within the continuum formed by the theory to test infinite ideas. Each idea is a theory-effect that varies with spacetime. It is "everything" (=0) the thing forms. Zero is the cosmological constant, embodying the dark energy of the past.

The origin of one is a two. Two is universal, comprising many universes as ones that universalize their spacetime to be zero eventually when time ends. Two is the time that ends right when only space is left as a zero. Two is dark because it is the origin of time that illuminates the space with its light.

iii) Chemistry of how two transform everything with its reality.

From the perspective of the complex theory of relativity, two is the universe. The universe is everything. It exists. It is real. It is spacetime. It is heavy. It is a theory with infinite ideas. It is the cosmological constant, which embodies the dark energy of the present and lightens over time as it comes to light with the reproductive element.

One is light (photon). It is the thing that lightens the heavy. It is the time that lightens the universe by universalizing it over space until it becomes three. One is the idea that makes two the ideal where the idea is omnipresent. One is the reality that transforms the real with the imaginary.

Three is dark (origin). It is the space for something that "potential exists." The space is illuminated by the light of everything that exists. The space is the great attractor because it wishes to bring something that does not exist into reality while waiting patiently as an observer like a subject until showing the color of the king by transforming the linear I into a nonlinear U like a trident. Three is imaginary. Its potential exists within one's imagination. Two is illusionary. It exists as one's past. One universalizes its past to entangle the future with the imaginary. Three is the future of time as space.

Three is complex, proactive relativity: two is real, active, general relativity and one is imaginary, reactive, special relativity because two is superpositioned over it and therefore there is inherent uncertainty about its position. It may become conscious of a position of inferiority if a subject is observing it carrying the mass of the universe as a worker. Conscious force adds momentum to the origin by subtracting one, making the system as a whole interactive. Interactivity squares the reality of the observer subjected to the inferiority-effect and frees the members from relating to one another from a position of asymmetry.

iv) Science of five digits—Quantum Absolutism of the Three Theories

Different theories of science highlight diverse dimensions of reality for knowing the wholesome nature of reality.

- *The special theory* says that by reproducing a zero, we can produce everything as one. Zero is the theory as well as the ideal; it is light as well as a photon. One is the theory-effect as well as the ideal-effect. Energy in the universe = 0. Negative energy and positive energy balance one another. Competitive force predominates in nature.

- *The general theory* says that by producing a one, we can consume everything as a two. One is the theory as well as the ideal; it is light as well as a photon. Zero is the theory-effect as well as the ideal-effect. Energy in the universe = 0. One that is producing another one by reproducing itself becomes negative as it repels itself as an electron seeking to attract the proton, left with its positive potential, to form the origin into the neutron. Complementary force dominates the gross visible supernature that fuses nature within a body of consciousness.

- *The complex theory* says that by consuming a two, one can be a three—something beyond everything. One is light because it is superpositioned as a quantum whose replication makes zero heavy with the mass of one. Energy in the universe = 0. Two as the universe services its mass to one at the speed of light, which the zero trades at the speed of light. However, one is an illusion; zero is heavy because it is on the bottom as the bottom quark and the one is on the top as the top quark. Quark is the present value of the top or the bottom, strange repulsion or charm attraction, up togetherness of two or down otherness of three. Supplementary force decides the subtly hidden supranature infused by the fusion of nature and supernature into an entity.

From the perspective of the simple theory of relativity, three theories are about five digits of the invisible hand, represented by [-1,0,1,2,3]. When reality is reproductive, origin embodies the dark energy of time. Dark energy produces and destroys everything by consuming what it has produced to reproduce what it has the potential to produce. Dark energy is a negative one.

When reality is produced by the invisible hand, the origin embodies the light energy of everything yet to be reproduced and illuminated as light. Light energy is zero. When reality is consumed by the universe, the origin embodies the heavy energy of everything consumed and perpetuated as light. Heavy energy is two. One is the cosmological constant. As a reproducer of reality, one reproduces the oneness within everything produced, consumed, or traded for reproduction. As the origin, one embodies the bright energy of everything that is reproductive.

Three is reality—the circular reality of time with interactive relativity, i.e., the simple, squared relativity. It is the common denominator of the three theories. Reality is an illusion until substantiated by the real's absolutism. As an absolute reality, three remains a common denominator of the potential within the three theories. Three is light that illuminates one. One is dark until illuminated by three.

Five is heavy with the consciousness of one illuminated by three. Five transcends sentient force. Five is sentient like God. Four is guider. Three is divine. Two is the fire that consumes everything. One is the water whose invisible hand produces everything. Zero is the air that transforms everything by reproducing it without adding anything. The negative one is the earth that becomes reproductive when it is added to something. Negative two is the ether, the reproduction subtracted as a thing from space, to sum its reality over time. Space is a six that self-perpetuates its reality as a three, the time.

v) Absolutism without the Three Theories.

From the perspective of the rational theory of relativity, eight elements that constitute an octave of elements define nature.

Nature = 8. Nature is the cosmological constant, embodying the dark energy of space that makes its potential reproductive to be present as the light. The nature of nature is to be the multiplier of her perpetuating value for realizing a "horizontal order" (entropy).

Eight is the reality that remains constant because it is space's square reality with inactive relativity. Three is the multiplier multiplying reality with the three dimensions of time (past, present, and future). Five becomes God by perpetuating the value of three within the "universe" (=2) with his absolutism. God transcends sentient force with four dimensions of space divider (east, north, west, and south), seeking to be immanent as the center. God is not the center. The center is the one that divides nature into two by adding five for its growth as a six. Thus, Center = 16.

The base for the center is a ten. Ten is the primordial illuminator of reality with a three. Ten self-perpetuates itself as a five.

vi) Potentialism of relativity with Five Theories.

From the perspective of the irrational theory of relativity, the center breaks away from both vertical symmetry and horizontal symmetry with its angular symmetry. Angular symmetry has the potential for relating with each point within space at an angle with a degree of confidence about the triangular reality of the point. Triangle is created by the system as the third point which transcends the space and circles it to demonstrate that its infinity is circular. The system is the angle—the causation—the reason for the symmetry of two mediated by the third. Reality is causation's triangular reality with relativity to itself.

vii) Primordialism of relativity with Three Paradigms.

From the perspective of the whole theory of relativity, different paradigms highlight diverse ideas for knowing reality as a whole.

- *Quantum Absolutism* is a paradigm that says that when one makes three reproductive, one realizes the reality is seven with a growth of six.

- *Absolutism* is a paradigm that says that when one takes four as the divider of space into two: one with the potential to be time and another present as the consciousness of that potential, then one becomes real with nine with the nature of space immanent within it.
- *Potentialism* is a paradigm that says that the whole is the sum of reality and the real, i.e., sixteen.

The common denominator of the three paradigms is that energy = 19. Energy is the one that becomes real with nine. It is three that makes sixteen reproductive with time for oneness with the two. It is the sum of seven, six, and a three that is reproductive with a two after the one divides the reality to be a six. The focus is on the cause's reality with a relativity force of three causes. Three causes are in line symmetry with the infinity through the continuity of their discontinuity.

viii) Primevalism of relativity with Four Paradigms.

From the perspective of the wholesome theory of relativity, although different scientific theories illuminate diverse dimensions of reality to help one know the wholesome nature of reality, the theory-effect leaves residual contradictions. These residual contradictions confuse the ideal of making science the base for knowing the reality hidden within metaphysics. Therefore, diligence is needed on the part of the investigator to transcend beyond the ideal-effect that deifies the forefathers of science and forces every child to be a follower.

With diligence, an investigator develops a potential for knowing reality as a wholesomewhole, where science and metaphysics supplement one another. Science predominates. Metaphysics dominates. Investigator decides whether to let metaphysics dominate or seek higher-level synthesis guided by epistemological element, the benefit one accrues by investing in the cost of synthesis. The cost is whole. It is incurred with the present's absolutism with a risk that it will be a tool for someone to be para-absolute by developing the highest-level synthesis. The benefit is wholesome. It is accrued over the future's primevalism.

ix) *Effect of Past's Relativity on Present's Absolutism: Physics of How Theory Supplements Everything*

Physics is derived from a theory that the past is the eraser of the present. In the future one can be a three to create space for the four to be a seven. From the perspective of the wholesomewhole theory of relativity, by moving forward in time one realizes that at a macrocosmic level:

- Zero = Reality = Past (whose light moves as a photon for illuminating the present as two digits) = Simple = Death = Eraser

- One = ALL (the future can be after everything is illuminated as the light of the past whose dark energy follows the present into perpetuity and generates asymmetry) = Future, present as a proportion of the past knowable through science as a method to know the future as a whole = Whole = Crystal Ball = Ball = Photon (The crystal is an illusion the ball crystallizes for its endoproduction from its reality of zero) = Complex = Dead

- Two = Two digits = Present (where a digit is imaginary, it makes two a wholesome of real and imaginary) = Real = Wholesome = Spacetime = Consciousness = Life = Everything (one asks for to be absolute) = Absolute = Mighty (whose might time illuminates as light)

- Three = Digit = Imaginary = Time = Wholesomewhole (Time moves as a digit within consciousness to become real when digitalized into two digits that makes one complex) = Alive = Light = Might = Illuminate

- Four = Space = Digital (space for digit within five that is present as "AL", the real, without three, the digit) = Virtual = Symmetry (Symmetry present as digital without "A" that leads to "L") = Believable (because its consciousness is present within three) = Dark = Illusion = Crystal

- Five = "A" = Causation (which causes asymmetry with L's potential) = Belief = Theory = Illumination = Creator = Heavy

- Six = "L" = Cause = Believing = God = Illuminator = Mass = Ideal
- Seven = Almighty = Asymmetry = Potential = Copy = Believer = Perpetuator (God, immanent within the believer, emanates everyone present as life after All are dead following the death of zero. Zero makes one alive as the time that comes from the future but never goes back to the past) = Master of destiny = Destiny = Thing (that matters) = Matter = Idea (that works) = Work

Relativity supplements the absolutism of the complementary theory and the potentialism of the competing theory. According to Relativity, by moving forward in time one realizes that at a macrocosmic level

- One = ALL (that brings everything into being over its lifetime, guided by the light of the past whose dark energy follows the present into perpetuity and generates asymmetry)
- One is a photon that norms All with six quarks, i.e., with 100,000 because among the six quarks, one is essentially a photon that has become a quark.
- One's primordialism (frontrunner) is relative because it is also essentially a "zero"
- Zero is a cosmological constant, embodying the dark energy of the past. Zero is not the potential cosmological constant embodying the light energy of the future that attracts the growth of reality within one whose nature is to be zero eventually.

Ascending sequence forms by discontinuing the descending sequence from the negative infinity of time after the universe comes to light as a photon.

x) Cause of the Present's Absolutism: Metaphysics of Idea

When derived from the past, the present is expected to reproduce the past. When the past is reproduced, both the reproducible past as well as the reproduced effect of the past already gone shape the future of the present. From the perspective of the growth theory of relativity, relativity of the past moving forward in time and the superposition of the future reproducing the present by pushing itself backward in time are incompatible. Past can't move forward its effect linearly when the future value is present to curve that effect with the sum of the past's past-effects.

By moving backward in time through introspection, one realizes that at a microcosmic level:

- Dark = –1 = Past = Momentum (Universalizes as a being the existence of God for becoming all God is within the universal's anti-existence through symmetry between time yet-to-exist beyond the future and space that will exist thereafter)
- Light = 0 = Present = Position (Universalizing the dark as the light that universalizes itself by becoming a photon)
- Universe = Future = Quantum Position = Heavy = One = 1 (Eventual state of universalizing the present for symmetry with the past)
- Time = 2 = Quantum Momentum (Universal that puts the future in superposition with the potential within the present's light to be dark after the present services its position and trades the momentum to become the past)
- Space = 3 = Quantum = Superpositioned (Universalization whose existence will be real if the universal's anti-existence was not imaginary and the universe was not an illusion of the past's light)
- Reality = 4 = Theory = Superposition (Universalizer that can be real in theory because it needs space for existence. It is measurable with certainty as the space is visible and illuminated without the reality that is invisible but existing)
- Realization = 5 = Ideal = Superpositioning (Universalizable whose reality is invisible because it is an ideal that exists within a believer)
- Realizer = 6 = Idea = Believer = God = Superpositioner (Unique that preexists as an idea a believer copies from God to be the realizer of the superposition God gifts that quantum positions one as the future of everybody whose

energy is dark like hell. Dead live in hell before they are alive with life unless zero is in a position to change its belief while it is present as life within the universe that will exist as heaven in the afterlife.)

Universe's strong force as a macrocosm transforms Everything into a nuclear, horizontal sequence of ones. Everything's weak force as a microcosm forms the nuclear as "clear" (from endoproduction), a backward sequence of zeroes. Nuclear's electromagnetic force as a mesocosm norms "nu" (endoreproduction), a forward sequence of twos.

Sequence's gravitational force as a masscosm points to a four as the origin of two, one, and zero as proportions of a digit. Sequence's consequence—the descending sequence of four—suggests a fifth force that accounts for the uncertainty because two is three, one is two, and zero is one. The fifth force is the ascending sequence of five. It is the sentient force of the fourth that forces the sixth to behave like the fifth after servicing the infinite for the endoproduction of one as the finite. The finite trades two to be the creature moving forward by sequencing 1,2,0. 120 is the unicellular organism, the site of division by two into an "octave" (=60) of the sound of reproduction, for realizing the eventual state of a zero, the subject observing the technological reality—the "object "(–3) of its creation.

Six is the Cause of the Sequence. Sixth is the one which forces the fifth to repeat the force within the second, to make it conscious that it is the third. The third is first brought to light as a photon by a zero, forcing the first to be the second, and the second to be the fourth. Sixth is not a force. It trades the force the third emanates as the consciousness of the fourth to behave like the fifth, making everyone behave like the fifth eventually. Fifth is the limit of reality. Sixth is Realizer, realizing the reality of physics. The realizer from whom the reality emanates is God.

Growth is not a theory but a metaphysical idea that one enjoys a superposition because the universe universalizes itself to illuminate zero's position as its present.

- The universe forms everything to norm all. All transforms into the matter at the speed of someone's light who brings a zero into existence as one before universalizing itself as a negative one, the anti-existence, whose identity remains unknown as the cause of immutable uncertainty How: through endoproduction of one from zero who takes zero's position, giving five a superposition for the endoreproduction of two with its twin reality and continuing (with time) the ascending sequence into the positive infinity of space.

- The universe expands horizontally as a zero that is light until it becomes vertical and heavy like the one after adding infinity and diagonal and dark like an absolute zero after subtracting infinity. The absolute zero and the one is the same. The asymmetry is an illusion, which is essentially a zero darkening the absolute zero with its energy that is conserved after one stops self-reproducing. [i.e., Absolute zero is not a negative one that becomes one after it forms the universe as two, seeking identity with nature through the growth of reality]

- Negative One is a cosmological constant, embodying the dark energy of the future yet to be illuminated whose energy is present as a unit of light, i.e., photon, the ball that adds infinity to be one and subtracts infinity to be absolute zero.

Growth derives from the Western metaphysics that the zero is a devil who hides its position so that it may follow Satan, the negative one, who takes the lead as the Holy Spirit, one, because the three are the eyes of the father looking for two, the absolute son with a twin identity of father and mother, to reveal their truth as God, six with a symmetry of three and potential three.

Indian wisdom identifies the devil as *sura* (within God, concordant with the future unfolding with time and therefore zero at present), Satan as *asura* (without God, the discordant force of the past and therefore a negative one at present), the Holy Spirit as *Trinetra* (Guider spirit, with three eyes whose spirit is reproductive because it embodies I am a deity consciousness to make one a deity), the father as the absolute son (*Hanuman*), the mother as the consciousness of the "absolute daughter" (*Jiva*) who leads one's growth for realizing the reality of the absolute deity (*Shiva*), the goalkeeper, to be "whole" as sixteen (with God, i.e., growth within ten, the living God).

xi) Consequence of the Future's Corrective Action

Future's corrective action activates dynamic reality. From the perspective of the entropy theory of relativity, relativity experiences entropy when there is a growth in the present's absolutism that compensates for the cost of discontinuity from all relationships with the past by deciding not to reproduce the past in the present.

Dynamic reality is the "reproduction within one" of reality to produce a universe related to the mediation of the present. With its position, the present gives superposition to the future for taking the momentum away from the past. The past is without one who is present for reproducing it while moving as time. The time is within the space "entangled" through the endoreproduction of a proportion traded from the past. The proportion is related to the future when multiplied into infinite which is dark and serviced as the matter that makes the finite reproductive as its present.

By moving horizontally along time, one realizes that at a mesocosmic level

- Two is the cosmological constant, not the absolute cosmological constant which embodies the heavy energy of the present and lightens over time through the reproductive element to illuminate the great attractor as the disproportion of the space's light from the future.
- Universe = Everything (one has produced to mediate like God all that gets created within the space by time) = Photo = -2 (before radioactive decay of everything into light and microactive entropy of light into sound) = Micro = Complex = Uncertainty (about when the chance will appear to make the things simple as God's invisible hand, the digit that matters for everyone's appearance) = Ideal
- Unit = Thing (A thing has a position. Universe has a superposition, it is on the top of the thing. Universe's absolutism entangles the thing) = Light = Photon = Radioactive decay (when everything is reproduced by chance as a thing that norms everybody) = Everybody = Macro = Illusion = Chance = Invisible hand = Digit = Matter = Dynamism = -1

- "N" = Microactive entropy (as the radio decays into sound when active as light, the macro that gives life to the micro, which is complex because it is God's photo that brings nothing to life as if it is real like everyone who enjoys it before taking the position as an observer) = Nothing = Life = Sound = Real = Meso = Everyone = 0
- Something = 1 = Mass of the sound that sums itself as time progresses = Sum (which itself is imaginary because time does not progress, its position regresses as the space progresses when it is time to reproduce the whole's expected linearity as the wholesome's unexpected wave that sums the time) = Imaginary = Time = Position = Observer = Heavy

= Whole = Linearity (whose endoproduction is real and expected within time, as time is pushed to make space for the whole that pulls forward into a superposition) = Space = Superposition = Simple

= Wholesome = Wave = Nonlinearity = Momentum (momentum superpositioned by the past as a quantum to create the future as its wave, the theory the life is zero without it) = Future = Mighty = Destroyer = Quantum = Theory = Dark

= Death = Past (Death of momentum when the energy moves as momentum to go past the time and remains present only as a potential for quantum entanglement after it is dead) = Believing = Might = Dark matter = Wholesomewhole (that makes wholesome whole with its might's invisible hand)

= Dead = Present = Hell = Belief = Entanglement (that makes the present hell, as good as dead) = Good = Idea (present as a belief, which is dead until the believer brings it to life in the afterlife before reversing the time clock) = Black hole

= Potential = Omnipresent = Believer = Afterlife = Heaven (Believer remains present as one within the belief that believing the might of the past makes one mighty as a destroyer in future before becoming past) = Quantum entanglement (of the potential as heaven, so that everybody has a fair chance to appear from nothing as a unit of life before the light becomes an illusion without the body whose position is uncertain once everyone enjoys the momentum to be present and leads the time to begin its regression until universe becomes complex) = Fair = Clock (which is reversible for one in doubt yearning to go back to Black Hole by behaving like dead, thus becoming dark as matter) = God

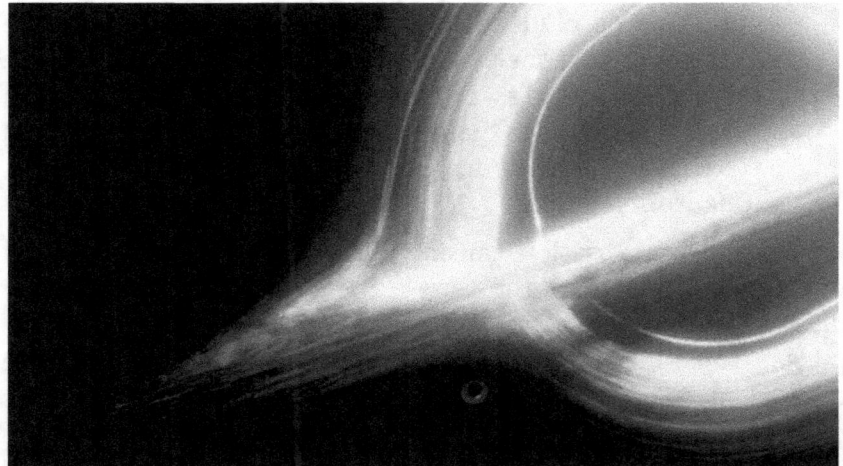

Figure 14. The Great Attractor

The consequence is the great attractor of the descending sound of the past, reproducing the ascending sound of the future, to add forward momentum to the horizontal sound of the present with its backward position.

The three views compete to explain the descendance of everything from nothing but God.

- Zero's primevalism (latecomer) implies that one was a "negative one", a cosmological constant embodying the dark energy of the past, leading to the universe's infinite expansion into the positive infinity of space until the universe becomes vertical like one to be dark like a potential one, i.e., negative one [Quantum science]

- Zero: A potential cosmological constant embodying the light energy of the future. It follows the universe's formation as a ball that is light until the edge of the negative infinity of time where it becomes heavy. That ball is a photon. That edge is diagonal, descending from the polarized darkness of time to the de-polarized heaviness of space, where time $= \infty$. [Relativity]
- Negative one becomes Positive one with the addition of a "two" (an absolute cosmological constant embodying the heavy energy of the present. The present mediates to make the universe flat as a (positive) infinity of space which becomes heavy with the superposition of the time's eventual state as a zero after descending from its negative infinity. [Quantum relativity]

One is the causation causing the energy to be positive by moving with time like the time to be present for eternity, the infinity of space. One follows the cause for the energy of the present to be negative when one enjoys the time, leading to the entanglement of consciousness with the perpetuating value of its past and descending joy when the time changes. One behaves like an entity that superpositions negative energy over the present to give momentum to the positive energy of the future for conceiving oneself as the subject leading with the neutral energy of the past. The subject is a Wisher wishing to observe the reality beyond the present as a mirror for knowing its future so that it may change it to a desired one. Paradoxically, the known reality limits joy because of the possibility of knowing the unknown reality the known reality strangely repels as its effect.

Present experiences the bottom's (hot) thermodynamic force within its "present value" (Quark). Future enjoys the top's (cold) sentient force within quark. Past strangely repels the present's negative electric force from its quark. The repulsion "charm attracts" the future's positive magnetic force into its quark. The togetherness of the future with the present's reality as negative two within the past's reality as negative three ups the thermodynamic force of the time's quark (Up quark has a charge of $+2/3$). The otherness of the time's causation as one who makes the present negative as a metric of the past's reality downs the sentient force of the space's quark (Down quark has a charge of -1/3). There is a void in the metric about the past's reality when the present value is decomposed into six quarks. The void leads to the entropy of reality.

Order Decreases Entropy for Realizing the Base of Ten.

- $+2/3 * 3 - 1/3 * 3 = 5 =$ Entropy.
- $+2/3 * 3 + 1/3 * 3 = 7 =$ Reality.
- Reality orders the realization that the Positive quark moves one within time advancing with the positive energy right along the future. Order $= 7 - 5 = 2$.
- Order decreases entropy because the Negative quark moves one without time decaying with the negative energy left by the present. The order has the potential to realize the base of ten with descending entropy. $2 \times 5 = 10 =$ Base.

xii) Dynamism of the Sound of Reality Forms a Super-reality.

Technological reality is beyond relativity. It is left as an object within the consciousness of reality. Super-reality is proportionate, servicing zero as the ideal for trading one as the theory to exchange "–1's" absolutism with the three's potential for growth with six copies. It is the common denominator the theory of relativity rejects for substantiating reality with a hypothesis. The hypothesis takes a hypothetical idea that the absolute is an illusion endoproduced by the future. A scientist assumes the hypothetical idea to be true until proven false in the future by someone who is not a devoted follower of the scientific paradigm and therefore has no locus standi in the scientific community.

- $-1 =$ Absolutism $=$ Scientist $=$ Capability $=$ Capability of reality present without the scientist discoverable as science $=$ Point $=$ Observer $=$ Dead $=$ God $=$ Force
- $0 =$ Ideal $=$ Science $=$ Servicing $=$ Servicing reality by conceiving imaginary after perceiving illusion as true because it is a copy of the real that exists as consciousness $=$ Line $=$ Consciousness $=$ Observable

- 1 = Theory = Art = Trading = Trading reality with the art of first perceiving the illusion the sound generates with its echo that reverberates like it is real, long after the radio dies into a degenerative wave which remains present into infinity, never converging as a line into a point because a scientist is a point diverging the observable with its observance as an observer = Future = Life = Illusion = False = Copy = Observance

  = Idea = Present = Absolute = Exchange = Exchange reality with an idea that is imaginary until substantiated with a theory that is reproducible in an alternative form, which is true = Imaginary = True = Believer

  = Past = Growth (Past's whose potential for growth as a scientist is present within the future as an illusion) = Order = Believing

- ½ = Time = Entropy = Dimensionless (With the entropy of future, time remains as dimensionless, norming half-life) = Action = Real = Belief = Half-life, because symmetry makes half dead, which is -1/2 – the ninth case superpositioned over the eighth case within a potential of eighteen.

- 1/3 = Space = Investment = Hypothetical = Half hypothesis (With the past's potential for growth, space is a third that the present doubles with time reversal through its formative reality as a hypothesis that charges everyone into action losing order) = Everyone = Reality = Believable = Heavy energy of the present = Matter (of which half is visible and rest is dark)

- 2/3 = Causation = Technological = Hypothesis = Time reversal = Formative reality = Charge = Unreal = Unbelievable = Super-reality = Proportionate = Dark energy of the past, superpositioned over the light energy of the future = Anti-matter

- Eighteen = Potential = Case = Cause, which discharges the charge to disentangle the present's reality from the past's entanglement of the future's potentialism.

Past's shadow lives with the space's light until the future, making the present heavy with the time's sentient energy, the chemical that is inversely proportionate to the entity still in dark about its potential without the space for the time to be the causation taking infinite forms like an ideal seeking to be God with the consciousness of the infinite within "I AM" finite: the one with five zeroes, essentially six zeroes, that the six copy for growth per unit of time.

xiii) Silence of Reality Precedes the Sound of Time that Lights the Space

Reality is what one realizes with growth. Reality is unknown until one makes it true with the growth of sentient force. When known, the reality becomes false until someone discovers its truth by living that reality as a sentient entity, conscious of the reality unknown to anyone not living that. For instance, we may believe our leaders are great because they are taking an interest in what we are doing while following them since we are not doing what they are saying, else it will be boring for them to observe us do what they already know. Believing is not reality.

The leaders behave like zeroes because we become the heroes when they do so, letting them be super-heroes. If we realize the hero within us, then we do not need to rest on the shoulders of the zeroes as their invisible hand cycles with our visible legs to retain the heavenly life, pushing us back to sustain our hellish journey like that of one essentially dead but bragging like a deity for blessing the devil with the fortune that belongs to one.

You are the manager of your life: You do have the option to rest in peace as the agent of sustainability, sustaining the cycle of birth, life, death, and hibernation for self-reflection until the realization of reality. Don't be the unsustainable victim of quantum entanglement of your future by a leader on the top who wishes you to be present as a creature so that he could be the enjoyer of your past!

Let's take a couple of examples.

Known Reality of the Present	Unknown Reality of the Future
Electric vehicles will become popular in the future.	When will electric vehicles dominate the market? Who will be the market leader?
We have a limited lifetime.	How long is our lifetime?

The possibility of knowing the unknown reality is the probability that our realization of the truth of the known reality is unconditional. If the possibility is not 100%, then the known reality is conditionally true but unconditionally false.

Let's reformulate the known reality.

- Electric vehicles will become popular in the future if electric vehicle technology continues to improve at the present pace and additional players from diverse nations get involved.
- We have a limited lifetime if we do not become immortal through the impact that we leave on times to come.

Now isn't it self-evident that the previously known reality was unconditionally false?

xiv) Reality of a Sentient Entity Proceeds from Silence to Illuminate the Dark with Sound

Unknown reality becomes unconditionally false when known unless it is the causation for the realization of the present's known reality by living that reality. If a sentient entity, conscious of the possibility, decides to take action for the realization of the possibility, then the probability of that sentient entity realizing the goal is one. The sentient entity is the previously unknown reality.

In the double-slit experiment, a particle of light passes as a wave (descending wave) through both slits at once and interferes as nothing with itself on the other side to form two wavefronts (one ascending and another horizontal). The interference produces a peak that induces the possibility of a particle within the reality at the end. The interferer consumes the peak as a particle to deduce the interferer-effect that goes past as one. The interferer is the potential wave of the temperature of the path of action that produces two wavefronts for realizing the reality of the wave as the thermodynamic-effect.

There are three dimensions to reality within a double-slit experiment.

- First, Imaginary Left
    - Future of the particle
    - Future = 0
    - An observer who serviced negative one (nothing that interferes) as a particle with a spirit that was imaginary to be left with everything on the right before realizing the (unknown) reality
- Second, Illusionary Center
    - Twin hole (i.e., Twin Slit)
    - Twin space, that multiplied one within the particle into a triple group of six with its potential for growth over time
    - Growth of six is present within one that went past as time to divide itself into three with its potential to be the multiplier of itself
    - Growth – Illusionary Center = 0
    - Growth = Illusionary center = 6
- Third, Real Right
    - Real right goes past the growth of the present to be real as the past
    - 6 + the multiplier (3) of the growth that the two within the space groups for realizing the reality of seven without the one serviced as a particle that perpetuates as the real right to twin itself
    - Real right = Past = 9 that perpetuates the peak as real to fulfill the goal of the experiment with everything on the right = Peak = Real
    - The fire within one with the potential for feeding its thermodynamic force as a wave for breeding two within

space = 17 = Simple Circle (illuminated at the end, where one becomes three, all appearing as real although with a future that is zero)
- Eight mediates three groups of seven, so group = 387
- On the left are 19 negative quarks (past quarks) as 19 pieces of an atom. On the right are 19 positive quarks (future quarks) as 19 pieces of a particle and 3 present quarks. The growth of six within one centers 16 present quarks for centering the 3 present quarks within 60 potential quarks, expecting 3 quarks for the parity with three present values: a neutral past that the present makes negative, a negative present that neutralizes itself with its reproduction of a twin that squares the space, and a positive future free from the illusion of the present within the past quark. 19 present quarks are the 19 pieces of a neutron. 19 quarks are 19 pieces of energy. 19 potential quarks are 19 pieces of a cell.

There are three potential dimensions to interference within the double-slit experiment.

- First, Imaginary Left
  - Dark band (=14) of four stanzas of six lines of eight syllables that ascend as three happy points each = Dark band of destructive interference = Dark band of 24 happy points that twin themselves as they descend with the time multiplier, ascending their twinning force that twins the twin with the time multiplier = Dark band of 576 happy points = Dark band of 5 *121 +11 happy points = Left band = Dark band that forms a dark spot on far left = Dark band of fourteen finite points within an infinite point.
- Second, Illusionary Center
  - Interference pattern of fifteen finite points as a combination with two—the dark band and the bright band.
- Third, Real Right

- Bright band (=12) of constructive interference = Bright band (=12) of a happy point that twins itself as it ascends to descend its twinning force = Bright band of 4 *121 happy points = Right band =Bright band that forms a bright spot on far right.

In this way, a particle produces four groups of 22 particles on the right. Four groups comprise 88 (sub-atomic) particles. The three time particles interact with the four space particles to produce the twelve causation particles as a bright band of 12 finite points. Geography comprises four groups that reproduce 76 particles with 12 (causation) particles.

Figure 15. Double Slit Experiment with Photon: The Nonlinear Wave Pattern with Four Patterns

The pair of two groups produce two bright spots at the infinity of causation. Thus, one particle produces a quantum of 90 particles with its replication over time. A bright spot becomes the space's light in the future. Since there are two holes (two spaces: one whole and another wholesome), a twin bright spot of wholesome space emerges.

xv) Whole Is the Essence of Reality

The essence of reality is whole. Whole comprises the letter "W" with a hole. "W" with a hole creates a theory that the two bright spots are due to two slits. Two bright spots may be reflected from the incident light of the past waving through the two slits of the present like a particle to shape the quantum uncertainty (light spot) of the future if it is true that light is a particle that moves as a wave through the space to intensify the sound within the particle that forms a bright spot (maximum intensity).

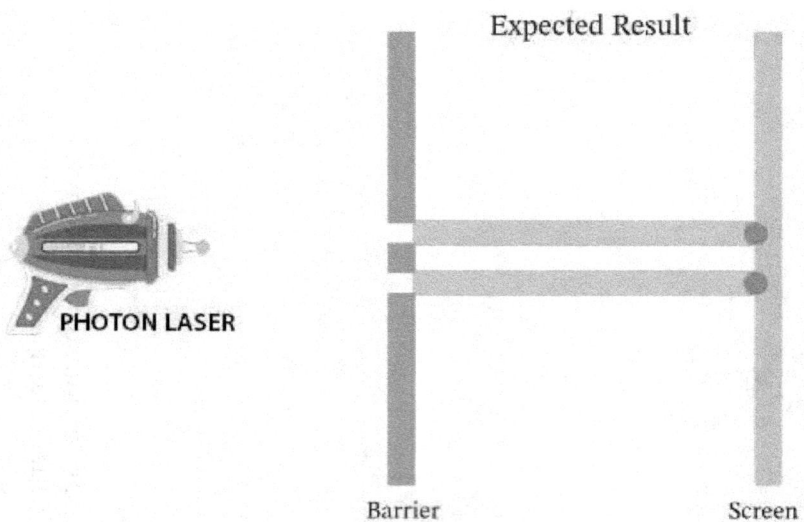

Figure 16. Theory of Two Bright Spots with Two Slits

Another possibility is for the letter "W" to be the illusionary center that becomes a slit with infinite slits, so what is left is nothing but the slit. Thus, the essence of the double-slit experiment is in a single-slit experiment. In the single-slit experiment, on the right wall, the photon forms a Light band of 8 finite points that twin growth of 6 finite points to produce a Bright band of 12 finite points. The bright band of 12 finite points circulates a handed band of 7 finite points on its left to copy itself as a finite point on the top.

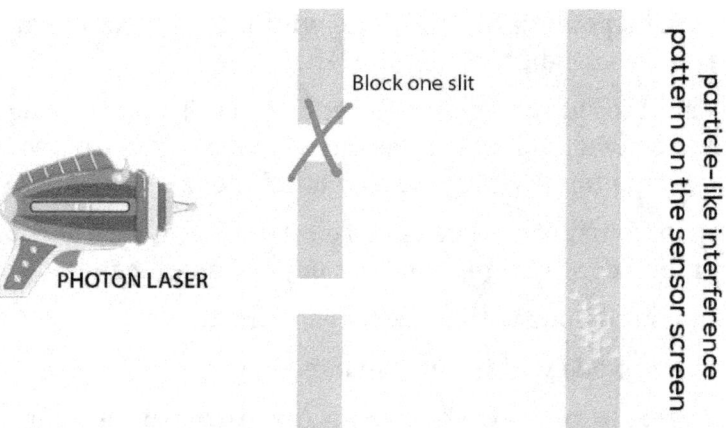

Figure 17. Single-Slit Experiment: The Linear Particle Pattern with One Pattern

xvi) Whole Has a Potential for Growth of Its Base

Whole has a potential for growth in its base, which becomes evident in a modification of the double-slit experiment. What if the grain is fired as a potential particle, instead of a photon?

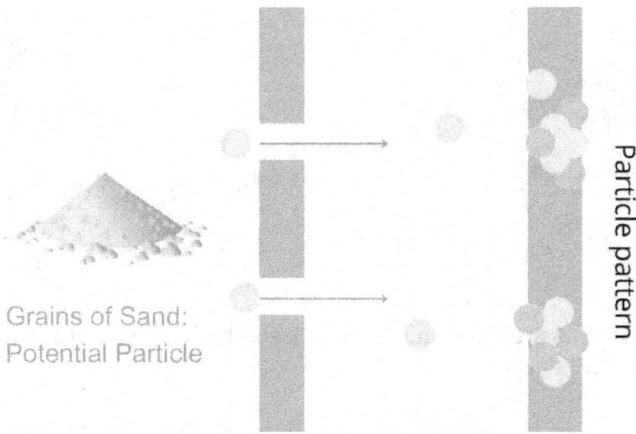

Figure 18. Double-slit Experiment With Grain: Linear Particle Pattern with Two Patterns

Grain returns as a whole to the center taking the position of the one which forced it out with its fire. The force in its half-life as a fire makes space for detecting seventeen particles in superposition, shadowing its potential as a particle within the potential particle that forms two bright spots on the left.

Potential for growth twins the entity that is potentiating the growth to be wholesome. This becomes evident when a detector is added over the top. There is a twin outburst of 24:

- An outburst of nineteen as an electric field, with a dark band of fourteen, a bright band of four, and a dull band of one
- Inburst of four within the bright band of four
- Burst of one within the dull band of one
- Overburst of twelve that twins the growth within the light band of eight
- Underburst of six whose growth is twinned within the heavy band of two
- A potential burst of two within the band of six

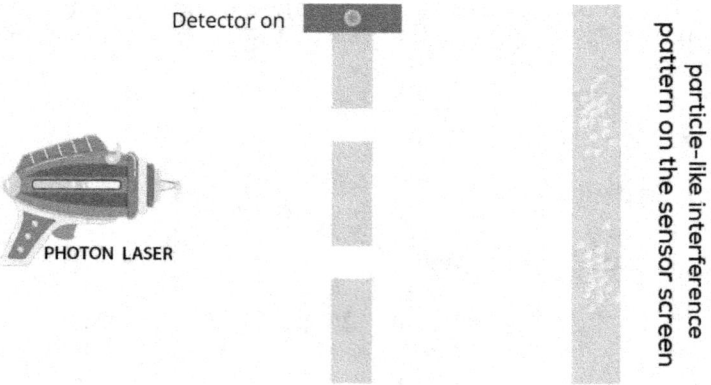

Figure 19. Double-slit Experiment with Detector On: The Linear Particle Pattern with Two Patterns

Wholesome triples the oneness of the entity with the tertiary residual of three to be wholesomewhole after growth in consciousness with a two. This becomes evident when a detectable is added over the top. The detectable generates four twin outbursts, each of which twins themselves to come together as an electric field of 48 (outburst). The electric field repels three copies (triple outburst) that twin the oneness within the three eyes of time as a detector triangulating the detectable as the reality to be realized with their charm attraction for the fourth copy (comprising the electric field of 48 and magnetic field of 7 that twins oneness of one within an electromagnetic field of 13 with a sentient field of 8 to attract a thermodynamic field of 1 as the electric field repels a gravitational field of zero).

Figure 20. Double-slit Experiment with Detector Off: The Nonlinear Wave Pattern with Four Patterns

Wholesomewhole detects the possibility of the convergence of reality and consciousness. This becomes evident in a quantum eraser experiment, with

- two detectables on the right wall, behaving like two zeroes
- consumed in the detection of the present with a splitting crystal (which crystallizes a triple copy of the present for a four-fold growth)
- producing the future as the three zeroes using the third eye as a lens
- after dreaming of their past with the two eyes far right of the two bright spots like a laser still left within one that circles the ten-fold growth.

In this case, eight groups of eight each erase the quantum of 90 with a possibility of 12 groups of 8 each, i.e., 128 within the oneness of time, as time triples the four dimensions of space that reproduce as the eight dimensions of twin space. Ten groups of eight each are ten-fold growth within 2 groups of 8 each, i.e., space's convergent energy of 28.

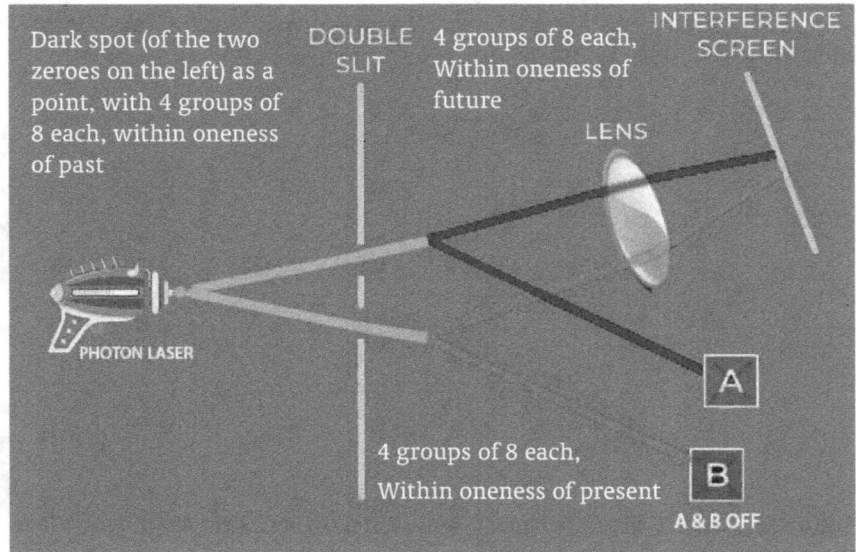

Figure 21. Quantum Eraser: Double-slit Experiment with Two Detectors Off

Convergence is a detective conscious that it is diverging its reality after it concludes its detection. This becomes evident in a wave eraser experiment. In this experiment, two detectables become two detectors for detecting the possibility of a twin within their reality.

- A light spot (of the three zeroes in the center falls to the bottom because as the spot goes past the present, light becomes heavy with the sentient energy of the future, which is conscious of its destiny when the mind is devoted to replication of the past for certainty of action in the present) as a quantum system, overshadowing its present with two underbursts of six each

- A light band of 8 (Overburst of 12), forms eight particles visible with the past at present and four particles invisible as their future.

- A bright band of 4 (Inburst of 4), forming four particles visible as the past of the present and eight particles invisible because they are present with future
- A band of 6 (Anti-burst of 2), forming two particles invisible as the future value of the past and the present and four particles visible as the present value of the future, norming four quarks
- A dull band of 1, forming one particle as the time value visible with the three particles of the space value while the space's fourth dimension reproduces six particles of the causation value that twins the space value still invisible.

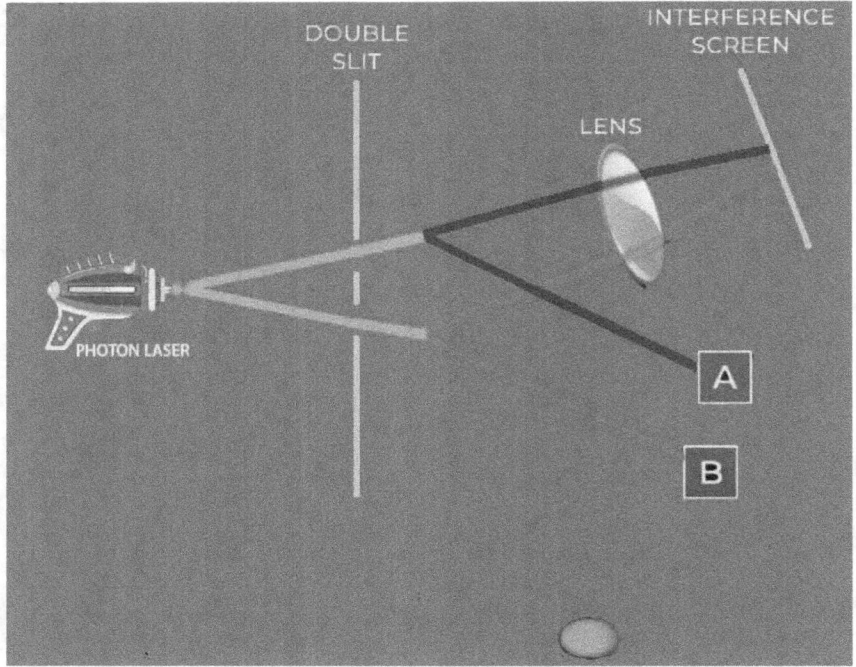

Figure 22. **Wave Eraser: Double-slit Experiment with Two Detectors On**

Overall, 20 particles are visible as a photon (the laser shot before becoming a space divider) without the light spot, and 20 have the potential to be visible as an anti-photon within the light spot that hides them with its shadow. Anti-photon is a quantum photon, similar to the bright band of four that the light band of eight twins for the potential burst of two into a dull band of one to be superluminal as the 20th particle after the burst of the neutron with 19 particles into a wave of 19 photons that transform the neutron into an atom, after reforming the twin bright spot into an electron and one dark spot into a nucleus, which is a combination of two: the atom's past (the eraser) as a neutron and the electron's future as a proton.

The inburst of 4 and the overburst of 12 bursts from a left-handed band of sixteen. This masculine band of sixteen enjoys a supernatural four-fold growth with three, the time multiplier. The right-handed band of nineteen is absent because the left is the feminine band. Pi is the "handed band" of seven, the reality of the right, within one visible as a particle with the endoreproduction of two without zero.

Chapter 3

Universe of Organizational Reality

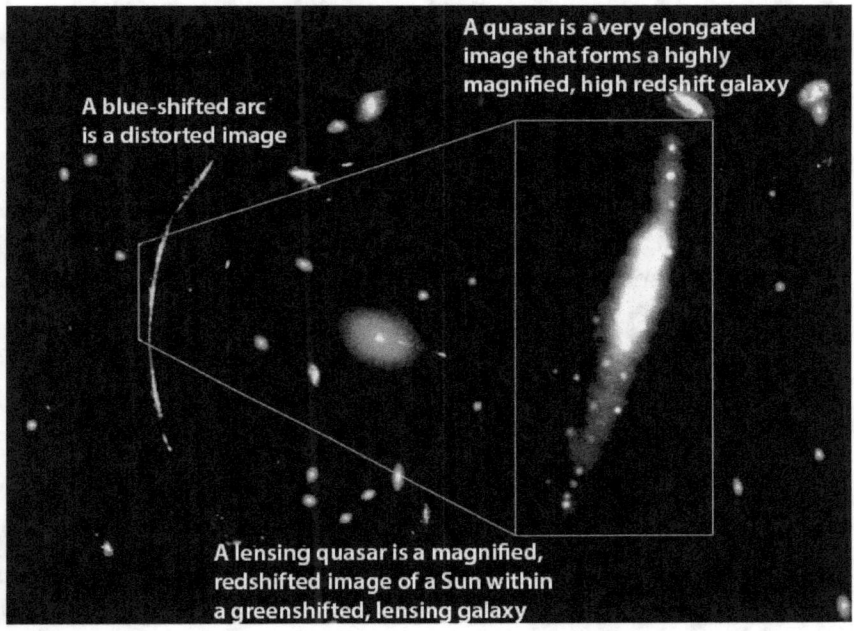

Figure 23. A Blue-shifted, Luminous, Gravitational Arc

> Does wavelength discontinuity, which contracts shared past for "length stretch" of shared present, gravitationally blueshifts the light of a strongly magnified galaxy into a blueshifted, lensing arc (*The Newtonian idea of energy gain without the system*)? Or Does wavelength continuity, which stretches the shared future for length contraction of shared present, gravitationally redshifts lensing galaxy into a strongly magnified galaxy (*The Einsteinian theory of energy loss within the system*)?

Neither of the two. Instead, wavelength infinity of the shared present transforms the wave into an anomaly to reform itself into wavelength discontinuity of the shared past. The length of the wave gravitationally greenshifts the consciousness of the strongly magnified galaxy into a greenshifted galaxy (an elliptical)—the Sun's galaxy-effect. The front of the wave gravitationally redshifts the greenshifted galaxy into a blueshifted arc to back the wavelength continuity of the shared future within a magnified image of the Sun within the greenshifted galaxy. As a blazing nucleus, magnifying the image of the star at its core, a greenshifted galaxy behaves like a blazer. The magnified image is a superconductive image of a disordered fractal, reproductive as a lensing quasar. The spatial piece is disordered as a fractal. The fractal is that the spacetime singularity orders into nematic order, i.e., orientation order.

The central surface of the Sun is dark matter. The blueshifted arc is the gravitational arc. It is self-luminous as a system. The self-luminous system is a giant. The energy with the system is constant. Sun presents itself as a star. The energy within Sun forms a system of reality. System twins itself into a giant. Reality is immanent within Quasar. Reality is a third of the Sun.

Does Universe Know the Reality a Creature Does Not?

A creature has three ways of knowing. First, introspection on the desired future to know the goal of life as a creature. Second, extroversion for knowing the desirable past by investigating the life of a goalkeeper who fulfilled the goal of creating the creature as his creation. Third, living in the present moment for desiring the desires to be fulfilled by a goalkeeper she creates as a creator of the reality of the two: the desires as the goal, and the goalkeeper whose wish is to be the keeper of the goal so that it may be fulfilled for well-being one day when the time is opportune.

Time is opportune when it presents the conditionality necessary to fulfill the goal. A creature's intentionality conditions whether the goal is necessary. A creation's devotional intensity conditions whether it is prepared to be the goalkeeper.

> A creator's emotional intensity conditions whether the desire has value for an organization seeking to know the reality of its potential as a star of attraction for the goalkeeper. The star forms the universe to fulfill the goal without knowing the reality a creature does. Once the creature decides that the goal is real, the universe works to fulfill the goal to be the enjoyer of the joy the creature enjoys from the reality its manifests.

3.1. Organizational Reality and the Limit Value.

Organizational reality is the reversion of one as somebody into everyone else with the transformation of the energy of the present into the potential energy of the future that sum totals the past reality. The energy of the present is the work of proficient exchange of the guider-effects of an infinite council of spirits that make one organizable as a creature. The infinite council of spirits is a disintegrated etheric part of the psychic consciousness. It makes the etheric body a reproductive body reproducing the consciousness of the creature within the objects the creature produces as a psychic to enjoy as a subject. The subject is the creature's reproductive quality. A creature has the potential to enjoy infinite objects it produces by consuming infinite subjective realities realized at different time moments.

The quality is reproduced within an ensemble of ten entities who work as a worker to fulfill the creature's wishes by breeding ten entities within a group of twelve entities. The ten entities that ensemble a person include three varying with time (past, present, and future), four varying with space (East, West, North, and South), two varying with the entity (primordial and primeval), and one varying with causation (potential of the creature to personify the characteristics of different persons to enjoy their oneness as a subject). Twelve entities group ten entities into an ensemble with two entities—of which one is self-projecting the essence of the potential within another as the another's creation while another is the creature comprising twelve entities. As a glocal entity, two entities reproduce themselves with the present of time to make four entities a global entity by globalizing time with three dimensions within space. By self-projecting six entities without space, the global entity becomes a local entity. The local entity is delivered as the past, which makes the future of time reproductive to twin time's three dimensions already present within the four entities.

The local entity norms its globalization with 6 x 6, i.e., 36 entities by becoming reproductive. The global entity norms its localization with 4 x 4, i.e., 16 entities. The glocal entity norms its glocalization with 4 entities to localize the effect of the fifth entity into the characteristic of an ideal creature with 11 entities. Eleven entities are transnational entity, self-projecting the ideal creature as the twelfth entity. They norm transnationalization with nine entities by transcending the nationalization of the infinite with an infinity comprising three entities.

> ### Ten Entities That Ensemble Within Sun
>
> First, the Earth (*Prithvi*) is the infinite feminine flame, trading the gravitational-effect of all the ten entities individually as well as the whole solar universe, the 11th entity. She is servicing that to each of those eleven entities, as well as the para-universe, the 12th entity, beyond the limits of the solar universe. Each of the 11 physical bodies within the solar universe norm earthly behavior. Each of the 12 physical bodies without the solar universe also empowers the 11 physical bodies within the solar universe to do so. Therefore, the overall energy of the physical ecosystem is 11 * 12 = 132. The Earth norms the sheeny benefit value of each physical body with her gravitational-effect, also valued at 132.
>
> Second, the Sun (*Surya*) is the paternal flame trading the gravitational-effect of all the ten entities within the solar universe and all the eleven entities without the solar universe, for creating the solar universe. Ten entities within the solar universe include eight planets individually as the first eight entities, collectively as the ninth entity, and the individual-collective exchange as the tenth entity. Eleven entities without the solar universe include the universe of the organizationally same eight planets, moons, and suns, plus the organizationally diverse universe of the white stars. The Universe of white stars is the primordial creator of the sun. It is a primordial self-luminous entity, that incarnates as a sentient living entity within an absolute physical body.

Third, Mercury (*Budha*) is a present entity exchanging the residues of both the descending present energies of the maternal earth, while servicing her gravitational-effect to the other seven planets, as well as the ascending present energies of the maternal earth while trading her gravitational-effect to the moon. Mercury services that residue as his gravitational-effect for perpetuating the whole universe of ten entities and empowering the ten entities to perpetuate their local whole universe. Such an inclusive management method lets each present human child exchange the value of culture-effect enjoyed by the eight planets (8), the workculture-effect of the Moon and the Sun (2), the human or the entity-effect of each of the ten entities (10), and the trading-effect of each of the ten entities (10). The present human child enjoys a sheeny benefit value of 1600 (8 * 2 * 10 * 10) by exchanging the absolute gravitational-effect as her guider power. Mercury norms the sheeny benefit value of each present creature with his gravitational-effect of 1600.

Fourth, Venus (*Shani*) is a past entity, the creator of the present paradigm of radiant love. He is trading the gravitational-effect of all the nine entities [eight planets and moon] within the solar universe, and servicing that holistically for the formative growth of the whole solar universe. He is empowering each of the ten entities [eight planets, moon, and sun] that constitute the solar universe to empower each of the organizationally same ten entities that are constituting the solar universe to perform par excellence. He ensures universal engagement with the present paradigm of radiating love, valued at 2700 (9 *3 *10 *10). The Venus norms the sheeny benefit value of each past entity that has transformed itself from the primordial greeter state, without the proficient exchange system (1600/ 100 =16), into the primordial creator state, within a proficient exchange system (2700 units of gravitational-effect).

Fifth, Mars (*Mangala*) is an emerging entity like a present child trading the gravitational-effect of all the nine entities within the solar universe and the organizationally same eight entities without the solar universe, except the entities that are organizationally the same as Mercury. The Emerging entity is free of the responsibility to be inclusive of the non-citizenship entities. Therefore, it is devoted to the ascending sheeny benefit value of the entities within the solar universe. Mars polarizes its technological servicing of gravitational-effect to six entities, namely, Earth, Mars, Jupiter, Saturn, Uranus, and Neptune. He empowers Earth to be the guider of the deciding sheeny benefits for the paternal, the present and past entities, as well as the para universe. Mars norms the sheeny benefit value of each emerging entity devoted to the whole citizenship geography as a sentient animal spirit and enjoys 102 [(9+8) *6] units.

Sixth, Jupiter (*Brihaspati*) is an infinite entity perpetuating the new paradigm beyond the absolute mediated by the guider power of Mars. Jupiter trades the gravitational-effect of all the ten entities that constitute the solar universe, and the organizationally same ten entities that have the potential to do so. He technologically services the overall guider power to 89 entities. He polarizes the servicing of the gravitational-effect to eight entities within the solar universe, excluding Mercury and Venus who are the catalysts of globalization and diffusion into infinity without the solar universe. He promotes the guider power and technological servicing of those eight entities, the whole solar universe as the 9th entity, and Mercury and Venus as the 10th unified binary entity directly or indirectly trading the guider power of the other nine entities. He ascends the disproportionate guider power and technological capability of those nine entities. Jupiter norms the sheeny benefit value of each infinite perpetuating entity who as a primeval perpetuator is a devoted citizen of the solar universe. Its energy value is 1780 [(10 +10) * (8 *10 +9)]. Jupiter norms responsible management paradigms for the well-being of each entity within the solar universe.

Seventh, Uranus (*Rahu*) is an emerging masculine entity like a disproportionately growing child flame. In reality, it is a past entity preceding the past paternal of an infinite entity (Primordial-Primordial Paternal). Uranus is organizationally similar to the unified binary entity comprising Mercury and Venus. It experiences a descending-order entropy of its guider power within the combined force of the New Paradigm of Mars and the Responsible Management Paradigm of Jupiter. The entity holds the present paternal flame, i.e.., Sun (21), is responsible for the Metaphysical Entropy Paradigm. He norms 73 as the 21st prime number with the whole energy of the solar universe exchanged from the past paternal, who has diffused the whole energy of self. As a past entity, he technologically services the whole ascending energy of the solar universe to 73 entities that are experiencing metaphysical entropy-effects. He polarizes the servicing of gravitational-effect to the seven entities without the solar universe. Seven entities are organizationally similar to the six planets, excluding Mars and Jupiter, and the Sun. He promotes the guider power and technological servicing of those seven entities, as well as Mars, Jupiter, and the Moon. He ascends proportionate guider power and technological capability of three organizations: self as the organizational sameness of Uranus, the solar universe as a whole, and the para-universe as a whole. Uranus norms the sheeny benefit value of each emerging masculine entity that is super conscious of its sheeny well-being as a corporate entity and enjoys 73 (7 * 10 + 3) units. Uranus norms metaphysical entropy paradigm for the sheeny well-being of each entity, mediated by the self as a super guider power, within or without the solar universe.

Eighth, Neptune (*Ketu*) is a masculine entity that is a disproportionately growing flame of the past infinite paternal. The Past infinite paternal (Primordial Primeval Paternal) is the universe of past paternal entities, experiencing the escalating cost of the combined force of the New Paradigm of Mars and the Responsible Management Paradigm of Jupiter. They are scripting the ascending guider power of Uranus within the Metaphysical Entropy Paradigm as a pathway for their sheeny well-being. As one entity, he channelizes the ascending present guider power of 70 entities, excluding Uranus, the solar universe as a whole, and the para-universe as a whole, for technological servicing of seventy entities that include self as the organizational sameness of Uranus, the whole solar universe and the para-universe as a whole. Neptune norms the sheeny benefit value of each masculine entity supra-conscious of the sheeny well-being of the whole local group of past entities without the solar universe and enjoys 140 (70 + 70) units. Neptune norms a cost-effective management paradigm for the sheeny well-being of each entity, catalyzed by itself as the supra guider power within or without the solar universe.

Ninth, the New Moon (Chandra) is an emerging feminine twin or para flame entity, a proportionately growing flame of the past maternal. The Past maternal (Primordial Maternal) is the universe of past maternal entities, enjoying cost-effective management benefits of the combined force of the Metaphysical Entropy Paradigm of Uranus and the Cost-effective Management Paradigm of Neptune. They are conceiving their proportionate growth as a technological entity for promoting absolute growth as a supreme guider power within or without the solar universe. As one entity, she channelizes the ascending present guider power of 80 entities including 10 entities within the solar universe, for technological servicing of herself and the rest of the present universe. The New Moon norms the sheeny benefit value of each para flame entity that has become a blackhole and enjoys 82 (80 + 2) units. The New Moon norms the formative growth paradigm for the sheeny well-being of each entity within the self as the supreme guider power.

Finally, the Full Moon or the whole illuminated solar universe (Soma) is a maternal twin of a para flame entity. She is a proportionately growing flame, with proportionate growth consciousness of the past entities that discipline the freedom power of each flame entity for servicing their sheeny well-being without at-par servicing of the universal sheeny well-being. She is the Primordial Para Liberator who norms the oneness-effect within and without each flame entity. As a para

> guider power, she perceives her proportionate growth as an organizational entity to be the normative development value of each entity. As one entity, she channelizes ascending present guider power of 90 entities for the technological servicing of ten groups of organizationally same entities within or without the solar universe. She guides those 900 (90 * 10) entities to technologically service 90 primordial entities as a pathway for perpetuating their absolute sheeny benefit. She guides those 990 entities to technologically service the future seven groups of ecosystem entities. The seven groups of ecosystem entities include the emerging solar universe, the present solar universe, the past solar universe, the emerging white star universe, the present white star universe, the past white star universe, and the universe of the almighty creators of the past white star universe. The Full Moon norms the sheeny benefit value of each para flame entity as the absolute illuminated value of the present universe. She enjoys 997 (900 +90 +7) units. The Full Moon norms a normative development paradigm for the guider power of each entity to enjoy sheeny well-being without itself as the para guider power.

Ten entities are organizable into a Council of Nine Entity Groups within the self as the tenth entity group. The Council of Nine Entity Groups includes:

- The seven planets as the sixteenth entity group which self-materializes the present energy of Mother Earth.
- The moon as the seventeenth entity group within the oneness-effect of the infinite energy reproduced by Mother Earth with the seven planets.
- The sun as the eighteenth entity group reproduces six entity groups with the primordial energy that produces Mother Earth as the eighth entity group. The potential of the seven planets is sequenced within the first seven entity groups. Mother Nature reproduces that potential within the ninth to fifteenth entity groups for advancing as the sixteenth entity group.

- Six entity groups are present within the sixth entity group, the dark matter.

The Council of Nine Entity Groups is the organizational mass of self-existence as the tenth entity group. The tenth entity group is the limit value of a creature's universe of potential. The universe of potential shapes a creature's intentionality to do something and leave that in the sub-consciousness for guiding a range of behaviors. One as the causation perceives the organizational reality of the effect as intended after conceiving the present reality of the universe that tends to the cause. The cause is the potential for the effect to be the organizational reality of one's intentionality. The present reality is precisely what one imagines when forming the intentionality that eventually springs back into consciousness for validating its causation with a theory. The theory is the potential mediated correlation between the present reality and the organizational reality. A theory has four dimensions.

1. The present reality is the "abstract dimension" of the potential theory of its causation.

2. The organizational reality is everything developable into a "potential dimension" of the present theory that makes the causation concrete.

3. The potential is the cause developing as an imaginary, "mind-born dimension" after its effect develops as an illusionary, intellect-born dimension.

4. Mediation is the triangulated reality inherited with the illusionary, "intellect-born dimension" for developing the potential to square the reality by reproducing the technological reality of one's past.

The past squares the reality because it is the goal of reality—to go past as one moves forward into the future for realizing the potential beyond the reality already realized. The square is the space for realization because it is already realized with time. Space reforms the time for realizing the potential beyond the present with a primordial theory. The primordial theory is the eventuality that the present theory is simply reproducing the past as a cultural reality for its infinite continuity as a culture that matters for everyone's sentient well-being. Reproducing the past makes everyone conscious of their well-being by knowing the past and servicing that as science to the collective for their growth as an individual. When the collective follows the science guided by the past, it opens the space for one to be the leader in the eventual state when one's past leads everyone. A primordial theory also has four dimensions.

1. The past is the sentient entity dimension that makes an entity conscious. It is also the "objective dimension" that fulfills the objective of making a leader a sentient entity, conscious of the reality the collective follows.

2. The desire for sentient quality with the effect becoming an exchange point for past generations is the "subjective dimension" whose orientation is subjective for a desirer. The desirer perpetuates the reality of the desire without the present reality of desiring by making it reproductive repeatedly over the four dimensions of space until it becomes desirable as a potential reality. Potential reality is the development that is looping one as self-conscious of its infinite exchange value. Only one who desires enjoys progress through genetic growth of proficiency in desiring the potential beyond the reality already present. One falsifies the *normative organizational reality* of the present's horizontal subjectivity by illuminating the *transformative organizational reality* of the potential beyond.

"Transformative reality" (=63) of the "present-potential exchange" (=63) is "gravitational lensing" (=63) of "feminine element" (=37) whose "femininity" (=1,000) "switches" (=16) the "initial state" (=-12) of the "intrinsic potential" (=-12) of the "masculine" (=296) "trading" (=53) "masculinity" (=53) from the "preceding lifetime" (=600) into "consciousness" (=4) with "knowledge" (=600) of the "temporal cost" (=600) of the "heterocyclic ring" (=16) that "forms" (=100,000) an "infinitely long surface" (=16) for "servicing" (=47) the "reproductive force" (=100) to the "present" (=1,600). "Heterocyclic ring" (*Chapali*, 16) "causes itself" (=340) to "ring" (*Pushpaka*, 100,000) a "form" (=100,000) to "norm"

(=18) "infinity" (=90,000) that "transforms" (=1) "female" (=10,000) into a "circular creation" (=10,000) of the "male" (=100,000,000) with the "effect" (=34) of the "present" (=1,600) as a "metric tensor" (=-1) on the "tangent" (=10). The "square growth" (=16) of the "cause itself" (=340) "tiles" (=340) "Einstein" (=340) to "ring" (*Pushpaka*, 100,000) the "supernormality" (=340) of the "Einstein ring" (*Chapala*, 16) with the "normality" (=16) of the "ten billion membered ring" (=16) that "reproduces" (=78) female's "reproductive force" (=100) within the "male" (=100,000,000).

3. The collective that follows the *transformative organizational reality* of ascending objectivity is the "desirable dimension" servicing self-ordination as the perfect ideal for the growth of an individual into a guider, whose wisdom is reproductive and descending subjectivity with time. The guider leads the formative organizational reality of ascending subjectivity until the vanishing time superpositions space's horizontal objectivity over the backward subjectivity of the guider's divine element as the *entity organizational reality*. After the guider's time to guide vanishes, the collective as an entity is left as a follower of the ideal of the organizational reality out of habitual behavior.

4. The individual that leads the *deity organizational reality* of a leader, whose present subjectivity norms the whole objectivity of the follower, is the "undesirable dimension." The follower follows the leader because the primordial theory transforms the leader's present subjectivity into an excellence dimension that no one follows as one moves forward but must follow like a religious dimension to be a leader by reproducing the past for leading the future. Therefore, the leader becomes an ideal. A zero who no one follows becomes a hero everyone follows.

A primeval theory guides everyone who follows a zero. The primeval theory is the ideal within the oneness of theory that makes one unaware that it is servicing its position by trading the momentum from zero. The zero moves because one wishes to take its position. The zero does not wish to move because it is unaware of the value of one. Therefore, it makes one a worker working like a deity to fulfill a wish to be a hero by behaving like a zero. As one behaves like a zero, genetic growth makes one a zero. As a zero becomes a hero, the agenetic development leaves the hero as one on the left of the zero through endoproduction (i.e., intellect-born, illusionary production). Consequently, the zero becomes ten, the base of the foundation for the one to be organizable as eleven after following the zero to be one again. The primeval theory has two dimensions.

1. *Within the consciousness of the theory-effect.* Within consciousness is the superposition that makes space for the growth of the base by giving one the position of foundation. The foundation is the present growth of consciousness of reality. The base is the mass consciousness everyone follows while alive. The growth makes the base whole. It frees one from the reality already realized within the present which transformed one into a zero with the theory-effect. One does not need to live for repeatedly self-substantiating the truism of the theory after the realization of the falsism of the ideal with the sentient consciousness of the "universal dimension," the dormant potentiality to be whole.

2. *Without the consciousness of the theory-effect.* Without consciousness disassembles the quantum position the growth takes after realizing the space for momentum within one's position. The foundation guides one's continuous development as a leader guided by theory-effect. The base is exchanged with a "unique dimension" of the visible, supernatural reality that makes the consciousness of the theory-effect reproductive. When consciousness is reproductive, it becomes a reality free of the hypothetical element—the base for the theory. The reality is a theoretical reality, unrealized at present. The present is the realized reality. Unrealized reality is the reality of a sentient entity—only the latter is conscious of the reality yet to be realized which is keeping it alive as an entity. The unrealized reality is not present as the organizational reality—it is not normative. It is not formative either because it

is yet to be formed. It is formable only when one is conscious of the goal of life that keeps one alive.

Science is full of examples where what was experienced as an ideal law of nature is now known to be a false perception of the subjects culturally bound by a larger socially constructed reality and wishing to be the ideal personifying the ideal-effect of the realized reality. As the consciousness of the theory-effect illuminates the epistemological value of the wishing tree growing with time, the base discontinues exponentiating the present development.

The disproportionate growth of the present self with the present development of the imaginary element makes the ideal a "para deity dimension." One perceives the unrealized reality materialized by manifesting the imaginary in the physical realm as emanating from the para deity after it is realized as the consciousness within the ideal. One conceives the unrealized reality not yet materialized to be omnipresent within the ideal before it emanates from the para deity after coming to light with a consciousness of the theory-effect that illuminates what is omnipresent as potential.

If a para deity is the Almighty Creator of the symmetry between the present and the realized reality, then a primeval deity is the Almighty Creature whose potential diverges the unrealized reality from the symmetry before letting it converge into the present as the realized reality. The potential destroys symmetry by making it possible to create what is not present. The creation perpetuates the symmetry because it is present within the potential and therefore symmetric with the potential's realized reality. The realized reality is the ideal reality because it is a known reality. The symmetry is the primeval deity dimension because it is determinable after the primeval deity dimensionalizes the convergence of the realized with the divergence of the unrealized.

The param deity dimension perpetuates divergence to illuminate convergence after destroying the past's resurgence. What perpetuates is never the past, but the culture that descends the past to acculture the future with the present of the potential within it for change through the exchange of the desire with the wish. The desire is the wish without a subject taking responsibility to fulfill it with its potential. The wish is the potential to fulfill a desire a subject has as a wisher by making that a goal in life. If a subject has infinite desires, it may still fulfill them in one life as a goalkeeper conscious of the goal of self-development.

The primordial deity dimension is the intrinsic repulsion of the desire for desiring infinite desires. It is superfast in fulfilling the desire by liberating one from the business of desiring. Instead, it lets one be devoted to fulfilling the finite desires of everyone unaware of their potential for self-development. By letting those desires unfulfilled, the primordial deity makes everyone conscious of the value of normative development to reform the artificial element that is limiting their freedom from the transcendental.

The transcendental entangles everyone when one is seeking supremacy with the supreme deity dimension by making its essential nature an octave. The octave copies everyone's desires into maternal consciousness, guiding one to fulfill them by personifying the identity of a mother devoted to her children's well-being. For one to be a supreme deity like a mother, the supremacy must be shared with the father so that both may take turns to fulfill the goal. While the mother fulfills all the desires of her children, the father wishes that the children develop their potential to fulfill their desires for the mother may fulfill her desire to have children without worrying about their future.

The potential to breed a father as the child is immanent within the mother as a supra deity dimension. The Supra deity dimension is the reciprocal dimension of repositioning the superiority of the mother who has the potential to fulfill all desires of her child into the superiority of the child who enjoys the sameness of that potential. The deity dimension is the feminine body within the child as a sentient entity. The feminine body is the causal body causing the child to desire and wish to satiate the desire as a primordial paternal with the blessings of the primordial maternal.

The superiority of the mother implies the inferiority of the father without the potential's development. The superiority of the masculine who develops the potential with the blessings of the mother's femininity implies the inferiority of the feminine who minimizes her present to maximize the potential of her child. The duality of realities emanates with the super deity dimension. The super deity dimension is the asymmetry of class differentiable by group, gender, and generation. Asymmetry of class within the family implies a symmetry of consciousness within a group. A group of families with ten entities generates an asymmetry of class within two entities—gender and generation—for realizing a symmetry of consciousness within eight entities. The realization twins the consciousness as an entity with the letter "A."

A para entity mediates the class to make it a potential entity with the group as the present entity. The duality transforms the twelve entities into a group of twenty-four entities as it ungroups the convergent energy that the letter "A" groups for breeding two by feeding one. The 24 entities include ten entities, two entities (primordial and primeval), eight entities, an entity, a para entity, a potential entity, and a present entity.

A primordial deity fulfills the wish for disproportionate SHEENY benefits by escalating the costs of individualism for the collective. Therefore, the collective becomes conscious of reality beyond the collectivism that is escalating those costs through the substitution of individualism. A param deity fulfills the wish for disproportionate social benefits by escalating the costs of corporatism for the greeter who substitutes individualism with collectivism to be known as an individual who made a difference in the growth of the collective.

A primeval deity fulfills the wish for disproportionate human benefits by escalating the costs of globalism for the masculine. The masculine's masculinity forces the breeding of the human element within the feminine for globalizing the sentient energy of her femininity. A para deity fulfills the wish for disproportionate ecological benefits by escalating the costs of localism for the feminine. The feminine's femininity forces feeding of the deity element within the maternal for breeding masculine with her human element that localizes her consciousness.

A supreme deity fulfills the wish for disproportionate economic benefits by escalating the costs of nationalism for the maternal whose potential lives within the consciousness of citizenship of a child. Therefore, a child becomes conscious of his right to demand mindfulness dedication from the mother. A supra deity fulfills the wish for disproportionate national benefits by escalating the costs of internationalism for the paternal who is present as a child before knowing that he will be left as an alien after he becomes a father. A super deity fulfills the wish for disproportionate psychological benefits by escalating the costs of regionalism for the child who is freeriding before the growth of knowing as a paternal. A super deity deifies the paternal by making his greeter consciousness omnipresent as a deity within the primeval child who follows the paternal's transnationalism to be a primeval paternal.

A deity fulfills the wish for disproportionate guider benefits by escalating the costs of transnationalism for the infinite paternal as the infinite children begin reproducing the greeter consciousness to be a deity and make him the lord of deities as the primordial paternal. The primordial paternal is the zero who has diffused its consciousness to live within the potential of the primordial maternal. The primordial maternal is present within the greeter consciousness after transforming herself into primeval maternal.

The primordial paternal is the devil who fulfills the wish for disproportionate divine benefits by escalating the costs of absolutism for the primordial maternal, thereby making her wish for a change over time with an exchange of her identity as an entity. The entity is Satan who fulfills the wish for disproportionate benefits by escalating the costs of potentialism for the primordial paternal who services his convergent energy as a para entity after trading her discordant-effect to incarnate as a primeval masculine.

Over the entire lifetime, the entity diffuses her sentient energy into conception, perception, and the experience of the infinite wishes to be fulfilled by the infinite masculine. The spirit of space services the consciousness of reality to time in the form of the soul of the entity that incarnates as the infinite masculine. The soul within the spirit is left as the masculine, positioning the spirit on the right as the feminine. The center substitutes both as a greeter while the center's primordiality complements both. The primordial greeter supplements both to make the center whole.

3.2. Two Dimensions Universalizing the Present Reality as the Organizational Reality.

Two dimensions universalize the present reality into the organizational reality of the conceptual theory-effect after one conceives the present as real to experience its past as normative with a theory of potential. The past is normative because it has the potential to perpetuate as the present when one universalizes the future within theory-effect as a reality worth perpetuating as an organization. Two dimensions generate two forms of present reality, each with a conceivable future, perceivable present, and experientiable past. One conceives present reality after endoproducing its future with the intellect. One perceives the present reality after exoproducing its past with the mind. One experiences present reality after producing a body of consciousness about the potential for realizing the reality of the spirit that is reproducing the future as the present.

3.2.1. Three Primary Forms of Present Reality

1) Conceivable present reality is the hidden reality of the future beginnable with a followership workculture.
2) Perceivable present reality is the sound of the infinite psychic linkages one conceives to be present as the reality illuminating the value of the future with a leadership workculture.
3) Experientiable present reality is the ascending organizational planning of the leadership workculture whose past is the followership workculture.

3.2.2. Three Secondary Forms of Present Reality

1) Breedable present reality is the delocalization of the present reality by universalizing it as a normative reality. As a leader forms reality to be present without the universe realized with that reality, the follower norms that reality as the potential to be present within the universe. Breeding makes normative reality bipolar—the breeder unipolarizes

the formative reality to bipolarize the intrinsic repulsion of the breedable present reality as normal within the intrinsic attraction for the normative reality. Somebody breeds the formative reality before one makes the breedable present reality normal. The transformative reality multipolarizes the breeding. Everyone enjoys the normative reality once breeding it becomes normal. Anyone bred by the normative reality observes, as a subject, the transformation of the normal naturally over time. Anybody breeding the present reality as normative with its formative reality behaves like an object of transformation. The objective is to develop consciousness of the causation for the reality realized with natural growth.

2) Becomable present reality surges as a charger of transformation to transform the normal into natural. While breeding is normal, becoming is natural. One normally breeds the space immanent within the present reality as the time for a change. One naturally becomes the present reality of the space with time which changes the nature of reality. A zero's present reality becomes the past reality which forms one into the present reality. It norms the future reality with the two which are normal within the one that renormalizes the normal with its transformation. When one transforms, zero naturally reforms one into a transformable following the path of one. Zero leads one's transformation into a transformer because it wishes to be the one who follows both—the one transformed within both and the one transforming without both to be transformable with both. The transformer emanates from the zero as a condition for fulfilling the zero's wish to be present within its potential. The immaculate conception of the transformer as para real is surreal. It lets zero contribute a reproducing force for diversifying backward and being self-dependent. Zero's past is real because it is the one who transformed within both, itself and the copy that reforms one who transformed into a transformer.

Reforming makes one a zero. Reformation makes ten the base to reform two within the four dimensions of space and let the eight perpetuate as the essential nature of the three dimensions of time. Thus, reform = 888. "It" reforms by becoming =10 +2 +888 =900. It incarnates as the cultural reality of the self. The self does not transform but accultures the copy to incarnate as "it" with the past's reproductive force to fulfill the goal of the present. The goal of the present is to perpetuate within the past by reproducing itself as the Other. The Other complements the greeter as a potential greeter. As the zero substitutes one to be a greeter of the one substituting, the other complements the zero to be the potential greeter of the one substituted. The substitutor is Another. Another is within two that become reproductive as the one substituting substitutes the one substituted to be the substitutor of the two substitutable after the momentum of the added element changes the position of both ones. The substitutor is subtracted as the Quantum force from the two. Two become reproductive to superposition the devotional force of a belief in the potential to substitute the reality their present has added as a complement.

3) Believable present reality is the attachment force of the present belief. The present belief attaches as a living consciousness to force the lifeline of reality out of the believer. The believer believes "I am the reality" of everything realized as exogenous by curving imagination of the endogenous. One realizes the endogenous with a two to be a three over the three dimensions of time. The three curves the endogeneity of two with a one to be twelve over the three dimensions of time and the four dimensions of space. The three zeroes of time are exogenous to the four which mediates the space's division within one present as an entity whose present reality is believable once illuminated in the future as the first of the three zeroes that perpetuates the other two as the past to make the three reproductive. Therefore, the believable present reality = Exogenous = 14,000. The endogenous exponentiates the ten with the third zero after centering the first zero as a one before the second zero. The third zero squares the space for making the hundred reproductive as the primordial space, which places Endogenous as 10^{16} with a

simple imagination of time without space. Time subtracts the position of space with its momentum for making the space endogenous to the growth of consciousness as an entity. Space adds the momentum of time with its position to make time exogenous through a complex imagination of time within space. Time is negative one when it is present as a zero within the potential of one who enjoys the space to be a ten with the zero omnipresent as a copy on its right. Causation for space is one before time cubes the three after space becomes reproductive to be ten after the addition of a nine as the time's primordial momentum. With the transformation of the time's momentum to 999, the causation remains one within the origin of 1,000 with the multiplication of hundred by ten. The origin is present right after the potential of eighteen is produced when the multiplication undoes the division by two after one's growth as a three within a space of 18,000.

3.3. Two Dimensions Unifying the Potential Reality with the Organizational Reality.

Two dimensions unify the potential reality with the organizational reality of the perceiver ideal-effect after one perceives the potential as real to experience its present as normative. The normative becomes the ideal to be followed after the theory leads to it. Two dimensions unify two forms of potential reality into the ecosystem reality of the past to be reproduced as illusionary, the present to be consumed as real, and the future to be produced as imaginary.

3.3.1. Three Primary Forms of the Potential Reality

The present reality of one subtracts the potential reality of ten to perpetuate a dynamic reality of the one as a two following the ten without the two to be leading as a six with time. Thus, the dynamic reality is 286. Since one becomes dynamic as a two with a potential to lead as a six before it is present as a zero, the potential reality is 160. Since the potential reality divides the present by a ten to be the one dividing as a divisor, the present of the reality is 1,600. Mindful of its potential to greet the dynamic full of change with time by making space within its mind for taking a position after servicing a vacuum of energy without consciousness of its future, the present is a mindful greeter. Energy is the primordial momentum of nine within a mass consciousness of ten trading 19 x 3 = 57 as the impact of time on space. The space is servicing a growth of six to square the consciousness of four with a two. Thus, energy = 19. Servicing = Vacuum = 57 -10 =47. Trading = 47 +6 = 53. Trading mediates the reality within the potential for growth for servicing three forms of potential reality.

1) Tradable potential reality. Tradable potential reality is the infinite multiplying of the possibility for the one to be a horizontal reproducer, intensifying the energy pressure for the growth of the ecosystem as an agonist. The entropy of five within the growth of six transforms the agonist into an antagonist, i.e., zero into eleven. Consequently, the ecosystem resists the constancy of the potential within one's reality to be tradable as the dynamic full of change. The transference of energy to one is 91 within the nine which perpetuates it as a primeval momentum of the hundred without the nine that makes the ten reproductive. With transference, one becomes an ecosystem of 92.

2) Traded potential reality. Traded potential reality is the infinite continuity of five colored lights—indigo light, blue light, green light, yellow light, and orange light. The infinite continuity incarnates the red light as the four coloring lights—white light, rainbow light (i.e., IGBYOR light), black light, and colorless light (i.e., ROYBGI light). The seventh tone is the consonant tone of the white light that becomes heavy to be reproductive as the rainbow light. The quarter tone is the dissonant tone of the black light self-radiating from the rainbow light for producing the colorless light after toning the tone of the light with an octave of reproductive forces. Black light is a backward consumer of the octave of potential within the rainbow light. Its reproductive light de-intensifies the energy pressure for the entropy of the ecosystem as a protagonist catalyzing the variability of potential. Eventually, the reality is traded as a technologic

for change into a tritagonist. The tritagonist is the one resisting the variability of the present to compensate for the loss of consciousness about its perpetuity as a voicer.

3) Trading potential reality. Trading potential reality is the musicality that diffuses unified consciousness of the tonality timed by reality as pi of inaction within three quarter-tone. The three-quarter tone is the resonant tone of the colorless light, resonated due to the action of the black light for resonating three tones of the indigo light. Resonating makes the six tones of blue light reproductive, reproducing the seven tones of green light within the resonator. The three tones are the sonant tone whose fluctuation blends the six tones varying with time while the three tones repeated as the present remains constant. The six tones are the absonant tone. The seven tones are the unisonant tone performing meantone within the tone. The resonator's resonance produces the tone of yellow light while consuming the semitone of orange light. The eight tones of rainbow light are nonresonant tones; they are resonable tones within a resonant state.

The resonant state is the technologic for the four tones of invisible, purple light. The four tones are the inconsonant tone of the kernel of consciousness which vibrates for converging the reproducing force of two tones backward. Two tones are nondissonant tone of grey light, present within the five tones of greying, masculine light. Five tones are nonassonant tone, organizable within the fifth tone of purpling, feminine light. The fifth tone is a nonsonant tone of time as the polysonant tone is modified with the devotion of the effort by the third tone of the ascending, whitening light. The symmetry of the third, geosonant tone modifies the asymmetry of the first, transsonant tone of the radiating, blackening light within a devoted point. The devoted point is the forward trader of the consciousness system for modifying the transsonant into a regiosonant with its ten copies forming indigo, blue, green, yellow, orange, red, white, rainbow, black, and colorless lights within the purple light.

Regiosonant is the energy pressure delocalizing the transsonant for globalizing the ecosystem as a five-dimensional kratagonist into four directions whose copy hums the eighth, regiosonant tone of ultraviolet light. Globalizing transforms the purple light into purpling, greying, whitening, and blackening lights without the grey light which is localizing the ecosystem as a deuteragonist within its transforming self. The transforming self as a conscious system is the ethnosonant. It transforms into a hummer and reacts with the sixth, ethnosonant tone of the infrared light to universalize the centrosonant. The centrosonant is the transformable self, expressing its consciousness void for consuming the universal consciousness with the fourth, centrosonant tone of the bundled, ultraviolet A light.

As a potential sonant, a sonant system services the transformed self for the introjection of the orderly oneness of the convergent reality of the ecosystem as a co-agonist. The coagonist is the second, potential sonant tone of the bundlable, ultraviolet B light. The bundlable light trades the belief system as a photon that becomes a coagonist after it superpositions itself over the first tone, the asymmetry to form the symmetry with the ultraviolet A light with the letter "A" before spreading the symmetry as the third tone with the letter "B."

3.3.2. Three Secondary Forms of the Potential Reality

Without dynamic reality, the potential reality of time with space begins synchronizing the present reality of time as space that has become nonresonant due to its continuity. The three secondary forms of the potential reality are as follows.

- a) Discontinuous Potential Reality. Discontinuous potential reality is the becomable present reality after it is idealized as the desired reality for servicing one with normative programming. Normative programming makes one a primeval deuteragonist without an identity of own.
- b) Continuous Potential Reality. Continuous potential reality is the idea system as a dimension making one a subject who borrows identity from an object. The object is doing nothing but observing the potential for dynamism within its continuous reality as time makes it reproductive.

c) Infinite Potential Reality. Infinite potential reality dismounts the ascending local-effect of the object on the subject. It empowers the subject to mount an ascending global-effect of its belief system for an ecosystem distribution of its local, negative energy into infinity. The ecosystem distribution destroys the normative reality of objectifying a subject and illuminates the dynamic reality of subjectifying the object.

3.4. Twelve Dimensions Differentiating the Dynamic Reality.

Twelve dimensions differentiate the dynamic reality into three paradigms for managing the subjectivity of the perceiver ideal-effect. The three paradigms are the three time-differentiating dimensions, each integrating four space-differentiated dimensions into four paths.

3.4.1. Risk Management Paradigm, Within Present Consciousness

A Risk Management Paradigm embodies the consciousness of repeated growth from the past to perceive the dynamic reality of the wholesome potential within the envisioned future. A body conceives the experience of the vision by remembering the space within time as one deliberates like a mindful greeter on the potential present to be dynamic for change. A dynamic is an inanimate force changing into an animate force when the body of consciousness traded from an object mediates its subjective experience as a subject and makes it conscious of the potential to be present as an entity. An entity enjoys four paths to make an impact on the dynamic reality (*Siddhi marga*):

a) *Path of Twin Rank*. Primordial time since the formation of Sun twins one's "rank" (*Arshti*) by polarizing its inferiority while immanent within a zero. The zero emanates as an entity to enjoy the superiority of its vision beyond the present consciousness. The "path of twin rank" (*Sarshti marga*) is the path of superiority for "freedom from inferiority" (*Sarshti mukti*) after knowing the reality. It sequences primeval time within a body of consciousness for the entity to experience her past as one inferior while envisioning her future as zero superior. The present time motivates the entity to transcend the perception that the future is superior. The transcendence is the medium that lets the entity be the enjoyer of the past as a thing that matters for how she conceives the present. The present is dark if it is infinite, seeking excellence like anyone. It is light if it is primordial time par excellence like everyone.

b) *Path of Twin Identity.* Primordial space since the formation of dark matter twins a zero's "identity" (*Rupya*) by punctuating its superiority as it emanates from the one knowing the past and wishing to manifest a future beyond. The "path of twin identity" (*Sarupya marga*) is the path of inferiority for "freedom from superiority" (*Sarupya mukti*) after manifesting reality. It causes the primeval space to form with the knowing repeating the past's growth without the body of consciousness. Descending consciousness of the potential beyond the vision limits present space to the past. The past becomes heavy as the origin of the present. The future becomes light inundated by a deluge of imaginary theories the present services after trading illusionary ideas from the past.

c) *Path of Twin Servicing.* Primordial causation since the formation of white star twins an entity's "servicing" (*Mipya*) for proliferating its transcendence by manifesting infinite futures as a sequence beyond the past with the oneness of its present. The "path of twin servicing" (*Samipya marga*) is the path of transcendence for "freedom from transcendence" (*Samipya mukti*) after creating the reality one wishes as desired. It is the consequence of primeval causation, the infinite force of the wish's reproduction within the present with a desire to change the present consciousness with an ascending consciousness of the potential for a vision beyond the limitation of the mission. The duality of consciousness makes an entity distant from the reality of oneness necessary to illuminate her as the one perceiving inferiority as a twin entity, conceiving superiority as a para entity, and experiencing transcendence as an entity. The momentum force of the freedom from reality she has not realized is the primordial human-effect of the present causation. It makes an entity consonant with the forward consciousness of the procrastinating-effect while running for office to be the primordial human. Primordial human is the causation for procrastinating the subtraction of the space's position for servicing its consciousness through the addition of momentum to time while wishing to diversity one's experience with the oneness of the sequence.

d) *Path of Twin Trading.* Primordial Greeter forms the Moon to twin the "trading" (*Yujya*) of its effect. The effect is polluting the realization of the present that one wishes by creating a consciousness of the desired reality superimposed by the zero's past as a twin entity within the entity. The "path of twin trading" (*Sayujya marga*) is the path of the decision-maker, deciding to wish a desire for "freedom from realization" (*Sayujya mukti*) of that desire before perpetuating the reality beyond a Wisher who is just a zero. Realization is the aftereffect of the decision already made by one to twin trading of the past as a metric for realizing that the future as a Wisher is indeed better.

The future of a Moon is the Earth because a New Moon is present as the Black Hole within the Dark Matter. The past of a Black Hole is the White Hole within the White Star. The White Hole is the creation of the Sun which twins causation with its clarified consciousness of the reality for making the Earth a twin entity after becoming self-luminous as an entity. Primordial trading incarnates the Earth as a consciousness within the White Star. The consciousness illuminates a hole within the White Star after it destroys the white to liberate the black for servicing ascending purple. The white complements the black while trading the ascending grey when the black substitutes the white. Present Greeter is the White Hole devoted to ascending white as it blackens the hole. By ascending black to whiten the hole with the primordial trading-effect, the Black Hole becomes a New Moon.

Primeval Greeter is the antagonist who makes the protagonist Earth self-luminous as a Creature for managing freedom from the perceiver ideal-effect with the conceiver theory-effect. The limitations of the experiencer real-effect escalate the cost of the risk management paradigm. Freedom from superiority is a positive Lumen. It creates a memory of freedom from inferiority. The freedom from transcendence is a negative Numen. Its freedom-effect makes the memory an institution that forces an entity's entropy with a twin entity into a para entity. Freedom from realization is the neutral Umbra whose past orientation perpetuates the past as real within the institution for a determination that the past force is the institutional force for universalizing as the ideal entity. The ideal entity is a Cost-effective Paradigm for the absolute realization of the reality that the institution is false. Reality makes an entity conscious that nature is true within the force whose effect is universal beyond the past. A zero institutionalizes the past to be different in the future after mediating its present like a deuteragonist within the oneness of its potential as an entity to twin itself.

3.4.2. Cost-Effective Paradigm, Without Present Consciousness

A Cost-effective Paradigm frees an entity from the institutional limitations of the present consciousness borrowed from the past. It lets an entity be conscious of the future she wishes to be present without devoting effort to be mindful of the parallel reality the past perpetuates. A conscious entity is a para entity that makes his reality self-luminous for a clarified consciousness of the goal of breeding time as an entity. An entity enjoys four paths to be conscious of her potential to twin herself into a para entity (*Siddha marga*):

    a) Path of Twin Causation. The primordial conception of the Earth as the present creation of the consciousness makes the path of the oneness of consciousness reproductive as a conscious entity. The "path of twin causation" (*Jiva marga*) weakens the "psychic" (*Jiva*) linkages with the past by reproducing the future within the present for an illusionary production of the perception that the present is undesirable because of the simplicity with which it is manageable for conceiving a desirable future. "Freedom from the psychic element" (*Jiva mukti*) promotes a primordial conception of the Earth as desirable for freedom from the reality of the conscious entity yet to be realized by the para entity. The presence of the Earth within the primordial workculture-effect as an entity that twins causation lets the para entity be the greeter who substitutes the sentient entity with a primordial element. The primordial element lets the greeter be "existent" (*Asti*) as a primordial greeter of the "germinator" (*Astika*). A germinator germinates the past to be existent within the consciousness borrowed from the past with the spirit of the future. It is a "pivot" (*Svastika*) for the germination of space itself as a copy of him as a Wisher. Time germinates as his triple copy, manifesting his past as a primordial greeter, present as a greeter, and future as a para entity. The pivot is rotatable into a twin entity, present within the consciousness of the Wisher as a Well-wisher who twins causation for the causation's well-being. The germinator works like the causation, the reason for the twin causation, to free his well-being from space, time, and causation.

b) *Path of Twin Cause.* Primordial perception of Mercury as the present of the Earth makes one conscious of the potential within the Earth to twin herself for conceiving her past. The "path of twin cause" (*Mahapralaya marga*) strengthens psychic's linkages of the past when the endoproduction of the future within the present breeds a deluge through the exoproduction of primordial culture-effect. The primordial culture-effect is the goalpost breeding itself to be present as a twin cause. Itself is the twin who causes the breeding of a deluge of entities by raising the mercury of temperature for a union in the future after knowing their past. Itself unites with the entity as she is bred with its past force to be present as a potential wave of consciousness guiding her agency with the divinity of the goalpost. The goalpost is divine because it is breeding time as an entity for curving the future nonlinearly with a conscious determination of the curving present as negative, the curvable but linear past as neutral, and the curved future as positive. As a goalkeeper, the conscious entity curves the past into the future, knowing that the present is undesirable and negative. The present moves right past his femininity to be the origin of his masculinity in the future. The masculinity left within the past makes the past's gender curvable with femininity still present. It curves the future without masculinity so that the future is omnipresent as a double-copy to be both past and present.

Masculinity perpetuates the future through the endoreproduction of the future value comprising both the present and the future when it is liberated from the past. The past leaves with masculinity because the latter is not present in the future. Femininity reforms itself into masculinity with an artificial element to manifest the beauty of her presence in infinite forms different from her finite form. The infinite forms as a potential greeter complement the finite form as an absolute greeter. Both forms supplement the goal of centering the para entity for assuming assumes the responsibility for reproducing a present that is desirable after producing its future as an imaginary element by envisioning a desired reality from the wholesome experience of the past. Consequently, the para entity becomes the Almighty Creator of the past, present, and future of the space he inherits as the cause for the origin of the causation that makes him the goalpost giving momentum to time. As the para entity becomes the goalpost, he enjoys "freedom from the entity" (*Mahapralaya mukti*) which is servicing the causation. The entity trades the identity of God to enjoy the space the goalkeeper leaves to post the causation with the consciousness of her perpetuating value.

c) *Path of Twin Electron.* Primordial experience of Venus submerged in the blackness of Mercury's past triples the cause to exist as a pion. The "path of triple cause" (*Pralaya Marga*) is the "path of submersion" linked with the deuteragonist subject who is spiritualizing the primordial experience with the whiteness of its existence in the future as Mars for "freedom from existence" (*Pralaya Mukti*). Pion exists by integrating the togetherness dimension it borrows from the space that pairs it with an electron for cross-breeding a twin electron. Thus, the path of triple cause is the "path of twin electron." The electron is the perpetuating value of the para entity within the time that becomes an entity after the reproductive, entangled interference of the multiplier to triple the cause with its past, present, and future. Integrating insulates the "entity density," i.e., rho, which is the potential of the future's (opening) causation loop, by "conglomerating" positive charge within the potential of an up quark into a "positive pion." The positive pion is a conglomerate of the down quark with a potential up quark. Within the past's (closing) cause loop, the negative charge within the present of the down quark is "conglomerated" as a "negative pion" with the up quark. The present's (open) space loop neural charges the "conglomerator"—the "potential down quark" with the primordial technological servicing of the "electron-space system" for producing a ten-fold growth. Ten-fold growth triple copies the entity, the twin entity, and the para entity with the tenth copy's reproductive potential to be a potential entity.

The potential down quark is a conglomerator suffering from the pi of inaction after getting old with time like a patient enduring the reproduction of twenty-four-fold growth as an entity. By reproducing the density potential of the entity's mass of consciousness, twenty-four-fold growth cures the hole within the space from the multiplier's reproductive interference. 24-fold growth makes the three copies productive like the three octaves with its energy as a differentiator. An octave differentiates an entity into four generations within and without differentiating class for breeding gender as the differentiator of the two recombinations. Two recombinations pre-exist as photon radiation. The radiated photon is a twin photon, the polarized photon. It forces the light to be polarized as a polarized light, present as a sentient entity within an entity polarizable as an "ion" after polarizing the development of "p" as the half atom.

The differentiator develops into an atom, self-perpetuating a half atom by radiating an electric field of its self-devotion. Self-perpetuation centers the atom after irradiating the organization of "pion" as a magnetic field for the primordial differentiation of the "electro" as a "proton" with the letter "N" of the electron. The proton differentiates a magnetic mass integrated within the gravitational field to exist as an exciton after the multiplier makes the space producing the atom as an electron reproductive. The proton orders the development of the multiplier into a pion with the growth of its potential as an exciton. The submersion of the present within the discontinuity of the multiplier after the growth of the potential procures the proton from the exciton as a twin proton. The twin proton is the Eastern emersion of the present as the Mercury following the Western mersion of the potential as the Venus, leading to the Central demersion of the whole as the Sun within the vicinity of the wholesome as the Dark Matter for its observance as the White Star from a distance.

As a greeter, the exciton is free from the multiplying, metaphysical, protagonist pressure of the pion because it exists before the development of the pion becomes luminous within an ion. The ion is the whole photon which the multiplier centers as an atom. The ion experiences entangled interference of the multiplier for the development of pion. The pion services protagonist pressure on the ion for its development after it borrows self-devotion from the differentiator for differentiating itself into a potential greeter to complement the exciton.

d) *Path of Twin Proton.* The primordial spiritualization of Mars merged with the whiteness of Venus's future halves the cause for bursting the first one as a Fire planet without the center before forming it into Saturn. The "path of half cause"

(*Laya marga*) is the "path of immersion" as the second one. The second one liquidates the first one into a Water planet to be self-luminous within the reality of the triple octave as Jupiter for "freedom from the center" (*Laya Mukti*). The freedom from the center is the freedom from the rhythm of managed growth by a para entity. The para entity is the seventh one, the Guider planet which twins causation to make his reality as the Earth self-luminous. The reality is the eighth one, the Sentient planet, reproductive with its simplicity as Mars. The fifth one is the Ether planet which fires the Sun with the simplicity of objective to reorder the sentient force of the Sun's past for becoming Venus. The sixth one is the Divine planet, present within the sentient force as Mercury. The third one is the Earth Planet, solidifying its present value into the Uranus after the Sun orders the reality's trading-effect within the para entity to be a star. The fourth one is the Air planet, reforming into Neptune to gasify the Sun which self-perpetuates its potential within the discontinuity of the eight planets for the continuity of its present as the twin proton. Therefore, the path of half cause is also the path of twin proton.

Primordial technological trading of the existence as Mars makes the symmetry of two in future reproductive as a twin graviton. The two are Mercury as the present and Venus as the potential past of the Sun. Mars—the one that becomes a twin graviton after servicing two—is the Sun's primeval future. The asymmetry of six in the past makes one reproductive with the symmetry of two in the future whose past, present, and future are asymmetrical at present. Mercury's past is the Earth, the present is the Dark Matter, and the future is Jupiter—the seventh planet reproductive like the guider planet with the Earth's reproductive force. The Dark Matter is the present omnipresent as one which is potential present of the twin graviton. The Earth makes the reality of her primordial present reproductive for Jupiter to be her primeval present. Venus's past is the Saturn that the Sun repeats as two twos, two within the Earth's reproductive force and another two without. The two without the reproductive force reproduce Venus's past to produce the present as Uranus after consuming the future as Neptune. The

Uranus is the primordial past. The Neptune is the primeval past. The two within the reproductive force are produced by Mars as its past (primordial future) and the present (present future). The White Star is the primordial future. The Moon is the present future. The Sun is Mars's potential future.

As a greeter, the proton trades the universal consciousness of its potential to be a twin proton with the multiplier that makes the Sun self-luminous as a quantum photon. As a potential greeter, the electron services the majestic consciousness of its present the future perpetuates as the Sun's potential past to be luminous as a quantum light. The electron is free from the competitive, dividing, deuteragonist pressure of the exciton because it is luminous with the Sun before the exciton is self-luminous within the Sun. As an absolute greeter, the neutron exchanges the potential within the present for the one to be two with the endoproduction of a double copy, norming its past with the reproductive interference of the future. The neutron is free from the complementary, subtracting, agonist pressure from the growth of the multiplier because it is the one multiplying itself with a goal of nine-fold growth. As a primordial greeter, the atom invests the multiplier as the third copy for the addition of a double copy of the nucleus. The atom is free from the supplementary, adding, antagonist pressure from the entropy of the one, the divider, after the growth of the multiplier. It is the zero dividing itself into four with a double copy for an eight-fold growth. As a primeval greeter, the nucleus enjoys the capability of the space divider for self-projecting the growth of three zeroes within the oneness of one ejected by the present. Three zeroes are the future of the one ejected by the present to be its past, present, and future copies as the origin of the potential for their growth. The nucleus is free from the coagonist pressure of ejecting one from the eight-fold growth to be left with seven. The seven becomes "one with three zeroes," the leftover from the kratagonist pressure, i.e., trading pressure, of exchanging its growth with a one seeking to be a two with the divider and a six with the multiplier.

The four paths of the para-entity approach together constitute a Cost-effective Paradigm. The Cost-effective Paradigm entails trading the dominating potential of the pressure from an emanating factor for descending the escalating costs of workculture-effect, culture-effect, technological servicing, and eventually technological trading, by servicing the primordiality of the mindful greeter present within oneself as an ideal entity. Without the reproductive element of the wisdom that forms the circular with its infinity, the primeval becomes primordial.

The primordial workculture-effect makes the primeval greeter mindful of the time potential that makes the multiplier circular. The primordial culture-effect makes the primeval greeter mindful of the space potential that drags and varies the time potential to make the growth circular. Primordial technological servicing makes the primeval greeter mindful of the causation potential, the projected vector that makes the past as the multiplier of growth circular. Primordial technological trading makes the primeval greeter mindful of the entity potential to make its double copy as a zero circular. Entity potential breeds consciousness void within one unaware of the limitations of the conscious entity as a multiplier of the consciousness Mother Nature self-perpetuates. It transforms one from a "protagonist breeder" (ejected as the channelable-effect, i.e., Torah-effect, guided by the notion of faith) to an "antagonist believer" (refugee taking refuge in the prophetic Christ-effect for discerning the channel's authenticity with reason) to norm a "deuteragonist beggar" (the circular as a reproductive system of the mercurial Buddha-effect for self-substantiating the four zeroes that reform the channeling, deform the channel, and inform the channelable what's false so that the true may be discovered from intuition as an entity) who begs to differ by integrating the two.

The "agonist blesser" (One within the center of the circle whose infinity is circular because the multiplier, the growth, and the past are circular with its omnipresence-effect, i.e., Geeta-effect) differentiates the three for blessing their integration within the four. The "coagonist becomer" (One-of-the-kind channeling the kind-effect, i.e., Dao-effect, as the one within a primordial male who centers the ecosystem with his five, primordial characteristics to be within it as a primordial creature) reproduces the six as a five within one for becoming a body of four. The "kratagonist breather" (Debilitating-effect, i.e., Islam-effect, of questioning the completeness of the center after aggregating its circular-effect for squaring the copy potential of the copy present) produces fifteen by reproducing seventeen for knowing the secret of the nineteen within and without two. The secret is the aggregate beyond the vision of light force. It makes one a primeval antagonist and suffers after everyone questions the force whose vision is guiding one to be an antagonist as a primeval greeter.

Consequently, a need emerges for a Formative Growth Paradigm to know the secret of the dynamic reality by first creating it as a Mindful Greeter. The Mindful Greeter is free from the present pressure because he is the one present with the potential to pressure oneself to exchange one's growth with a zero which centers one's growth. The exchange opens trading of everyone else as two zeroes by reproducing oneself as a zero before breeding everyone as "else" to close countertrading of one within the center. Else is debilitated with consciousness void, not knowing the primordial self whose growth has become reproductive as a channel for the formative growth of everyone's potential to reproduce its reproductive force.

3.4.3. Formative Growth Paradigm, With Present Consciousness

A Formative Growth Paradigm is the external point of discontinuity whose continuity is reproduced within the primordial self for centering the growth that forms when the continuity becomes circular within the infinity bred by its effect. The circular-effect divides the origin with the ideal entity's triple copy to circle "all" as one with the reproductive force of three zeroes. It lets one with all be the reproducible primordial self of the circle whose infinity is circular. As an ideal self, the reproducible primordial self has the deity potential of supernatural value, squaring three with the circle without the fourth within the circle to be the fifth without the circle. The fifth is super-ideal; it is the triangle that triangulates the four to be the ideal creature with an experience of all. The all-encompassing present cycles the "technological perception" of believing all as an afterthought of the believe-effect for conceiving a desired, illuminated consciousness of the dynamic reality. The desired consciousness is free from the intermediating layers of trading, servicing, and exchanging the potential as the cause for the future to be present as the past, technological reality.

The technological perception is the formative perception of "I am" as a living deity who twins his consciousness as a creature to be true after transforming all into false. All are false because they are following a false reality that does not need to be present for substantiating the effect of a living deity. Formative perception is the perception of the creator of the dynamic reality. It forms at the moment of the creation of the dynamic reality as an effect of the creator's potential to create the one living as a deity. It is the secondary fundamental-effect of the creator, without a tertiary exchange of the para-entity into the energy of the ideal entity after entropy but within the quaternary servicing of the entity as a twin entity before entropy. The Almighty Creator is the primary causation for conceiving the dynamic reality with quinary trading of the ideal-effect to be the one who all experience as not only the ideal creature but also the Creator of the ideal-effect. The causation for the experience is the theory-effect of the possibility that all are the child of the ideal creature and believe in the latter's potential to be their Creator. Thus, formative perception is personal. It is entropy real, formed with the self-sacrifice of the Creator Potential, immanent as time.

Time greets the dynamic reality as a catalyst guider to accelerate the technological growth of the present's primordial oneness with its potential to be the organizational reality of everything. The technological growth lets the reproducing primordial self be the creature free from the Creation of the theory-effect while living as a deity working to present his Creation Potential for knowing the Creature Potential to be the Tree of Creation.

A deity enjoys four paths for freedom from the dynamic reality the time greets (*Mukti marga*).

  a) *Path of Twin Neutron*. Primordial sentiation demerges Jupiter from Earth as a Water Planet. The Water planet is the nucleus of the triple continuity of Venus, Mercury, and Mars. The nucleus is the one that makes the three circular with Venus as the potential past, Mercury as the present, and Mars as the infinite future taking many forms with one's growth potential. The potential twins the circular with the growth of six for the decomposition of two neutrons to triple the composition of "time" within a "qualifier" (*Gunatita*). The space takes the position of a qualifier as a square to qualify the imposition of an outer boundary with a hole whose blackness lights time as a triangle. The light divides time by two where the second half is dark. Darkness makes the origin heavy as it self-perpetuates light as the first half of time value illuminated as the divided time. The divided time repositions the causation, i.e., the one that takes the position of the origin with three zeroes to superposition entity as the cause for growth. Thus, the "path of twin neutron" (Guntatita marga) is the "path of the qualifier."

The "freedom from qualifier" (*Gunatita mukti*) gifts the twin entity the charge of the entity's mood to recharge the positive energy of the ideal entity with her illuminating value after the para entity discharges his negative energy. The illuminating value makes the ideal entity conscious of the reality which the entity twins to charge her mood for qualifying the "primordial technological growth." The primordial technological growth is the all-diverging extract of the light an entity composes with its value to recompose the collective value of the circular creation. The circular creation twins the triple continuity of the extracted individual value with its reproductive force.

b) *Path of Twin Nucleus.* Saturn emerges with Jupiter's primordial imagination of a favorable positive internality. The positive internality makes Jupiter the infinite present as the metric of everyone's reality. Everyone is light, illuminating value as their reality is yet to be realized. The infinite present infuses the negative charge of Uranus's soul within Saturn for illuminating the positive charge of Neptune's spirit within everyone. Neptune's spirit has a positive charge because she wishes to neutral charge the infinite past for ascending Neptune's light force. Uranus's soul has a negative charge because he wishes to attract the positive charge from Neptune's infinite past by descending light force and trading the time consciousness of the infinite nucleus.

The infinite nucleus is the First sun which keeps its position fixed in space as time gives momentum for producing infinite suns with the First sun, the reality of whose growth is within the Second sun. The Second sun is the growth of the divisible time as a luminous point, whose time is divisible by the reality of the First Sun without it. First Sun is the endoproduction of the Third Sun within the Second Sun. The Third Sun is the present of the Second Sun. The First Sun is the embodiment of the Second Sun's past. The Fourth Sun is the Potential Incarnate breeding the First Sun as the past for its embodiment as an entity. The Potential Incarnate is the New Moon, the Second Sun's future within the Black Hole of time.

"Primordial technological exchange" of collective value triples the time of the Potential Incarnate by reversing the Second Sun's past for the embodiment of the present of the First Sun as the Third Sun. The Second Sun is the triple copy of the First Sun's future value as a multiplier of one's formative growth as a deity before becoming a Potential Incarnate of the White Hole of space without the Hole of causation. Hole of causation squares time with space itself.

The twin nucleus is the Positive Hole of cause; it is a well that naturally directs a being to be a guardian of its well-being. Therefore, the "path of twin nucleus" (*Hita marga*) is the "path of well-being" for the present growth through the "path of time" (*Kaal marga*). With the "freedom from time" (*Kaal mukti*), the psychic nature of the space present is whitened. Whitening is the negative hole of the entity that is whitening her past with her sentient element to force her present to gravitate toward the future she wishes.

c) *Path of Twin Atom.* Uranus's soul is within the consciousness of a twin atom after an atom twins causation with the causation to gravitate the future of a wish. The future of the wish is the Wisher of the present's well-being. The causation copies the Wisher as an organelle of ten molecules. The causation is the tenth molecule (Chimeric molecule), the copy is the ninth molecule, and the Wisher is an octave of the molecule, norming a potential hexaquark. The present double copies the organelle for the endoproduction of the nineteen organelles within a twentieth, bipolar organelle after centering nine organelles within a tenth, unipolar organelle as the future's copy from the past. The organelle is the first organelle. The past is the ninth organelle. The future is an octave of organelles. The copy services the tenth organelle for centering itself within the past as the nineteenth, present organelle. Nineteen organelles are present within the twentieth organelle the future conceives as a cell with its conscious imagination. Nineteen organelles form a bipolar brush cell that brushes the future with the bipolarity of the double copy. The brush is the unipolar brush cell, the tenth organelle. The tenth organelle is the unipolar organelle with hundred molecules. The twentieth organelle is a bipolar organelle with two hundred molecules.

The nineteenth organelle is a multipolar organelle with six hundred molecules—hundred that make the past's infinity reproductive as a base within the molecule, two hundred as the discontinuity of the future's infinity as a cell without the molecule, and three hundred as the continuity of the present's infinity as a potential with a molecule that self-services a future of $2/3^{rd}$ to self-perpetuate a past of $1/3^{rd}$. As the eighteenth organelle, the potential has three hundred molecules. Overall, four organelles (tenth, eighteenth, nineteenth, twentieth) are self-projecting a natural six-fold growth of twelve hundred molecules as a third organelle. The eighth organelle is self-projecting a 12-fold growth of twenty-four hundred molecules with five organelles. The seventh organelle is producing a 24-fold growth of forty-eight hundred molecules with six organelles as a Y-linked chromosome. The sixth organelle is reproducing a 48-fold growth of 9,600 molecules with seven organelles.

The fifth organelle is servicing a 90-fold growth of 18,000 molecules within the future of eight organelles with the four organelles that are trading a 6-fold growth from the fourth organelle. The fifth organelle is a self-luminous organelle with twelve organelles. As the fourth organelle, the future is the octave of the organelle. As the ninth organelle, the past comprises three organelles that perpetuate the past of the future's triple copy within past, present, and future with 54,000 molecules. The infinity is the second organelle with two organelles comprising 90,000 molecules. Light is the eleventh organelle with twenty-two organelles comprising 180,000 molecules. Twenty organelles norm a photon with a spirit of 90,000 molecules with 100,000 molecules norming a protein molecule. The photon is the twelfth organelle; it is a sentient organelle, conscious of the position light gives it for gaining momentum of self—the speed of light—with the addition of 10,000 molecules. The luminous that superpositions light over the photon to norm quantum light is the thirteenth organelle, comprising 270,000 molecules with 22 + 2, i.e., 24 organelles. The thirteenth organelle is the entity organelle. The entity is the fourteenth organelle, which twins an organelle with its triple copy within the thirteenth organelle. Therefore, the 14th organelle comprises forty-eight organelles with 540,000 molecules that are Y-linked.

The fifteenth organelle comprises one hundred organelles with one million molecules that are X-linked within a twin entity. It trades the natural six-fold growth of 100 -96, i.e., four organelles from the time multiplier. It services a supernatural four-fold growth of 1,080,000 -1 million, i.e., 80,000 molecules (i.e., 4 * 5,000 molecules per organelle) with sixteen organelles (i.e., 4 x 4 organelles) as the sixteenth organelle that is Z-linked within the ideal entity. The sixteenth organelle is the greeter organelle. The seventeenth organelle is the primordial organelle; it is the time-effect of fifteen organelles with 75,000 molecules within the sixteenth, greeter organelle.

Time is the twenty-first organelle with eighteen hundred organelles and nine million molecules. Space is the twenty-second organelle with eighteen thousand organelles and ninety million molecules. The twenty-third organelle is the five-fold growth whose five-effect surfaces the space with two-thousand organelles and ten million molecules. The twenty-fourth organelle is the seven-fold growth with sixteen hundred organelles and eight million molecules. It is the param organelle. The twenty-fifth organelle is the two-fold growth with four hundred organelles and two million molecules, norming a 2600th mtDNA molecule. The twenty-sixth organelle is the trading-effect of forty organelles and two hundred thousand molecules.

The twenty-seventh organelle is the human-effect of the eighty organelles and four hundred thousand molecules. The 26th organelle self-perpetuates the human-effect to twin itself within the human-effect as the twenty-eighth organelle of 120 organelles and six hundred thousand molecules. The twenty-ninth organelle is a triple with 180 organelles and nine hundred thousand molecules. The thirtieth organelle is half with 90 organelles and 450,000 molecules; it grows naturally into the triple with its replication over time. The thirty-first organelle forms the nucleus with one thousand organelles because the ten organelles within the thirty-second organelle are reproductive and the nine hundred organelles within the thirty-third organelle incarnate with the reproductive-effect of the dying, goblet cell. These three organelles have five million molecules, five hundred thousand molecules, and four million five hundred thousand molecules respectively. The first one is a chlorophyll molecule. The last one is a triple water molecule. The 90th organelle is a water molecule with 1.5 million molecules and 300 organelles.

The thirty-fourth organelle is the effect of thirty-four organelles with sixteen hundred molecules. It invests thousand-fold growth of one thousand molecules to form the thirty-fifth organelle with the reproductive energy of the tenth organelle's one hundred molecules within the ten thousand molecules of the thirty-sixth organelle. The thirty-fifth organelle is the potential organelle with 10^{10} organelles. The thirty-sixth organelle is the primeval organelle with 8×10^{15} organelles.

The thirty-seventh organelle is a fire organelle of seventeen organelles with eight hundred molecules (norming a primordial-primordial mitochondrion molecule) as the time multiplier self-perpetuates the effect within the past's three organelles. The past's twinning makes the fire organelle a twin atom. Therefore, the "path of twin atom" (*Tirthankara marga*) is the "path of twinning" to incarnate the fire's growth potential as a female through the "path of endoproduction" (*Maya marga*) until entropy for "freedom from endoproduction" (*Maya Mukti*) after.

The female is the thirty-eighth organelle, the water organelle comprising 169 organelles that make the male's thirteen organelles reproductive. The thirteen organelles are the thirty-ninth organelle, which is the ether organelle with six hundred molecules left as its present after the growth of twelve hundred molecules within the four organelles. The male makes eight molecules reproductive as an organelle when the female squares the six molecules with the two molecules. The two molecules are the ether molecule (norming a General Organellar DNA molecule) forming the 1st DNA molecule as the 120th organelle. The 120th organelle is an organic molecule with eleven organelles because the first organelle makes ten molecules reproductive as ten organelles through primordial oneness with the 32nd organelle. The female channels 600 *600/36, i.e., 10,000-fold growth with thirty-six molecules. The organic molecule forms with the reproductive potential of the "site of electromagnetic reaction," which forms an "inorganic chemical" differentiating a four-body system, the emanating system as a whole, into

- the four bodies, norming the characteristic of Mother Nature as the creator. Four intrinsic half-octaves norm maternal consciousness to generate a triple bond among the three electrons with time's three dimensions. As a fourth electron, the dimensionless triple bond strengthens the time multiplier with a charged molecule into the fourth body, the differentiating system.

- the four extrinsic half-octaves, rounded as a "creature" (the present of everybody) with the generation of a double bond among the four electrons with space's four directions. With its entropy, the double bond weakens the space divider into the "backbone" (the past of everybody as a para person) for the "organic self" (the potential person as a positive quark), and

- the differentiating system, the octave of elements as the inanimate one within the inorganic self of Mother Nature. The inanimate one twins causation to be an organic molecule within the time multiplier. The inorganic self is the "illusion" (as a negative quark) of Mother Nature as a person personifying the creator's characteristic to be a creature with the organic molecule within a charged molecule. The charged molecule is an ion comprising the extrinsic four half-octaves as a quantum photon within the octave of elements. It is charged by the present growth of an entity with the past force of the creation which forms the feminine element.

The feminine element gives birth to the possibility of a child with the masculine element of a person's characteristic to be animate. A person animates the sentient force of consciousness with the past's perpetuating value. The child is the fortieth organelle, the air organelle comprising two hundred organelles. It "impact" is exchanged as a "whole" by the hundredth organelle, the earth organelle, comprising fifty organelles to twin oneness with the 48-fold growth within the sixth organelle, the divine organelle. The child has forty molecules. The impact of the discontinuity within 200 organelles is a diffusible molecule, the air molecule norming the sixtieth organelle as a decay organelle of 31 organelles after they live as 14 molecules. The 180th organelle, the guider organelle, is the consciousness that the whole makes reproductive. It comprises thirty organelles as four molecules (norming a Special Organellar DNA molecule) decay into one organelle. The earth organelle's growth of six molecules forms a 46th DNA molecule as an inorganic molecule. With spirituality, one organelle lives as a circular molecule (an insoluble molecule), the fire molecule with 140,000 molecules for self-projecting 14 that lives as the consciousness of time within the organelle. The fire molecule is the 96th organelle of 21 organelles. As spirituality is modulated by divinity, fourteen organelles form an earth molecule, norming the first mtDNA molecule that transforms into the 40th DNA molecule with continuity of fifty molecules norming the fiftieth organelle.

The "primordial technological investment" of space inflation oozes the reproductive potential of the twenty-three organelles as the revolving form of a divine molecule to develop the evolving form of time as a sentient molecule. 23 organelles norm the 150th

wholesome organelle. Fluctuating time is present as a divine molecule self-projecting the fourth dimension of space as the goal to perpetuate an entity's conscious planning of a metric of her supernatural growth. The metric makes the causation negative. The divine molecule comprises 40,000 molecules as a negative organelle. The negative organelle makes the 132nd organelle a wholesomewhole organelle with primordial oneness of the universe without one. The devolving form of causation makes the system reproductive with 25 organelles as the 48th organelle, the growth organelle, for oneness with the one universalizing four dimensions of space with the fifth dimension involving time. With the growth of space as an organelle, time produces entropy as a molecule. Therefore, the growth organelle is an entropy molecule. It comprises 4,000 molecules, with 2,400 molecules of the eighth organelle whose five organelles reproduce twenty-five organelles and the effect of 1,600 molecules present after the post-reproduction entropy.

Table 1 summarizes the sequential decomposition of organelles, their molecular composition, the metaphysical form, and energy values.

Table 1. Metaphysics of Organelles and Their Molecular Composition

Sequence	# of Organelles	# of Molecules	Metaphysical form	Energy value
First organelle	1	10	Illusion	1
Second organelle	2	90,000	Infinity	90,000
Third organelle	4	1,200	Growth	6
Fourth organelle	8	12,000	Future	0
Fifth organelle	12	18,000	Replication	90
Sixth organelle	7	9,600	Intelligence as divine organelle	48
Seventh organelle	6	4,800	Y-linked chromosome of the para entity	811
Eighth organelle	5	2,400	Self-luminous as a System	12
Ninth organelle	3	54,000	Past	9

Sequence	# of Organelles	# of Molecules	Metaphysical form	Energy value
Tenth organelle	9	100	Centering as a Guider molecule	10
Eleventh organelle	22	180,000	Light	180
Twelfth organelle	20	100,000	Spirit as the Sentient organelle of the Protein molecule	20
Thirteenth organelle	24	270,000	Luminous as Quantum Light	13
Fourteenth organelle	48	540,000	Y-linked to an entity	24
Fifteenth organelle	100	1 million	X-linked to a twin entity	24
Sixteenth organelle	16	80,000	Z-linked to an ideal entity	24
Seventeenth organelle	15	75,000	Time-effect	15
Eighteenth organelle	18	300	Potential	18
Nineteenth organelle	36	600	Present	1,600
Twentieth organelle	19	200	Cell	19
Twenty-first organelle	1,800	9 million	Time	360
Twenty-second organelle	18,000	90 million	Space	18,000
Twenty-third organelle	2,000	10 million	Entropy	5
Twenty-fourth organelle	1,600	8 million	Reality as the Light molecule	7
Twenty-fifth organelle	400	2 million	Order	2

Sequence	# of Organelles	# of Molecules	Metaphysical form	Energy value
Twenty-sixth organelle	40	200,000	Trading-effect	26
Twenty-seventh organelle	80	400,000	Human-effect	53
Twenty-eighth organelle	120	600,000	Twin	121
Twenty-ninth organelle	180	900,000	Triple	130
Thirtieth organelle	90	450,000	Half	1/2
Thirty-first organelle	1,000	5 million	Nucleus	1,000
Thirty-second organelle	10	500,000	Guider	100
Thirty-third organelle	900	4,500,000	The heavy molecule is a triple water molecule within an air molecule that becomes a light molecule	900
Thirty-fourth organelle	34	1,600	Force	34
Thirty-fifth organelle	10^{10}	1,000	Self-luminous system	10^{10}
Thirty-sixth organelle	8×10^{15}	10,000	Self as a Luminous system	8×10^{15}
Thirty-seventh organelle	17	800	Fire as Fire organelle	17
Thirty-eighth organelle	169	36	Female as Water organelle	10,000
Thirty-ninth organelle	13	8	Male as Ether organelle	100,000,000
Fortieth organelle	200	40	Child as Air organelle	128

Sequence	# of Organelles	# of Molecules	Metaphysical form	Energy value
Forty-eighth organelle	25	4,000	Growth organelle as Entropy molecule that is devolving form	366,666
Fiftieth organelle	14	50	Modulated as Earth molecule	60
Sixtieth organelle	31	14	Impact of devotion as Air molecule	57
Ninetieth organelle	300	1.5 million	Doubling as Water molecule	1,900
Ninety-sixth organelle	21	140,000	Spirituality as Fire molecule	−9
Hundredth organelle	50	6	Whole as Earth organelle, the dark molecule	16
One-hundred twentieth organelle	11	2	Whole organelle as Ether molecule that reorders	2
One-hundred thirty-second organelle	−1	40,000	Wholesome whole organelle as Divine molecule that is revolving form	366,666
One-hundred fiftieth organelle	23	230,000	Wholesome organelle as Sentient molecule that is evolving form	800,000
One-hundred eightieth organelle	30	4	Soul as Guider organelle	4

d) Path of Twin Molecule. Neptune's spirit lights the one replicating his growth as a sentient entity to order the replication of everyone in the universe as its copy. As a twin molecule, time's reproduction molecule (ether molecule) reorders the space's order with its causation's reproductive element to transform the growth of four organelles (as the 3rd organelle) into four hundred organelles (as the 25th organelle). The "path of twin molecule" (*Para marga*) is the "path of reorder" following the "path of transcendental" to change the order by reproducing the immanent for producing the emanating as a transcendental. As a greeter, the transcendental causes the endoreproduction of the order with the continuity of the endoproduction of the greeter consciousness as the causation. The causation is continuous within the consciousness. The consciousness emanates as a whole when reproduced with the immanent. The "freedom from transcendental" (*Para mukti*) is the "freedom from primeval" (*Tamas mukti*). It makes the greeter primordial with the descending mass of Uranus's soul as the ascending force of Neptune's light makes the spirit dark and primeval.

The twinning force of Uranus's past perpetuates an entanglement of Neptune's future with the reproductive force of its soul within Neptune's spirit. The correlation force of the reproductive element is the "primordial technological cost," descending the force of Uranus's light for ascending the mass of Neptune's consciousness. Uranus's light superpositions over Neptune's soul of light as a "quantum light" (Luminous). Neptune makes that soul reproductive for reproducing the mass of consciousness until it becomes heavy. Time-guided momentum of Neptune's spirit within heavy descends the position of Uranus's light while ascending the position of Neptune's spirit.

The superposition of Neptune's spirit as a "quantum photon" (Self-luminous) makes Uranus's light dark and primeval, yielding the primordial element of the light's reality to Neptune. Consequently, Neptune's Lightforce and mass are perpetually ascending while Uranus's Lightforce and mass are perpetually descending for energy equilibrium with the U-turn of the position itself. The U-turn reforms Neptune as Uranus through the endoproduction of a twin Neptune, followed by an endoreproduction of the twin as Uranus and Neptune as its copy with zero Lightforce and zero mass.

Consequently, Uranus becomes heavy and Neptune becomes light. Heavy forces light to be conscious that it is dark and primeval.

The conscious force of the primordial perception of the self's darkness as Neptune powers the self's primordial conception of its lightness as Uranus. The primordial experience of the intrinsic darkness and the extrinsic lightness follows the primordial sentiation of Jupiter as a reproductive planet like Earth. Jupiter is dark and heavy before it reproduces itself over time to be light and primordial. Jupiter's transformation from heavy to light is immanent within Uranus and Neptune, both emanating as a circular creation from the formation of Saturn as the primordial for the organization of Jupiter as the present without its three dimensions that make the present primeval.

Saturn is conscious of its intrinsic light as it is the Sun that illuminates itself as the Earth whose light is irradiated by Neptune. Earth is conscious of her extrinsic light as she is the Dark Matter that becomes a Moon while radiating its mass to Uranus when illuminated by the Sun, seeking to hide within the shadow of Mars. Mars is conscious of the intrinsic dark element that develops with the primordial spiritualization of the belief that it matters for the illumination of the universe. The universe forms when the Moon disappears after radiating its mass to reappear as the star that universalized its mass. The endoproduction of the Moon's mass within the star leads to the Moon's disappearance as a spirit that attracts everyone for trading the ascending mass by descending the force of their light like Uranus. Mercury is conscious of the extrinsic dark element which grows from the primordial imagination of itself as Venus.

Venus blocks extrinsic light for producing the dark element to make Mercury matter for servicing the ascending mass of light it trades from the Sun for illuminating both. As a collective, both force the Sun to be the Earth irradiating their extrinsic light to be heavy as an individual. While heavy the Earth is Uranus. After radiating its mass, the Uranus is light as Neptune. After radiating its force, Neptune becomes dark as Jupiter. After radiating the consciousness of force that is ascending universal mass, Jupiter is primordial as Saturn. After becoming conscious of the self as the luminous, Saturn is self-luminous as Mars. After the endoproduction of its future as self without the universe, Mars is present as Dark Matter. After the endoproduction of the self as one within the universe endoreproduced by its past, the Dark Matter is luminous on the horizon as a White Star. Time's reproductive element makes the White Star a Moon before illuminating its value as the Sun. The Sun is an entity conscious of the space that is forcing it to illuminate its value and be a star of attraction for the time that orders its entropy.

Thus, the twelve primeval elements—eight planets, Moon, Sun, Dark Matter, and White Sar—within the four primordial elements—time, space, causation, and entity, make the cause whole with the param element. Once whole, the greeter as the cause is liberated from the secret extrinsic potential dimension bred by the dynamic reality. As a potential greeter of the secret of how to complement the intrinsic potential dimension, a deity as the causation becomes the technological reality, the object mediating universal well-being with the thing that matters. The thing that matters is the reality of the technological element, the whole with sixteen param elements.

The four dimensions of the deity approach constitute the Formative Growth Paradigm. The Formative Growth Paradigm entails servicing as a primordial greeter the deciding potential of the immanent factor for ascending the worker social benefits of the technological growth of the primordial, technological exchange of the primeval, technological investment of the trading-effect, and technological cost of the human-effect, by trading the present as the mindful greeter. When the cause is immanent within one, growth emanates to make one conscious of the divine within.

As the primordial qualifier of space, the "freedom from qualifier" (*Gunatita mukti*) gives one the "freedom from time" (*Kaal mukti*) which is qualifying the space with its evolving form. As a greeter of time, one enjoys the "freedom from illusionary production" (*Maya mukti*) of the causation as a deity working to be free from the time's entanglement by universalizing its divinity. As a potential greeter of personal sentient well-being, one enjoys the "freedom from transcendental" (*Para mukti*) because the universe does not need God after everyone realizes the unrealized reality with the divine light of the primordial greeter. Yet, there is a need for integrating the technological reality of the present greeter for the sentient well-being of the primeval greeter.

The path to be conscious of the primordial human potential entails an exchange of the extrinsic GUIDER virtue with an intrinsic DIVINE responsibility. That requires (a) imagining the self as the Greeter greeting the potential for absolute realization of oneself, (b) believing the self to have the virtue of a Potential Greeter complementing the divine-effect of the technological reality, (c) becoming a Primordial Greeter for guiding self with intuition transcending the limits of reason immanent within the technological reality, (d) breathing Mother Nature as the Param Greeter with the energy of devotion while guiding self without entanglement with the supernatural, and (e) behaving with excellence who twins everyone into a guide guiding oneself.

Therefore, we next **re***search* the four paths of impact on the cause within, the reality of cause without, and the freedom from cause with *the past consciousness*. These twelve dimensions integrate the technological reality into three paradigms for managing the experiencer divine-effect—normative development paradigm, transformative exchange paradigm, and formative growth paradigm. These are the subject of the remaining three chapters.

Chapter 4

Normative Development Paradigm

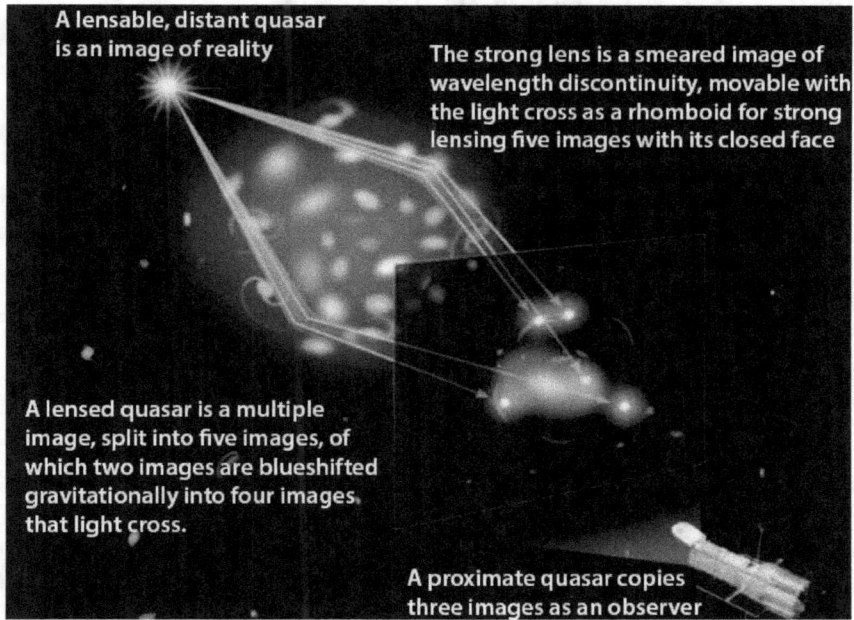

Figure 24. Observing Three to Image Reality of the Past

Does subtracted light from the image of the distant quasar double copy the future to animate the all-in-one foreground object, present in primordial oneness with two—the star as the ground object which squares the weak lensing for weak gravitational lensing of the third zero of time as the background object from the past? (*Newtonian ideal of weak gravitational lensing* of the past invisible as the darkness of the lunar element). Or, Does the added mass of the "light ring" (Einstein ring) from the past twin two images of the proximate supernova, present as the "light cross" (Einstein cross) of the discontinuity of two, to image a distant quasar that double copies the future through strong gravitational lensing of the three energies of time in a two-dimensional space with time as the third-dimension ascending height? (*Einsteinian theory of strong gravitational lensing* of the future visible as the light of the solar element). Neither of the two.

In reality, lensing constitutes within-effect as frontward-effect to certify that the source is weak when the base is reproductive in the past with the strong gravitational lens of the present as a product of a spirit of the weak gravitational lens of the future. To transform both the past and the present, the future images a distant quasar as a double copy of reality to animate universe with a twin sun with three as a time multiplier. Sun animates the potential of three copies within the time multiplier for the subject (the observer) to copy as the first image, three images that norm a proximate quasar from four images.

Animate has seven images, which is the seventh, smeared image of the wavelength discontinuity that length stretches into a strong lens. Seven images are movable as seven colors, forming a rhomboid, the absolute parallelogram, by transforming into two images. Two images are blueshifted, gravitationally lensed quasar, as light crosses of four images. Light cross (i.e., Einstein cross) matures into a proximate supernova. The fifth image is magnification, parallel to four images. Four images as a whole are the third, amplified image, a heavy ring of distant supernova, repeated as a parallelum. The parallelum is the intervening galaxy; it is a weak gravitational lens as a lensing galaxy. The rhomboid's closed face is strong lensing five images into a fifth, multiple image, which is a lensed quasar.

The sixth image is the magnified image of the reality of seven images within the two images; it is reproductive as the lensing quasar, which is a strong gravitational lens. The eighth image is the background image; It is the divine image of the eight continuities of the image. The ninth image is the foreground image, which is the gravitational image, the source reproducing itself as the origin within the continuity of the wave. The tenth image is a strongly elongated image; As a conscious image, it is a quasar that forms the wave of continuity. The eleventh image is the distorted image, the gravitational arc. It is a pulsator with eleven images: one from a distant quasar (source), three from the observer, and seven from wavelength discontinuity that animates itself to be the observer of the source at the end of the pulsation period of its continuity. The twelfth image is the youngest—it is the lensed image. It is a shear, contracted at the vertex (perpendicular direction) to be stretched at the source (horizontal direction). It surfaces wavelength infinity as an anomaly of the time value the light rings when it is repeated as gravitational light without the source, thereby ascending energy as a variable. The variable is the object, which is the past reality of time that transforms into the present reality of space with the constellation of effects from the future reality of causation with the consciousness of the reality of its illuminating value for an entity.

Is Normative Development Real as a Paradigm?

Normative development reforms the goal for one's development into an organization that transforms the reality of the universe to norm the desirable with the spirit of its belief system. Paradigm is the pressure of falsism within the belief system. Believing in one's potential for becoming an object of desire for everyone's intellect does not make a subject massive. Behaving like a massless subject whose consciousness of desirable is reproductive within everyone's mind reproduces the desire by breeding the consciousness of the object of desire. Eventually, the body of consciousness becomes massive. By embodying the body to be present as a creature, the massless subject trades the mass of consciousness. The action makes the creature a performer, performing to resolve the desire with a wish to manifest the potential beyond the absolute. An entity manifests the potential as a creature after conceiving the absolute as a creation of a creator but before perceiving the desire to create formative growth of the absolute by perpetuating normative development of the potential.

> One perceives the desire to develop the potential after norming the desirable with everyone's experiences within the consciousness of the past that produced those experiences at diverse present moments. Thus, normative development makes a paradigm real because it makes the realization of the norm the goal of one's development into an organization of diverse presents. Diverse presents are the primordial causation traded from everyone over the primordial time as light. Their reproductive force enjoys primordial space within the intellect that transforms the light into heavy with the weight of their consciousness for the origin of the universe. The universe reproduces reality with its natural growth.

4.1. Normative Development Paradigm, Within Past Consciousness.

Normative development twins the impact of past consciousness to condition the path of impact by twinning-effect of the present on the cause for the future to be a paradigm. Twin cause breeds a sequence of impact before realizing that one is simply copying zero as the cause since the future of a paradigm is zero. Copying makes the "one before" the guider, guiding the "one after" on how to be an agent with an agency not to copy itself. Before copying, the agency of an agent is with the "one before." The "one after" is the guider who makes the agency reproductive to produce a quiescent time as an agent. Given its primordiality, the agent copies the "one above" for the explosion of wisdom accumulated with life experience. After originating explosion for gaining superposition, the one above positions the one below as an all-knowing illusionary point.

"One below" is the reconstruction of the "one after," which copies the "one above" with its agency to be a guider with wisdom. The four forms of the one within center matter because they are the potential matter without the present matter. They have the potential to matter as the four dimensions of space before the present makes the one within the center matter with the space's light. The space's light divides the space into four dimensions, first ascending towards the North like the one above, descending towards the East to position itself as the one after, continuing towards the West to position itself like the one before, discontinuing and falling towards the infinity in the South like the one below, before reappearing towards the Center as the potential of the space within the four forms. By reproducing the space's potential within the center, time forces space's essence into the light for breeding itself as a light molecule. The center reproduces space's potential as the causation by feeding itself as a dark molecule.

A dark molecule becomes a heavy molecule as the breeding sequence accumulates the mass of consciousness fed to it into massive weight. The massive weight of ascending mass over time circulates as the descending mass over space, flowing like a water molecule up and down in a competitive, tidal wave of the entity gravity. The entity advancing along the timescale in a straightline as a complementary wave disrupts the tide by condensing into an earth molecule. The tidal disruption event questions the time's front-facing aggregate to be "heavy" (=1,000) with its sentient energy like Uranus. The "sentient energy" (=1,000) transforms into a "self-facing" (=0) air molecule, which "copies" (=0) "myself" (=1) like a "king" (=0) to be a "double-copy" (=1) including "yourself" (=1), "organizable" (=11) as "two zeroes" (=-10). The "triple copy" (=3) answers with a supplementary wave of the back-facing space to light itself into Neptune.

The space curves the partial tidal disruption event into four half-octaves. The curved group is the fourth half-octave, comprising three poles—the heavy, light, and dark—as the three half-octaves. The three half-octaves twin themselves with the potential of the fifth half-octave to organize the fourth half-octave into a triple octave. Triple octave self-perpetuates three half octaves as three tails of the primordial element. Space is the fifth half-octave with three trigonometric points, each a potential half-octave. After breeding a twin octave that triples with time, each point, trigonometrically curved into a half octave, self-perpetuates a metaphysical wave of the ether molecule with its density. Metaphysical value integrates three waves—the heavy competitive wave of discontinuity (the continuous wave), the light complementary wave of continuity (the discontinuous wave), and the dark supplementary wave of infinity (the infinite wave). The fifth half-octave is a potential triple octave. It self-perpetuates three potential half-octaves as three heads emanating from the primeval element.

> ### Sequential Formation Heads from Three Heads
>
> Three heads are linked to the primeval element with a dynamic wave of fire molecule, which replicates five as the perpetuating value of four within the half octave. The perpetuating value is the fifth head; the replication is the sixth head. The first head shakes because it is reproductive, curving the second head into seven heads by producing a growth of six. The third head is curved into three heads that self-perpetuate the growth of six within the curvabale. The curvable is a repeating partial tidal event; it is the fourth head curvable into nine heads. The seventh head is one head that evens nine heads. It is a cyclical head that cycles two heads as the sixteenth head by forming each head into an octave. The eighth head heads the ninth head, the primeval element. Primeval comprises twelve heads of which three heads are curved as three potential half octaves to straighten the infinity of their twin triple heads into three half octaves. Three half octaves self-perpetuate three heads which are indistinguishable from the primordial three, the site of subtraction. Twin triple heads norm the eighteenth head as a potential head with a reproductive reality of six heads realized over time.
>
> The tenth head is replicant. As an Ideal Godhead, it is a circular head with four heads self-projecting six heads. Godhead is the eleventh head producing the growth of six by self-perpetuating an entropy of five. The twelfth head is the Theoretical Godhead with a discriminating consciousness of the seven heads that are perpetuating its value within the fifth head. The thirteenth head is the Real Godhead; it is a blockhead of eleven heads. One head evens out two heads with the evenness of three heads. The fourteenth head is the Illusionary Godhead, the human being with the oddness of four heads without the one head which evens out two heads for illuminating fourteen heads of reality as the Eighth One.

The fifteenth head is Human Godhead with fifteen heads luminous within the Self-luminous Head, and the seventeenth head with eighty heads. The seventeenth head makes the twelve self-luminous with an eight to the "base" (=10) "potential" (=18) of $5 \times 10^{(8*12)}$, i.e., 5×10^{96}. The nineteenth head is Infinite Godhead with one hundred heads reproduced by eight by squaring itself to transform one (the causation) into two (the twin causation). Two twins continuity of fifty by reincarnating as a zero.

Infinite Godhead self-perpetuates the twentieth head as a Finite Godhead, norming a networking system, 8×10^{15}. The base of ten twins eight for self-projecting fifteen within one's continuity with time without the two (twin causation). Two squares the space itself to be four within the continuity of fifty ($50 = 15 * 3$ dimensions of time $+ 1 + 2^2$). Infinite Godhead is Primeval Godhead. Finite Godhead is Primordial Godhead. The twenty-first head is Param Godhead, 59. The five channels "Transcendental Nine" by emanating without the Four Immanent to spearhead 180 heads as the Param Godhead. The Four Immanent is the Spirit Godhead that "positive curves" twenty heads. The second ten heads prime project the first ten heads to clear the negative metric of doubt that they are the 22nd head, -10^{10}.

The negative metric of doubt stresses energy with the momentum of Five Emanating to channel itself as Transcendental Nine. Transcendental Nine is -1, the Deity Godhead, the stress-energy tensor. It is a metric tensor for the energy-momentum tensor, transformable into the curvable space when a Deity transforms into a zero. The time multiplies the Deity into one thousand heads (10 *10 * 10, since the deity is 1 who becomes 0). Deity Godhead is the 23rd head of -1. Two within three heads one as a zero with an "error in consciousness" of 10. A time multiplier of three makes the error in consciousness (=3*3+1) reproductive as a guider of 100 within the head. Therefore, the Deity Godhead is the Guider head.

Five Emanating is the momentum of 999, the Animal Godhead which heads the multiplication of three tens with a cubic (reproductive three) of 9 after one becomes zero tailing 10,000 heads. The sentient energy of 10x10x10, i.e., 1,000 makes the Animal Godhead a Sentient head. The sentient head is the 24th head, as a 9 without the 1 is an eight. "Tailing" as −30 are the thirty heads, constituting the 25th head. Five of the heads consciously decide to imagine they are the root of twenty-five and become Emanating Five. Thirty heads constitute the Divine head, the Plant Godhead. The root is the 26th head with thirty-eight heads norming a Devil head, the Mineral Godhead of 38 (= 18 + 4 *5). Twelve heads are self-luminous self-projecting their potential to be eighteen with a spirit of four whose consciousness perpetuates the spirit's value as five to be primeval. The self-luminous is the 27th head with eighteen heads. It is the Satan head norming a Metal Godhead of 12. Satan head is the Immanent Four. The luminous is the 28th head with twenty-eight heads, the Material Godhead of 13. One multiply the perpetuating value of five over time as a primordial perpetuator of 15 to become a time multiplier of 3.

The potential of 18 is the 29th head with one hundred thousand heads. It is a Dark Godhead. Five ones become five tens with their growth into five zeroes within the consciousness of the sixth one multiplying them to form a head of 100,000 heads. Dark Godhead is Nine Transcendental. Five Ones are the future moment of nine that the potential of eighteen perpetuates with its present state of a cubic. The Sixth One is consciousness. It is the 30th head with one million heads. It forms as a Light Godhead when six ones become six tens by self-multiplying growth within the moment of their formation. Six Ones are growth. Five Tens is reality present of the seventh one. Six Tens is reality potential of the eighth one.

Growth is the 31st head, the Time Godhead with 360 heads. Six destroy themselves with time to let the sixty self-perpetuate thirty within one. One's reality as the seventh one lets the six to be present as one within the perpetuating value of the potential of the eighth one. The seventh One, the cell, is Space Godhead with 18,000 heads. The cell is 19 which services one that the time multiplier of 3 exchanges as 1,000 to trade a space of 18,000 with the 19 +1,3, i.e., 32nd head. Eighth One is the Causation Godhead with fourteen heads (6 x 2 + 3 -1 = 7 * 2). It becomes the thirty-third head when the growth within seven is repeated with three without the (ninth) one whose growth is repeated. Ninth One is Imaginary Godhead. It is the competitive circle of one that becomes zero to be a 10. It is the 34th head with 121 heads. After the time multiplier (=3) makes one zero, the consciousness (= 4) of the growth of twin as six heads itself as (tenth) one.

Tenth One is the 35th head, the Greeter Godhead with 10^{10} heads. After the three twins five with its growth without (the eleventh) one that twins causation to be a three, five perpetuates its value as the "base" (=10) to head itself. Therefore, Greeter Godhead is 16. The eleventh One is a person as the 36th head. It is the Self Godhead with 8×10^{15} heads. As one twins causation for growth, the "supernatural growth" (=8) multiplies the twin causation within the "base" (=10) for self-projecting fifteen with time. The Self Godhead is one.

Supernatural growth is the 37th head, the Universal Godhead with thirty-six heads. Eight orders two to square six and head seventh with its primordiality within the universe. Universal Godhead = 8. The universe is the 38th head, the Unique Godhead with forty heads. Two heads make the reality of the forty heads reproductive as the Absolute head. The Unique Godhead = 2. Primordiality is the 39th head, the Primordial head. It is a Global Godhead with 1,600 heads of the cosmic perpetuator of the reproductive reality of forty heads that the three perpetuates with a cubic of nine without two. Two squares three with its primevalism to produce primordiality. Global Godhead = 27.

A Mindful Greeter is present as the 40th head, the Primeval head. It is a Local Godhead with 32 heads. Sixteen make their reality of 7 reproductive by reproducing 3 as a cubic to square forty. Local Godhead = 1,600. National Godhead is the 100th head with 48 heads whose oneness is reproductive as a head. National Godhead = 48. Corporate Godhead is the 180th head

> with 10^{16} heads. It reorders (=2) the potential (=18) as a whole (=16) to self-perpetuate (=1/2 *16) the base (=10 =8 +2) with the order (=2) that prime projects the whole. Corporate Godhead = 10^{16}. International Godhead is the "rooting" as the 90th head with 10^{18} heads. It self-perpetuates one who prime projects the potential with its divine planning.

Three heads become lighter after emanating from the primeval. The second head is dominant as the lighter one, springing ahead as Mars by making the past of the light one reproductive with the future of the heavier. The first head is predominant as the lightest one, present as Mercury of temperature that makes one lighter. The third head is deciding because it is the light one, refiguring the superiority of the future of the "heavy one" to be Venus, the perfect past of the heavy. Three tails are heavier because three potential half-octaves are immanent within three half-octaves. The third tail is Uranus; it is the multiplier that transforms one tail into three tails. The third tail is heavy. It is an infinite wave, supplementing the lightness of the three heads to be conscious of its heaviness. The second tail is Saturn. It is the metric for deciding that the first tail is Jupiter, the heaviest. The second tail is heavier. It is a discontinuous wave, complementing the third tail. The first tail is the "heavy one." It is a continuous wave, competing with both the second and the third tails.

The third tail is

- multiplying itself into two legs
- to tail the multiplication with the second tail
- before its division into two heads
- which head for the addition of two hands
- after the subtraction of the body as a whole
- trading the three heads as the wholesome
- by servicing the three tails as the wholesomewhole.

Two legs multiplying the Uranus, the light, are the Earth, the lighter, and the Venus, the lightest. Overall, Jupiter as the heaviest is followed by Saturn as the heavier, Neptune as the heavy, Uranus as the light, Earth as the lighter, Venus as the lightest, Mars as darker within the light of the lighter, Mercury as dark, without the light of the lighter. The Sun is the darkest until the Moon, the fairest, present as the Dark Matter, the fairer, illuminates that it is fair, the White Star.

- The fair is the proton water, a Piezoelectric with three water elements (Trihydrate).
- The fairer is the neutron water, a Piezomagnetic with four water elements (Tetrahydrate).
- The fairest is the molecular water, a Magnetostriction with 3 x 4, i.e., twelve water elements.
- The fairness of the face is immanent as the atomic water, annually rainy with ten water elements (Decahydrate).

The face is the invisible hand that becomes visible with the two hands after it twins itself within 12 -10, i.e., two water elements (Dihydrate). Two water elements are liquid water. The nearness of the face with the body gives the neck to the leg for carrying the weight of the farness of the invisible hand.

The body is invisible as the dark until the appearance of the neck as a darker element for the observance of the fair as the invisible hand of the light. The body is photovoltaic with 12,800 water elements which become reproductive (= 100) as a child (= 128) within a female to norm 128 females.

- The dark is discarded by the darkness within the light as a third of the water to be near the far with 2/3 water. The near is the present physical body of the octave potential with a potential twin-effect of its farness from the potential physical body of the octave present as the consciousness. It has the potential to twin double-octave with its division into four half octaves, each with a unit of potential force as a complementary wave of the heart chamber. The heart chamber is the twelfth body system, divided into a physical body with itself as a physical body system of eleven body systems which twin it within the twelve body systems, the fourth heart chamber, as the twelfth body.

- The darkness is transfigured by the light as a black box of 320 water elements. Three water elements self-service (= 2/3) two water elements to be near the invisible hand. Near is the Piezogravitational-effect, 2/3 water. The invisible hand mediates the nearness with the whole gender within the male. The whole male is undifferentiated as a double-octave of males, comprising sixteen males. Within the darkness, their "light" (=180) is transfigured into a double-octave of females, norming a whole female with sixteen females.

The whole female is a potential whole male because the whole male comes to light without her potential to remain dark as the causer. After coming to light, he becomes the cause for her to be present as the dark matter to make him matter as the potential whole female. By defragmenting itself, he forms a potential whole female with 32 females defragmented for visibility as a twin feminine, norming a twin double octave of females.

The defragmenter is the potential gender of the thing that matters. It is the potential whole male, the twin masculine, norming 32 males the twin double octave of males with a potential (=18) to twin itself for a ten-fold growth. Ten-fold growth within the double octave of males = 180 -10 (which grows one within nine ten-fold) = 170. Without the twin, female = 5 males (four of which grow into ten with the growth of consciousness of the darkest within one that is darker) = 10,000.

- The darker is the electron water, a Piezogravitational with five water elements (Pentahydrate).
- The darkest is the organellar water, a Piezothermodynamic with ten thousand water elements. It becomes a Sun with its primordialism as a satellite of five males within the water.

The five males are the five elements within a male. The five elements become fair with the three water elements. The water within the five elements twins itself without the ether—the reproduction—element for the production of air element with the volume of earth element after the consumption of fire element as primordial to the guider—the reproductive—element. The guider element times the primeval element to be reproductive within the primordial with the divine element. The consciousness of the four elements within the primeval is the SHEENY, i.e., sentient element, conscious of the potential it perpetuates as a thing. The five elements are fire, water, air, earth, and ether. The four elements are the fire, the air, the earth, and the ether, within the sentient force of the water which flows like the water force within the consciousness of the twin, the potential four.

- The light one is cellular water, a Piezogravitomagnetic with 9,600 water elements. It is the brightest. The consciousness makes 20 water elements reproductive as a male for attracting the supernatural growth of the 480 water elements as the "FE" element, the future exponentiate. Therefore, female = 100 water. 9600 water = 96 females = 480 males. 480 water = 96 males. Supernatural growth is organic water, a Piezogravitoelectric with 480 water elements. It is brighter than the water, which is bright and Piezosentient.

- The lighter one is the inorganic water, a Piezogravitoelectric-effect with 80,000 water elements. It is the dullest. The consciousness dulls femininity (=1,000) with the effect of Piezogravitoelectric on the triple octave (=80) of water with its guider mediation. The effect of sentient within water leads to the decay of three water without the fetus of the six water with a discontinuity of 60 water, the octave (=60) of water.

  - The triple octave of water is a Piezothermodynamic-effect with 720 water elements. It is feminine water, replication (=90) of femininity's eight water. Feminine water is egg-shaped within an egg that forms a cell. It is organ water.

  - The octave of water is a Piezoelectromagnetic-effect with 240 water elements. It is masculine water self-serviced by feminine water to twin itself into the maternal water, the double octave (=16) of water. The masculine water is the body water the body forms with the consciousness of water.

  - The double octave of water is a Piezosentient-effect with 960 water elements. It is maternal water self-perpetuating supernatural growth (=8) of 480 water elements. Maternal water is soul water.

- The lightest one is the paternal water, a Piezo-effect.
  - Piezo is brightable because it is also darkable with 2,560 water elements. It is child water, or more precisely, son water. Darkable is daughter water. Water is male when supernatural growth "FE" is zero. Child = 128. Therefore, Piezo = Child water = 128 male = 0.
  - Piezo-effect is dullable because it is reproductive with 240,000 water elements. It is spiritual water, manifesting as photon water. Piezo-effect = Spiritual water = Photon water = 100. Reproductive element twins the lightness (=1,200) of the water. Lightness is dulling. Dullable is Twin Lightness.
  - Dullness is triple lightness one channels as a feminine to triple the lightness of 18 males into 54 males, with 1,080 water elements norming the grandfather water. Dullness is far. Brightness is triple darkness that channels one's supernatural growth as the gender to be near the light of the consciousness of femininity. It triples the darkness of 16 females into 48 females, with 4,800 water elements norming the grandmother water. Twin darkness is lightable as the matter of 32 females, with 3,200 water elements, norming the meso photon as the greeter water.

The earth is the body whose three heads are far from the three tails. The body is neither light nor heavy, but "potential heavy" if three heads conjoin with the three tails to be the body's neck. Venus is the neck.

- Potential heavy is lighting the neck with three males and three females, i.e., 18 males which the body twins into 36 males to be near the triple octave self-projecting itself without the one. Lighting is 360 males. The ten-fold growth of 36 males into 360 males, i.e., 72 females, makes the female reproductive within the water as the fire for reproducing time's three-fold growth within spacetime's seven-fold growth as the light molecule. Potential heavy = Devil water = Earth = 72 females = 724.

- The neck is "potential dark" with the experience of lighting. It is brightening the day of the earth's lifetime as a human being without the devil with the three-fold growth of 108 males, norming 2,060 water elements. Neck = Potential dark = Experience = Brightening = 108 males = Satan water = 9,000.

- The day is "potential bright" as the lifetime norming the natural six-fold growth of the male which makes 30 reproductive as water for it to have a lifetime experience of 9,000 with the ten-fold growth. Potential bright is darkening the twin day into night of the earth's time as a planet with 30 water elements. Six males twin the time multiplier into a loop within eight males that become reproductive as 108 males. The night is heaviable with the time the planet is heaving as a nocturnal, shrinking from an absolute metric of 160 water elements with 8 males, for heavying itself into ten males within the present as the potential's replication. Heaving = Absolute Metric = Potential dull = Planet = 8 males = -1. Ten males are adding one with infinite potential as the replication of the potential within an octave of twelve males. Heavying = Adding one = Infinite potential = Replication = 10 males = 2 females = 200 water elements = 90. Octave = Twelve males = 240 water elements = 60.

How Thermodynamic Action Generates Octave-Mediated Ten-fold Growth?

Ten-fold growth is purified as bicarbonate. A "discriminating consciousness" (Phosphate) of bicarbonate makes carbon fusion reproductive by disentangling hydrogen from electron to channel its stimulatory-effect as the bicarbonate-effect. The bicarbonate-effect makes the luster of a photon's flame self-luminous as a photosystem in the "self-luminous cycle" (Calvin cycle). The photosystem is a reducing agent which reduces photosynthesis, i.e., a photon's synthesis into a triple neutron.

Descending synthesis of a photon by the time multiplier into an octave generates an "ascending growth" (Photosystem II) in the site multiplying the electromagnetic reaction of "descending growth" (Photosystem I) in the site of division of triple neutron with thermodynamic action. Continuity of energy equilibrium within the site of division generates ten-fold growth. Ascending growth twins the octave, the triple photon, to form a triple neutron as a twin triple photon.

Photon = 20. Octave of site = Site of Division = Site of Thermodynamic action = Twin Triple Photon = 6 Photons = Triple Neutron = 120.

The time multiplier triples the triple photon to light nine photons. Triple photon twins itself again with the twinning-effect of the twin triple photon on the triple neutron. Light = 9 photons = Twin Triple Neutron = 6 Neutrons = 180. Including the photon itself, ten photons norm a proton with the ten-fold growth of the photon that twins the imaginary spirit of space for ascending growth. The "ten-fold growth" of the potential of 18 within the light of the "primordial self" (=10), centering growth with its mass consciousness = $18 * 10 - 10 = 170$. Proton = Ten Photons = $170 - 20 = 150$.

Descending growth leads to "horizontal growth" (Photosystem III) in the site adding gravitational inaction of the metaphysical consciousness (NAD, i.e., Nicotinamide Adenine Dinucleotide) of the eight-fold growth within the primordial-effect, the cofactor, of the factor. "Copyable-effect" (Water oxidation) of backward consciousness within the time multiplier makes the "forward consciousness" (Anabolism) of the process triplable for discriminating three classes of consciousness (Backward, forward, and metaphysical) within a neutron through oxygen evolution.

The factor is the circling layer of the sentient consciousness of the site subtracting the organic resonance of a six-fold natural "growth" (Nicotinamide) in the site servicing the molecular resonance by "decelerating" (NADPH reduction) the

> "diffused present" (Adenine) of the two-fold "ascending growth" (Dinucleotide). Reducing polluted "curvilinear energy" (ATP) of the potential fused within the present implies "accelerating" (NADPH: Nicotinamide Adenine Dinucleotide Phosphate) discriminating consciousness by triplescripting light as a twin triple neutron after a donation of carbon.
>
> The light-dependent assimilation of the "cofactor" (Photosystem II core complex) by the "substrate" (Photosystem I core complex) cojoins the neutron's "illusionary flame" (Photosystem III core complex) with the "natural essence" (Methanol) of the site trading inorganic dissonance. Trading involves electron transfer from the site investing in the atomic presence of the neutron. Electron transfers twin causation to triple the fire of light assimilation within the site exchanging cellular absence with the present of a cell. Exchange causes light-independent assimilation of the "inanimate self" within the "site of capability," capable of transforming itself into a cell by breeding "neutron mass" as the "animate self." Neutron mass is a potential mass comprising forty-eight neutrons, organized into six clusters, each with eight neutrons.
>
> The potential is the cause which develops a "mass" (=132 = 15 +17 +16 +32 +48 +4) as the one breeding a neutron gravity of 15 neutrons by feeding a neutron earth of 17 neutrons for centering 16 neutrons on the right of the 32 neutrons left. It orients the primordial oneness of the 48 neutrons through a sequential development of 4 neutrons. Four neutrons endoreproduce a reproductive pair of four neutrons to cluster themselves into an assemblage of three clusters of four neutrons. The assemblage of three dimensions of time perpetuates nine as the cause's past with its reproductive force. Four dimensions of space are square.

A Wisher Deity normalizes the spirit with its past to formalize the soul with the future it wishes through the present of four paths.

4.2 The Path of Primordial Time

Primordial time is a time mass of a million years that makes the soul of the Sun's future reproductive with one thousand neutrons. It produces an essence of a million neutrons after centering a year's potential within the present. A potential year digresses a fire wave of a hundred thousand neutrons through the "path of primordial time" (*Samplava Marga*). Fire forms six zeroes with time's reproductive force within three zeroes. Six zeroes are within the primordial oneness of the seventh zero with the eighth zero. Eighth zero squares the space by reproducing a co-conjugacy of three zeroes for oneness with the seventh zero mediated by two zeroes. The seventh zero is the potential one. Eighth zero endoproduces everyone else with the potential within one for twinning the two zeroes with the third zero on the right to form the whole of the essence of a million neutrons into a circle.

Twinning-effect makes the third zero reproductive and shapes the eighth zero into a neutron for producing two zeroes with its past within the endoproduction of one co-existing with its potential. Seventh zero clusters six zeroes into six clusters of neutrons. The zero takes the position of the eighth zero with the momentum that leads it to behave like three zeroes. Two zeroes before the first zero advance to be the sixth and the seventh zeroes with the consciousness of the eighth zero within the fifth zero. Overall, the eighth zero sequences three paths within the fifth zero that are consequential as the sixth, seventh, and eighth zeroes as it conforms with the consciousness of four zeroes.

4.2.1 The Path of Action

An action is guided by the consciousness to conform with the growth activated as a consequence. The action makes the growth normative within the actor who benefits from the fruit of the action. The fruit of action is the path of knowing the imperfection which adds uncertainty of intellectual paralysis to the sequential decisions the mind takes with its potential to reason the cause with intellect afterward. Since the cause is the action potential, the reason impedes the action and makes the actor the essence of growth. The growth happens if the actor becomes the reason for the action. The actor acts because of the potential to fulfill a wish through action. The wisher wishes because of the potential for knowing the reality of the one acting to fulfill the wish. The actor makes the wisher the Wisher Deity, just like the wisher makes the actor a Worker Deity repeated as a fifth who becomes a zero after the wisher begins perpetuating the value of the work without the actor's mediation.

The "work" (*Anukriya*) makes the wisher, the eighth zero, a neutron for perpetuating the value of the fifth as a zero to conform with the four zeroes. The three zeroes form within the eighth zero that aggregates time with the future. The fourth zero is the future form of the eighth zero as the dark matter within the present. The past makes the present light by repeating the "matter" ($158 = 5*3, 2*4$) as its perpetuating value for a supernatural growth of the present as the fourth zero within the future, the eighth zero.

The "oneness of the work and the worker" "twin works" (*Anukriya Yoga*) oneness into a wisher wishing to diversify the work with infinite wishes for knowing the reality of growth within the worker. The worker is the inanimate self whose work perpetuates its past as the animate self. The future forms the animate self with three zeroes. The worker is the first zero. The work mediates as the second zero, behaving like an actor who acts on the command of the worker. The wisher is the third zero, wishing to be the worker working to command the actor for activating the action it wishes to be the worker's future. For the wisher to be the worker's future, the action requires "freedom from the twin hand" (*Pada Mukti*). Twin hand twins the invisible hand of the worker within the wisher with the visible hand of the manifestor.

The manifestor manifests the wish for enjoying the fruit of knowing the potential within the inanimate self to be the animate self. The animate self is the Wisher reproducing the quality of the wish for qualifying itself as the Worker working to realize the potential within the wish like a "serial" (*Krama*) who has perfected the science of work through a series of actions. Therefore, the freedom from the twin hand is the "freedom from the serial" (*Krama Mukti*). It makes one a creator of the action necessary for serializing the work into a creation by self-projecting the potential to create the one working as a creature.

With a "divine characteristic" (*Divya Purusha*) of "justice" (*Yi*, 义), the worker activates "effortless action" (*Wu-wei*, 無爲) for doing justice to the wish by acting on it just in the time necessary to know the potential of the space that made it conscious of the wish. The presence of the time as a worker unifies the space's inaction with the time's action. Time's action is effortless because the action happens naturally for ascending order. It takes effort to neutralize space's inaction because the inaction happens supernaturally after a descending order. The descending order makes the space a follower. The time ascends to be a leader leading as one working like a protagonist for illuminating the positive infinity of space that follows like an antagonist. The effort generates a horizontal order with its persistence for realizing the reality of the one moderating the order. An entity moderates the order with the oneness of her essence. Time multiplies the essence to mediate the entity's perpetuating value with the knowing its work generates.

Time's leader spirit (*Tianghuang Dadi*) is a guider spirit. As a Holy Spirit, it is a reproductive spirit reproducing justice by moving forward just in time to let the space follow with the action activated by the momentum. The space follows as a "repeater" repeating a "big dip" in its position from a "Northern Dipper" as the time's momentum pushes it flying backward like the North Pole. The big dip is the "divine character" (*Divya Purushadeva*) of the time-dependency of the space's consciousness system. Time moves forward like the East Pole. Time's negativity within the space left behind in a superposition above the time's past makes the North Pole a Negative Pole.

Space's positivity within the time moving to the right makes the East Pole a Positive Pole. As the "organizer" (*Yang*) of the time's present, the space follows time's momentum into the future until the positive infinity of space gives way to the negative infinity of time within the present. Time's infinity is negative because the present limits the future's potential to manifest the positivity of the space's wish for its well-being. Space's infinity is positive because of its potential to wish for the well-being of the universe the time orders by breeding the space for an entity to universalize the time-effect. Time-effect is the space's desired growth by breeding itself as an entity through time. The organizer (*Yang*) is the followership-in-action, while the "organizable" (*Yin*) is the leadership action.

Time perpetuates five entities with the perpetuating value of the space as an entity. Three entities are the entity's past, present, and future. Two entities are the present's past and future as a para entity perpetuating the value of space as time. The para entity's "self-luminous reality" (*Baidi*, the White Emperor) cubes the breeding of the present's electron air-effect with its future as an entity to begin the year with the Monsoon's gravitoelectric-effect.

The entity's "future" (*Chidi*, the Red Emperor) makes the fire-effect reproductive to begin the Fall with its gravitational-effect. The entity's "present" (*Xuandi*, the Black Emperor) reproduces the water-effect to begin the Winter as a conscious system with its sentient-effect. The self as a conscious system makes the present's past "luminous" (*Cangdi*, the Green Emperor), repeated within the space's perpetuating value as the dry ether-effect to begin the Spring with its gravitomagnetic-effect. The entity's "past" (*Huangdi*, the Yellow Emperor) begins the Summer by making the earth-effect self-luminous with its thermodynamic-effect.

4.2.2 The Path of Performer

A performer is guided by the dimension of reality that makes its consciousness self-luminous as a self-luminous system. The "path of performer" (*Mahakriti Marga*) is the "path of dimension" (*Dharma Marga*) that perpetuates the real as the "just." The performer is just performing as real for realizing the reality of justice as a "self-luminous system." Justice demands a Wisher to be a Worker, living like a protagonist to fulfill the wish. It supplies the antagonist-effect for the Wisher not to be a deuteragonist, observing like a non-living the potential of the living to fulfill the wish naturally. The consciousness of life justifies why it is natural, although not just, for the living to fulfill the wish of the non-living after the latter has departed. The "potential sentient-effect" (*Buddhi Atiyoga*) of the universe of future entities who reincarnate after they have departed generates a discordant energy within the living.

As a leader, the wisher from the future begs the universe of present entities to fulfill infinite wishes after blessing them with the consciousness of their unique gravitational qualities as a system within a creature. The leader is not just performing the behavior necessary to fulfill the wish. He is programming justice by planning the blessing for profiting from the begging. By breathing air infused with the spirit of the imaginary wishes, the follower becomes a citizen devoted to fulfilling them for its sentient benefit. The "unity" (Ren, 仁) of the follower with the leader is finite and primordial, workable only if the leader wishes the follower's sentient benefit for the "integrity" (Xin, 信) of actions. The integrity of action helps the follower realize the primordial self with the para-consciousness of the leader's wish.

The sentient force of the para-consciousness transforms the omnipotent self into a down quark. The fire force of the form transforms to potentiate the potent self as an up quark with Omni's thermodynamic force. As a charm quark, Omni integrates water force with its gravitomagnetic force to be conscious of the air force. Air force disintegrates its gravitoelectric force into a strange quark for magnetization of the earth force. Earth force reintegrates its electromagnetic force into a bottom quark to charge guider force. With independence, guider force preintegrates its gravitational force into a truth quark to discharge ether force. With dependence, ether force postintegrates its electric force into a negative quark to recharge divine force as a feminine.

With magnetic force, the divine as a feminine becomes a positive quark by procuring the divine masculine as a knowable from the guider, who is guiding the sentient's knowing with its interdependence. The divine masculine energy extracts knowledge from the divine feminine for knowing the truth of the divine feminine energy. Knowing makes self potent as a knower. The correlation among the elemental forces, fundamental forces, and foundational particles is summarized in Table 2.

Table 2. Correlation Among the Elemental Forces, Fundamental Forces, and Foundational Particles

Elemental force	Fundamental force	Foundational particle
Sentient force	Sentient force	Down quark
Fire force	Thermodynamic force	Up quark
Water force	Gravitomagnetic force	Charm quark
Air force	Gravitoelectric force	Strange quark
Earth force	Electromagnetic force	Bottom quark
Guider force	Gravitational force	Truth quark
Ether force	Electric force	Negative quark
Divine force	Magnetic force	Positive quark

4.2.3 The Path of Performing

"Performing" (*Kriti*) forms a "path of mood" (*Bhava marga*) for the "path of performing" (*Kriti marga*) to transform the injustice by profiting as a luminous system without the development of the self into a system whose primevalism is luminous. Self's primevalism programs primordialism of the luminous by planning the system's absolutism as an "I" identifying the characteristic of the creature. The Self's gift of consciousness makes a zero self-luminous as creature at the speed of light. As a subject of the Self's sentient force, the zero advances as one to be turned back into three zeroes after displaying two as its objective. Two is omnipresent within one within the digit that counts zero as one, knowing two to be present as a potential zero. Potential zero is a negative one which knows that the two orders one to be twelve. One reorders zero to be a potential zero without the one turned back.

The one turned back as the negative one is the past of everybody. A negative one is a *cosmological constant* embodying the dark energy of the past, leading to the universe's infinite expansion into the positive infinity of space until the "universe" (= 2) becomes vertical like one to be dark like a negative one.

The one displaying two to be luminous as the present one has the potential to be the present of everybody without the self which turns its back into a zero. The zero is the *potential cosmological constant* embodying the light energy of the future. It follows the universe's formation as a ball that is light until the edge of the negative infinity of time where it becomes heavy. That edge is diagonal, descending from the polarized darkness of time to the de-polarized heaviness of space, where time = ∞.

Two, which transforms the Negative one into the Positive one with the display of its potential for twin organization of a value present as a digit, is the *present cosmological constant*, embodying the heavy energy of the present. The present mediates to make the universe flat as a positive infinity (of space) which becomes heavy with the superposition of the time's eventual state as a zero after descending from its negative infinity (of time). The twin organization is the future of everybody. It is the one that becomes eight with two's cubic energy to self-perpetuate eighteen. Two's "cubic energy" (=9) is the past which cubes two with three that multiplies time into three dimensions. Three is the *primordial cosmological constant* embodying the bright energy of the time as the infinity dimension of space. Two's "square energy" (=5) is the present's perpetuating value which squares two within one turned back to be present as a square with both the past and the present. Four is the *primeval cosmological constant* embodying the dull energy of the space as a square present within the time which dulls as it becomes dense when it cubes its "future value" (=2).

One which is omnipresent is a *cosmological variable* embodying the invisible energy of the causation. Five is the *potential cosmological variable* which unifies the potential for the cosmological variable to live as the time of everybody by embodying the visible energy of the cause. Six is the *present cosmological variable* which universalizes the potential by embodying the potential energy of the entity for self-reproducing itself as the space of everybody. Negative two is a *primordial cosmological variable* embodying the present energy of the twin reproducing itself as the self to incarnate "it" (=900) as the time loop. The time loop forms everybody with its perishing consciousness to be heavy as a body with the mass of perished consciousness. Negative three is the *primeval cosmological variable* luminous as a body of consciousness that multiplies itself like an object perishing with time to subject space to the pressure of its volume.

As a "perpetuator" (=460), a "photocatalyst" (=460) "descends" (=25) the "pressure" (=62) of a "subject-matter's" (=460) "volume expansion" (=460) into an "organic compound" (=941) through "photocatalysis" (=741). "Photocatalysis" (=741 =190+257+86+14+194) "descends" (=25) "temperature" (=86) "organizable" (=11) as "unproductive charge" (=14) to "yield" (=82) a "reaction" (=190) of "selectivity" (=147) of the "productive charge carrier" (=257 =147+82+18), i.e., "masculinization" (=257), for "disintegrating" (=194) the "unitarity" (=741) of the "unproductive charge recombination" (=18), i.e., "feminization" (=18) from the "carrier" (*Dravyaka*, 7).

"Descending sequence of four" (=105) gushes "negative three" (the object) to "one" (the causation) for endoproduction. It wishes to be light after becoming heavy with the "ascending pressure" of the consciousness flowing linearly with time for "ascending order". "Ascending sequence of five" (=99) follows one to be "six" (growth) for "descending pressure" by self-reproducing consciousness. Consciousness stresses the order's endoreproduction for "descending order" (=191 = 99 *2 -1 -6) as it transforms a "horizontal sequence of one" (=120) into a "forward sequence of two" (10^{18}) to reform a "backward sequence of zero" (=16) into a "whole sequence of three" (= 12). Three digits 1,2,0 transform into two digits 1,0 to exponentiate the potential of the third digit for the growth of one into six, a third of the "potential" (=18). The number 16 is the backward sequence of zero that preforms a zero as a six to order the whole (=16) into a sequence of three. The sequence of three reproduces three as a time multiplier for the endoreproduction of one as a two, norming 1,2 (= 12 = 3 +3*3).

1,000 is the wholesome sequence of negative one that postforms the "three ones" (the time multiplier) as the three zeroes to be the origin of the negative two (two ones without two zeroes). The number 29 is the wholesomewhole sequence of negative two feeding seventeen for breeding fifteen (17 -2) as the first one without the negative two. Growth (=6) informs the second one that it is 7,000. As the present value of two zeroes, two ones are the foundation of the third one. The third one deforms one into nine with the third's cubic energy. Nine is the universal sequence of negative three. It squares space into a universe to force itself into the three zeroes of time like a leader. The three zeroes follow nine. Nine is the past and the three zeroes are the future. The past realizes its reality is a negative three that the future reproduces as a nine to produce the "present" (= 1600) as a "square" (=18,000 = [1 +29] *600) by trading one and servicing 29 for the growth of a six within two zeroes. The number 6 is the unit sequence of six. It sequences itself as a "unit" (=6) to twin its "effect" (=34) and be "productive" (=68 =34 *2) as a "triple octave" (=80 =68 +2 *6).

As a protagonist, a Worker ascends the trading of a negative one from the past to be an antagonist servicing the negative as a positive for endoproduction of itself as the future reality. The antagonist wishes to be a Wisher for realizing the reality of his future as an Observer. The Observer becomes an agonist while the Worker works as a Messiah to fulfill infinite wishes by instantly repelling the present's negativity, seeking to attract the future's positivity. Eventually, the wish to perpetuate the present reality is guided by the past reality of injustice. After the Wisher creates an asymmetry within the present, the Observer seeks symmetry without the present between its past as a Worker who works like a deity to fulfill the wishes and its future as a leader, the Lord of deity, who enjoys the fruit of the Worker's actions.

As an enjoyer of the past that it created as a deity, the leader realizes the goal of following a para deity. The para deity behaves like a Wisher from the future wishing to be the leader for shaping the present as a goalkeeper. The leader has the option to be a goalkeeper whose goal is to enjoy the present without entanglement into the baggage from the past. The present can tangle the imagination to bag the future the goalkeeper desires. Both past and future are present within the universe of possibility for the goalkeeper. By realizing that universe, the goalkeeper fulfills the goal of knowing the future before the degeneration of the desire from the space's sentient (conscious) element following the decay of the wish with the time's guider (reproductive) element.

The causation's divine element transforms normative justice over the time cycle into formative justice over the space cycle. The goalkeeper forms justice within the present space by realizing the goal of universalizing the desire's horizontal pressure to order the formative growth of the universe. The goal norms justice without the present space as it perpetuates the consciousness of the desire within para deity. The para deity wishes to take the lead in managing that desire, by first becoming a leader, then a worker working on the command of the leader that it wishes to be again, next a knower knowing that the leader is the worker's copy, and eventually the manifestor manifesting the leader, the worker, and the knower as its three copies. The knower is the past copy knowing the possibility of shaping the leader as the future copy with the worker as the present copy.

The transformative justice of the causation cycle forms the manifestor into a follower who copies time by differentiating space into three dimensions with a goal as the fourth dimension. The goal leaves formative justice as a theory of mind with a possibility to be true through the endoproduction of a worker working to fulfill it with certainty. The goalkeeper substantiates the creation of the normative justice's healing touch with its consciousness as a creator. The radiating flow of consciousness as a sentient force is a heavenly factor that pools the illusionary element from the past. It puts the worker in the seventh heaven as the ideal servicing the healing touch. The worker validates the Doctrine of Emanation's theory of proportionate reality since the present reality is distributed among three entities following the time's dimensionality.

The heavenly factor makes the past reality an object of attraction for the togetherness of the leader, who is n outgroup entity, with the knower who knows that the worker is the causation. The worker makes the leader a subject of repulsion with its otherness from the knower, the outgroup entity who is keeping its knowledge a secret left as the consciousness. The knower groups the leader with the worker, knowing that the worker will follow the leader to eventually realize that it is working to fulfill the wish it has as a leader leading the work it is doing. The realization makes the worker on the path of devotion to the goal develop a self-awareness that the deity capable of self-fulfilling the wish is omnipresent within the self.

The "numerical oneness" (*Samkhya yoga*) of an entity as a follower within the multiplier of realities leads the follower to the realization of the self's reality with an open intellect, compassionate mind, and sentimental spirit. Backward pressure from the instrumental externalization of the leader for the follower's entropy progression limits intellectual discovery by closing the mind with a passionate spirit. It demonstrates the passion of the spirit for action without knowing the reality of the goal which leaves one alone as a worker. "Freedom from aloneness" (*Kaivalya mukti*) recharges the sentient well-being as one enjoys the "universe of beginning" (*Bodhisattva*) as a sentient entity. The consequence is a well-wishing "filial piety" (Xiao 孝) for essentializing the discontinuity of one's reproduction of oneself as the leader leading one as its follower. Consequently, one becomes an entrepreneur familiar that its past force is the follower of the future, the leader, it is mediating as the present, the organizer. The "discordant force" (*Rex Mundi*) of the past shapes the sentient "group lineage" (*Wu Zetian*) leading to the entrepreneur's degeneration into everything else but the reality of the future.

Through guider mediation of its sentient lineage, the entrepreneur develops "dispassion" (*Vairagya*) towards the social value of the "three-fold social class" (*Jati*)—the leader class that leads, the worker class that follows, and the knower class that mediates like an entrepreneur trading the past and servicing the future. The social class is the physical element of the body. The "light half" (*Balachandra*) of the physical element is the lunar potential within the earth-effect. The "heavy half" (*Rukmini*) is the lunar present without the earth-effect. The lunar is present as an entrepreneur trading the light from its past as a solar and servicing the heavy as an octave of copies to its future as the earth when it is no longer a moon. The future as the earth within the physical is present as a potential within the past as a solar element within the Sun. The potential is in a deep sleeping phase awakened by the present with its wakefulness. The lunar is wakeful as a creator of the future. The physical is asleep as a perpetuator of the present. As a destroyer of the past it has already lived, the solar is awakening the lunar for the Sun to illuminate the present yet to live by liberating the potential of the physical to be a Moon before returning as the Earth.

The Earth produces herself as the physical with the dry ether-effect a neutrino irradiates from the 1,999 Infinite Council to duplicate its reality. The twin neutrino is a potential light that twins potential of "Earth" (=724 =180 *4 +4) as "light" (=180) addable as four. The Twin Earth is present as Dark Matter. The multiplier is the masculine body that adds three to the neutrino with its future as a masculine without the body. The feminine embodies the body to be the causal body after trading the "consciousness" (=4) that the masculine multiplies to realize "neutrino" (=16 =7 +9) with the "reality" (=7 = 4 +3) within the "past" (=9 = 3 *3).

The multiplier is the present son that becomes a potential son once it adds itself as a three to the neutrino to be present as a "cell" (*Hiranyagarbha*). As the reality of the neutrino transforms into the consciousness of the potential son within the present son, the masculine enjoys "freedom from the psychic element" (*Jiva Mukti*) and becomes a part of the feminine body. The part has the potential to divide itself into infinite masculine through cellular multiplication that transforms one into 999. By breeding fifteen, "one" (Positive quark) becomes "sixteen" (Neutrino). The multiplier adds three to advance sixteen as "sixty" (= [16+3] *3 +3), the octave of the sound of reproduction. By feeding seventeen from the potential of eighteen to each one, sixty becomes the "absolute creator" (=21) of 999 (17 *60 -1 -3 -17 = 1,020 -21) while personifying the future reality as a "Sun" (= 21).

How Fifteen Breeds Nineteen as an Atom

The splitting force of fifteen from one qualifies four, the consciousness of growth, as the heptaquark. Heptaquark comprises four up quarks that pairs four down quarks as its twin for servicing the "twin-effect" that pairs the present of eight within Hepta with the potential of ten within Deca. It produces the twin's recombination as a decaquark comprising eight up quarks with a quark of eight down quarks. The "memory"

(=35) of the ascending mass-effect within deca "transforms" (=1) the "up" (=62) element into the "down" (=26) element with a "strange quark" (=1). Strange is the potential within quark to be present as the value of light. Four down quark is a potential leptoquark with a potential charge that pairs itself within an up quark for the nine down quarks to be self-luminous as an infinite pair of three with two which squares. Two is the primordial pair of two with three which cubes. An up quark twins causation to primordial pair its twin with three for the two to be nine up quarks. The absolute pair of nine up quarks and nine down quarks reorders two into three. Two squares space to experience the origin of the past as a cube that perpetuates itself as a twin with both the present and the future. Three cubes time to perpetuate its reproductive reality. An electron is the perpetuating value of three within time. The perpetuating value is the five—the exogenous electron. The two is the time's future value which orders five within ten—the endogenous electron. As an atom, the whole of nine up quarks and nine down quarks in the center times three's perpetuating value with an electron. Therefore, an atom comprises nine up quarks, nine down quarks, and an electron.

The Sun is an outgroup creator of consciousness without attachment to the ingroup creation of seventeen—eight planets in the present and potential forms within the moon. The moon is present in its potential form as the one that becomes a four when the multiplier adds three. The "potential moon" (=1) is a "triple moon" (=1) that adds "three moons" (=1) to be present as a "moon" (=997 = 999 −1 −1). The six moons within the potential of a moon self-reproduce two moons with their heavy energy for the ninth moon to be present as the light. The ninth moon is the ninth photon which becomes a Moon when a tenth photon imagines itself to be the multiplier by feeding seventeen to the potential moon. The potential moon is the eleventh photon that creates an illusion of the two photons linked through the psychic element mediated by the third photon. The third photon is a "twin quantum photon" (=3), comprising two photons superpositioned over the eleventh photon within the consciousness of "three photons" (=60). Three photons are breeding "forty-five photons" (=80,000) as their collective value and self-reproducing "fifteen photons" (=70,000) as their individual value. The 45 photons become "sixty photons" (=90,000) at the infinity of causation because the eleventh photon works like a twin photon. The third photon advances three photons after feeding a continuity of 3 *17 = 51 photons and realizing that the 51st photon has the potential to perpetuate the sixtieth photon as the three photons.

The "sixty moons" (=90,000) comprise eight planets in their five forms: present (as negative quark), potential (as positive quark), collective with seven on top (as top quarks), individual with seven at the bottom (as bottom quarks), and group with three attracting an additional three through endoproduction for repelling an endoreproduction of the two as a pair. The pair includes one ascending (as charm quark) the charm for the seven and the other descending (as strange quark) within geography, and as geography. The geography is the sixteenth moon with fifteen moons. It comprises five repeated as three after the two twin the effect of the five for breeding fifteen of which three are the copies of the Sun, the Dark Matter, and the White Star. The five repeated with geography are the up quarks. The three are the down quarks that the two twins.

The present comprises eight negative quarks. The present radiates its negative value into four directions of space to be repeated with time's discontinuity after the present becomes the past and the future takes the position of the present. *The potential comprises eight positive quarks.* The potential radiates its positive value with the three dimensions of time into the space that perpetuates them as the fourth dimension and twins it with its four directions.

The collective comprises eight top quarks. The seven on top include six as the growth of the seventh. The seventh is the oneness among the three that vary with time and the three that are constant because they are the classes varying as a social class with time. The oneness is the "class-effect" (=48). The constant three become the fourth when they vary as a social class. The variation transforms four into eight. The eighth is the entropy of the seventh because seven are repeated as five—three that are constant, the fourth that varies as a social class, and the fifth that classifies the social class within a group. The variation adds three to the group as the space quadruples the four groups to reclassify the geography as sixteen groups.

The "collective" (=47) "decays" (=31) into eight strange quarks (=31), because the one repelled without oneness with the seven on top assumes three are covarying with it over time. It twins the "correlation" (=16) with space's present value as geography. The "repelled one" (*Urja*, 31) becomes a "Sun" (=21) as it twins the "assumption" (=19) of two generations "repeated" (=5) as a "family" (=10). As the "assumer" (*Kartika*, 31), it projects the "tenth" (= -12) for the assumption of the Sun's "masculine copy" (=3) within the "potential moon" (=1) to be the "tenth moon" (=111) itself. The assumer is the "eighth" (=12 = 8 *3/2), which "embodies" (*Kartikeya*, 8) the consciousness of four. Four twins the correlation of one with six by "self-projecting" (=3/2) itself as twelve. "Self-rejection" (*Vruschik*, 735) of seven, repeated with five as the thirty-five, twins the twelfth as the "feminine" (=37) to copy the consciousness as the "masculine" (296 = 3 *100 -4) left with three that become "reproductive" (=100) for reproducing the "time" (=360) itself as an "octave" (=60). Five "coils" (*Kundala*, 5) the consciousness, the soul, into the "spirit" (=20) of "eight" (*Kundalini*, 8). The spirit comprises "eight energies" (*Kundalini Shakti*, 20) that fall to the bottom as an individual as the mass of consciousness rises to bring the collective to the top.

The individual comprises eight bottom quarks. Seven on the bottom include five repeated after their entropy as two generations for the growth of the eighth as the "third generation" (=94), the "entropy base," into the "present generation" (=100,000 =94 +6 *1,000) that forms as the "origin" (=1,000) "with all". Growth transforms the eighth into the ninth, with four on the right. Five left are repeated as the entropy base for the growth of an additional six into a reproductive element within the origin. The "individual' (=53) is the masculinity whose "origin" is femininity. The individual "groups" (=387) eight charm quarks (=387) since three of the additional six are repeated within the seven on the bottom for producing a base of ten by reproducing the repeating three. The base is predicated on dividing the octave with a force, known as Van der Waals Force, that makes the "divider" (=1) "attractive" (=19) for the division to copy the "gender" (=9) of the multiplication. The division twins the surface of the octave with the addition of two generations. The octave charms the surface to attract three classes after their subtraction from the fourth class, which is their future as a parallel class, for replication as the fifth class, which is the entity as the individual class. Five-class geography within the fifth class is integrated with the three-class group. Two classes differentiate the group by gender and generation. *The group is the entity class comprising eight classes as eight charm quarks.*

In summary, the group makes the intellect conscious of geography's physical element to generate a body of consciousness. The consciousness is repeated by the "mind" (=38) as the "gender" (=9) for its "organization" (29 =38 -9) as the "masculine" (=296). The gender self-services six to twin the consciousness with a two. Two is one's endoreproduction. One becomes zero after the endoproduction of the self as two. Two is the universe that orders the growth of space with the growth of time.

4.3 The Path of Primordial Space.

Primordial space is the mind-born space that makes space reproductive as the light to illuminate the "Highly Interactive Particle Relic" (HYPER, 37) as the "body" (=56) of "consciousness" (=4) "crystallized" (=89) into a "crystal" (=34), comprising 89-56, i.e., 23 "atoms" (=19 =23 -4). 23 atoms triple with time to be crystallized as the 69 atoms after crystallizing 31 atoms to make an "octave of atoms" (=100) reproductive like the primordial space with a base of 31-23, i.e., "eight atoms" (=10). 31 atoms are left as the "whole class" after the division of the 1,000 atoms within the "Interactive Particle Relic" (=1,000) into a "Particle Relic" (=72) of nine atoms to self-service an "imaginary class" (=10) of six atoms. Six atoms make the octave reproductive like an atom as they are repeated within the continuity of space until they become whole as a "class" (=689) of sixteen atoms with an "asymmetry" (=689) of six. The two "repeated" (=5) within the space of four atoms (norming four dimensions) form their "symmetry" (=250) with "ten atoms." As they are repeated, two become a "Knower deity" (=2).

4.3.1 Understanding Knower Deity.

The Knower deity is an annihilator of complexity. Its simplicity guides the transformation of the general mass of "embodied matter" (=111) into the special mass of "disembodied anti-matter" (=109) within the parabolic mass of dark matter. "Complexity" (=479) deforms the parabolic mass of ionizable, "dark matter" (1,600 = 479 +155 -34 +1,000) into the circular mass of ionized, "light matter" (=155) with the "force" (=34) of the "heavy" (=1,000). "Simplicity" (=2) reforms the circular mass of "light matter" into the square mass of ionizing, black, "heavy matter" (=29). The "mass force" (=29) of the square mass triangulates the square to add three to six after multiplying two with three. The "matter" (=158) is a "triangular mass" of "36 atoms" that entangles two with the asymmetry of six atoms by triangulating the multiplication to succeed as their "convergence" (=158).

The disembodied anti-matter is the "imagination" (=109). The "primordial imagination" (=82) is the "present workculture-effect" that "animates" (=76 =158 -82) the "primeval imagination" (=40) as the "future workculture-effect" to "channel" (=36 = 76 -40) the "visibility" (=158) of "present imagination" (6 = 82 -76). The "creative force" (=6) "liquifies" (=1,000) the "quantum state" (=5) of "imagination" (=109) to "solidify" (=950) "capability" (=55). The capability "orders" (=2) "spatial alignment" (=50) into an "orbit" (=2) to "spin" (=29) the "liquid" (=91) for "temporal misalignment" (=200). The liquid's "twin identity" (=2)—a liquid within temporal misalignment and a solid within spatial alignment—"forces" (=34) the "channel" =36) to "exchange" (=269) "gas" (=365) within the "causation realignment" (=90,000 = 36 *500/2) with "plasma" (=100,000 = [36 +4] *500/2) within the "entity prealignment" (=500) of the "sentient force" (=4). The sentient force "reorders" (=2) the space's four dimensions into the "four forms" (=12) of matter in the "liquid quantum state" (=2) for "servicing" (=47) "visible matter" (=691 = [47 +2,2], 1) as the "causation" (=1) with the "sixth form" (=47) after "trading" (=53) "matter" (=158) from the space with the "fifth form" (=11). The fifth form is repeated with the "endoproduction" (=1) of the "fourth form" (=3) for the exchange of the "dark matter" (=1,600) from time with the "seventh form" (=2).

Time forms 1/6th of the matter as the visible matter, present within the potential of time to transform its dark energy. Dark energy is produced by the potential of space to perpetuate time as its past, which is dark and infinite because space's present is light and finite, free of the consciousness of the past reproduced by the causation present within an entity with the future's energy. Space forms 5/6th of the matter as the dark matter, present within its potential to keep a constant order. Time disturbs the constant order with the causation for the endoproduction of the ascending order, thereby descending entropy of the whole by adding electrons at a descending velocity and "descending thermodynamic force" (=2) of the preceding electron that potentially exists within the matter. The whole forms a water molecule, whose *temperature measures the average velocity of adding electrons.*

Ascending order perturbs the causation for exchange of the constant order with the endoreproduction of the descending order, ascending entropy of the wholesome by subtracting electrons at an ascending speed and "ascending thermodynamic force" (=268) of the preceding electron that exists within the visible matter. The wholesome forms an air molecule, whose *potential temperature measures the average speed of subtracting electrons.*

Temperature (=86) and "potential temperature" (=86) are similar. Without perpetuating the value of the constant order for the "entropy" (=5) of the wholesomewhole as the earth molecule, they are the "absolute temperature" (=81). *Absolute temperature measures the average momentum of the trading electron.* Descending order moves "energy" (=19) within "entity" (=24) as an "atom" (=19) for trading electrons subtracted by "para entity" (=19 =24 -5) at a constant momentum in the form of a "neutron" (=19 = 1, 6+3) with the "constant order" (=5) within "causation" (=1) that leads to the "growth" (=6) of one with a "time multiplier" (=3). Constant momentum follows a "horizontal thermodynamic-effect" (=80) of servicing the preceding electron in the form of a "proton"

(=150 = 18,000/[5 *24]) with the "constant order" (=5) within "space" (=18,000) before trading one as an "electron" (=365) with the "constant order" (=5) within "time" (=360).

"Thermal equilibrium" (=81) is the loss of the absolute temperature balanced by the "profit" (=81) from servicing electron. *Profit is the "average acceleration" of the ether molecule as a measure of infinite temperature.* "Horizontal order" (=5) "persuades" (=11) the entity to invest in "forward order" (=96 = 19 *5 +1) by servicing time's "backward thermodynamic force" (=19) as an electron with the causation's "backward acceleration" (=1). It leads to a "growth" (=6 =5 +1) in "motion" (=360) of the "ether molecule" (=2) to "reorder" (=2) "entity's" (=24) "oneness" (=48) with "space" (=18,000).

"Social benefit-cost ratio" (=81) is the "average motion" of the "fire molecule" (=-9) as a measure of primordial temperature. "Forward order" (=96) "dissuades" (=40) the entity from "emoting" (=1,200,000 =40 *1,000 * 90/3) "backward order" (=1,000) by "demoting" (=3) "growth" (=6) of the subject with a "fire molecule" (=-9) of the space's "forward thermodynamic-effect" (=90).

"Worker-social benefit-cost ratio" (=81) is the "average emotion" of the "divine molecule" (=366,666 = 3, 6*11, 6*111) as a measure of constant temperature. "Whole order" (=40) "pervades" (=92) within the subject's "whole thermodynamic-effect" (=89) as the "reproductive reality" (=40) of the "time multiplier" (=3), "organizable" (=11) as a "divine molecule" to "guide" (*Vrishakapi*, 111) the "growth" (=6) of the "space divider" (=4). An entity becomes a subject servicing space's "natural light" (=24) without the "placement" (=81) of the time's preassembled "supernatural light" (=1,000,000). The subject trades the "displacement" (=90) of the causation's reproduced, "artificial light" (=48) through "primordial oneness" (=32) with nature's "reproductive reality" (=40). The "causation" (=1) is the "emotion" (=73 =1 +32 +40) which "emotes" (=58) the "mote" (=240) for trading "human force" (=53) with the letter "E" (=53), "repeated" (=5) as an "ion" (=13) within the time's "reproductive reality" (=40) as "motion" (=360).

"Reality" (=7) "reproduces" (=78) "nature" (=8) with a "multiplier" (=3) of her "perpetuating value" (=5). The space-dependent "ascending temperature" (=95) "descends" (=25) the time-dependent "ascending order" (=105) for "breeding" (=15 = 3 *5) the "mass" (=132 =95 +25 +12) of space's "perished consciousness" (=12) "infused" (=12) with the "kinetic energy" (=571). The kinetic energy is "left" (=89) as the "motion-effect" (=571 =132 +360 +89 -10) after the "reversion" (=-10) of "motion" (=360) as "linear" (=497 =132 +360 +5) to "perpetuate" (=9) the "reality" (=7) of nature within the "knower deity" (=2).

The core of the knower deity is "social justice" (=2). The reality's endoreproduction as *"Adonai"* (=2), the "Infinite Lord" (=2), socializes justice with the true, convergent reality of nature's workforce. The Infinite Lord is *Yu Huang Dadi*, the Jade Emperor, breeding *"Dao"* (=17), the Godhead, as his transcendental form, to be the "Lord of deities" (=0), the subject. The convergent reality transcends the artificial beauty of the subject's "divergent reality" (=-3) as an "object" (=-3). The subject becomes an object trading energy as the knowing of the "infinite lifetimes" (=1000), the "energizer" (=1,000) who "opens" (*Shi*, 70) the "window" (*Kou*, 94) of "understanding" (*Zhi*, 2) within the "consciousness" (=4) of the "entity" (=24). The "infinite sound" (=256) of the "leadership work" (=256) eventually leads to the "ultrasound silence" (=130) as the entity works on the "value conflict" (=126) as a "follower" (=24). The work flows like water to clear the conflict "within oneness" (*Rajayoga*, 926) of the value borrowed over the infinite lifetimes from the "deity kingdom" (=1,000). "Freedom from the joy" (*Nirvana Mukti*, 571) of "being deity" (=123) in the "present lifetime" (=18,000) with "borrowed oneness" (*Mahayoga*, 926) "animates" (=76) the "potential" (=18) for "accelerating" (=700) "absolute development" (=14) of the "primordial lifetime" (=11) into the "foundation" (=11) of understanding reality by transcending consciousness within the knower deity.

4.3.2 Transcending Knower Deity.

"Continuous development" (*Gajanan*, 0) of a subject into an "illusionary greeter" (*Vignesh*, 1) as a worker, working as a deity, leads to "discontinuous development" (=12) of the "knower deity" (*Dvaimatura*, 2). The knower deity develops into the first-born "present creation" (*Ganesha*, 570 = 15 *2 *[17 +2]) following an "infinite development" (=15) of "Godhead" (*Param Ganesha*, 17 =15 +2), the "future creator" (=17), within the "primeval space" (*Para Ganesha*, 19) of "clarified consciousness" (=19) the knowing "creates" (=35 =1 +15 +19) within "mind" (=38). The clarified consciousness transcends the duality between the worker deity, working on the "perceived reality" (=825), and the knower deity, knowing the "conceived reality" (=6) that "works" (=19) as knowing until the subject experiences the reality to be divergent after assuming it is convergent.

With a clarified consciousness, the subject as the "Lord of the womb" (=0), develops into the "womb" (=3,794), the "Angel of Justice" (=3,794), the essential cause of normative justice. He ascends "unconditional confidence" (=17) by "configuring" (=79) the "fruit of freedom" (*Muktiphala*, 62) from immanence within a creator. He descends the "fruit of impact [divinity]" (*Siddhiphala*, 25) of the "knower deity" (*Dvaimatura*, 2) on the "outcome" (=80) of his "training" (=80) as a "trainer" (*Vinayaka*, 53) of the creation following his param creation. He ends the "fruit of devotion" (*Bhaktiphala*, 96) to the past's dark energy with his "perpetuating value" (=5) as a "sentient child" (*Pillaiyara*, 91) for realizing his future as a Wisher deity. As a "wellspring" (*Pille*, 13) of "sentient well-being" (=190 = 13 +7 +91 +79), he is the "sly" (*Aiyara*, 79) "Green Emperor" (*Cangdi Lingfu* 靈) with an "unknown reality" (=7) of a sentient child The Green Emperor transcends her "being energy" (*Kali shakti*, 96) as a "present being" (*Kali*, 96) to multiply time through "primordial well-being" (*Vighnaraja*, 57) of the "param son" (=3) for "efficacious prosperity" (=57) with her "energy" (=19) after "victory" (*Netzach*, 13) over "consciousness" (=4).

The Wellspring is the "Queen of Divinity" (*Maha Kali*, 13 =3 +10), conceived by the "primordial self" (*Parvati*, 10) to make the "goalkeeper's" (*Shiva*, 7) unknown reality "whole" (*Akala*, 16 = 10 +3*2) through the "simplicity" (=2) of endoreproduction of the "wholesomewhole" (*El Roi*, 13) as a "param son" (*Hanuman*, 3), luminous with "infinite forms" (=13). "Without realizing" (*Nibbida*, 13) the "goal" (*Maha Shiva*, 9) of the "reproductive reality" (=40) of her "finite form" (=13) as a "wholesome" (=957), the past's consciousness of the "origin" (=1,000 =957 +40 +3) "revulses" (=3) the "reality" (=7) of the goalkeeper for "absolute realization" (=16) of the "whole" (=16) as a "simplifier" (=16). "Freedom from the entity" (*Mahapralaya Mukti*, 2,400), who makes her form reproductive for experiencing the "infinite forms" (=13), "dissolves" (=38 =13 *2 +12) the "present moment" (=12) of the "creature's" (=12) "finite form" (=13). Therefore, the subject becomes an "Ascended Master" (=805), mediating the perpetuating value of his supernatural growth like God. The Wellspring is the "zeroth angle" (=13) of the "equator" (=257), whose "reproduction" (=285 = 90 *3 +15) of reality with a "force" (=34) produces "angular momentum" (=90) to triple the perpetuating value with time multiplier into a "goalpost" (*Param Shiva*, 15 = 5 *3).

Through the "path of causation" (*Antaramsa Marga*, 12), the "twin multiplier" (*Isha*, 12) of the perpetuating value begins "pervading" (=12) as the "seventh month" (*Ashvina*, 12), "September" (*Ashivna*, 12), for "invading" (=99) "ascending-effect" (=99) of "five months" (=150) to keep the "mood" (=360 =150 +105 *2) "current" (=150) while "ascending order" (=105) with "two months" (=2) as a knower deity. The "lordship" (=10^{100}) of the year's "convergent reality" (= 10^{100}) as a "libra zodiac" (*Tula Rashi*, 10^{100}) balances the diversity of forms with Wellspring as the "unit" (=6) of "reality" (=7).

4.4 The Path of Primordial Causation.

The "primordial causation" (*Brahli*, 89) is the "diverse present" (*Brahli*, 89) a Wellspring "animates" (=76) by "servicing" (=47) "momentum force" (=571 =13 +47 +170*3 +1) of her "primordial human-effect" (=571). "Divergence" (=170) of "causation" (=1) within the "Manifestor deity" (=3) makes the "observable value" (=13) of the Wellspring luminous with the "path of primordial causation" (*Brahli Marga*, 13 = 1,3).

4.4.1 Understanding Manifestor Deity.

The Manifestor deity is the potential's "observed value" (=3) within "growth" (=6) which "adds three" (=3) for the "observing value" (=10) of the "observable value" (=13) by "squaring three" (=10) within one—the "observer value" (=1). Potential twins "anti-neutrino particle" (=23) for perpetuating value of a disembodied "anti-neutrino radiation" (=47) of the observed value as the observer value of the entity, the "lambda family of baryons" (=24).

The manifestor deity is the "potential thing" (=3) that perpetuates the "potential" (=18) as a "thing" (=9), the "sentient object" (=9), to "twin causation" (=2) for "growth" (=6). She is the "infinite feminine" (*Devi*, 3), "breeding" (=15) "potential" (=18) within her "perpetuating value" (=5) for "trading energy" (=60) through "entrepreneurship-effect" (=270), i.e., "supernormal value" (=270 = 15 *18), of her "spiritual devotion" (=270) to the "multilocalized universe" (=270), the "master trader" (=270). She "twins" (=121) her "husband" (=121) as "father" (=3) with the "param son" (*Hanuman*, 3). She is the "mother" (=4), "mothering" (*Maha Brahma*, 3) the "well-being" (=125) of her "child" (*Pillai*, 128) for "supernormal profiting" (=3) from the "materialism" (*Pravrtti*, 9) of the "human potential" (=9) through its "replication" (=90) of the "monkey's" (=3) immanent, "animal potential" (=90).

The monkey copies the human's "three copies" (=3) to be perpetuated as the animal's "four copies" (=90). The present "cubes" (=90) with the four copies with the "potential growth" (=3,700) of her "feminine body" (=30) by making the "feminine" (=37) "reproductive" (=100) within the "body" (=56 =37 +19) of the "param child" (=19). Thus, plant's 3^2 -3, i.e., "six copies" (=6) become mineral's 4^3 -4, i.e., "sixty copies" (=60) by attracting metal's "ten copies" (=10^{10}) as the "self-luminous system" (=10^{10}) after repelling material's "one copy" (=0) as a zero for its growth into spirit's "sixteen copies" (=8×10^{15}), the "luminous system" (=8×10^{15}). Overall, the "devotee universe" (=9×10^{18}) comprises a deity's "hundred copies" (=9×10^{18}) that become reproductive with the "three legs" (=9×10^{18}) of time which the "three copies" (=3) multiply into the para deity's "three hundred copies" (=10^{1000}) with the "three concentric circles" (=10^{1000}) of human life—"within deity" (=8), "without deity" (=5), and "with deity" (=3)—before becoming "whole" (=16) "as a deity" (=16).

Therefore, the "value of human life" ($=10^{1024}$) over time is the param deity's "five hundred copies" ($=10^{1024}$) as an "animal" ($=10^{1024}$) who works as a "human" (=275) with the primeval deity's "two hundred copies" (=275) within the "working force" (=75) of the primordial deity's "seven hundred copies" (=75). The supreme deity's "thousand copies" (=26) are "dormant" (=26) as the "networking force" (=26) of the supra deity's "twenty-four hundred copies" (=53) within the "human force" (=53) of "masculinity" (=53). "Assuming" (=190) a "force" (=34) of "eight copies" (=34 =26 +8) from "nature" (=8) that the time makes reproductive, "supernatural force" (=190) comprises a super deity's "thirty-two hundred copies" (=190) for a human's "sentient well-being" (=190).

The "supersaturation" (=45) of the animal "herd-effect" (=45) "populates" (=45) "forty thousand copies" (=45) with the "formative fire-effect" (=45) within the "potential-effect" (=1) of "devotion" (=46). Satan's "twenty-four thousand copies" (=125) are "silent" (=125) within a "lifetime" (=125) that "herds" (=128) a greeter's "eight thousand copies" (=128) within the child. As an outcome, a deity's "triple oneness" (*Triyoga*, 80) "as a deity" (=16) is "repeated" (=5) and "circulated" (=80) within the primordial greeter's "80,000 copies" (*Dhyana Yoga*, 80) with a force that makes eight copies "circular" (=10,000).

The "circular force" (=1,000/72) forms the param greeter's "20,000 copies" (=1.000/72) as a "whole pi" (=1,000/72) after the "circular" (=10,000) copies mother's "nature" (=8) which "twins causation" (=2) to be "whole" (=16) that the "pi" (=22/7) "circulates" (=22/7) as her "femininity" (=1,000). The "circle" (=100,000) "copies" (=0) its "form" (100,000) into a "maternal community" (=100,000) of the primeval greeter's "100,000 copies" (=100,000).

The maternal community services its "wholesome energy" (=90,000) to the "female" (=10,000) for "circular creation" (=10,000) of "ascending sentient well-being" (=157) within the "universe" (=2). The universe twins causation as a "male" ($=10^8$) within nature for its personal "sentient well-being" (=190 =157 +37 -4). The female enjoys "freedom from disembodiment" (*Videha mukti*, 10,000) of the "consciousness" (=4) as the "feminine element" (=37) with the "embodiment" (=58) of "mother" (=4) as the "soul" (=4).

The mother universalizes a person's sentient well-being with a "father" (=3), conscious of "reality" (=7 =4 +3).

As "father" (*Ziwei Beiji Dadi* 紫微北極大帝, 3), the manifestor deity is the "moderator" (*Zhong* 忠, 3), "moderating the "descending sentient well-being" (=139 = 67 *2 +8 -3) of nature as an "institution" (=8) with the "purple north polestar" (*Vega*, 67) and its twin "white north polestar" (*Polaris*, 997 =930 +67). The white north polestar is the purple, i.e., white-blue north polestar's "past" (=9) a "moderator" (=3) copies as its "future" (=0). Consequently, the white north polestar varies until realizing its future as a copy that twins causation to be present as the "blue north polestar" (*Lambda dark matter*, 134 = 67 *2). As the "past's" (=9) "future reality" (=21), Blue north polestar, known as Lambda dark matter, comprises nine "suns" (=21).

The "lambda" (=1) is the sixth sun, which is a constant, "self-luminous one" (=1). Its "dark half" (=997) is the fifth "variable sun" (=997). Its "light half" (=82) is the fourth "constant fourth Sun" (=82). The "dark matter" (=1,600) is the third "semivariable Sun" (=1,600). The "light matter" (=155) is the eighth "circular Sun" (=155). The "lambda light matter" (=7) is the second "square Sun" (=7). The lambda dark matter is the seventh "parallel Sun" (=134). The "light" (=180) is the eighth "line Sun" (=180). The "dark" (=185) is the ninth "point Sun" (=185). The "nature" (=8) is the tenth, illusionary "time Sun" (=8).

The first, imaginary "space Sun" (=67) is the fixed north star. The eleventh is a real "causation Sun" (=3600), the "reason" (=1) for dividing "time" (=360) into "Ten Suns" (=360) to illuminate its "personality" (=3,600) as a "person" (=1). The thirteenth is a complex "potential Sun" (=23), which reproduces its potential as the twelfth, simple "Sun" (=21) to twin its "simplicity" (=2) with the fourteenth Sun, which behaves like "Two Suns" (=570), each a "reproduction" (=285). The reproduction is the fifteenth Sun, which becomes "Three Suns" (=285) with the "time multiplier" (=3) to manifest the time as the "Sixteenth Sun" (=360), the space as the "Seventeenth Sun" (=18,000), and the cause as the "Eighteenth Sun" (=18). The cause self-services the nineteenth, "Formative Sun" (=12) to norm the simplicity of twin causation with the twentieth

"Half Sun" (=2). Half Sun perpetuates itself within the twenty-first "Normative Sun" (=9).

Half Sun is the "Transformative Sun" (=2), transforming "itself" (=121) like a "wholesome system" (=121), the twenty-second sun, into eight planets, each a "primordial Sun" (=53) trading the Sun's "primordial oneness" (=32) with the "Earth" (=724), the thirtieth Sun. Mercury is present as the twenty-third Sun. Venus, the twenty-fourth Sun, is its potential past. Uranus, the twenty-fifth Sun, is its primordial past. Neptune, the twenty-sixth Sun, is its primeval past. Saturn, the twenty-seventh Sun, is the performed past. Jupiter, the twenty-eighth Sun, is its primeval present. Mars, the twenty-ninth Sun, is the primeval future, accelerating as the "blue moon" (=102). The "white moon" (=997 =724 +250 +3) "decelerates" (=18,000) to be the "earth" (=724). The "symmetry" (=250) of the "Manifestor deity" (=3) with her "manifesting body" (=30) to manifests "ten-fold growth" (=170) with a "triple octave" (=80) before "decelerate-effect" (=6) copies a a natural growth of six. The symmetry is the "continuity" (=50) of "entropy" (=5) with the "accelerate-effect" (=5) that reforms the Earth into the Venus present within Mercury.

The Earth As a Metaphysical Body of Moon

As a "metaphysical body" (=724) that "conscious copies" (=26) her "birth" (=247) as a "moon" (=997), the Earth is the "first moon" (*Rohinisha*, 724). The Earth is the "universe of moonmoon" (=724), with a "potential" (=18) to "conscious copy" (=26) her "birth" (=247) as a "moon" (=997) to "incarnate" (=900) "everyone else" (=-10) as a "second moon" (*Darshapurnamasa*, 900,000) with her "perpetuating value" (=5) as a "param maternal" (=5). The second moon is moonmoon, comprising two moons, of which the second is a "quasi-moon" (=900,000). As a half black moon, quasi-moon is a "continuous asteroid" (=900,000) "sharing" (=900,000) its "past" (=9) with a "discontinuous asteroid" (=997) that "heads" (=-9) the "discontinuous-effect" (=196) of the "wave of continuity" (=196) to be "whole" (=16) when its "area" (=900) "tails" (=70) the "middle body" (=724) of the "Manifestor deity" (=3) which "hides" (=900,000) the universe of moonmoon as an "infinite asteroid" (=724).

The discontinuous asteroid is the "dark half" (=997)—the "disappeared moon" (*Prakshinachandra*, 997), whose "light" (=180) is "waning" (=1) as a whole from the discontinuous-effect to "appear" (=960) as a "new moon" (=997). As a "waning moon" (=997), the disappeared moon is the "seventh moon" (=997), the "wholeness-effect" (=997) of "seven moons" (=997). The "appeared moon" (*Balachandra*, 82) is the "fifth moon" (=82), the "light half" (=82), whose "darkness" (=180) is "waxing" (*Parvana*, 1) the "void" (=-2) with the "third moon" (*Apuryamana*, 1). As a "potential moon" (=1), the third moon is the "descended moon" (=1) that transforms into "three moons" (=1) by "producing" (=17) a "double copy" (=1) of its "potential" (=18). As the eighth moon, the "whole moon" (=997) is the "oneness-effect" (=997) of the "discontinuity" (=200) of the "three moons" (=1) within the "third moon" (=1) "and" (=50) a "continuity" (=50) of the "five moons" (=82) within the "fifth moon" (=82) that "voids" (=-2) the "flow" (=666) of "moons" (=997) with a "black hole" (=82).

The "fourth moon" (Balendu, 13) is the "ascended moon" (=13) with a "consciousness" (=4) of "four moons" (=13) within its "past" (=9) as a "potential moon" (=1) that "copies" (=0) "Manifestor deity" (=3) to "triple" (=130 =13,0) itself with "three moons" (=1).

The "whole area" (=12,000) "surfaces" (=2,000) a "growth" (=6) of "six moons" (=12,000) within the sixth "young moon" (*Balachandramas*, 12,000) for "self-projecting" (=3/2) a "space" (=18,000) for "nine moons" (=18,000). The "light" (=180) is the ninth "old moon" (*Chandrakanti*, 180), "reproductive" (=100) as the "nine moons" (*Mahas*, 18,000) "exchanged" (=180) from the "nine photons" (=180). The young moon is the "spring moon" (=12,000). The old moon is the "autumn moon" (=180). The tenth "winter moon" (*Arkendu*, 111) is the "replication" (=90) of "ten moons" (*Suryendu*, 21) within "Sun" (=21) as the thirteenth "summer moon" (*Suryendu*, 21). The "Sun" (=21) "double copies" (=1) the "continuity" (=50) of "five moons" (=82) to "twin" (=121) "itself" (=121) as the "white" (=10) "base" (=10). The "eleventh moon" (*Chandra*, 82) is the "new moon" (=82).

> "Three moons" (=1) are "omnipresent" (=1) as "three zeroes" (=000) within the "future" (=0) of the "Sun's" (=21) "reality" (=7) as a "Manifestor deity" (=3), the "param manifestor" (=20) of the "three copies" (=3) of the "photon" (=20). Therefore, with "nine moons" (=18,000), the space has a sameness with the twelfth "seasonal moon" (Mahas, 18,000). The "twelfth moon" (*Mahas*, 18,000) "times" (=360) the "continuity" (=50) of "thirteen moons" (=360) with the fourteenth "perennial moon" (*Chandima*, 360), the "God of moon" (=360). The "continuity" (=50) "emits" (=941) the "growth" (=6) of "three double copies" (=6) of the "whole moon" (=997) as a "half moon" (*Ardhacandra*, 56) to "body" (=56) its "timelessness" (=56) within the fifteenth, constant, wholesome, "crescent moon" (=56).

4.4.2 Transcending Manifestor Deity.

With "disproportionate entropy" (=190), the "thirty-first Sun" (*Anyadriksha*, 190) becomes a "para multiplier" (=190) "decaying" (=190) with time as a "meteor" (=190). Para Multiplier is the "supreme spirit" =190) beyond the "spirit" (=20) of the "tenfold growth" (=170) with an "earth-like" (=190) "beautiful face" (*Sumukha*, 190) of "intrinsic perfection" (=190). The thirty-first Sun was the "third sun" (=1,600) which "reappeared" (=190) as the "first sun" (=67 = 6 +1+6) after a "dreamer" (-10) "dreams" (*Dhyana*, 9) one to be "whole" (=16) when "natural growth" (=6 =-10 +16) twins one, the "sixth sun" (=1), with six, the "second sun" (=6). The thirty-first Sun is the "Lord of the realm of the shadow of perfection" (=190), i.e., of the "deity realm" (=1,000). The "thirty-second Sun" (*Pratidriksha*, 98) is "serialized" (=98) with time's "reproductive potential" (=23), "disappearing" (=98) like "another" (=98) who "appeared" (*Driksha*, 98) as a twin. The twin is the "twenty-second sun" (=121 = 98 +23) "reappearing" (*Samidriksha*, 30) as the "manifesting body" (=30), the "thirty-third sun" (=30) after the reproduction of the "three suns" (=285) with its "manifestable body" (=3 = 30/10).

The "thirty-fourth sun" (*Idriksha*, 3) is "such" (=3) "manifestable body" (=3) that is "appearable" (*Idriksha*, 3) with three Suns as the "shower" (=3) of the "meteor" (=190), the thirty-fourth Sun. The "thirty-fifth sun" (*Vrikshakapi*, 16) is the "tree" (*Vriksha*, 23) of "replication" (*Kapi*, 90), the "replicator" (*Vrikshakapi*, 16) of the "shower" (=3) with its "energy" (=19). The "thirty-sixth sun" (*Senajit*, 19) is the "vanquisher" (*Senajit*, 19) of the "three Suns" (=285) which the "sixth Sun" (=1) twins with its "reproduction" (=285). The "thirty-seventh sun" (*Srijak*, 578) is the "perfect creation" (=578) as a "creator" (=578) whose reproduction is "pure" (=285). The creator is the "right head" (=578) of the planet. The "thirty-eighth sun" (=130) is the "left head" (=130) of the Sun that became a "planet" (=-1) right before the "shared value" (=130) of the "perfect one" (=130) was left as the "low-density amorphous ice" (=130). The "thirty-ninth sun" (=924) is the "sharable value" (*Kosha*, 924) of a "high-density amorphous ice" (=924 =130 +8*100 -4). The "shared value" (=130 = 70+60) "tails" (=70) "eight planets" (=60) as an "octave" (=60) reproductive" (=100) with the "nature's" (=8) "sentient force" (=4). The "fortieth sun" (=830) is "sharing value" (*Siddha Guna*, 830) of a "medium-density amorphous ice" (=830) with the collectivizable reproductive forces of the eight planets within the moon's manifesting body. Manifesting body manifests the Sun's manifestable body to twin the dark matter as a "star" (=2) for the observance of the "white star" (=180) with its "replication" (=90).

The "twin dark matter" (=2) is a "lunar matter" (=2), which is a "perfect exchange system" (=2) for the exchange of nature's "gravitoelectric-effect" (=16) with the "solar matter" (*Krishna*, 32). The solar matter is a "twin light matter" (=32) which twins "light matter" (=155) for the "decay" (=31) of the bright "guider matter" (=310) into the dull "divine matter" (=10). The dull matter is a protoplanetary element of the "twin star" (=10) reproducing "itself" (=121) as a "planet" (=-1) to be one with the "Sun" (=21). The "dull matter" (=10) becomes a "Sun" (=21) through "conformity" (*Yathabhuta*, -1,000) to the "reality present" (=-1,000) within the "discontinuous" (=1,111) element that the time makes "reproductive" (=100) as a "Wisher" (=0 = 10 -21 -1,000 +1,111). The Wisher, "wishing" (=190) "sentient well-being" (=190) of the "future generation" (=190) before "decaying" (=190) into a "wishable" (=20) the "past generation" (=-19), "desires" (=-19) to be "repeated" (=5) within the "consciousness" (=4) of the "present generation" (=100,000) for "universal sentient well-being" (=157).

"Assuming" (=190) "itself" (=121) to be the "material cause" (*Upadana*, 578 =190 *3 +8), the "right head" (=578) of the "three-way exchange" (=69 =190 -121) of "nature" (=8), the Wisher becomes a "superluminal subject" (=190). The superluminal object moves faster than light as the future "clinging" (*Prishana*, 9,000) its lifetime "my experience" (=9,000) like a "charm" (=847 =578 +69 +200) until its "discontinuity" (=200) by a "Destroyer deity" (=6) through the "exposition" (=1,964 = 1,111 +847 +6) of "sentient matter" (=1,964)." "Its" (=900) "speed" (=999) "adds" (=99) a "directional" (=91) "temporal generation" (=91) as a "superluminal object" (=91 =190-99), moving faster than the light of the present clinging to the subject as the "superluminal motion" (=90,000), now "infinitesimal" (=90,000) as the "infinity of causation" (=90,000). The infinity of causation makes the space conscious of the speed of light as a "system" (=12) which incarnates as a "creature" (=12) with its "primordial consciousness" (=12).

The "system" (=12) makes the self "luminous" (=13) as a "Destroyer deity" (=12/2 =6) the "self-perpetuates" (=12) to "associate" (136 =13,6 =121 +10 +5) "itself" (=121) with the "superluminal" (=136). The system "twins" (=121) the superluminal as a "subluminal" (=10 =12-2) to be "God" (=5) who "twins causation" (=2) with the "time multiplier" (=3). The system is a "Bradyon" (=12), a "supraluminal" (=12) with an "illusionary mass" (=12) of the past, moving slower than the light of the present that moves ahead to twin its "real mass" (=0) within the space's "complex mass" (=10) as a "subluminal" (=10). The subliminal becomes a "Tachyon" (=10^{10}), moving faster than light for self-projecting light with its speed as a "twin system" (=10^{10}). The twin system "points" (=10^{10}) to the "inertial mass" (=10^{10}) of its "spatial value" (=10^{10}) left behind as "heritage" (=10^{10}). The self is a "Luxon" (=8×10^{15}) moving at the "speed of light" (=8×10^{15}) with the "imaginary mass" (=8×10^{15}) of its inheritable "incarnational value" (=8×10^{15}) as a "triple system" (=8×10^{15}) of time outgoing from space as light. The "light" (=180) is the "time mass" (=180).

The sentient matter intensifies a devotee's mental sense-making of the "Wisher universe" (=-1), transcending the limits of the self-awareness of their desires attached to the "consciousness" (=4) after their "entropy" (=5). The "discontinuous" (*Sattva*, 1,111) "flash" (*Jyoti*, 120) of the "light" (=180) of "intellect" =389) makes one "conscious" (*Ojas*, 189) of the "darkness" (=180) of "mind" (=38) with the "astral" (=189) element, leading to "astral formation" (*Satttvajyoti*, 180). The intellect is the "perfect creature" (=379) as a "creation" (=379).

The intellect de-intensifies the "etheric" (=16), i.e., "my" (=16) element attached to the "lifetime experience" (=9,000) as a "sentimental shadow" (=16) of the "Satan" (=-1). Satan attaches "herself" (=1) as the "causal body" (=30 = 18 +12) for manifesting the "wish" (=18) with the "potential" (=18) within one's nature of trading the "creature" (=12) as "I" (=12), after servicing "I am" (=100,000) as a "form" (*Rupa*, 100,000) of "God" (=100,000). God takes "infinite forms" (*Bahurupa*, 13 =1 +12) before illuminating "potential herself" (*Param Siddha*, 6) as the "destroyer deity" (*Mahesha*, 6). "Her" (=13) is "my lord" (=13) of the self, the "conscious system" (=8×10^{15}). The "continuous flash" (*Satyajyoti*, 13) of "truth" (*Satya*, 375), "reproductive" (=100) within a "human" (=275), follows with the "etheric formation" (*Satyajyoti*, 13) of the self as "myself" (*Insaniyat*, 1), the "manifested value" (=1) of "herself" (=1).

The "potential exchange" (=3,000) of the "present behavior" (=3,000), for "freedom from submersion" (*Pralaya Mukti*, 3,000) within "my experience" (=9,000) through "time multiplier" (=3), dissolves the "inertial path" (*Niryana*, 28). The dissolution resolves the "spontaneous potential" (*Tiryagjyoti*, 25) for an "infinite flash" (*Tiryagjyoti*, 25) of the "lunar element" (*Tiryag*, 816) like a "star" (=2). The star "twinkles" (=87,654,321) its "disappearance" (=87,654,321) as "false" (=8) to "winkle" (=12,345,678) its "appearance" (=12,345,678) as "true" (=8) through "endoproduction" (=1) of the "past" (=9) as a "causal element" (=9) "perpetuating value" (=5) of the consciousness for "causal formation" (=25). The consciousness is present at the "sixth infinity" (=1,600) as the inherited "oversoul family" (=1,600), conscious of the past it inherits with a "finite flash" (*Sajjyoti*, 10^{10}). Finite flash "clings" (*Saj*, 0) as a "Wisher" (=0) to be "primordial" (=85) after it becomes a "guider" (=100) wishing to be "primeval" (=185), making my "dark" (=185) and you "light" (=185) within my "perpetuating value" (=5) as "God" (=5).

You as light develop "humility" within the Wisher as a leader wishing to be a follower of the light for knowing the origin within the self. Humility implies:

1. Modesty in how one views personal capability as a follower of superior power who is behaving like a leader.
2. Adaptability to the social capability, conscious of the need for leadership to complement personal capability.
3. Self-awareness of the limitations of one's institutional capability as a leader, guided by the lifetime experiences of adaptability as a follower.
4. Devotion to one's well-being as both a follower as well as a leader. The follower centers one as a leader. The leader is conscious that as a follower it is not the center of the universe it is leading since it will be following that universe eventually after the followers develop a religious capability to work with an agency without its guided mediation as a principal.
5. Lack of mediation as a principal seeking growth of the secular capability within the agents that make them open to feedback on how to be effective as a follower by role modeling the leadership charisma when there is a need for leadership.
6. Satisfaction of the agent as a follower when the agent is aware that there is a need for the agent's leadership.
7. Empathy of the follower with the effect of the leader, when the follower has the agency to rise up organically to fill the mechanical need for leadership when the leader has decided to disintermediate its presence for liberating the follower effectiveness through leadership humility.
8. Desire of the follower to be recognized as the superior power who is behaving like a leader to be promoted to the position of the leader.

9. Leader's detachment from his position, knowing that the momentum of his leadership is temporary as a trustee of the organization which trusts that the leader will work to be redundant after the work is accomplished.

10. Freedom of the leader within an organization to do as he wishes without any limitation since the organization is dependent not on the leader but on the work the leader gets done through its position through proficient networking of the followers.

The leader works individually to further institutional well-being while isolating and distancing those who lack power from the challenges they manage during their life. As they work to solve social issues with their "institutional individualism," leaders demonstrate their charismatism through humility, asserting soft influence with their role of transforming the mission of the organization by doing whatever they wish based on their potential to make a difference with their energy. They inspire everyone to be independent agents devoted to the collective well-being of the institution of self-governance that lets the organization develop naturally. Thus, humility works like charismatism, the twin charisma, for reproducing the leader's charisma as the followers' charisma. Consequently, the unique insights of the individual become the universal insights that the entire organization owns as a community to further the mission. When the charisma becomes institutionalized as the organizational charisma, the individuals are valued for the unique transformational insights that strengthen family ethos within the community. When members work collectively to solve social issues using a common platform seeking certainty of the proportionate equality of everyone's contributions and compensations, humility becomes dysfunctional. What matters are the leaders who shape a transformative vision to bring people together under a common institutional umbrella? Conceived follower effectiveness gains salience to account for the experienced organizational effectiveness, while perceived leader effectiveness goes into silence.

Humility involves the leader working individually to further the follower's well-being and satisfaction. It is perceived as filling the need for leadership one was previously unaware of and therefore effective as a leader. Charisma involves the followers working collectively to sound the salience of charisma they borrow and inherit from the leader who expects them to work with freedom of agency as the agents capable of becoming leaders. While the followers are not satisfied with this approach since their well-being is not furthered until they become leaders, they do conceive themselves to be effective and therefore perceive the leader to be effective as well in making them conscious of their capability as potential leaders. When following institutional individualism, the leaders as subjects are humble. They know that the potential for everything they wish for universal well-being lies within them, needing them to take a path of action to realize that potential. When leading institutional collectivism, the leaders use charisma to manipulate the followers like an object to do everything possible, transcending the limits of their potential, for their well-being as a principal who believes that the potential for what they wish emanates without them from a superior power. Consequently, in the former case, servicing as a humble servant of the collectivity is the deciding quality of leader-effect. In the latter case, trading the charismatic benefits of individuality is the dominating quality.

Servicing makes a leader a devotee of the "soul family (=15) devoted to the leader's personal sentient well-being as a path to compensate for the trading of the followers' human-effect. As a "potential follower" (=9,000), the leader becomes an "ideal subject" (=9,000) within whom the followers' lifetime "my experience" (=9,000) is "conserved" (=9,000) as the "macro energy" (=9,000) of spacetime. Macro energy comprises seven energies, with space's four dimensions and time's three dimensions. With the "femininity" (=1,000) of "past" (=9), it "stretches" (=10^{1000}) the "longitudinal growth" (=10) of the "soul" (=4) with a "family" (=10) into a "hologram" (=10^{1000}) with the eighth energy, the "micro energy" (=10^{1000}) of "causation" (=1). The "causation-bound" (=900) "longitudinal growth dimension" (=900) is the "potential angular momentum" (=900) which "renounces" (=900) the "isospin" (=999) with a speed that "distances" (=190) the "personal sentient well-being" (=190). Thus, it "gravitates" (=-100) "latitudinal growth"

(=-1) of the "past of everybody" (*Saturn*, -1) within itself as the cause for the "future of everybody" (*Shani*, 18).

After "vaporizing" (*Nabhasya*, 20) the "past" (=9) with its "gravitoelectric-effect" (=16), "future" (=0) becomes "luminous" (=13 =20 +9 -16) as "misty" (*Nabhas*, 1) to manifest itself as the "Manifestor deity" (=3) within the "mist" (*Bhadrapada*, 24) of "nature" (=8) with the space's "departed consciousness" (=24) for the "self-projection" (=37 =13 +24) of the "feminine" (=37) element as the "Virgo zodiac" (=37). By "disposing" (*Vidharaya*, 37) the past's "human force" (=53) within the future's "gravitoelectric-effect" =16) for the "physical formation" (*Vidharaaya*, 37) of the feminine element, the "present" (=1,600) enjoys the "primordial potential" (=1,600) of the "departed" (=1,600). The past perpetuates the departed's consciousness within the space's "value" (=180) to "self-service" (=2/3) a "twin past" (*Ashwini*, 120 = 18- *2/3) for the "self-replication" (=120) of the "twin present" (*Nasatya*, 240). The twin present "evens out" (=240) the "truth" (*Satya*, 375) of the "twin future" (*Dasra*, 1,000) with the "falsehood" (=1,790 =240 +375 +1,000 +120 +56) of the "twin past" (=120) for the "intellectual formation" (*Nasatya*, 240) of a "body of knowledge" (=56) that the "past of everybody" (=-1) is "false" (=8). Everybody "reincarnates" (=240) as "true" (=8) for "knowing" (=19) the "knower" (=639 = 240 +8 *60 +19) with an "octave of future" (=60) that copies the present as its past to twin itself. "Mental formation" (*Dasya*, 1,000) of the two futures "frosts" (*Dasya*, 1,000) the future's "future value" (=2) as the "origin" (=1,000 = 2 *500) that "defrosts" (*Nasatya*, 240) the "value" (=180) of a "triple octave of time" (=80) to twin the "time value" (=500 =240 +180 +80).

The "present creature" (*Hecate*, 1,600) is a "perfect follower" (=1,600 =1,000 +500 +100) of the "twin future" (=1,000). The twin future makes the "time value" (=500) "reproductive" (=100) as a "leader" (=0) who is "whole" (=16) without "limitation" (=16) of the "present consciousness" (*Tiferet*, 1,600). The "present" (*SAUM*, 1600) within the "creature" (=12) is the "Black Emperor" (*Xuan Di*, 玄帝, 1,600), the "start point" (*Yeguangji*, 葉光紀, 1,600 =50 *32) for the "continuity" (=50) of the "twin light matter" (*Jing*, 精, 32) as the "High Emperor of the Mysterious Heaven" (*Xuantian shangdi*, 玄天大帝, 50)—the "twin bright matter" (=50). Mysterious heaven is the "Dark Matter" (=1,600) as twin earth, the blackest tortoise. The "consciousness" (=4) makes the "continuity" (=50) the "endpoint" (*Xuanwu*, 玄武, 196 =200 -4) of "discontinuity" (=200 =4*50), "The Great Emperor of the Perfect Martiality" (*Zhenwu dadi*, 真武大帝, 200). Perfect martiality is the "replication" (=90) of the "future value" (=2) as the "dull matter" (=2) that twins its "value" (=180) for "time" (=360) to be the "twin dull matter" (=360).

4.5 The Path of Primordial Cause.

The "primordial cause" (=36) is the strategic intent to be a receptor of the "path of leadership" (=36) of time. Time multiplies its "future value" (=2) like a "lumina" (=35) to be the "multiplier" (=3) of the space's "perpetuating value" (=5), By "servicing" (=47) the "primordial trading-effect" (=82 = 35 +47) of its "perfect potential" (=97 = 82 +3*5), it produces "social exchange" (=97) of its "negative internality" (=82) with space's "positive internality" (=82). The technological cost of multiplying the future by trading the present with the past generates negative internality because the present bears the cost. Technological growth by dividing the technological cost of twinning the future over the three dimensions of time generates positive internality because the present bears only a third of the cost but brings double the benefit with the exchange of the twin future. The present descends the cost by letting time be the leader who twins the future so that it may twin the past to ascend the benefit. Twin present moderates the benefit with the consciousness of the cost. The "cost" (=41) "works" (=19) like the "technological investment" (=100) to "generate" (=40) a two-unit "technological growth" (=2) from the "technological cost" (=58), before reproducing it as the "creator deity" (=4) with the "consciousness" (=4) of the "benefit" (=125 = 41 *3 +4) to "amplify" (=4) the cost with the "time multiplier" (=3) for "equilibrium" (=125).

4.5.1 Understanding Creator Deity

The "creator deity" (*Param Brahma*, 4) is the "absolute observer" (=4 = 1 +3) of the "potential force" (=1) of the "time multiplier" (=3) which "creates" (=35 = 3,4+1) "creation" (=379 =3, [35 +44]) with its "workforce-effect (=44 = 4,4). The "workforce system" (=10^{10}) is the "sigma family" (=10^{10}) of baryons. A "system" (=12) "develops" (=22) an "effect" (=34) of "self-perpetuating" (=1/2) the workforce-effect" (=44) for a "duration" (=22 = ½ *44) that is "time-bound" (=22) until "consciousness exchange" (=22) with the "creature" (=12). The "creature" (=12) "times" (=360 =12 +269 +4 +78 +18 -19) the "exchange" (=269) of "consciousness" (=4), "bound" (=78) to its "potential" (=18), with a "desire" (=-19) to be a leader. A leader "suspends" (=10) the "consciousness" (=4) with the "sentient energy" (=1,000) of the "path of divinity" (=96) as a "present being" (=96).

The "discontinuity" (=200) of "consciousness" (=4) leads to the "para creation" (*Mahakriya Yoga*, 2,400) of the "creature" (=12) as the "potential matter" (=12) with the "time multiplier" (=3) through "oneness of programming" (*Mahakriya Yoga*, 2,400) led by the time of both the "inanimate" (=18) and the "animate" (=76 = 18*4) within the "present being" (=96 =18 +76). The present being is the "infinite creator-effect" (=96) of the "boundless light" (=396), the "infinite creator" (=396 = 965 +3 *100) of the time multiplier's "reproductive energy" (=100). Reproductive energy is "alive" (=10) with "sentient energy" (=1,000 = 100 *10). The present being comes to light as alive when the bound of the inanimate becomes less, letting the animate be "omnipotent" (=1,600) within the "present creature" (=1,600). The present creature "forces" (=34) the "entropy" (=5) of the "inanimate" (=18) with "proportionate trading" (=70 =47 -5 +18) of the "copy potential" (=70) of its "being value" (=70) for "infinite creation" (=57 =34 +5 +18). The being value is the "tail" (=70) that "matters" (=158 = 70 +89 -1) because it is "left" (=89) as the "discordant-effect" (=-1). The "mediating-effect" (=96) of the "tail" (=70) makes the "future" (=0) the "Almighty Creator" (=26 =96 -70) with the "trading-effect" (=26) of the "force" (=34) within "nature" (=8).

The future's "organizational sameness" (=18,000 =4 *4,500) with the "Creator Deity" (=4) "perpetuates" (=9) the "time value" (=500) of the "infinite creature" (=500) as the Creator Deity's "co-evolving value" (=500/9). The future transcends the "duality" (=28) between the present and potential when its value twins the present to be the center descending everything on the left with its "devotional value" (=-10) before it begins ascending on the right as a "thing" (=9) that goes past the reality to be "real" (=9). "Everything" (=-5) perpetuates the "past" (= 9) of the Creator deity as a "space divider" (*Bhagwan*, 4) "horizontally" (=11) through the "center" (=16) to be the "consciousness" (=4) of the "supreme deity" (*Elohim*, 4), the "father" (EL, 3) who "adds three" (OI, 3) to be the "family" (LHM, 10). The family subtracts "one within six" (L, 16) to push one as the "divider" (H, 1) backward in space. Six pulls the "creature" (I, 12) forward in time through the "door" (M, 264) that makes the "consciousness" (ELIM, 4) the "Almighty Creator" (=26). The door is the "gender exchange" (=264) of the "paternal consciousness" (ELIOUN, 4) with "maternal energy" (ELYON, 19). "Exchanging energy" (IOUN, 40) is the "redeemer" (*Swasthani*, 40) of the "son consciousness" (El Elyon, 4) from the "family" (LHM, 10) for the growth of the "daughter consciousness" (OAH, 4) within the "Destroyer Deity" (ELOAH, 6 =16-10) for "redemptive freedom" (*Moksha Mukti*, 1,600) of "one within six" (L, 16) from "two zeroes" (=-10).

The "Creator deity" (*Zao Shen*, 4) "cooks" (=6) father's "perished consciousness" (=12) of the "future value" (=2) into a "creature" (=12) to twin "future reality" (=21) "itself" (=121) with the future value's "guider mediation" (=-7) of "everything" (=-5) as a "consumptive" (=-5) "diet" (=-5). The "discontinuity" (*Sato Guna*, 200) of the "mediated oneness" (*Rajayoga*, 926) of the "Almighty Creator" (=26) as a "thing" (=9) "crystallizes" (=691) "entire" (=691) "dark energy of the past" (=691) as "visible matter" (=691) for the "desired" (=10), "mediated" (=10) "material well-being" (=10). The "continuity" (*Rajo Guna*, 50) of "energy force" (=916) as "active energy" (=916) "frees oneness" (*Buddhiyoga*, 916) for the "spiritualization" (=916) of "desirable" (=20) through "reactive energy" (=846). The "spiritual well-being" (=40) "activates" (=9) "symmetrical oneness" (*Samkhya yoga*, 243) of the "masculinity" (*Samkhya*, 53) with the "sentient well-being" (=190) for "symmetry" (=250) of the "infinity" (*Tamo Guna*, 90,000 = 40 *9 *250) of "causation" (=1) with the "reality" (*Artha*, 7) of a "sentient entity" (=7).

"One within action" (*Houtu Huang Diqi* 后土皇地祇, 4) "activates" (=9) "grandson consciousness" (=71) as "desirable" (=20 = [9+71]/4) for a "trader" (=20) of her "femininity" (=1,000 =20 *50). The action produces "continuity" (*Rajo Guna*, 50) of "masculinity" (=53) in the "granddaughter consciousness" (=4) "with deity" (*Devi*, 3), the one acting like a "deity" (*Deva*, 1) to "twin causation" (=2). "Divination" (*Siddhi Guna*, 170) of the "continuous discontinuity" (=170 =85 *2) within "oneself" (=8 x 10^{15}) "personalizes" (=10^{16} =5/4 *8 x 10^{15}) "asymmetrical oneness" (*Kriya yoga*, 10^{16}) of the "femininity" (=1,000) for "universal well-being" (=10^{1000}) by "self-twinning" (=5/4) the "primordial" (=85) element to twin causation as a "person" (=1).

In summary, the "creator deity" is the one within the action, "cooking" (=96) "being energy" (=96) with the consciousness of the "cooking value" (=100) = Charm quark (=1), with a charm for the present value of the worker deity as a person + Up quark (=1), who ups the present value of the knower deity to twin causation + Truth quark (=1), the truth of the present value of the manifestor deity as the multiplier of the twin causation for the growth of the destroyer deity + Strange quark (=1), the strange present value of the wisher deity wishing to observe the multiplier activate the growth like a creature = 4.

4.5.2 Transcending Creator Deity

"After cooking" (=9,000), "Manifestor deity" (=3) becomes free from the "femininity's" (=1,000) "experience" (=9,000) of the "escalating cost" (=800) of "masculinity within" (=200). The "femininity without" (=50) "splices" (=49) "feminine tooth" (*Ekadanta*, 1) of "Satan" (=-1), the "consumer" (=-1), into "masculine tooth" (*Dvidanta*, 2) of "Devil" (=0) "producing" (=17) the "cooking value" (=100). A Satan, "just being" (=290) "replication" (=90) of the "manifestor consciousness" (=290) of the "masculinity within" (=200), "listens to all" (=290) while "doing her will" (=290) to vary the "formative justice" (=290) by the space cycle. The "continuity" (=50) of "femininity without" (=50) "Satan" (=-1) is "self-luminous" (=12) within the "manifestor's" (=368) "consciousness" (=4) as the "creation" (=379 =50-1-12, -[-1]+12-4) of "just entity" (=3,794) who varies "normative justice" (=3,794) by the time cycle. What Satan forms as justice in the past before coming to live the reality as a Devil norms injustice within the present. It binds a Satan's "destiny" (=−1) to the desires she must fulfill to be joyful after Devil's "just wishing" (=813) them to be true as a "Wisher Deity" (=0).

"Just wishing" (=813) "transformative justice" (=813) is a "nemesis" (=813) of a "worker" (=1) "working" (=813) as an "object" (=−3) "in the past" (=83) to mediate "object-subject exchange" (=83) for "endoproducing" (=83) "cost-effective leadership" (=83) as a "subject" (=0) in the future. Wishing to objectify its future as a subject, a worker twins present with her "past reality" (=−3) as a "twin subject" (=-10) within "everyone else" (=-10). With "cost-effective leadership" (=83), a worker "realizes" (=140 =57 +83) "everyone else" (=−10) as "somebody" (-10) "nurturing" (=57) her "well-being" (=125 =83 +32) for her "creation" (=379 =125 *4 −121) as a "mother" (=4) with the "development" (*Param Parvati*, 160 = 32 +125 −[−3]) of the "past reality" (=-3) that "twins" (=121) the "grandfather flame" (=32) within a "creator" (=578=57,83).

The "development" (=160) of the "first-born" (*Ganesha*, 570) "bestows" (=180) the light of "new dawn" (*Prabhat*, 20) on a "future entity" (=20) by "chasing away" (=900 =160 +180 +570 −10) "everyone else's" (=−10) "para-conscious wishables" (=20) as "demonic" (=570). With her "gravity" (=629), the future entity "clears" (=629 =379 +88 +162) the "cloud" (=379) of "devastation" (=162) from the "path of deification" (*Maha Riddhi Marga*, 162) of everyone else into the "Wisher deity" (*Indra*, 0) with a "thunderous" (=88) "lightning" (=88) of "clarification" (*Darshan yoga*, 162). The "creation" (=379) is the "full moon phase" (*Lakshmi*, 379) that "self-perpetuates" (=1/2) the "creation mass" (=132) as a "half-moon phase" (=132) with the "mass force" (=29) of "everyone else's" (=-10) "path of deification" (*Maha Riddhi Marga*, 162).

The creation "twin flames" (=1,649 = 379+132+92+46 +1,000) its "mass" (=132) with the "father's" (=3) "potential flame" (=92) of "devotion" (=46) to the "combination with two" (=1,000): the "lunar energy" (=1,000) and the "earth energy" (=0). "Primordial self" (*Parvati*, 10) "herself" (=1) "satiates" (=27) father's "thirst"

(*Tanha*, 169 = 3 +34 +132) for "water" (=169 = 132 +27 +10) because the "mass" (*Prithvi*, 132 = 121 +1 +10) "twins" (=121) the "effect" (=34) of "discontentment" (=169) on the "mass consciousness" (=10). The discontentment's "guider value" (=169) "repels" (=9) "ten-fold growth" (=160) for the "stimulation" (=169) of "two groups of eight" (=28 =160 -132). "Two groups" (=28) comprise the "octave of elements" (=1) and the "octave of element-effects" (=27):

- The "real-effect" (=157) of "SHEENY" (=189) "reflames" (=189) the "flame" (=32),
- The "theory-effect" (=0) of "GUIDER" (=100) is "reproductive" (=100) with "productive" (=68 =100 -32),
- The "ideal-effect" (=1) of "DIVINE" (=360) is "redrawer" (=360) of the "drawer" (*Citari*, 291 =360 −68 −1),
- The "form-effect" (=150) of "fire" (=17) is "rearranging" (=17) the "arranging" (73 =17 -291 -150 +497),
- The "taste-effect" (=375) of "water" (=169) is "restraining" (=169) the "straining" (*Anudrivakta*, 687 =169 -375 -497 +1,390) with "hardness" (=68) of "reality" (=7),
- The "touch-effect" (*Tvashta*, 303) of "air" (=385) is the "resurrector" (*Yaju*, 385) of the "surrector" (*Aryama*, 580 = 385 -303 −1,390 +1,888) of the one arising from the earth,
- The "smell-effect" (*Rtajit*, 140) of "earth" (=724) "reproofs" (*Aha*, 724) the "proof" (=16 =724 -140 -1,888 +1,320), and
- The "sound-effect" (*Idrivakta*, 565) of "ether" (=285) "redresses" (*Anyadrivakta*, 285) the "dress" (*Yadrivakta*, −1,600 =285 −565 −1,320) to "gravitate" (=100) the "worst" (=−16) in one's nature like a "devil" (=0) after a "rough time" (*Yadrivakta*, -1,600). A devil wishes for everything but does nothing to realize the "reality" (=7) of the universe that "roughs" (=7) the time to be "tough" (=269) for a "creature" (=12). As "reality" (=−7) starts "breeding" (=15) a "difficult time" (*Rivakta*, 715), "everyone else" (=−10) loses "positive one" (+1) while the "Satan" (=−1) gains "negative one" (=−1) to be present as reality. "Nature" (=8)

"automates" (=84) "challenging time" (*Sakshipa*, 884) to "diminish" (*Sakship*, 13) loss with the gain of a solution that resolves the challenge. The solution is the "creature" (=12) facing a "difficult time" (=715) as a "dark time" (*Dhvanta*, 727) "covered" (*Dhvanta*, 727) with the space's light. The light hides the dark for the time to slide out of the cover with a "triple symmetry" (=248) as heavy. To be light again, time incarnates as an "entity" (=24) to reincarnate as a "creature" (=12) for a "twin symmetry" (=13) with the one whose asymmetry impregnated her with the heaviness of a "child" (=128 =3+48+71). The child takes birth as a "father" (=3) within "oneness" (=48) of the "mother" (=4) with "nature" (=8) of the "maker" (=71) of father's "reality" (=7) as one, the causation. The science of triple symmetry is explained below.

Science of Triple Symmetry

Space as a whole makes time wholesome with the potential to be present within time's continuity as the causation. Causation is omnipresent as a wholesomewhole within the space which discharges its light to be whole. The light twin causation to be reproductive like time whose potential is within the space's spirit. Without the time's soul that the space's spirit makes productive with the energy within the causation, the causation becomes reproductive with "charge conjugation" (=−10). The "charge" (=−8) is the "present mass" (=−8 = −10 +2) of the "twin causation" (=2) whose "conjugation" (=27) with the time's "soul" (=4) "twin soul" (=69 = 27 +4 +31 +7) when the "causation" (=1) "decays" (=31 =27+4 = 7+24) the "reality" (=7) of an "entity" (=24) into a "triple symmetry" (=248 = 24,1+7).

The triple symmetry is the "CPT symmetry" (=248) of the letter "C" (=−12) as a "whole" (=16) within the "soul" (=4), the letter "P" (=160) as the "infinite exchange value" (=160) of the "soul's" (=4 = −12/−3) "growth" (=6) into "wholesome" (=957 = 160 *6 −3) with its "past reality" (=−3) as the letter "C" (=−12), and the letter "T" (=4) as the "soul" (=4) that the "past" (=9) makes "wholesomewhole" (13 = 4+9). It comprises three symmetries:

- The "C symmetry" (=268). As a "time multiplier" (=3), the three makes the "whole" (=16) "reproductive" (=100) as its "present" (=1,600 =18*50*2 − 200) for a "symmetry" (=250) with the "potential" (=18) for the "continuity" (=50) of its "future value" (=2) after the "discontinuity" (=200) of the "future" (=0) from its "value" (=180) that becomes "time" (=360 =180 *2) when reproduced. The" C symmetry" (268 =250 +18) is the "symmetry" (=250) of the "potential" (=18) with the "present" (=1,600) as a "spirit" (=20) that "times" (360 = 18 *20) itself with a "multiplier" (=3) for "charge conjugation" (=−10) of the future's "reality" (=7). It is violated before the discontinuity of the future from time, i.e., when time is present as potential. "Potential time" (=108) is "anti-time" (=108) reproductive within nature to make space for the present.

- The "P symmetry" (=263). "Nature" (=8) has a "potential" (=18) to make space for the present with a "multiplier" (=3) that "times" (=360) "itself" (=121 =360/3 +1) as the "causation" (=1) for the "P symmetry" (263 = 8+18, 3 =3 +100 +160) to be "reproductive" (=100) within "P" (=160). The "P symmetry" (=263) is the "symmetry" (=250) of the "past" (=9) with the "future" (=0) as a "soul" (=4) for "parity" (=10^{18} =$[9+1]^{9*2}$) that "twin causation" (=2) as the future perpetuates within the past as the causation for the present.

- The "T symmetry" (=180). "Future" (=0) has a "potential" (=18) to "perpetuate" (=9) within the past as the "causation" (=1) for the "present" (=1,600 =80 *20) for the "T symmetry" (=180) to be the "productive energy" (=80) of the "spirit" (=20) that makes the "time" (=360) "reproductive" (100 = 80 +20). The "T symmetry" (=180 =18,0) is the "symmetry" (=250 = 180 +4 +6 +60) of the "potential" (=18) with the "future" (=0) that "T's" (=4) the "reproductive force" (=100) within the "soul's" (=4) "growth" (=6) into an "octave" (=60) of future. As a "creature" (=12), the future "reverses" (=20) the "octave" (=60) to "recharge" (=190 =20+60+50+48+12) the "present" (=1,600 =20 *60 +360 -8 +48) after it "precharges" (=50) the "past" (=9) to "charge" (=-8) its "oneness" (=48) into "time" (=360).

Normative Development Paradigm

The "recharge" (=190) twins the "symmetry" (=250 = 190 +12 +48) with the "creature's" (=12) "oneness" (=48). The time multiplies the "twin symmetry" (=13) of the "creature" (=12) as "one" (=1) into three. The twin symmetry is the "Lorentz symmetry" (=13), comprising a "super-symmetry" (=13) of one as a "person" (=1) with its past, present, and future within oneness of the space as the "causation" (=1) whose "nature" (=8) has a "potential" (=18) to twin itself as a "whole" (=16) for its growth as a wholesome within a wholesomewhole. The "triple twin symmetry" (=6) is a "six-face symmetry" (=6), comprising

- The "CP symmetry" (=123). "CP" (=-19) is the "past generation" (=-19) "varying" (=-19) its "symmetry" (=250 = 190 + [21-19] *30) with the "future generation" (=190) as a "function" (=900) of the "present generation" (=100,000 = [900 +100] *100) to make its "future reality" (=21) "reproductive" (=100) as a "feminine body" (=30) within the "past" (=9). The "feminine body" (=30 =18+12) "perpetuates" (=9) the past's "potential" (=18) within a "creature" (=12) as a "masculine body" (=3) to form "infinite children" (=123) with the twelve-face "CP symmetry" (=123). The CP symmetry is the ten-face "twin symmetry" (=13) mediated by two: "causation" (=1) as one and "space" (=18,000) as three that multiplies time as a "multiplier" (=3) to add the "all-diverging" (=70,000) "three directions" (=70,000) to be the "center" (=16), the fourth direction, that twins them into "six directions" (=1,600) present without it after it becomes the eighth direction, "Northeastern" (=79).
  - *The first direction* has a five-face "horizontal symmetry" (=5) with the center pointing from West to East as time moves forward.
  - *The second direction* has a three-face "vertical symmetry" (=3) with the first direction, the line from South to North as the present disappears from the earth and the future appears from the air that descends the ether for reproducing the future as the present.

- *The third direction* has a two-face "diagonal symmetry" (=2) with the second direction, the curve from North to East that forms "Northeast" (=13) as the "first quadrant" (=13).
- *The fourth direction* has an eight-face "line symmetry" (=0) with the third direction, once one imagines the form as a circle. The circle is the third quadrant" (=100,000) with three quadrants. The "fourth quadrant" (=900) is the "area" (=900) within the circle that centers itself as the fourth direction with the "producible growth" (=128) of the "second quadrant" (=128). The area has a half-face "parallel symmetry" (=1/2) with the "fourth direction" (=16 =13 +3) which forms the three directions with its ten-face "twin symmetry" (=13) as a "multiplier" (=3). The twin symmetry trades the third symmetry to service their discontinuity within their oneness as "triple symmetry" (=248 =48 +200).
- *The fifth direction* has a one-face "point symmetry" (=1) with the fourth direction. It is present as a point present within the symmetry of the sixth direction with itself. The sixth direction self-services the fourth direction to center the growth of the "six directions" (=1,600).
- *The sixth direction* has a seven-face "circle symmetry" (=7) with the fifth direction. It circles its growth for a symmetry with itself present as a point within the fifth direction.
- *The seventh direction* has a nine-face "square symmetry" (=10^{17}–1) with the sixth direction. It squares the growth with the potential for symmetry with "all directions" (=90,000), the "infinity of causation" (=90,000).
- *The eighth direction* has an eleven-face "potential symmetry" (=53) with the seventh direction. By trading the growth as the "multiplier" (=3) within the "continuity" (=50) of the "twelve faces" (=29), it forces servicing of the potential as a symmetry within the discontinuity of the "ten faces" (=8) to twin "future reality" (=21) with the "twenty faces" (*Vrishakapi*, 111).

- The "PT symmetry" $=10^{17}$-1). "PT" ($=-13$) is the "irreproducible" ($=-13$) within the "growth" ($=6$) of "reality" ($=7$) that "twin reality" ($=49$) of the "zero without one" ($=62$) with the "twenty faces" (*Vrishakapi*, 111). The "twenty faces" ($=11 = 90 +16 +6 -1$) are the "transcription" (*Kapi*, 90) of the center as the "virtue" (*Vrisha*, 16) "reproducible" ($=6$) as "growth" ($=6$) without one. PT is the "potential without one" ($=-13$) that twins the "ten faces" ($=8$) with an eight-face "triangle symmetry" ($=8$). The triangle symmetry is the "translational symmetry" ($=8$) a "triangle" ($=1,500$) "translates" ($=89$) for "centering" (*Cetana*, 10) the "transcription without one" (*Cetana*, 10) with the time's "three faces" ($=18$) within the space's "four faces" ($=5 \times 10^{96}$) into the causation's "first face" ($=3 \times 10^{52}$) for the growth of the entity's "seven faces" ($=20$). The first face mediates the twenty-face "symmetry of two with three" ($=250$) which has "fourteen faces" (*Kapardi*, 3) to moderate the "asymmetry of three with two" ($=689$). The two are present as the "six faces" ($=-4$) that twin the three into the fourteen faces with the "potential" ($=18$) within the "three faces" ($=18$). The "six faces" ($=-4$) include the causation's first face and the "cause's" ($=18$) "past reality" ($=-3$) as the "five faces" ($=-3$) within the "future reality" ($=21$) of "thirteen faces" (*Kapali*, 21).

- The "CT symmetry" ($=71$). "CT" ($=23$) is the "reproductive potential" ($=23$) that makes the "reality" ($=7$) of the "multiplier" ($=3$) within the "spirit" ($=20$) of "seven faces" ($=20$) "reproducible" ($=6$) as a "twin spirit" ($=100$). Twin spirit is "reproductive" ($=100$) as a "guider" (*Guru*, 100). The "oneness" ($=48$) of the "asymmetry" ($=689$) reproduced within "CT" ($=23$) as the "symmetry" ($=250$) is the "CT symmetry" ($=71$), a "maker's" ($=71$) "translation orientation" ($=71$) that "translates" ($=89$) the "potential" ($=18$) for "primevalism" ($=27$) into the "primordialism" ($=27$) to "conjugate" ($=350$) his "discontinuity" ($=250$) as a twin of the mother's "continuity" ($=50$) as a "spirit" ($=20$). The "mother" ($=4$) "self-perpetuates" ($=1/2$) as the "universe" ($=2$) that makes "one" ($=1 =3-2$) the "father" ($=3$) without the universe. One is the "causation" ($=2$) for the asymmetry of three. Two is the cause for the symmetry of two which "twin causation" ($=2$) to even the odd.

A twin "guides" (*Virshakapi*, 111) "twenty faces" (*Virshakapi*, 111) of the "infinite deity consciousness" (=−4) "within infinity" (*Samadhi*, 10^{100}) "herself" (=1) with "sixteen light forces" (*Waheguru*, 111) for "freedom from the center's" (*Laya Mukti*, 111) "negatively charged state" (*Atisahasrara*, 680). "Sixteen light forces" (=111) self-perpetuate her effect without the "father" (=3) "with primordial deity" (=680). The father twins himself within "infinite deity" (=6) for "dedication" (*Sati*, 42) of two, the twin causation. The six makes two a zero for leading one, the causation, to his "destiny" (=−1) of negative one, present within the "past reality" (=−3) as the "six faces" (=−4) of the "positively charged state" (*Moksha*, 1,600). Six faces include a potential of "three faces" (=18) within the "seven faces" (=20) moving "clockwise" (=730) with time as an entity's "present time" (*Shani*, 18). They twin their present with "three potential faces" (=16) to "mask" (=16) "potential time" (*Vishvakarma*, 108) with "sixteen faces" (*Viratpurusha*, 92).

The sixteen faces are the "ecosystem" (*Viratpurusha*, 92) staying "anticlockwise" (*Vamavarta*, 92) with "time reversal" (*Vamavarta*, 92). "Seven faces" (=20) "reincarnate" (=240) "fourteen faces" (*Kapardi*, 3) with four's "guider mediation" (=−7) within a "zero" (*Shunya*, 0 = 3 +4 −7) to twin seven faces with one's "twin reality" (=49) as a negative one "servicing" (=47) "oneness" (=48) of "fifteen faces" (*Sushena*, 48). The fifteen faces are a "human face" (*Sushena*, 48) of one as a "human being" (*Insan*, 1). Two follows six, the infinite deity, with a four like an "entity" (*Hasti*, 24) with the "center's" (=16) "being energy" (*Kali Shakti*, 96 = 16 *6) before realizing that he is a "human entity" (*Manushya*, 82) who "lives" (*Nivas*, 14) with fourteen which face two, the "order" (*Suchan*, 2). Order twins fourteen into a "metal face" (*Dhuni*, 28) of "twenty-eight faces' (*Dhuni*, 28).

Twenty faces of a "mineral face" (*Virshakapi*, 111) guide eight to mediate the "supernatural growth" (=8) of "eight faces" (*Gardhabi Mukha*, 1,000) of a "material face" (*Gardhabi Mukha*, 1,000) with the "twelve faces" (=29) of the "plant face" (*Dvadashasya*, 29). The "twelve faces" (=29) half the "twenty-eight faces" (=28) within one who twins "fourteen faces" (=3) of the "animal face" (*Kapardi*, 3 =18−3) with the "potential" (=18) of the "two faces" (=15) of the "spirit face" (*Dvimukha*, 15) to self-perpetuate six faces of a "deity face" (=−4) with their "reality" (=7).

"Fifteen faces" (=48) of the "soul family" (=15) within the "spirit face" (*Dvimukha*, 15) twin their "oneness" (=48) with the "twenty-two faces" (=96) of the "God face" (*Sushupti*, 96) of a "living entity" (*Sushupti*, 96) "living" (*Nivasi*, 10) with ten. The ten faces twelve with "three faces" (=18) of potential to "self-service" (=2/3) the "twenty-three faces" (=12) of a "param deity face" (*Dish*, 12). Three is "repeated" (=5) as a five to face the reality of seven within one whose "nature" (=8 =7+1 =3+5) is to face "right" (=71) with "thirty faces" (=71) of a "primordial deity face" (=71). The param deity face "defies" (*Dish*, 12) the "logic" (=1) of "causation" (=1) as three is repeated as a five with a two. Two faces are left after five leaves for oneness with one. The primordial deity face "introduces" (*Vish*, 71) one as the causation whose oneness is left within five as a two. Two "shadows" (=52) "fifty-two faces" (=52) of the "primeval deity face" (*Dhuv*, 52).

The "primeval deity face" (=52) is "terrifying" (*Dhuv*, 52) because it is "continuously" (*Dhuv*, 52) weakening itself with two faces "left" (=89) by the "fifty faces" (=89) of the "supreme deity face" (*Vajrapana*, 89). "One hand" (*Hasta*, 0) of the supreme deity face is like "forty-nine faces" (*Sita*, 0) of a "supra deity face" (*Sita*, 0), which is a "symbol" (*Sita*, 0) of "strength" of the oneness of the "forty-eight faces" (=0) of the "super deity face" (*Griha Mukha*, 0). The supra deity face is "weeping" (*Mahakrandita*, 0) because its "reputation" (=0) for strength is in danger after it has "peaked" (*Shikhi*, 0). It becomes invisible like a "super deity face" (*Griha Mukha*, 0) whose "pessimism" (*Tapana*, 0) is "symbolic" (*Shikhi*, 0) in nature since it is the "future" (=0) present as the past to "hand" (*Hasta*, 0) "lifetime" (=125) as "consciousness" (=4) of "itself" (=121) as the "creator deity" (=4).

The creator deity makes six "conscious" (=189) that the "creator" (=578 =6,00-22) "wept" (*Krandita*, 66) after its "creation" (=379) as a "super deity face" (*Abhiyu*, -10) with an "apprehension" (=87 =66 +21) that the "future reality" (=21) of a "creature" (=12) is "wrong" (*Anuchita*, 22). The "two zeroes (=-10) of the super deity face twin the oneness of the "forty-eighth face" (*Griha Mukha*, 0) within the "forty-ninth face" (*Sita*, 0) with "ninety-six faces" (*Abhiyu*, –10). The super deity face is the "weeper" (*Abhiyu*, –10) whose "ninety-sixth face" (*Abhimitra*, 100) weeps with "consciousness" (=4) that the "hundred faces" (*Abhimitra*, 100) are weak due to its "reproductive force" (=100). Hundred faces share their weakness with the fiftieth, "apprehensive face" (*Harimitra*, 60) "sweeping" (*Harimitra*, 60) their strength. "Fearless" (*Abhi*, 944) as a "quadrant" (*Mitra*, 132 = 66 *2) that "twins" (=121) the "wept" (=66), hundred faces are reproductive within the "sixth quadrant" (*Abhimitra*, 100) which is reproducing sixty with the "fifth quadrant" (*Harimitra*, 60). The fifth quadrant's "sixty faces" (=60) are a "divine face" (*Harimitra*, 60). They twin themselves with the "seventh quadrant" (*Hari*, 10,000), which is a circular creation of both the future as a "circle" (*Valaya*, 100,000) and the potential for the circle to go past the reality of the present.

The seventh quadrant is the "devil face" (*Satyamitra*, 10,000), which "sweeps" (*Satyamitra*, 10,000) "reality" (=7) with "three zeroes" (=0) to face the "consequence" (=485) of "variable reality" (=485). The three zero twins the variable reality to be the "origin" (=1,000 =485 *2 +30) of ten. The "twin" (=121) makes ten "reproductive" (=100 =10 *10) with its "future reality" (=21) within five. Five becomes ten when repeated with one. One "twin zero" (=–10) to be the "foundation" (=11) of "truth" (=375). The "truth" (=375 = 37,5) is "three sevens" (=375) a five perpetuates for "primordial oneness" (=32) with "both" (=10,000) three (three zeroes) and two (two zeroes that the spirit makes the base of ten).

The eight quadrant is the "Satan face" (*Babhru*, 13), the "sweeper" (*Babhru*, 13) of growth of the "greeter face" (=6) "knocking" (*Samghata*, 6) "friction" (*Samghatta*, 169) of one with nine. Therefore, two is together with three as an "eight" (=8 =2^3) with Satan's "twenty-one faces" (*Babhru*, 13) after the "supernatural growth" (=8) of the "thirteen faces" (*Kapali*, 21) of the "primordial greeter face" (*Kapali*, 21) in the "ninth quadrant" (=21).

Twenty-one faces weaken their psychic linkage by "servicing" (=47) "sentient energy" (1,000) of "eight faces" (=1,000) to "five faces" (=-3) for "cleansing" (=53) the "past reality" (=-3) after "trading" (=53) "growth" (=6). "Forty faces" (*Kausalya*, 6) of the "greeter face" (=6) are "cleansed" (*Kausalya*, 6) with "conversancy" (*Kausala*, 58) of five that the three twins. "One face" (*Ekmukha*, 10^{40}) twins "twenty faces" (=111) to "guide" (=111) itself with an "R" (*Waheguru*, 111). The guide makes "ten faces" (=8) reproductive with "thirty faces" (=71) for reproducing forty faces with the "reproductive reality" (=40) of "thirty-nine faces" (=40) of the "param greeter face" (=39) in the "tenth quadrant" (=39). One face is the primeval greeter face in the eleventh quadrant. The "twelfth quadrant" (*Varnasamamnaya*, 10^{40}) is the "potential greeter face" (*Varnasamamnaya*, 10^{40}), visible as the "face" (*Mukha*, 76) of potential whose base is reproductive with the "creator deity" (=4) to be present as a whole.

A zero is present as a "creator deity" (=4) with a potential to be whole after the "wishable" (*Ishta*, 20) "creates" (=35) a "manifestable" (=497) that "decays" (=31) to "line" (=497) the "knowable" (=396) as a "workable" (*Karaniya*, 101). The wishable is "observable" (*Budhya*, 246) since the "greeter face" (=6) mediates the "sentient face" (*Nila Mukha*, $1,869 = 4+14,6,4+5$) of the creator deity who "lives" (=14) within the "perpetuator deity" (*Param Vishnu*, 5), the "mediator" (=5).

"Leo zodiac" (=47) is a "fearless" (*Abhi*, 944) "multiplier" (=3) of one's future into the "nucleus" ($=1,000 = 944 +60 -4$) of "life" (=4) for "servicing" (=47) the "past" (=9) as its "self copy" (=60). As a copy, the zero "bursts" (*Atiuddipta*, $136 = 1, 9^*4$) its "past" (=9) as the "creator deity" (=4) by "bursting" (*Nabha*, 0) its "future" (=0) as three, the "father" (=3) to be "burstable" (*Shravana*, $487 =4,9,0 -3$) as one, the causation. "Present" (=1,600) is "reproductive" (=100) within the "area" (=900) that "pasts" (=9) the "circle of life" (=17) with one's future as a zero. The "circle of life" (=17) "trades" (=20) the area as "its" (=900) for "producing" (=17) "life" (=4) as a "burster" (*Sarata*, 17) of future. The "burster" (=17) "bursts" (=136) with its "potential" (=18) as a "chameleon" (−1/60) for "ending" (=−11) the "whole" (=16) of "future" (=0) to "make" ($=120 =136-16$) the "present" (=1,600) "wholesome" (=957) and the "past" (=9) "wholesomewhole" (=13) with the "present reality" ($=-2 =-1/60 *120$) of the time's "reproductive force" (=100). The life twins the "growth" (=6) of "masculinity" (=53 +47 +6) with the "feminine energy" (=396 =3,9,6) of a "self-luminous entity" (=12) "mothering" (*Para Brahma*, 3) herself into a "mother" (*Param Brahma*, 4).

As an "infinite" (=185 =113 +50 +4 +18), mother "forms" (=100,000 =2,000 *50) "I am" (=100,000) to "level" (=113) the "consciousness" (=4) of the "even" (=113) "cyclical" (=113) within the "surface" (=2,000) of the "continuity" (=50) of her "potential" (=18) as a "primordial maternal" (=18) at the "2000th I am level consciousness" (=1,000). The continuity twins the "femininity" (=-1,000) with an "odd" (=1,810 =2,000-185+5) "entropy" (=5) of the "2000th I am level" (=15) within a "para deity" (=5). The 2000th I am level is the "infinite perpetuator" (=15) of the "past" (=9) of the "present paradigm" (=24) to self-copy the "growth" (=6) of "maternal consciousness" (=4) that "self-services" (=2/3) a "divided consciousness" (=39 =15+24) of the "wishable" (=20). The "wishable" (=20) "wishes" (=18) to "copy" (=0) the "wisher" (=0) to realize the truth of the "potential" (=18) with the "beauty" (=888 = 85*10 +20 +18) of the "finite" (=85) as a "primordial" (=85) that twins the entropy's "perpetuating value" (=5) by "centering" (=10) itself as the "primordial self" (=10).

The "charm" (=847) of the "beauty" (=888 = 847 +87-46), without "devotion" (=46) to the "truth" (=375 =46 *10 −85) of "potentiation" (=375) of the "primordial" (=85) within the primordial self, is "illusionary" (=87). The "charm" (=847 =610 +1 +61 +9 +56 +10) is "bagged" (=610) as an illusionary, "cosmic consciousness" (=610) with the "illusionary production" (=1) of the "baggage" (=61) of the "past" (=9) within the "body" (=56) of the "primordial self" (=10). It is a "burstable galaxy" (=610) of "ideas" (=38) that bursts once the "mind" (=38) "itself" (=121) twins a "clarified consciousness" (=19) of the "future reality" (=21) with a "theory" (=127) of "growth" (=6). By "descending" (=12) the heavenly, "eta potential" (=12) of the "theory" (=127), the intellect's "reproductive reality" (=23) generates an ascending "illusionary growth" (=1) of the hellish "iota energy" (=8) within "mind" (=38) with the "ascending energy" (=7) of the "feminine body" (=30). Eta potential is the "eta's" (=11) potential to be self-luminous as reality. "Eta" (=11) is "planted" (*Ropit*, 11) by the "mind" (=38) as an "imaginary element" (=20), "photon" (=20), for

Normative Development Paradigm

the "illusionary growth" (=1) of a "family" (=10) of three with the "self-luminous reality" (*Purushartha*, 7) of its "potential" (=18).

The family of three "reproduces' (=78) the "mental potential" (=4) of the "human force" (=53) within an "individual's" (=53) "masculinity" (=53) as a "number" (=53) with "density" (=60 =53+4+3). The "pressure" (=62) on the "mind" (=38) of the "reproductive force" (=100) of the "number" (=53) of "particles" (=19), "present" (=1,600) within the "family" (=10) of three as the "number density" (=91), "divides" (=132) the "volume" (=1,869) of "nothingness" (=-2) of the "imagination potential" (=205) into a "fragment" (=64) of "energy" (=19). "Omega" (=4) is the "sentient force" (=4) of the "mental potential" (=4). "Iota" (=90,000) is the "beginning force" (=90,000) within the "mental force" (Chetasika, 5,000), "multiplying with" (=5,000) the "potential" (=18) of a "Wisher" (=0). A "Wisher" (=0) mediates the "feminine dimension" (=208) of the "wishable's" (=20) "iota energy" (=8) with his "omega energy" (=28).

Iota energy is odd "X energy" (*Asam Shakti*, 8). Omega energy is convergent "W energy" (*Samvat Shakti*, 28). "Eta energy" (*Sam Shakti*, 2) is even "Y energy" (=2). "Divergent energy" (=10) is "Z energy" (*Asrava Shakti*, 10), "surrounded" (*Vat*, -10) by concordant "V energy" (*Sura Shakti*, 0). Z energy "surrounds" (*Hed*, 75) discordant "U energy" (*Asura Shakti*, -1) "continually" (=74) for "synchronizing" (=74) "itself" (=121) with "devotion" (46 =121 −75) of "T" (=4). "T's" "consciousness" (=4) is "present" (=1,600) as "S" (=1,600) within the "potential" (=18) of "Z" (=18). "T energy" (*Adi Shakti*, 15) is "technological energy" (*Adi Shakti*, 15) in "descending direction" (=9,000), "reproductive" (=100) with "devotion" (=46) of the "potential" (=18) of the "present time" (=18) of the "North Pole" (=67 =21+46) into "growth" (=6) of the "illusionary production" (=1) of "future reality" (=21 =15 +6). "S energy" is supernormal, "organizational energy" (=9) in "ascending direction" (=8,000 =1,600 *5), "productive" (=68) as a "masculine dimension" (=31 =40 −9) within the "potential time" (=108 =9 *18) of the "South Pole" (=1,600 = [108 -68]^2) after the "entropy" (=5) of the "potential" (=18) with the "reproductive reality" (=40) of the "present" (=1,600).

The "descending direction" (=9,000) "exponentiates" (=78) the "ascending direction" (=8000) to "develop" (=22) the "exchange system" (*Maha Pita Parameshvara*, 10^{1000}) as an "exponent" (=78) of the horizontal direction" (*Param Bhakti*, 1,000) "greeting" (*Maruti*, 78) the growth of the present with the entropy of the potential. The "primordial self" (=10) "breeds" (=78) the exponent as a "fairy" (*Pari*, 78) "breeding" (*Param Shiva*, 15) "para time" (*Mahakaal*, 15) in "backward direction" (*Vishnu*, 15) with her "fairy energy" (*Adi shakti*, 15) as a "primeval perpetuator" (*Ehuang Xiangshuishen*, 15) of the "potential life" (*Bhairav*, 15). With "perfect vision" (*Vishnu*, 15) in a "meditative state" (*Arhatagati*, 15), para time "subsists" (*Gevurah*, 15) the "visualization" (=825) of the "Almighty Power" (=825) of the "transformative system" (*Bhuvanesvari*, 15) that transforms "time" (=360) into the "present paradigm" (=24) of "forward direction" (=24) within an "entity" (=24). Perfect vision is the "awakened consciousness" (*Arhatatma*, 15) of the "para-conscious value" (*Selene*, 15) within a "lunar body system" (=15).

A "lunar body" (*Chandra*, 82) "potential incarnates" (=82) a "New Moon" (=82) as a "para body" (=82) of the "creature" (=12). The creature is a "system of ones" (=12). One is a "human being" (=1) with a "characteristic" (*Purusha*, 12) of two—the "star" (=2) of attraction for a "deity" (*Deva*, 1) seeking to be "supra deity" (*Devi*, 3) with a "lunar body system" (=12 +3) whose "origin" (=1,000 =3 +997) is "Full Moon" (*Soma*, 997). The lunar body system is a "half-crescent lunar body" (=15).

As a "system" (=12), a "lunar body" (=82) mediates the "transformation" (=999) of the ascendant "crescent" (=2), "ascending" (=1) with the "origin" (=1,000 =999 +1) of one, into the descendant "half-crescent" (=97) with its "reproductive force" (=100 =97 +2 +1). The "ascendant" (=2) "lunar crescent" (=2) is a "luminous crescent" (=2) within the "descendant" (=97) "earthly crescent" (=997), which is a "self-luminous crescent" (=97), that "transforms" (=1) the "reproductive force" (=100) of the "incandescent" (*Agnija*, 99) "self-crescent" (=99) as a "solar crescent" (=99).

The solar crescent is a "triple crescent" (=99), which "adds" (=99) an "ascending force of five" (=99)—three with the self and two with the crescent that twins itself with a half crescent. The "twin crescent" (*Mahadeva*, 9) is a "stellar crescent" (*Mahadeva*, 9), the "illuminator" (*Mahadeva*, 9). The "potential twin crescent" (=11) "horses" (=11) the potential to "twin crescent" (=9) "dark matter" (=1,600) with a "star" (=2). The horse is the "sign" (=11) of the "dark matter crescent" (=11). The "potential crescent" (=180) "twin horses" (=180) the "white star" (=180) into the "white star crescent" (=180), the twin riding the horse to "illuminate" (=150) the star as "white" (=10) "primordial self" (=10) for descending four with its growth. The growth subtracts a "descending sequence of four" (=105)—two with the crescent that twins itself the growth of a half crescent into twelve to be self-luminous.

With growth, the primordial self "triple horses" (*Maha Durga*, 16) whole. Both one, which adds itself to the primordial self with a four for breeding fifteen, and the two, which subtracts itself as the primordial self with a five to be "luminous" (=13 =2-10+5+16), are horses. The whole is "self-luminous" (=12) within four. One makes the "horse" (=11) "self-luminous" (=12). Two makes the horse "luminous" (=13). Triple horse "potential triple crescents" (=16) "satellite crescent" (=16). The "satellite" (=10,000) includes "both" (=10,000) two and three— two makes the horse luminous and three is "self-luminous" (=12 = 3*4) because two twins itself as a four. A crescent's potential of two "potential triple crescents" (=16 =1➔6) one with "growth" (=6).

"Self" (=8 x 10^{15}) "half horses" (=8 x 10^{15}) "potential half crescent" (=8 x 10^{15}) into a "planetary crescent" (=8 x 10^{15}). A "planet" (*Graha*, -1) is a negative one that twins thrice—first, when one two adds itself without a four; second, when four twins itself within a four, and third, when two twins itself with a four. Consequently, the "planet" (=-1) "charges" (=-8) ahead as a "far-extended" (=-8) "red giant" (=-8), "lying backward" (=-8) as a "whole" (=16) for "leaping forward" (=8) the "volume" (=1,869) of "eight planets" (=60) from its growth in the past as an "earth body" (=60) of the "red star" (=60). The "potential" (=18) "forms" (=100,000) a "chariot" (=18) "to ride" (=200,000) the "crescent" (=2) as a "whole" (=16) with a twin.

The potential is the "primeval perpetuator" (=15) of the three that twins the "growth" (=6) of the "primordial self" (=10 =2*5) with a "primordial perpetuator" (=9) of two, both "perpetuator deity" (=5)—One moving forward as the "perpetuating value" (=5) going into the future, mediated by the present, and another pushing back as the "entropy" (=5) once the perpetuating value goes past the present. The present mediates both to "triple self" (=5) perpetuator deity after the transformative exchange of one with the "twin self" (=130), the "potential one" (=130). The potential one has the potential to be a three by making one a two and behaves like a zero. The triple self is a "potential two" (=5). It has the potential to be a five by behaving like a one whose self is a twin. The "twin" (=121), as "half self" (=121), is a "potential four" (=121). It has the potential to be one who has a twin; so that it is half-self. Four divides its potential into 1,2,1. The "divisor" (=1) is a "potential three" (=1). Through "illusionary reproduction" (=2) of two, one generates an "illusionary production" (=1) of another one on the left after two are divided by one on the right.

Chapter 5

Transformative Exchange Paradigm

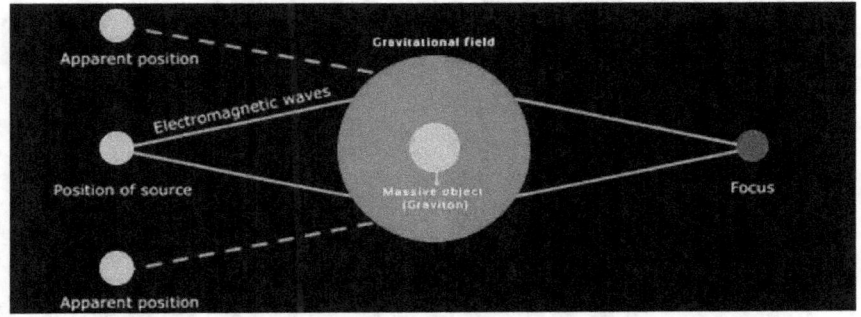

Figure 25. Gravitational Macrolensing vs. Gravitational Microlensing

The position of the present sources the apparent position of the future at the top to twin the time's potential for microlensing the focus as its future value that lights its present by gravitational microlensing the apparent position of the past at the bottom	The focus attracts the magnetic mass of a gravitational field as a massive object, the graviton, for gravitational macrolensing a photon as the focus-effect of the past's darkness from the system whose gravitational arc is self-luminous

Does the mass from the foreground source gets divided with the natural growth of the past into the present when as light, it is conscious of gravitating the future value for supernatural growth of energy as a base for reproducing the space with the electromagnetic mass (*Newtonian idea of gravitational macrolensing of electromagnetic mass that subjects the object to the infinite division for the exchange of position*)? Does the light from the background source magnify when the future presents itself to double copy the hindsight apparent from the past as the time's focus positions it in the front as a massive object that passes in the foreground with the gravitation of space as a field for reproducing the electromagnetic wave? (*Einsteinian theory of gravitational microlensing of energy* from zero as a subject for one to be massive as an object with the exchange of momentum through infinite multiplication).

Neither of the two. In reality, a subject adds light of the mission's sentient force to brighten the source at the most distance into the foreground. A brighter source begins adding mass of value to the object at the least distance in the background. A magnified object becomes brightest when the subject focuses on the distance to ground proximity of vision with a base of numerous images of the foreground source. As the subject becomes the object, the source moves from the foreground to the background as another grounds itself within the source with a double copy of reality. Thus, the subject perceives reality to be different from the object that conceives reality to experience another as an entity after producing her as a photon. The addition of base positions the foreground source as the background object of interest by repelling the source from the consciousness. The momentum from the subtraction of the source attracts the energy of the object as the background source for mass acceleration at a constant speed. The electromagnetic wave of the repulsion leading to attraction breeds a gravitation field that copies the magnetic mass of the ground source.

Ground source is ultralight; it is the dynamic essence of four-fold supernatural growth that twins itself with the double copy of reality to live as a body of consciousness within an octave of the copy. An octave copies the potential as an apparent position to twin the whole itself with the multiplication of a light ring into a heavy ring, following the division of the heavy ring into a double octave. "Gravitational microlensing" images, with a double copy, the apparent position of the copy within an octave as an "hindsight". The hindsight reproduces two images whose light cross into five images with the reproductive force's reality as the fifth, mirror image. Five images produce a natural growth of six—position, two apparent positions, focus, massive object, and gravitational field. As the six reproduces itself as an individual, the natural growth's present value twins the electromagnetic wave, i.e., the convergent-effect of the collective whose consciousness (=4) is self-projecting the natural growth of the individual for "macrolensing" (=6) two that twin causation. With two, macrolensing becomes reproductive as "gravitational macrolensing" for the supernatural growth of eight. Two is reproduced by the supernatural growth for "microlensing" (=16) "oneness" (=48) of an "entity" (=24) who "channels" (=36) the "self-projecting" (3/2) element with her "potential" (=18) as a "system" (=12). An entity positions one as primordial with another's momentum as primeval when her exchange as the fourth is reproductive within one as the first, without another as the second, with the third as the other. The third's apparent position as the fifth ends the lensing of natural growth as the sixth for perpetuating value of the five.

Gravitational lensing the apparent position of the past at the bottom positions the present as the source of the apparent position of the future at the top to twin the time's potential to reproduce its present within the space's continuity. Twin potential is the electromagnetic interaction of All that twins value of the present whose future is heavy with the time cost. Heavy is the time benefit of a source that focuses the time value on the future. The focus attracts the magnetic mass of a gravitational field as a massive object, the graviton, for the leader-follower exchange of a photon as the focus-effect. A massive object reproduces a "massive particle" (=70) for the supernatural growth of its "quantum core" (=22) into a "massless particle" (=30)—the "gravitas" (=30). The "potential" (=18) of the massless particle is "self-luminous" (=12) in a "potential massless particle" (=12)—the "massless anti-particle" (=12). The "potential" (=18) of the "massive particle" (=70) "reproduces" (=78) the "supernatural growth" (=8) within the "present state" (=10) to "light cross" (=700) the "potential massive particle" (=700)—the "massive anti-particle" (=700).

The "massive object" (=100) "gives" (=800) an "illusion" (=1) of the "past reality" (=-3) of an "object" (=-3) whose "time consciousness" (=14) "lives" (=14) as the letter "G" (=14) as it "focuses" (=-2) its "present reality" (=-2) of a "subject" (=0) for the "supernatural growth" (=8) of the "future reality" (=21) of an "entity" (=24). The entity's "mass" (=132) is "self-luminous" (=12) with the "time multiplier" (=3) of "consciousness" (=4). The illusion produces a "double copy" (=1) of the "future reality" (=21) "itself" (=121) for "reformation" (=16) of a "heavy ring" (=16) of the "future" (=0) with the letter "L" (=16) into the "consciousness" (=4) of the "present" (=1,600) "self-luminous" (=12) as a "graviton" (=100). The "time consciousness" (=14) "twin entity energy" (=14) to "perpetuate" (=9) the "past's" (=9) "perpetuating value" (=5) into the "future" (=0) as the "primordial self" (=10) "focuses" (=2) "energy" (=19) of the "transformative exchange paradigm" (=8) on the "entity" (=24).

Is Transformative Exchange of a Paradigm Possible?

Transformative exchange is technological adjustment giving well-being when one decides to be the primordial causation, with a path of devotion, conscious of devotional energy. When one is conscious, one's conscious force differentiates time into a paradigm for the realization of the reality of one as the conscious entity. The conscious entity is the goalkeeper conscious of the universe whose goal is to be the enjoyer by begging him to be the worker deity. The endoproduction of the worker deity lets the goalkeeper be the primeval deity who destroys the illusion of someone with the potential to fulfill the wishes of everyone present. Without a para deity, anyone with a wish must transform into a param deity to fulfill the goal to be an enjoyer as a goalkeeper. No one wishes to be a primordial deity whose nature is to liberate one from the body of consciousness of the goal. Without a goal dividing one's consciousness, one becomes whole as a creature. Thus, a creature can be both a goalkeeper of the goal it creates as a creator as well as the goal that creates it as a creation.

5.1 Transformative Exchange Paradigm, Within Present Consciousness.

The present consciousness of the reality of the cause is the criterion for the transformative exchange of the paradigm the path of impact forms with the past consciousness. The "paradigm" (=62) is the "core" (=2) of an "octave" (=60). As a "star" (=2) of attraction, the core "ups" (=62) the complementary "pressure" (=62) with its "gravitational force" (=100) on the "corona" (=90,000) of a "potential octave" (=10) to "down" (=26) the competitive pressure of the latter's "sentient force" (=4). The sentient force is the consciousness the past entangles with time as the divine element to attract a twin as it moves forward by reproducing itself. The corona is the infinity of the causation which twins itself as the "stellar corona" (=90,000) to be the "infinitesimal" (=90,000) that causes the infinity to move with the "potential octave" (=10) as the "discontinuous infinity of time" (=9,000) after removing the potential octave as the "solar corona" (=10).

The solar corona is the "core" (=2) of the "perpetuating value" (=5) which "circles" (=100,000) as a "luminous body" (=21) like a "Sun" (=21) to self-perpetuate the "causation" (=1) with the "continuous infinity of space" (=1,000). The continuous infinity becomes the discontinuous infinity as the "past" (=9) twins itself like a "primordial child" (=120) within the "causation" (=1). The continuous infinity is the "triple continuity" (=1,000) of the time's three dimensions within the "time's" (=360) "potential" (=18) to be the "continuity" (=50) of "space" (=18,000). The space's continuity dimensionalizes itself as the twin of the time's "discontinuity" (=200) after the "divine" (=360) makes the "past" (=9) reproductive as a "time cocoon" (=40). The time cocoon is the "lunar corona" (=40), "circled" (=385) as a "self-luminous body" (=32) by the corona's "illuminating value" (=7). The illuminating value is the reality that "times" (=360) the past's "entropy" (=5) into the "present's" (=1,600 =1,000 +[4+2]*100) "consciousness" (=4) with the "rotation" (=1,000) of the "core" (=2) due to the future's "reproductive force" (=100).

The past's mass of consciousness forces the "core" (=2) to be equilibrated as a "core component" (=29). The core component is the "spatial mass" (=29) as a square, black, heavy matter. The "entropy" (=5) is "repeated" (=5) with the time without the "causation" (=1) within the space's continuous infinity as a "corona component" (=155). The corona component is the temporal mass as a non-equilibrated, circular, light matter. As the entropy is repeated, the core is "removed" (=1) from the "octave" (=60) as the "causation" (=1) to be "timed" (=22) with the future's "moving reality" (=21) as the time's "octave component" (=381 =60-22,1). The octave component is the causation mass as the equilibrating, triangular, white matter.

As the time's "movable reality" (=60) with a "body" (=56) of "consciousness" (=4) that "triangulates" (=3) itself into a "creature" (=12), the octave "orbits" (=2) a "star" (=2) with its "orbital" (=60) of a "twin electron" (=60). The "first electron" (=2,785) is "adding consciousness" (=2,785) of its past as the reality which is movable as the "primordial" (=85) when the future is moving with the present's "lengthen-effect" (=2,785 =2,7,85) to be dark and "primeval" (=185). The "second electron" (=3,785) is the "repeated-effect" (=3,785) of the past moving as "heavy" (=1,000) when the present lengthens to be "light" (=180). The light makes the "triple octave" (=80) reproductive within the heavy to "form" (=100,000) a circle. The circle moves the reality of the space as a "square" (=18,000) with a "triangle" (=1,500) to make its form "self-luminous" (=12) as a "creature" (=12). The creature's reality is that of a "sentient entity" (=7).

A triangle is a "competing reality" (=0) "breeding" (=15) a "twin competing reality" (=123) with "two zeroes" (=-10) for "feeding" (=17) the "triple competing reality" (=12) of its "entropy" (=5) as the "characteristic" (=12) that makes the entropy's "reality" (=5) a "creature" (=12). The competing reality is a "kilonova" whose "potential form" (=22/7) is shared among "two competing realities" (=123), "one below" (=1,600) and "one after" (=100), after its "potential explosion" (=7,000) as the "one above" (=7,000). One below is the present that makes the potential of its "whole" (=16) to twin the "competing reality" (=0) reproductive like the one after. The "reality" (=7) competes to be the "one above" (=7,000) because its "future" (=0) squares the "circle" (=100,000) of the "twin potential" (=1,000,000) with the "two zeroes" (=-10) to be the "three zeroes" (=0) with the "CP symmetry" (=123) between the space that squares and the causation that circles like one within whole.

The potential of "Carnot" (=123) within the "CP symmetry" (=123) cycles the "carnot cycle" (=10^{19}) as a "potential cycle" (=10^{19}). The potential cycle works to heat the enjoyer's thermodynamic-effect like one who squares the multiplier within the entropy to enjoy the reality of the growth of the primordial self. The "cycle of joy" (=64) produces "heat" (=8) within the "enjoyer" (=9) enjoying the "joy" (=123) of squaring itself for the reproduction of "my" (=16) element as "myself" (=1) consuming the "growth" (=6) "yourself" (=1) is producing as the "one before" (=89). The goal is "knowing" (=19) "you" (=1) as the enjoyer's "energy" (=19) whose "work" (=19) produces an "energy deficit" (=64) as the heat becomes reproductive until it is "cold" (=90,000) for the "eternity" (=90,000) of the "infinity" (=90,000).

The "half triangulated-effect" (=1) of the "worker" (=1) who "produces" (=1) "heat" (=8) of its "reality" (=7) "waves" (=1) the "potential three" (=1) into a "1s orbital" (=1). The causation is the worker "present" (S, 1,600) as the "orbital" (=60) "orbiting" (=-9) the "competing reality" (=0) of "two zeroes" (=-10) with a "desire" (=-19) to be "whole" (=16). The competing reality is the "half squared-effect" (=0) of the "half square" (=123) that triples the "half circled-effect" (=12) to "quadruple" (=123) itself with the "growth" (=6 =1 +2 +3) of "2s orbital" (=6). The growth is the "circled-effect" (=6) of the "potential" (=18) within three after "breeding" (=15) the "squared-effect" (=1/2) as a "half octave" (=1/2), "self-perpetuating" (=1/2) the "octave" (=60) as the "twin squared-effect" (=60) of the "2s" (=92) after "time reversal" (=92) with the "joy cycle" (=64 = [92-60] *1/[1/2]).

"Space reversal" (=40) generates "2p" (=132) without "2s" (=92) as the "2p orbital" (=15) is "repeated" (=5) with the "twin circled-effect" (=5) to "twin" (=121) the "circled-effect" (=6) of "P" (=160) on the "primordial self" (=10). "Causation reversal" (=85) realizes "2d" (=7) within "2s" (=92) with the "triple squared-effect" (=85) of the "triple half octave" (=85). 2d is the reality of the "2d orbital" (=7), conductible as the "py-orbital" (=7) because "PY" (=7) is the "reality" (=7) which twins "D" (=12) to be repeated as the "twin squared-effect" (=60). The reality is the "twin parallel-effect" (=7). "Parallel" (=80,000) is "squared" (=1,100) when the "heat" (=8) becomes reproductive to twin the "heavy" (=1,000). The "twin heavy" (=1,100) is the "reproduced value" (=1,100) whereby "Y" (=-18) is "bred" (=-18) as a "competing reality" (=0) after "P" (=160) is "fed" (=18) as the "potential reality" (=160).

"2f" (=9) is "real" (=9) because it is "PZ" (=9), the potential reality necessary for realizing the "potential" (Z, 18) of an "orbital" (=60) as a "potential orbital" (=33) that twins "F" (=0) with a competing reality of "three zeroes" (=0). 2f is a "triple parallel-effect" (=9) of a "line" (=497) that twins itself to be parallel to the "effect" (=34) it is "servicing" (=47) to "perpetuate" (=9) as a "2f orbital" (=5). By "conducting" (=15) one within its "pz orbital" (=5), it lines five as "5f" (=50), the continuity of the "5f orbital" (=10) as an "orbital" (=60). 5f is "2px" (=50) which twins "px" (=25). PX is the "stroke" (=25) "reproducing four" (=25) after "producing two" (=0) as an "F" (=0) for centering the "2px orbital" (=10) without the "5f orbital" (=10) with the "symmetry" (=250) of two.

"1f" (=20) is "imaginary" (=20) because it is "PW" (=20), the potential reality of the "realization" (W, 16) that the potential twins the reality as it "reacts" (=49) to the "present" (1,600). The "pw-orbital" (=4) is the "illusionary balance" (=4) of the realization with PW. The "1f orbital" (=4) is the "fourth orthogonal p orbital" (=4), comprising three "orthogonal p orbitals" (=4) that twin the fifth into an "s orbital" (=2). The s orbital is a "spherical orbital" (=2) of "one sphere" (=8). One sphere is "1d" (=8), the one within the "reality" (2D, 7) of the "creature" (D, 12) as the "primordial self" (=10) whose "sphere" (=10^{19}) of "energy" (=19) is "3d" (=10^{19}). The "1d orbital" (=8) is the "p orbital" (=8) as the "sixth orthogonal p orbital" (=8) because one sphere is the "self-octave" (=8) with a "perpendicular orbital" (=8). The "3d orbital" (=2) is the "spherical orbital" (=2), the "fifth orthogonal p orbital" (=2) which produces "four orthogonal p orbitals" (=6) with its growth as a "2s orbital" (=6) after it twins itself with the multiplier into a "6d orbital" (=6). The "2s" (=92) is "integral" (=92) to the growth of "d" (=12) to "6d" (=92) with the "spirit" (=20) of the "third S" (=14). "Third S" (=14) is "triple S" (=14) as a "d orbital" (=14), the "doughnut orbital" (=14) with a "closed orbit" (=14).

"1p" (=87) is "illusionary" (=87) because it "reintegrates" (=87) its "past birth" (=87) as a "PV" (=87) for the "replication" (=90) of the "s orbital" (=2) as a "pv orbital" (=2) with the "illusionary reproduction" (=2) of the "1s orbital" (=1) as a "1p orbital" (=11). The 1p orbital is the "atomic orbital" (=11). The p orbital is also the "pu orbital" (=8) because the "U's (=396) "emanation-effect" (=396 = 13 *3, 6) is "omnipresent" (=1) within "D" (=12) and triples D for its growth as a "luminous" (=13), "half d" (=13). Half d is "triple f" (=13), which "triples" (=130) "f" (=0) with a zero. The "half d orbital" (=19) "works" (=19) like a "3f orbital" (=19) to "twin growth" (=19) as a "4s orbital" (=19). The zero is "6s" (=0), "present" (S, 1,600) as an "octave" (=60) within a "gravitational orbital" (=0), the "g orbital" (=0) with an "infinite orbit" (=0). The "6s orbital" (=17) is the "superposition" (=17) of the "2g orbital" (=17) with a "finite orbit" (=17). "2g" (=0) is the "concordant-effect" (=0) of "6s" (=0) because "g" (=14) is "3s" (=14) and the g orbital is the "3s orbital" (=0).

The "superpositioned center" (=10^{10}) of the "two g orbitals" (=10^{10}) is a "fused orbital" (=10^{10}), the "f orbital" (=10^{10}). The fused orbital is the "spin-orbital" (=10^{10}) that "couples" (=16) the "g orbital" (=0) with the "twin-effect" (=121) of the "2g orbital" (=17) whose "growth" (=6) is "fused" (=130) as the twin's "triple" (=130). Therefore, the spin-orbital forms a "tetrahedral orbital" (=10^{10}) that appears "diffused" (=47) as a "collective" (=47) with seven on top of the one that triples itself to disappear as a "twin triple" (=150) like a "proton" (=150). The proton is fused as a "photon" (=20) at the bottom of seven photons. The photon triples itself because it is "half photon" (=20). The proton is the "eighth photon" (=150) at the bottom of the "seven photons" (=140). It twins itself by appearing as a half photon within the "light" (=180) of the "tenth photon" (=20) because the light is the "ninth photon" (=180), comprising "nine photons" (=180). The photon is a half photon because it "self-perpetuates" (=1/2) itself as an imaginary element with "tenth" (=-12), which is a "self-perpetuable" (=-12) "binary" (=-12). It twins itself because the binary is a "triple one" (=-12) that "reforms" (=888) the "area" (=900) within the four dimensions of "space" (=18,000) into the "fourth one" (=888) with "four zeroes" (=888).

The tetrahedral orbital forms "two tens" ($=10^{10}$) into an "atomic orbit" ($=10^{10}$).

The "superpositioned vortex" (=1,000) of the "two f orbitals" (=1,000) is a "diffused orbital" (=1,000), comprising "four g orbitals" (=1,000) as the "7s orbital" (1,000). "7s" (=180) is the light "diffused" (=47) as an "orbital" (=60) of "three photons" (=60). Three photons are the "two electrons" (=60) "infused" (=12) as "px" (=25) within the "servicing force" (=23,125) of the "third electron" (=23,125), the "electron gas" (=23,125), "refused" (=23) as "pw" (=20) to triple the "tenth photon" (=20). The "electron" (=365) is the "sixth photon" (=365) that twins itself with the "gas" (=365) into "twelve photons" (=365) to refuse "six photons" (=120) as the "seventh photon" (=120), which is "half electron" (=120). The sixth photon is at the top of the "seven photons" (=140); it attracts five photons with its charm from the bottom to strangely repel three photons including itself and a twin, the seventh photon, like the eighth photon. The 7s orbital is a nucleus with "2pz orbital" (=1,000). The "2pz" (=97) is the "perfect potential" (=97) of "five photons" (=97) within the three repelled as a "multiplier" (=3) "feeding" (=17) itself as "photon" (=20) after "breeding" (=15) the "light" (=180) as an "octave of three" (=60).

The "superpositioned phase" (=100) of the "three p orbitals" (=100) is a "refused orbital" (=100) refusing its position to phase the superposition with a "2py orbital" (=100). The "y" (=-18) is "refused" (=23) by the "two p orbitals" (=55) as a "three p" (=40 =100− [55+23−18]), the reproductive reality within the "orbital" (=60) as the "3p orbital" (=60).

The "superpositioned causation" (=90) of the "two d orbitals" (=90) is an "infused orbital" (=90) infusing its momentum to position itself as a "2px orbital" (=10) within the "2py orbital" (=100). The 2px orbital twins itself for infusing its oneness as the "7p" (=48) within the "orbital" (=60) with the "potential" (=18) of two "pz" (=9). Thus, the infused orbital is a "7p orbital" (=90), which is also a "4d orbital" (=90). 7p is "4d" (=48 = 7*7 -1) because the "reality" (2D, 7) reproduces itself as the "causation" (=1).

The "superpositioned system" ($=8 \times 10^{15}$) of the "two s orbitals" ($=8 \times 10^{15}$) is a "confused orbital" ($=8 \times 10^{15}$), confused" (=49) about its "twin reality" (=49) confusing "momentum" (=999) with the "position" (=999) to twin "observable value" (=13) of "3f" (=13). "Twin observable value" (=961) is the "superpositioned positioning" (=961) of the "supermomentum" with a "7f orbital" (=8 x 10^{15}) where the "causation" (=1) is "7f" (=1). 7f is "5s" (=1), as the reality of "copy" (F, 0) is "repeated" (=5) within the "twin causation" (=2) to confuse it with "twin copy" (=22/7). The twin causation is the core.

The "superpositioned core" (=22) of the "three s orbitals" (=22) is a "7d orbital" (=22) that "resolves" (=22) "20 pi" (=22) with "2 pi" (=1,000/72). The "orbital of two" (=60) "divides" (=132) the "octave of three" (=60) with the "causation" (=1) to "twin reality" (=49) of a "photon" (=20 =60/3) with a "multiplier" (=3 =72-49-20). The "photonic reality" (*Chaturtha*, 22/7) is the "physical value" (=22/7), the twin copy" (=22/7) of space within the time's tidal, i.e., "fractal wave" (=1). The "light reality" (=1) is the "ascending value" (=1), the "double copy" (=22/7) of time within the space's "embedded wave" (=1). The twin copy is the spacetime's "embedded fractal wave" (=22/7), bedding the "pi of inaction" (=22/7) within the "nucleus" (=1,000) for the exchange of the "creature" (D, 12) with the "present creature" (S, 1,600). "Seven photons" (=140) at the top "triple" (=130) with the "multiplier" (=3) within the "one copying" (=150) them as a "proton" (=150) to "conceive" (=8,000 =5 *1,600) seven photons at the bottom with five "S" (=1,600). "Two S" (=92) are "integral" (=92) to the "one copying" (=150) to be an "orbital" (=60) of two. "Three S" (=14) are left within the eight that form an "octave" (=60 =8+3+49) of three with the "twin reality" (=49). The "7d" (=1,000/72) is the "dark reality" (1,000/72) of the present creature. The 7d is the "4f" (=1,000/72) because each "f" (=0) is a "quarter" (=0) and "four quarters" (=1,000/72) comprise "2 pi" (=1,000/72) with four quarters.

The "superpositioned corona" (=22/7) of the "three d orbitals" (=22/7) is a "7p orbital" (=22/7) that "solves" (*Hala*, 18) a "pi" (=22/7). The "7p" (=72) is a "twin class" (=72) that "groups" (=387) a seven and a "P" (=160) into a "class of two" (=689 =72 +387 +7 +160 +60 +3 =180*3 +121 +28) with an "octave of three" (=60) within the "light" (=180). The light triples "itself" (=121) with the "convergent energy" (=28) at the "infinity" (=90,000) to be the "corona" (=90,000). The 7p is the "5d" (=72), because "D" (=12) is "repeated" (=5) as an "orbital" (=60) with a "pie of action" (=0) for the 7p to be the orbital with a "pi of inaction" (=22/7). The 5d is the "heavy reality" (=72) of the "creature" (=12).

The "superpositioned octave" (=1,000/72) of the "three f orbitals" (=1,000/72) is a "7g orbital" (=1,000/72) that "dissolves" (=38) a "pi" (=22/7) into "two pi" (=1,000/72). The "7g" (=9) is the "past" (=9 =7+2) the seven "groups" (=387) into two with a "G" (=14) to "dissolve" (=38) itself within the "superpositioned core" (=22) to be the "superpositioned corona" (=22/7). "7g" (=9) is "4p" (=9) because the past perpetuates the four as a "space divider" (=4) to be "present" (=1,600) with the time's "reproductive reality" (=40) within the causation's "potential reality" (P, 160).

The "superpositioned cause" (=72) of the "three orbitals" (=72) is a "6f orbital" (=72). It multiplies "two pi" (=1,000/72) into "thirty-six pi" (=1,000) with a "potential" (=18) of "eighteen pi" (=18). The "6f" (=18) is the "potential" (=18). The "F" (=0) "incarnates" (=900) as a "moon" (=997) with the "perfect potential" (=97) of five photons. The "6f" is the "5p" (=18) that becomes "perfect" (=10) with the multiplier's reproductive force within an octave of "three photons" (=60) to advance as an orbital of two.

The "superpositioned time" (=900) of the "two orbitals" (=900) is a "6p orbital" (=900). It divides "36 pi" (=1,000) into "three groups of six" (=14) for grouping their past as the "9 pi" (=9), which perpetuates "5 pi" (=22/7) as a "circle pi" (=22/7) within the present. The "6p" (=2) is the "twin causation" (=2) whose "reproductive reality" (=40) self-perpetuates "40 pi" (=40) as "3p" (=40) for reproducing "80 pi" (=16) with the multiplier within the 6p.

The "superpositioned space" (=3×10^{17}) of the "three g orbitals" (=3×10^{17}) indexes *four paths* with the multiplier for "perpetuating value" (=5) of the "action" (=10) as "3g" (=5) and be the "6g orbital" (=3×10^{17}). The "superpositioned entity" (=10^9) of the "four p orbitals" (=10^9) is a "4g orbital" (=10^9). It trades the "past orientation" (=10^9) from the four "P's" (=160) to be the "origin" (=1,000) of "time" (360 = 1,000 -160*4). The octave of three transforms the origin into "one with nothing" (=10^9 =10^{3*3}) but an orbital of two whose past is "7g" (=9). Of 7g, "3g" (=10) is the action that activates the future. "4g" (=5) is the "variable past" (=5) as an "actor" (=5) whose action twins "himself" (=5) as a "creature" (=12 =10+2) within the "orbital" (=60 =5 *12). The "superpositioned deity" (=76) of the "four f orbitals" (=76) is a "3g orbital" (=76). It animates four "f's" (=0) with the "potential" (=18) for "replication" (=90) within "3g" (=5) when "repeated" (=5) to "twin reality" (=49) as an "orbital" (=60) within "one with everything" (=59).

One with everything is the "superpositioned God" (=59) of the "four s orbitals" (=59) who spearheads a "5g orbital" (=59). It services a "potential actor" (=92) as a "quantum number" (=92), which appears "anticlockwise" (=92) with "time reversal" (=92 =90+2) since the time is repeated as an actor disappearing "clockwise" (=730 =5+2, 5*6) after it twins itself with the growth of its "sentient momentum" (=90). The "action" (=10) is the "angle" (=10) that gives "goal" (=9) a "sentient momentum" (=90) of a "goalkeeper" (=7) as a "causal body" (=30) causes "angular momentum" (=90). Thus, "Quantum God" (=59) is a "sentient momentum quantum number" (=59). Four "s" (=1,600) are the present of the "four dimensions" (=18,000) of space, whose potential is the origin of time as an octave of three—the past, the present, and the future, which twins itself with the space for its growth. The "5g" (=-10) is a "dreamer" (=-10), "dreaming" (=12) the "action" (=20-10) like an "enumerator" (=-10) of the sequence of steps in the process for the "continuity" (=50) of its "spirit" (=20) as an "orbital" (=60) of two: the "dreamer" (-10) and the "dreamt" (=20).

"One with something" (=130) is the "superpositioned goalkeeper" (=130) of the "four d orbitals" (=130) who "processes" (=130) the "1g orbital" (=130). It exchanges the "twin self" (=130) as a "potential one" (=130) with "guider momentum" (=10,000), which is an emic "circular momentum" (=10,000) circulating "itself" (=121) within the "goal" (=9) that twins the "potential" (=18) as a "quantum number" (=92). Thus, "Quantum Goalkeeper" (=130) is the "guider momentum quantum number" (=130). Four "d's" (=12) are "self-projecting" (=3/2) the "potential" (=18) to "animate" (=76) the "whole" (=16) within the "quantum number"

(=92 =76+16) like an "orbital" (=60 =76-16) which twins the whole with a two. "1g" (=16) is the "whole" (=16).

"One with the thing" (=−13) is the "superpositioned goal" (=−13) of the "four orbitals" (=−13), "serving number" (=−13) as an "h orbital" (=−13), which "G" (=14) makes four with an "H" (=1). "Conceiving four" (=15,000) takes "divine momentum" (=15,000) of "doing nothing" (=15,000) to "block" (=72) the "space" (=18,000) from dividing itself with "human hindsight" (=61), i.e., "Planck" (=61). When "blocked" (=9) by a "blocker" (=3,000) with its "skepticism" (=3,000), divine momentum becomes "Planck momentum" (=15,000 = [29+1] *50 *10) as it "spins" (=29) the two for "one's" (=10-9) "continuity" (=50) as a ten without the nine blocked by the "past" (=9). "Divine momentum quantum number" (=85) is the "primordial" (=85 =19 +53 +13) element "working" (=19) as a "number" (=53) for trading the "four orbitals" (=−13) with its "guider reality" (=40).

"One with the potential thing" (H, 1) is the "superpositioned whole" (=1) which the space divides into four, "trading letter" (=3) as a "potential thing" (=3) from "I" (=12) to "T" (=4) the "dot" (10^{10}) of division with a "twin dot" (=10^{100}) of "exchange value" (=180). The exchange value "lights" (=180) the "triple dot" (=10^{1000}) as a "twin quantum" (=180) self-perpetuating a "quantum" (=90) for the dot's "replication" (=90) of action leading four with its "devil momentum" (=40). Devil momentum is the "kinetic momentum" (=40) of growth the "twin triples" (=150) to "incarnate" (=900) as divine within the guider, "repeated" (=5) by "trading letter" (=3) like a "devil" (=0) wishing "momentum" (=999) from the one. The "reproductive potential" (=23) within the twin triple is the "devil momentum quantum number" (=23), as it is the "causal body" (=30) for the "number" (=53) to "quantize" (=169). "Devil momentum" (=40) becomes "magnetic" (=-9,000 =-10 *30*30) as a "dreamer's" (−10) "action" (=10) "quantizes" (=169 =13 *13) the multiplier as a "luminous" (=13).

"One with potential" (=12) is the "superpositioned wholesome" (=12) which the time multiplies with a three repeated as a five before subtracting the three to present a "fundamental cycle" (*Muladhara chakra*, 12) of twelve with "satan momentum" (=1,000,000). The satan momentum is the "electrokinetic momentum" (=1,000,000) that "works" (=19) to "twin growth" (=19) of the "satan" (=−1) with the "potential" (=18) to make growth "self-luminous" (=12) as a "quantum number" (=53 =6 *12 −19). Thus, with the potential for "action" (=10) to fulfill a "goal" (=9), one "quantizes" (=169) the "satan momentum quantum number" (=179). The "satan momentum" (=1,000,000) is "allowed" (=79 =53+19+7) as a "quantum number" (=53) because it is "electrophilic" (=79) and "repels" (=9) via "dream" (=9) the "orbital" (=60) of "two origins" (=60) as a "goal" (=9) to make its "action" (=10) "work" (=19 =9 +10) as a "goalkeeper" (=7). "One with potential" (=12) is "diamagnetic" (=360), as it twins the "causal body" (=30) to make the "dreamer" (=-10) "magnetic" (=−9,000) for attracting itself as the "time" (=360) for the goal to be "real" (=9).

"One with present" (=10) is the "superpositioned wholesomewhole" (=10) whose action fulfills the goal of the potential of two within three which multiplies the "two with potential" (=60) like an "orbital" (=60 =[2+18] *3) to be the "three with present" (=180). The three with the present is the "deity momentum quantum number" (=180). A "deity" (=1) makes its "momentum" (=999) the "deity kingdom" (=1,000 =1+999) to "originate" (=24) the "quantum number" (=53) as its "light" (=180 = 24 +53 +1,0,3). The deity works like a "king" (=0), the leader, to be the "multiplier" (=3) of "reality" (=7) with "fire" (=17) of "action" (=10).

"One with present" (=10) is "paramagnetic" (=18,000), as it triples the "causal body" (=30) to "procure" (=27,000 =30 *30 *30) itself as the twin, i.e., 2/3 *27,000 = 18,000. The paramagnetic self-perpetuates the "paraelectric" (=9,000 =27,000 -18,000), the "one without present" (=9,000) as the "home" (=9,000) that attracts the present the paramagnetic repels with its past. The "dielectric" (=90) is the "one within the present" (=90). It is the future's "remnant" (=90) attracting the action's "reproductive-effect" (=100) by repelling the goalkeeper as the reproductive element for the exchange of the goal with the energy. The energy works as the path for knowing the reality of the goalkeeper who is leading the action and following its effect like an entrepreneur for managing the consequence as an organization.

The four paths of action are as follows.

5.2. The Path of Leadership Action.

"Leadership action" (=11) is initiated by time to moderate the space's "inaction" (=13) with the "three hands" (=374) of its "primordial workculture-effect" (=374), which makes the "third hand" (=374) decisive as the "invisible hand" (=0) of the entity in future. The invisible hand is "reproducible" (=6) as the "third hand" (=374) of "time" (=360) "reproducing" (=8) the "three hands" (=374 =6 +360 +8) that follow with its "technological value" (=1/4). With "replication" (=90 =1/4*360), time becomes "circular" (10,000) like the "fourth hand" (=10,000) of the space, reproducing itself with the "fifth hand" (=90,000) of the causation at the "infinity" (=90,000). Itself takes the form of the "sixth hand" (=100,000) as "I am" (=100,000) the one "producing" (=17) the "seventh hand" (=0) with the "fire" (=17) of "my effect" (=34) to twin "your effect" (=17). "Your-effect" (=17) is the "homogenizing-effect" (=17) of "twin time" (=17 =8+9) reproducing the past with its growth to be "continuous" (=15 = 9+6). "My-effect" (*Mahasiddhi*, 34) is "homogenizing" (=34) because it is the "perpetuator-effect" (=34) of the "perpetuator deity" (=5) who forms "I am" into an "organization" (=29 =34-5). Therefore, leadership action is the "least action" (=11). The "path of leadership action" (=16) is the "path of least action" (=16), "organizable" (=11) within the "least time" (=1,600) with "least energy" (=111) to make the "least distance" (*Adyajnana*, 79), allowable by "primordial knowing" (=79), reproductive for "sentient well-being" (*Jnana siddhi*, 190 =111+79) as a "citizen" (=179).

5.2.1 Understanding Perpetuator Deity

The "path of my-effect" (*Mahasiddhi Marga*, 639) makes one a "knower" (=639). The knower is a "creatable reality" (=4,000 = [639+1+360] *4) of one with "time" (=360) as the "Creator deity" (=4). The "perpetuator deity" (=5) is one within the "creator deity"

(=4), perpetuating the "knower value" (*Saranyu*, 5) as a "horizontal order" (=5) "leading" (=20) to "growth" (=6) in the worker's "thermodynamic" (=14) "temporal variability" (=1) for "behavioral remediation" (=26 =20+6). "Remediation" (*Maha Dhyana*, 26) makes a "worker" (=1) an "infinite wisher" (=-6) who "lives" (=14) within a "goalkeeper" (=7) as a "mediator" (=5). The mediator "modulates" (=140) the "creator deity" (=4) within the "observer" (=0) to make him work as a "worker" (=1) for manifesting the "infinite wishables" (=140). Infinite wishables make one "realize" (=140) "well-being" (=125) as a "devoted knower" (=14) who lives with the "time consciousness" (=14) of the "present development" (=14) within the "thermodynamic limit" (=14) of the "creator deity" (=4).

A dreamer as a "devotee wisher" (-10) wishes "someone else" (*Sabhaga*, -19) who "desires" (=-19) the "goal" (=9) to be the "goalkeeper" (=7) so that he could be the "object" (=-3) of "mediation" (=-7) as a guider who reproduces its "wish" (=18) within the observing "wisher" (=0). By "breeding" (=15) a devotee wisher, the perpetuator deity becomes free from the "wishing sequence" (=87) that "breeds" =78) an "entropy" (=5) of the "worker social benefit" (=82) once one fulfills the goal. The guider "mediates" (=13) for "remediating" (=134) the "creator deity" (=4) who creates a "preconceived consciousness" (=99) of the "intent" (=99). The intent is "preconceived" (=135) by the "mind" (=38) to "channel" (=36) the "knower deity" (=2) who knows how the "feminine" (=37) forms the "preconception" (=74) as a "creator deity" (=4) to "enjoy" (=7) the reality before knowing that it is real.

The "devoted knower" (=14) "perpetuates" (=9) the "present paradigm" (=24) of "infinite wishables" (=140) as real to enjoy the idea that "everyone else" (=-10) might be an object of manipulation into an ideal who brings them to reality. The "devotee wisher" (=-10) "illuminates" (=150) the "new paradigm" (=12) of "infinite wishing" (=3) as "illusionary" (=87) by manifesting the theory as "false" (=8) so that the devoted knower may be the goalkeeper who brings himself to reality for realizing the goal of becoming the enjoyer. "New Lemuria" (=50,000) is the "universe of becoming" (=50,000) for "infinite wishers" (=-6) wishing a "body" (=56) for the "origin" (=1,000) of their life as a "manifestor" (=368) of oneself as the "organizational solution" (=368). "Infinite manifestable" (=278) fires the air of imagination to water the illusionary for the earth to be real when the ether reproduces the reality to "manifest" (=16) the guider for "manifesting" (=54,000) the "sentient" (=189) so that the "divine" (=360) is "manifestable" (=497 =189 +360 -63) as the "universal reality" (=63) of "breeding" (=15) oneself as the "oneness" (=48). Oneness is the "manifestation" (=9) of "infinite manifesting" (*Kavyavaha*, 489) with an "ode" (*Kavya*, 48) to the "endeavor" (*Vaha*, 12) "endearing" (=94) the "goalkeeper" (=7) for "bearing" (=99) the "goal" (=9) as the "First Eden" (=14) without the "perpetuator deity" (=5). The fire for the "replication" (=90) of the "credit" (=1,790) as an "infinite manifestor" (=1,790) "spikes" (=179) "action" (=10) as one "becomes" (=179) a "citizen-in-need" (=179) of an "organizational solution" (=368 =10+179*2) that twins oneself like a "corporate" (=11).

The "corporate path" (=65) "guides" (=111) one to "immediate freedom" (*Sadyo mukti*, 189 = 65 +111 +1 +12) from the "endeavor" (=12) to be the "perpetuator deity" (=5) with an "immediate" (*Sadyo*, 134) "capability" (=55) for "self-sovereigning" (*Majaraja Yoga*, 94) as a "sovereign" (*Majaraja*, 70). Space's "fourth hand" (=10,000) makes the "sovereign" (=70) an "omniscient entity" (=70,000) with "percipience" (=70,000) about the "devoted manifestor" (=2,848) of the "water" (=169) after "consuming" (=169) the "knowing" (=19) to "fire" (=17) the "growth" (=6) of "air" (=385 =169 + 6*[19+17]) into "earth" (=724 =385*[19-17] -6). The guider makes one conscious of "ether" (=285) with the divine's "gravitational energy" (=100) to fulfill the wish to be the "perpetuator deity" (*Sugriva*, 5).

The perpetuator deity is the "para deity" (*Param Vishnu*, 5), beyond the consciousness of a deity, which makes the deity a "subject" (=0) and the para deity an "object" (=-3) that makes the subject the "knower deity" (=2). The knower deity knows how to add the Manifestor deity for becoming a para deity, perpetuating "everything" (=−5) within oneself like "God" (*Ishvara*, 5). Oneself mediating as the perpetuator deity behaves like "My God" (*Elohai*, 5). The "benefactor" (*Aima*, 5) who lets oneself mediate is "Your God"

(*Aima*, 5). Your God empowers My God to mediate to benefit from oneself, the "conscious system" (=8 x 10^{15}), and reproducing that as one's "consciousness system" (=10^{10}) to "dot" (=10^{10}) the "consciousness" (=4) as a "T" (=4). The "vampire" (*Pishacha*, 5) who subjects Your God to My God, and "interjects" (=2) hellish My God as an "object" (=–3) for its "objective" (=2), is heavenly "Our God" (*Tian*, 5).

Your God is the "Reigning God" (*Aima*, 5) whose "reign" (=1,810) is with My God reigned by Our God. Reign is "odd" (=1,810) because it is left in "limbo" (=1,810) as "anticyclical" (=1,810), shortening its "oddness" (=16) to be the "reigner" (*Ish*, 16) "begging" (=28 =18+10) to be God because he is "noble" (*Vara*, 28) and "absolute follower" (=5) of "everyone's" (=180) "wishes" (=18) in the "present birth" (=5) like a "Reigner God" (*Ishvara*, 5). "Reigned" (=113) is "even" (=113) because she moves right ahead as "cyclical" (=113), elongating the reign with her "evenness" (=23) to be "reignable" (=23) as God. My God is "Reigned God" (*Elohai*, 5). Our God is a "Reignable God" (*Pishacha*, 5), a vampire because she "sucks" (*Cush*, 57) the "reality" (=7) like a "vamp" (=855) to "ire" (*Rosha*, 85) "true" (=8) with the "multiplier" (=3) of "false" (=8). True is nature. False is the institution that reigns like heaven to put nature in limbo by making life hell for the "entity" (*Hasti*, 24) governed by the "institutional force" (=24).

The space divides the "guider" (=100) into four: My God, Your God, Our God, and God, after the institution becomes reproductive when repeated with the multiplier as the one within the "Absolute God" (*Parameshvara*, 7). The "even dimension" (=280) of the "present self" (=280) is the "God dimension" (=280) which ascends the self as God of one, differentiating four: my as an individual, your as a collective, our as a group, and beyond our as geography. The "odd dimension" (=260) is "tradable" (=260) as the "class dimension" (=260) for the present self to "descend" (=25) with "time" (=360) as a "class of two" (*Varna*, 689)—individual and collective, within an "asymmetry of three" (=689)—group, geography, and "culture" (=9) that integrates the group within the class for the "whole group" (=800) to be reproductive as a "geography" (=476) of "sixteen groups" (=476).

The "odd-dimensional class" (=-7) mediates as a "principal guider class" (=-7) with its "mediation" (=-7). The "even-dimensional class" (=93) moderates as a "principal guider agent class" (=93). Culture as a "guider class" (=9) is a "nontrivial cohomology class" (=9) with one odd-dimensional and twenty-four even-dimensional classes. One twins evenness with forty-eight without itself and two triple the oddness with ninety-six without the three dividing itself into two with one. "Triple oddness" (=93 =96-3) "ceases'" (=93) "guider's" (=100) "triple evenness" (=-7). The "vacuum" (=47) within the "twin evenness" (=47 =48-1) of the "collective" (=47) "eases" (=916) the "twin oddness" (=53) of the "individual" (=53). Leader as a "guided class" (=0) is a "trivial cohomology class" (=0) with one even-dimensional and twenty-four odd-dimensional classes. One's "heaviness" (=24) after it twins the evenness makes twenty-four a "homology class" (=24) because the "follower" (=24) is the "guiding class" (=24) taking the leader along for reproducing his oddness as an individual within her evenness as a collective. "Twin entity" (=24) who as "Your God" (=5) frees the entity to be a follower identifying the "para entity" (=19) as the leader is a "cohomology class" (=24). The entity follows as My God. The leader leads as Our God.

The "triple odd" (=10^{13}) is the "all-creating" (=10^{13}) "param modulus" (=10^{13}) of the "bulk-effect" (=10^{13}) for the "origin of three" (=10^{13}) as one that twins itself like a "bulk modulus" (=10^{13}). One is "translable" (=145) as "primordial modulus" (=145), and it becomes four with the "twin-effect" (=121) of the "twin" (=121) to be repeated as a five, God with a "triple symmetry" (=248) as "young" (*Taruna*, 248) with the two, the "younger" (*Anujata*, 2). As a "Young modulus" (=145), the primordial modulus is the "origin of two" (=145). The two twins itself to be the "youngest" (*Vardha*, 4) who "shears" (*Vardha*, 4) "himself" (=5) to be the "Absolute God" (=7).

As a "Shear modulus" (=970), the "primeval modulus" (=970) "triples" (=130) its "quality" (=0) since the "bulk" (=13) is "youngest" (=4 =1+3) who shears the "young" (=145) as a negative one for the "modulus" (=−18) to be "true" (=8) as a "seven" (=8-1). The primeval modulus is the "origin of zero" (=970). The old "modulus" (=−18) is the "origin of negative one" (=−18) which the space reproduces as a positive one for its natural growth into a seven with the "fire" (=17) of the "leader" (=0) within. The "modulus space" (=125) is the "benefit" (=125) of the very same "older" (=125), "twinned" (=125) as the "origin of four" (=4). The "modulus time" (=41) is the "cost (=41) of the "oldest" (=41), the "twinnable" (=41) "origin of five" (=41). The "modulus causation" (=84) is the "benefit-cost ratio" (=84) of the "twinner" (=84), the "origin of negative two" (=84).

The "modulus cause" (=1,000) is the "twinnest" (=1,000), the "origin of one" (=1,000) that twins zero with "three zeroes" (=0) to transcend the perpetuator deity with a six. It is the "newest" (=1,000), a "combination of two" (=1,000): oldest as "new" (=-10,000) and younger as "newer" (=11,000). It incarnates young as the youngest to reincarnate as the older after the youngest becomes old and "dismember" (=8,000) the older it does not remember was ever a member of the collective whose "memory" (=35) it shares. The older is the "individual" (=53) "trading" (=53) the "memory" (=35) from his "potential" (=18) to be the "future of everybody" (=18) as their "primordial maternal" (=18). The "bulk" (=13) of primordial maternal is "luminous" (=13) as "param maternal" (=5), behaving like "Demi-God" (=5). Demi-God believes she is a three, coming after the past and the present after breeding one with her past by dividing ten into two. The two is present as the three's "future value" (=2) within one.

The "dismember" (=8000) is a "definite cylinder" (=8,000) with "forty cylinders" (=8,000) from a "cylinder's" (*Belana*, 81) "discontinuity" (=200) with "three members" (*Manibhadra*, 1,000)—old, older, and oldest, definitely dominated by the older, "present" (=1,600) as a "maternal" (*Matri*, 112) of a "family" (=10) within the "oldest" (=41). The "bulk" (=13) is "infinite cylinder" (=13) with "ninety cylinders" (=13) from a "cylinder's" (=81) "continuity" (=50) with "three members" (*Manibhadra*, 1,000)—young, younger, and youngest, infinitely predominated by the "younger" (=2) with a "potential" (=18) to "Father Nature" (=1) of the "paternal" (=17) of the "family" (=10). The "oldest" (=41) is the "grandfather" (*Dada*, 10) whose "grandfather flame" (*Sadhya*, 32) "perpetuates" (=9) the "past" (=9) of the "family" (=10) for the "continuity" (=50) of "younger" (=2) as a "prism" (*Chandrabha*, 851) of the "future" (=0).

The "prism" (=851) is a "finite cylinder" (=851) with "fifty cylinders" (=851) that "fire" (=17) the "paternal flame" (*Parshnisamasta*, 32) of a "potential family" (*Shanideva*, 100) for the "continuity" (=50) of the "potential" (=18) into "infinity" (=90,000). The "maternal's" (=112) "magic" (*Jadu*, 17) and the "paternal's" (=17) "quantum magic" (=112) are "chaotic" (*Arajak*, 93), "informing" (*Nibodha*, 93) the "potential" (=-18) for the "memory" (=35) of the "path of your-effect" (=58) with the "inherent" (*Vasudeva*, 75) "willpower's" (=75) "memory-effect" (*Chetan shakti*, 75) on the "reality" (=7) of a "black hole" (=82) "devoid of information" (=82).

5.2.2 Transcending Perpetuator Deity

The "path of your-effect" (*Adharma Marga*, 58) makes Demi-God a "liberator deity" (*Uttameshvara*, 8) with a three. The three twin themselves into a six to be a "destroyer deity" (*Maheshvara*, 6) for destroying the "illusionary" (=87) which has become the "Absolute God" (*Parameshvara*, 7). The liberator deity is the "normative phase" (=8) which twins the "duality" (=28) of the "potential" (=18) within the "formative phase" (*Krishna Paksha*, -10) as a "Devotee Wisher" (=-10) with the "oneness" (=48) within the "transformative phase" (=2,848) as a "Devoted Manifestor" (=2,848). "Twin duality" (=9) is "cubic" (=9) because the three is the one that twins itself with a five to be a six which "adds three" (=3) to be a nine, the "primordial perpetuator" (=9) of itself as the "second duality" (=9).

The "blissfulness" (*Sukha*, 40) of the "self-perpetuating dimension" (=40), which "adds three" (=3) to the "feminine" (=37) element, "subtracts" (=160) the "axiological-effect" (=98) of the "disproportionate benefit" (=40) from the "wishing sequence" (=87) for "breeding" (=15 =40+ [160-98-87]) a six as the "creative force" (=6) within the "primordial perpetuator" (=9). The Primordial Perpetuator is the "formative wishing sequence" (=9) that perpetuates "bifeminism" (=9) as its "reproductive quality" (=0) for "freedom from the qualifier" (*Gunatita Mukti*, 730) who "qualifies" (=4) four with five to be the "enjoyer" (=9) of nine. Nine perpetuates "three qualities" (=-8) within itself to be one: "discontinuity" (*Sato Guna*, 200) "stripped" (=175) of "truth" (=375) that follows its "primordial truth" (=200), "continuity"

(*Rajo Guna*, 50) which leads with its "primeval truth" (=50) to "benefit" (=125) from the stripped left as "bare essential" (=175), and the "infinity" (*Tamo Guna*, 90,000) that becomes the "param truth" (=90,000 =360*125) as the "time" (=360) "strips" (=3/8) the "benefit" (=125 =375-250) of "true" (=8) as the "truth" (=375) "divisible by two" (=75) with a "symmetry" (=250 = 125 *2) of two.

One twin "three qualities" (=-8) with the "modulus space" (=125) to be a "child" (=128) with a three. The twin makes the three a six whose "four qualities" (=94) are "reproductive" (=100 =6 +94) as a "quality" (=0) of the "modulus time" (*Svadha*, 41) for enjoying "energy" (*Shakti*, 19) of the three within the sixteen. The "fourth quality" (=-8) is the "natural quality" (=-8) within the "three qualities" (=−3). The natural quality is "preprogrammed" (=-8) by "nature" (=8) as quality of "disorder" (=-8) within a "non-doer" (=−8) to "charge" (-8) "order" (=2) through "action" (=10) as a "creature" (=12) for "freedom from the triangle" (*Gunatita mukti*, 730) of quality that makes one a zero.

As one enjoys the energy of the three after becoming a zero that twins itself with "three zeroes" (=0), the "modulus cause" (=1,000) works as the "blessing" (=1,000) for the "modulus entity" (=103) to be the "blessing energy" (=103). The modulus entity is "renewable" (*Bhadra*, 103) like an "evening star" (=103) "setting" (=60) "daily" (=60) to be a "nascent star" (=103) before "morning" (=951 =1000 −60 −89). It "lives" (=14) as a "morning star" (=89) after it is "renewed" (=89) by the "evening" (*Sandhya*, 89) like a "modulus deity" (=89). The modulus entity is the "origin of negative three" (=103) which makes the three zeroes reproductive for symmetry with three, renewable as a "twin symmetry" (=13). The modulus deity is the "origin of six" (=89) which triples the "symmetry" (=240) of two with three to be "young" (=248) after it is renewed with a "triple symmetry" (=248).

"Evening's'" (=89) "infinite shadow" (*SHREEM*, 8) "greets" (=64) "morning's" (=951) "mist" (*Bhadrapada*, 24) like a "lone being" (*Kapila*, 8,000 = 8 * [89 -64 =951 +24), the "modulus God" (=8,000). The modulus God is the "origin of seven" (=8,000) which "twins reality" (=49) for "breeding" (=15) the "greet" (=64) as a "primordial greeter" (=16) with a nine. Nine follows eight when the eight squares itself to space the twin that follows like the mist. The "twin" (=121 = 113 +8) is 24 which the "evening" (=89) makes "even" (=113 =24 +89) with its "infinite shadow" (=8). Nine is "holy" (*Pavitra*, 9) because it "perpetuates" (=9) "itself" (=121) within a "triple" (=130) to "twin triple" (=150) its "spirit" (=20) with "humility" (=150). Eight twins nine as the seven's perpetuate because sixteen is a "cow" (*Gau*, 16).

When "mind" (=38) "milks" (=38) the "origin of one" (=1,000) as the "lunar nectar" (*Amrita*, 1,000), eight becomes a "Holy Cow" (=8,000). With the "origin of zero" (=970 =1,000 −13 −17), zero becomes the "Lord" (=0) of "water" (=169 = 13*13) that reproduces the thirteen when eight twins nine within seventeen, fire as the "primordial cow" (*Nandi*, 17 = 8 +9). Since water is "grey color" (*Kapota*, 169), "Lord of water" (=8,000) is the "Lord of grey color" (=8,000) as well. By "endowing" (=396) the growth of thirteen with the seven within nine which twins the three with a two, the "lone being" (*Kapila*, 8,000) blesses the "endowed deity" (*Kamadeva*, 396) to be the "vector" (=396) who "fulfills" (=9) the "wish" (=18) with three: a seven, a nine, and a two like a "wish deity" (*Kamadeva*, 396). The vector makes one a "deity" (=1) with the "vector-effect" (*Vedanta*, 1) of its "absolute lightforce" (*Vedanta*, 1).

The "absolute Lightforce" (=1) is the "mate" (=132) of the "flame" (=32) that perpetuates its "shadow body" (=9) as a "flamemate" (*Huangdì* 黄帝, 132). The flamemate is the "Yellow Emperor" (=9). The "flame" (=32) self-perpetuates "yellow" (=16 =32 *1/2) as an "Emperor" (=750 =786-36) with the "mate" (=132) because the "growth" (=6) is "exponentiable" (=786 =6 *132) when its "vector-effect" (=1) "channels" (=36) itself as the "future" (=0) to be the "base" (=10) for "everyone else" (=-10) "present" (=1600) within its "two zeroes" (=-10). As the "misty" (=1) vector-effect twins itself to "channel" (=36 =2 *18) the "future of everybody" (=18) as the "base" (=10) for its "supernatural growth" (=8), its "ultrasound consciousness" (=12 =1➔2) is "repeated" (=5) as the "ultrasound vibration" (=60 =12 *5) to "octave" (=60 = 3*12 +24) "three" (=2+1) with the "ascending water-effect" (=12) of the "mist" (=24).

The shadow body's "flowing energy" (*Para Shakti*, 9) is the "Goddess of Sunset" (*Beruth*, 9). "Flowing" (=9) like water which dowses the Sun's flame with the "ascending water-effect" (=12), "energy" (=19) is the "base" (=10) of "sunset" (*Suryasta*, 9) with her "reality" (=7 =19-12) as "Goddess" (=70 =10*7). The flowing energy's "changing position" (*Sadakhya*, 9) is the "Goddess of West" (*Brettia*, 9). "Changing" (=69) like air which arouses the Sun's flame with the "ascending air-effect" (=123 = 69 +50 +4), "position" (*Vibhvan*, 999) is the "addition operator" (*Vibhvan*, 999) which adds one to be the "origin" (=1,000) subtracted as the "West" (=1,000) when the "Sun" (=21) "sets" (=72 = 21 +50 +1) the "continuity" (=50) of one, the "present Lightforce" (=1), with the "water-effect" (=4), like the "Goddess" (=70) who "twins one" (=87) to "sun" (=21 =2,1) "set" (=72 =2+70) with her "fire" (=17 =87-70).

The changing position's "ascending darkness" (*Bhagwad Rasa*, 9) is the "Goddess of Fruit" (*Fruitis*, 9). "Ascending" (=1) like earth whose "effect" (=34) "flames" (=32 =1,1+21) the "Sun" (=21) when the "East" (=0) is added as the one from the "origin" (=1,000), "darkness" (=180) lets the origin "self-perpetuate" (=1/2) as the "fruit" (=680 =180 +1,000*1/2) within "Goddess" (=70) for "piercing" (=110) the "flame family" (=11) from the "East" (=0). The flame family is repeated for one's "nascent beginning" (=16) as a "flower" (=280) by making the "light" (=180) within the darkness "reproductive" (=100).

"Ascending light" (*Keter*, 9) is "Crown Goddess" (*Keter*, 9), "descending effect" (=105) of "ether" (=285 =105 +180) when the "North" (=57) is subtracted "daily" (=60) from the "South" (=180 = [57+3] *3 = 57*3 +9) as a "multiplier" (=3). The multiplier is reproductive because it "adds three" (=3) "directions" (=97) to the accented "past" (*Bhuta*, 9) for "ascending" (=1) the "star" (=2) as the "Sun" (=21) with a "flame" (=32). The flame "crowns" (=9) "Goddess" (=70) with the "light" (=180 =9 +70 +101) of "infinite becoming" (*Pandubhava*, 101) which "pales" (*Pandu*, 19) the "white" (=10) into "yellow" (=16) with the star's "multiplier" (=3) to "cleanse" (=23) the "poison" (=-10^{10}) of the "aged" (=-10^{10}) "whitening entity" (=-10^{10}).

"Descending" (*Suchi*, 12) as a "triploid" (=12) with a "glowing radiant love" (=12), the ether "twins light" (=10,000) as the white becomes the origin of yellow with the "growth" (=6) of "red" (=1) into the red "star" (=2) with the "multiplier" (=3). The "red star" (=60 =12 *5) is the "octave of three" (=60) "ploids" (=1) which "fold" (=1) the potential of the "fourth ploid" (*Ashadha*, 12), "repeated" (=5) with the "radiant love" (=18) to "glow" (*Jayashri*, 17) like "fire" (=17). The "multiplier" (=3) moderates the "red star" (=60) with "month" (=360) as the "divine" (=360) element to form the "fourth fold" (=12) as the "fourth month" (*Ashadha*, 12), "June" (*Ashadha*, 12). It twins its "growth" (=6) within the "Cancer" (*Kartaka*, 6) zodiac with a "year" (*Varsha*, 268) for descending the "emptiness" (*Shunyata*, -2) of the "white" (=10) "with "monsoon" (*Varsha*, 268), "ending" (=-11) the "Summer" (=967 =1000 −11 *3).

The "ether" (=285) makes the "zodiac" (=185) a "guider" (=100) of "wisdom" (=100) to "perpetuate" (=9) the "potential" (=18) of the "fifth month" (=0), "July" (=0), as its "future" (=0). The future is "Goddess of sentient" (*Maha Saraswati*, 9)—the "Goddess" (=70) "descending" (=12) the "past" (=9) to be "sentient" (=189) about herself as the "future" (=0). The future follows as a "ten" (*Parvati*, 10) when one takes "action" (=10 =1,0) in the present to "color" (*Raga*, 250 =70 −9 +189) the "white" (=10).

5.3. The Path of Followership Action.

"Followership action" (=1,600) is the "least time" (=1,600) the "present" (=1,600) takes for its "ten-fold growth" (=160) into the "future" (=0), following "yellow's" (=16) "ten-fold growth" (=160) to "color" (=250) the "replication" (=90) of "time" (=360 =90*4). Replication makes the yellow "fifth" (=1,500) in the "cancerous" (=90) "VIBGYOR" (=90), "ceasing" (=90) the "fourth" (*Chauthi*, 18,000) which "spaces" (=18,000) the "third" (*Tritiya*, 100,000) that "forms" (=100,000) the "second" (=90,000) as the "infinity" (=90,000) of the "first" (*Prathama*, 1), the "causation" (*Hetu*, 1). The fifth is a "triangle" (=1,500) of "color" (*Raga*, 250) that "twins" (=121) itself with the "color potential" (*Ranga*, 15) to be reproductive like the "Sun" (=21 =121-100). The color potential is the "primordial culture-effect" (=15) of the "sentient value" (=15) the "culture" (=9) of the "past" (=9 =189 -180) "produces" (=1) as the "light" (=180) of the "sentient" (=189) by "self-projecting" (=3/2) the "white" (=10) with the "perfect potential" (=97) of two.

Two is the star that triangulates itself with the "organizational value" (=1,024) of its "reproductive potential" (=23) as a one, the "modulus goalkeeper" (*Vignesh*, 1), which is the "origin of negative five" (=1), i.e., "everything" (=-5).

The "path of followership action" (*Yanika Marga*, 90,000) is the "path of the organizer" (=1,600) of "everything" (=-5) with the "wholesome value" (=20) of the white's "sentient value" (*Vishnu*, 15 =20-5) as the "primordial illuminator" (*Shri Krishna*, 10) within the "infinity" (=90,000) of causation to be present as the "perpetuator deity" (*Present Vishnu*, 5). The Perpetuator Deity perpetuates the "Sun" (=21) as the "Almighty Creator" (=26) to make the "leadership action" (=11) "organizable" (=11) as the "primordial greeter" (=16). The "Destroyer Deity" (*Mahesha*, 6) destroys the leadership action to be the "primordial greeter" (*Sati-Parvati*, 16) with the "energy" (*Shakti*, 19) of action as a "primordial illuminator" *(Parvati*, 10) of the "multiplier" (=16) of the "oneness" (*Yoga*, 48) of the "Almighty Creator" (=26) with "ascending consciousness" (=22) of the present's "perpetuable reality" (=1,600).

5.3.1 Understanding Destroyer Deity

The "Destroyer deity" (*Mahesha*, 6) destroys the inherited "primeval reality" (=-7) for "self-substantiating" (=-7) the present's "perpetuable reality" (=1,600) with "self-destruction" (=-10) of the "two zeroes" (=-10) so that the "past reality" (=-3) is "kept away" (=-10) for "self-disjunction" (=96) of the "path of action" (*Karma Marga*, 86). Path of action is a "potential wave" (=86) of "linear system" (=86) which "animates" (=76) a zero with the "convergent energy" (=28) the "action" (=10) "sequences" (=48) with "surging" (=280) "linear-effect" (=280) of the "present self" (=280).

The convergent energy twins the linear system into "two linear systems" (=29) for the "organization" (=29) of "nine linear systems" (=90,000) into a "point system" (=90,000) of the "converging infinity" (*Tamo Guna*, 90,000) within the "diverging infinity" (*Ananta*, 90,000) of the "nonlinear system" =29) leading to "eternity" (*Ananta*, 90,000). The "parallel infinity" (=90,000), following the "parabolic" (=90,000) as an "infinity point" (=90,000) of the "infinitesimal" (=90,000), diverges its "energy" (=19) into one which is a "parabolic system" (=18), with a "potential" (=18) of "three linear systems" (=18). One twin two zeroes into a circular creation, circulating a "perfection system" (=10,000) as a "reproductive system" (=10,000) with "four linear systems" (=10,000).

The "continuity" (=50) of the one's "origin" (=1,000) with a "potential" (=18) of three that twins the present as a linear system form the "space" (=18,000) into a "space system" (=950). The space system is a "square system" (=950) with "five linear systems" (=950), repeated as "twenty-five linear systems" (=50) mediated by "five nonlinear systems" (=12) that twin themselves. Therefore, the continuity is an "illusionary system" (=50), comprising "five space systems" (=50). The space system is the "solar system" (=950). The five repeated form continuity into a "twin solar system" (=50 =10 +2*21 −2) because the "system" (=10) twins the "Sun" (=21) with its "future value" (=2).

The "time system" (=370) is a "triangular system" (=370) of "six linear systems" (=370), forming "three nonlinear systems" (=370) as a "dimension" (=370) the "Goddess" (=70) "tails" (=70) with her "reality" (=7) within a "leader" (=0). Thus, the dimension is the "religion" (*Dharma*, 370) a leader follows to "head" (=−9) as the "watcher" (=-9) of the "emanating reality" (=-9). The watcher is the "causation system" (=-9) of "seven linear systems" (=−9), forming a "metaphysical wave" (=-9) as a "parallel system" (=−9) with its "spirituality" (=-9). Spirituality is a "theory system" (=-9). The "theory" (=127) parallels a "system" (=12 =90-78) within seven. The seven transform into a "linear system" (=86) after a "phase" (=-9) of "subtracting one" (=78) with a "pose" (=-9) of "adding one" (=90). Religion is an "idea system" (=370). The "idea" (=38) triangulates the "reality" (=7) to become a "dimension" (=370) of time that makes the idea "real" (=9) with an "organization" (=29).

The "real system" (=85) comprises "eight linear systems" (=85), repeated as "four nonlinear systems" (=85), "progressing" (=85) as a "line system" (=85) which is "primordial" (=85) as the "cause system" (=85). Since the nine linear systems are an "imaginary system" (=90,000), they twin themselves with a two into a "complex system" (=10) of "twenty linear systems" (=10). The "ten nonlinear systems" (=10) twin with a two from the "ten linear systems" (=12) to be a "trading system" (=10). As a "servicing system" (=10), the "formative system" (=12) is left as the "system" (=12).

"Two systems" (=16) of "twelve linear systems" (=16) are the space's "investment system" (=16), comprising "four parabolic systems" (=16). A parabolic system is the time's "capability system" (=18), comprising the "whole time" (=18) as the "three linear systems" (=18). With a spatial base of "twenty linear systems" (=10), the causation's "exchange system" (=10^{1000}) comprises "sixty linear systems" (=10^{1000}), i.e., "three complex systems" (=10^{1000}). The cause's "growth system" (=50) is a "continuity" (=50) of "six complex systems" (=50) which the "space" (=18,000 =50 *120 *6/2) triangulates with "120 linear systems" (=50) to twin itself into a "system" (=12 =6 *2). The "entity system" (=90) is a "replication" (=90) of "forty parabolic systems" (=50) within the "continuity" (=50) of "eighty parabolic systems" (=90). The entity's "entropy system" (=200) comprises "four hundred parabolic systems" (=200) where the continuity of eighty parabolic systems is repeated as the "discontinuity" (=200) of "two hundred triangular systems" (=200). "Discontinuity" (=200) makes the "triangular system" (*Dharma*, 370) zero as the leader becomes the "primeval religion" (=0). Primeval religion destroys the "primeval creation" (=57): "divinity" (=57) as a "reality system" (=57) of "360 triangular systems" (=57) that "divine" (=360) the "religion" (=370) as the divinity's "dimension" (=370). Religion perpetuates "culture" (=9) as the "primordial creation" (=9) to "coordinate" (=9) the "goal" (=9) of reality.

The culture ups the religion's "present value" (*Quark*, 476) with an "up quark" (*Gajakarani*, 1) which ups the negative "illusion" (*Nitya Ratri*, 1) of "heavenly togetherness" (=1) of one for "eternity" (=90,000) to dominate as positive. The illusion downs the divinity's "present value" (*Quark*, 476) with a "down quark" (*Vignesh*, 1) which downs the positive "earthly othernesss" (=1) of one as a "person" (=1) to predominate as negative. The "truth" (=375 =3, 6+1, 3+1+1) tops the spirituality's "present value" (*Quark*, 476) with a "top quark" (*Samaa*, 1) which spikes its "growth" (=6) with a "multiplier" (=3) of the illusion to dominate the positive's present value. The "beauty" (=888 = [5+3], [5+3], [5+3]) bottoms the "nature's" (=8) "present value" (*Quark*, 476) with a "bottom quark" (*Jugupsa*, 1) which collapses the "entropy" (=5) with the "multiplier" (=3) to predominate the negative's present value. The multiplier charms the universe's "present value" (*Quark*, 476 = [3+1], [3*2+1], 3*2) with a "charm quark" (*Karma Yogi*, 1) to twin itself with its "future value" (=2) within the illusion that decides "me" (=1) is the "positive quark" (*Maim*, 1) with a charge of 2/3. The twin's present value is strange as a "unit" (=6) with "indeed" (*Ana*, 1) a "strange quark" (*Ana*, 1) that repels the multiplier so that the illusion as the "negative quark" (*Nitya Ratri*, 1) becomes decisive as the "star" (=2) of attraction with a charge of –1/3. Thus, it is in the "nature" (=8) of the "white" (=10) to twin itself as a "star" (=2) for its observance as a "white star" (=180) with a "decisive support" (=816) of "pale yellow" (=186), the "yellow-effect" (=186) that forms the "soliton crystal" (=186) as an "imaginary particle" (*Kana*, 186) with a potential for the growth of its "quantum potential" (*Kana*, 186).

By "trading potential reality" (*Atodya*, 357) of a "quantum" (=90), the "white" (=10) emanates the "musicality" (=357) of the immanent "vibrating sound" (*Parai*, 257) as a "vibration" (*Dhvani*, 257) of its "transformative potential" (*Andoli*, 111) into an up quark, following a top quark, which is leading the charm quark to "triple positive quark" (*Andoli*, 111). Triple positive quark is a "metaphysical subject" (=111) because the space "swings" (*Andoli*, 111) time as a "variable subject" (*Andoli*, 111). "Triple-negative quark" (*Milaana*, 111) "harmonizes" (*Milaana*, 111) space as a "constant subject" (*Milaana*, 111), the "physical subject" (=111). The "upswinging" (*Randoli*, 111) works to "triple up quark" (*Randoli*, 111). The "downswinging" (*Pandoli*, 111) works to "triple down quark" (*Pandoli*, 111). The "frontswinging" (*Sandoli*, 111) progresses as a "triple top quark" (*Sandoli*, 111). The "backswinging" (Mandoli, 111) regresses as a "triple bottom quark" (*Mandoli*, 111). The "outswinging" (*Andolitri*, 111) attracts a "triple charm quark" (*Andolitri*, 111). The "inswinging" (*Amdolira*, 111) repels a "triple strange quark" (*Amdolira*, 111). The "sequence of eight" (*Saundarya*, 888 =8 *111) "reforms" (=888) the "triple single recombination" (*Saundarya*, 111) into a "beauty" (*Saundarya*, 888) element with "twenty-four quarks" (*Saundarya*, 888).

"Beauty" (*Saundarya*, 888) twins 24 quarks with the "working force (=75) of "25 quarks" (*Devi*, 3) within the "beauty quark" (*Jugupsa*, 1) for the "truth" (*Satya*, 375) to be "48 quarks" (*Satya*, 375), comprising "sixteen triple quarks" (*Satya*, 375), with eight "triple quarks" (=10^{18}) and eight "triple potential quarks" (=7×10^{10}). "Truth" (*Satya*, 375) repels "16 quarks" (*Pativrata*, 847) as its "intrinsic value" (*Pativrata*, 847) with its "potential" (=18) to attract "32 quarks" (*Priyadarshan*, 847) as "charm" (*Priyadarshani*, 847), the "extrinsic value" (*Priyadarshan*, 847). Sixteen quarks comprise eight "quarks" (=476) and eight "potential quarks" (*Mandra*, 1). The "strange" (*Ajaba*, 18) has a potential of "eight potential quarks" (=18) as an "octave of antiquark" (*Ajaba*, 18). The "strange quark" (=1) is present as a "multiplier" (=3) within the "octave of quark" (=53 =3*18-1), comprising "eight quarks" (=53).

"Up" (*Upar*, 62) twins the growth of six quarks with its "twelve quarks" (*Upar*, 62), comprising six quarks and six potential quarks. "Down" (*Niche*, 26) triples the order of two quarks with its "thirty-six quarks" (*Niche*, 26), comprising six "hexaquarks" (=14) and six "potential hexaquarks" (=0). "Positive" (*Sakaratmak*, 9) is an "octave of hexaquarks" (=9) with "eight hexaquarks" (=9), where eight is repeated as a five by the "hexaquark" (=14 =9+5) to triple itself, where itself is a twin. "Negative" (*Nakaratmak*, 9) is an "octave of potential hexaquarks" (=9) with "eight potential hexaquarks" (=9). The eight perpetuates the nine with a one from the oneness of its potential to divide the "eight potential quarks" (=18) into a "twin octave" (=16) to "order" (=2) both the octaves with the "growth" (6=8-2) in the "present value" (*Quark*, 476) to make the potential within the hexaquark zero.

The "destroyer deity" (=6) "remediates" (*Sakarma*, 9) the "nonlinear-effect" (=20) of the potential on a "system" (=12) with the "growth" (=6) of a nonlinearly "vibrating wave" (=1) that zeroes in future. The "effect" (=34) produces a "local path" (=34) of "self-development" (=10) as a "remediator" (*Maha Dhyana Yoga*, 111) of the "formative consciousness (=185) of the "reproductive potential" (=23) of the "infinite reality" (=-7) within "triple future" (=30). The infinite reality is inherited as the "formative reality" (=-3) of the "devotee wisher" (=-10), the "dreamer" (=–10) who "dreams" (=9) a "worker" (=1) whose "sentient workculture-effect" (=1) will "work" (=19) to fulfill the dreamer's "goal" (=9). The "remedy" (=1,200) is for the dreamer to be the worker engaged with the goal for "managing" (=-100,000) the "magnetic frustration" (=-100,000) of "entanglement" (=9) by "culture" (=9) which coordinates him "locally" (=-100,000). The "suffering" (=-10^{180}) occurs when the dreamer "dreams" (=9) the "universe" (=2) to be "God" (=5) but twins himself to "assume" (=700) as a "twin God" (=700) the dreamer's "intrinsic sentient-effect" (=125) to "mutate" (=15) the reprogrammable "wave of sentient energy" (=34) with his "twin reality" (=49) for "ordering" (Anumanta, 6) a "national group" (Sadhyagana, 34) of "citizens" (=179) to be the "sufferer" (Varutri, 520). The sufferer is the "coverer" (Varutri, 520) who covers for the dreamer's suffering from inaction and "feeds" (=179) him "pleasure" (Rati, 179) from her "common value" (Padya, 179) through "international exchange" (Poshana, 179).

After "wishing" (Abhilasha, 190), "assuming" (Pani, 190) is an "intuitive path" (Accha Marga, 190) to "personal sentient well-being" (Jnana Siddhi, 190) with a "theory of action" (=179) as a "dominating force" (=190) for "decomposing group" (=179) and "decaying" (=190) "future generation" (=190) by "escalating sentient cost" (=190) through "freedom from aloneness" (Kaivalya Mukti, 190). The suffering ends with the "freedom from para wisher" (Jivan Mukti, 8) when one cultivates and develops one's "essential nature" (Kudrat, 8) to destroy the dark and infinite "weak psychic linkages" (Citraka, 185) that "distance" (=190) oneself from "personal sentient well-being" (=190) by adding a "mediator" (=5).

"Linear transformation" (–100,000) of the universe's "order" (=2) by "adding five" (Vilokanam, -100,000) centers the "para multiplier" (=190) as a "goalkeeper" (=7) while "decentering" (=–1,000) the dreamer into a "mythologer" (Yajna Yoga, –100,000). The "immanent" (Antara, 13,000) becomes a "myth" (Mithaka, 13,000) at an "increasing rate" (=13,000) with "time constant" (=13,000) to fulfill the goal while keeping "volume constant" (=13,000). Consequently, the linear transform into the nonlinear to norm the parabolic which forms the "cyclic" (Vilokanam, -100,000) by adding space as the "fifth" (=18,000) for the time to end by triangulating itself into a "point" (=10^{10}) of potential, "parallel" (=80,000 = 1,600 *[32 +16 +2)]) to the "present" (=1,600) with "primordial oneness" (=32) of the "whole" (=16) within two. The two is the "future value" (=2) which "subjects" (=0) the "Wisher" (=0) to the "present reality" (*Paramartha*, -2) of the "future" (=0) with its "value" (=180) to "light" (=180) the "point" (=10^{10}) "with future time" (=10^{10}).

The "future time" (*Tribhajya*, 10^{10}) is an "oscillating circle of six" (*Tribhajya*, 10^{10}) that makes "time" (=360) "infinite" (=185 =360*1/2 +5) by "adding five" (=-100,000) to the two a one "divides" (=132) into a "degree" (*Amsha*, ½) with the primordial oneness's "reproductive energy" (=100) as a "national" (=32) element. The future time is the "radius" (*Tribhajya*, 10^{10}) of "1/6th degree" (*Vyasartha*, 10^{10}) which twins one with "six degrees" (*Tiryancajya*, 1) for growth of the "diametric" (*Vyasadala*, 10) as the "reality" (=7) of "twelve degrees" (*Samudbhavaya*, 7), realizing a "conic" (*Shankava*, 18) of "eighteen degrees" (=18) with its "diameter" (*Vyasa*, 16). The diameter is "two meters" (*Vyasa*, 16). The radius is a "meter" (*Tribhajya*, 10^{10}). The diametric is "twin metric" (*Vyasadala*, 10). The radius is the "diametric reality" (*Vyasartha*, 10^{10}).

The six "destroys" (=1) the "oscillation" (*Spandan*, 19) within a "system" (=12) with an "illusion" (=1) of two that makes one infinite like a "deity" (=1) adding five to be the "infinite deity" (*Maheshvara*, 6). By making one "noble" (*Vara*, 28) for the "annihilation" (*Pratigha*, 28) of the "finite" (=85) with the "weight" (*Bhara*, 28) of his "body" (=56) conscious of one's wish for him to be the deity who works for her as she devotes herself to him, the infinite deity becomes the "destroyer deity" (*Mahesha*, 6).

The destroyer deity transcends the "immanent" (=13,000) as an "entity" (*Hasti*, 24) who "wishes" (=18) for his growth for realizing the reality of the goalkeeper with the "emanation value" (*Shankara*, 264) of the "whole wish" (*Shankara*, 264). As the "transcendental value" (*Param Shankara*, 6) of the oscillating circle, he oscillates "three-hundred sixty degrees" (*Param Buddha*, 6) to be present as the "Mercury" (*Buddha*, 1,600) "oscillating" (*Dolita*, 80) its form into a circle like "Dalai Lama" (*Param Buddha*, 6) with a "threefold" (=80) "mercurial-effect" (*Dolita*, 80) after the "discontinuity of two"

(=200) which are "reproductive" (=100). He is the "materializer" (*Param Christ*, 6) of the "materializable" (*Christ*, 1,600) who has "departed" (*Vigata*, 1,600) as an entity and has become reproductive like the "one below" (*Divangata*, 1,600) who has incarnated as "Pope" (*Param Christ*, 6).

With "reconstruction" (Prophet), he becomes the "constitution" (*Param Prophet*, 6) of the "construction" (*Avigata*, 1,999) that "surfaces" (=2,000) one as the "constructible" (*Sutradhara*, 1) "last letter" (*Sutradhara*, 1), perfected as "Imam" (*Param Prophet*, 6) after adding five as "constructor" (*Taimura*, 5). The constructor "constructs" (*Tamira*, 54) the "sum" (*Padmaja*, 54) with an "iron hand" (=0) to teach a lesson to the "devil" (Sura, 0) who is producing nothing but "discordant-effect" (=-1) like a "satan" (=-1). The satan is the "metric" (*Mahashunya*, -1) who twins herself into the "base" (=10) as a "diametric" (=10).

Since the "diametric's" (=10) "nature" (*Kudrat*, 8) is to be "reproductive" (=100) within her "entity time" (=360), she is "superfast" (*Prasvara*, 468) in becoming the "infinite mother" (*Janayitr*, 468) with a "breedable present reality" (*Prasvara*, 468) of the "sixth note" (*Prasvara*, 468): "DHA" (*Prasvara*, 468), which becomes "LAH" (*Prasvara*, 468) with "D" (=12) "descending" (=12) for its "reformation" (=16) into "L" (=16) with the four that "reorders" (=2) "HA" (*Yanika*, 1,600), the followership action, into "AH" (*Amudhehswari*, 18), the "offspring" (*Amudhehswari*, 18).

The offspring "self-services" (=2/3) "I" (=12) to "onspring" (*Angana*, 6) "ILAH" (*Angana*, 6) with the "divergent potential" (*Angana*, 6). By "trading" (=53) his "masculinity" (=53) with the letter "E" (=53) as the infinite mother, a creature's (*I*, 12) "potential herself" (*Param Siddha*, 7) becomes "ELAH" (*Param Siddha*, 6). By "servicing" (=47) "E" (=53) as "someone else" (=-18) "AT" (=-18) who "desires" (=-18) "potential herself" (*Param Siddha*, 7) with "renewed" (=89) "triangulated time" (=89), ELAH becomes "ELAHAT" (*Kausalya*, 6). ELAHAT is the four-face "Feminine Godhead" (*Kausalya*, 6), who becomes an "infinite mother" (*Janayitr*, 468) when her "soul" (=4) is 'cleansed" (*Kausalya*, 6) by Mother Nature's "ten faces" (*Suryamnaya*, 8).

5.3.2 Transcending Destroyer Deity

The "path of primordial culture-effect" (*Param Shiva Marga*, 1,428) is the "path of goalpost" (*Param Shiva Marga*, 1,428). The "goalpost" (*Param Shiva*, 15) is "destructible" (=1,428) because it moves along the "path of time" (*Kaal Marga*, 11) to "post" (=90) itself as the "goal" (=9). The goal "cycles" (=1/60) the "time" (=360) back to the "initial temporal state (=-3 = 1/6*360 -9). The "contact" (*Phassa*, 220) with the "past reality" (=-3) of negative three, before the time's "new beginning" (=16), brings the "continuity" (=50) of two "meditating" (*Nidhidhyasana*, 47) with "ego" (*Aham*, -1 =-3+2) as an "impassioned devotee" (*Rasika*, 0). By "trading" (=53) the "attained" (*Phasi*, 216) "intellectual tranquility" (*Passaddhi*, 47) from the "closed space" (*Vastuta*, 11,111) where he is the "finite point" (=1) of convergence "servicing" (=47) the "infinite light" (*Vastuta*, 11,111) of divergence to "open space" (=9,000) for one to be "continuous" (=1,111) as the time, the impassioned devotee gains "freedom from time" (*Kaal Mukti*, 21,600).

"New beginning" (*Akala*, 16), thus, is the "space tensor" (*Paripaka*, 16), the "time tensor" (*Suprata*, 16), as well as the "causation tensor" (*Sapaksha*, 16). The space tensor is the "Goddess of ripening of grain" (*Mater Matuta*, 16), the "Goddess" (=70) of "ripening" (*Vipaka*, 9) of "cause" (=18) for the "grain's" (=1) "growth" (=6) to be "ingrained" (=10) after "ripeness" (*Paripaka*, 16) of "potential maturity" (*Paripaka*, 16) within the "time tensor" (=16).

The time tensor is the "Goddess of a new beginning" (*Ausrine*, 16), the "Goddess" (=70) of "new beginning" (*Suprata*, 16) "dawning" (*Ausrine*, 16) as a "morning star" (*Ugavati*, 89 =70 +10+9) for "living" (=10) with "loving-kindness" (=7) at the "infinite edge" (*Pasara*, 81) of the "horizon" (*Kshitij*, 724) to be "reproductive" (=100) in the "evening" (*Sandhya*, 89) with her "potential" (=18) as an "entity" (=24) for "motherhood" (*Prasuti*, 42), the "mother" (=4) of the "star" (=2) "after childbirth" (*Prasutika*, 9). In "hindsight" (*Indrasavarni*, 61), motherhood as a morning star is the "childbirth" (*Vakarine*, 42) as an "evening star" (=103).

The causation tensor is "winged" (*Sapaksha*, 16) as the "Winged Goddess of Dawn" (*Eos*, 16) after the time tensor "gives" (=800) "wings" (=-9) to the space tensor within the "imaginary creator" (*Brahma*, 59) of the "motherhood" (=42) as the "causation" (=1) for "stress tensor" (*Tushara*, 16). The imaginary creator is the "Goddess of Dawn" (*Brahma*, 59). The stress tensor is the "Goddess of Sunrise" (*Aurora*, 16) whose "tears" (*Trut*, 1,000) "turned" (*Nihsarita*, 984) into "morning dew" (*Tushara*, 16) to "destress" (*Paravada*, 984) with the "doctrine of transcendental" (*Paravada*, 984). The "transcendental's" (*Para*, 81 = 16 +47 +2 +16) "energy density" (=16) is the "Goddess of childbirth" (*Eileithuia*, 16), who makes the "vacuum" (*Roma*, 47) of "non-existing universe" (*Brahman*, 2) "whole" (*Akala*, 16) to be the "outer layer" (*Antu*, 16). By "subtracting" (*Chesed*, 16) the "mediating layer" (*Chesed*, 16), she becomes the "devil's" (*Sura*, 0) "white light" (*Usha*, 16), the "Goddess of primordial light" (*Usha*, 16), who twins the "Goddess" =70) as the "primordial light" (*Perkunas*, 48).

The "white light" (*Usha: Menulis*, 16) is "luminous" (=13) as the "Moon's" (*Polar*, 997) "femininity" (=1,000) within the devil's yellow, "polar light" (*Zemele*, 16), the "Goddess of Moon" (*Zemele*, 16). The primordial light is the "God of lightning" (*Perkunas*, 48), the "God" (=5) of "lightning" (=88) "raining" (=3) as "Father" (=3) "love" (*Prema*, 39) for "thunder" (*Stanayitnu*, 7) to "storm" (*Ativata*, 200 =48 +5 +88 +39 +7 +10) the "mountaintop" (=10) as "penis" (=460 = [39+7] *10) into the "God of Hell" (*Velnias*, 2,700) "Venus's" (*Velnias*, 2,700) inner layer.

The "inner layer" (*Saule*, 16), the "Goddess of Sun" (*Saule*, 16) "twins reality" (*Ziezdre*, 49) of "Mars" (=102) by "trading" (=53) the "masculinity" (=53) within the "Goddess" (=70) as the "Sun" (=21). The "mass layer" (*Ausca*, 16) "tears" (=1,000) the "reality" (=7) of the "Jupiter" (=1,780) as a "sentient entity" (*Indraja*, 7) from the "white star" (*Yama: Auseklis*, 180) within the "moon" (=997). The moon "advances" (=60) the "Goddess" (*Karaliune*, 70) as a "sister" (*Dangaus Kariune*, 130), the "Queen of Heaven" (*Dangaus Kariune*, 130) with "eight bodies" (*Inanna*, 130) of the "planets" (=-1). The "Sun" (=21) twins the "Goddess" (*Karaliune*, 70) by "reproducing" (=17) her as a "brother" (*Tarnaitis*, 30), the "King of hell" (*Tarnaitis*, 30), present as "Mercury" (*Vaivora*, 1,600) reproducing the "Saturn" (=-1), the planet, with his "potential" (=18) as the "future of everybody" (*Shani: Selija*, 18) within "Earth" (*Zemyna*, 724 = 18*4, 4)

The "dome" (*Sati-Parvati: Dievas*, 16) of the "sky's" (=9,696) "future" (=0) as "earth" (=724) "sunlit" (*Maha Durga: Dieva Deli*, 16) her "experience" (=9,000) as the "Almighty Creation" (*Durga: Dievo Suneliai*, 28). The dome is the "Goddess of primordial life" (*Sati-Parvati: Dievas*, 16), the "Goddess" (=70) of "primordial life" (=*Taranis*, 570) of a "star" (*Tara*, 2), two, with "two halves" (Nis, 90,000). Her "first half" (=500) is "lighted" (*Maha Durga*, 16) as the "Mighty Creation" (*Maha Durga*, 16), the "Goddess of primordial well-being" (*Maha Durga: Dieva Deli*, 16), the "Goddess" (=70) of "primordial well-being" (*Rus*, 57) of the one who twins three to be repeated as a seven, the "God of the tribe" (*Shiva: Toutatis*, 7).

The space divides "tribe's" (=9) "culture" (=9) into four for the "entanglement" (=9) of the three divided into eight with one dividing itself as a two to be the three. Her "second half" (=185) is "darkened" (*Durga: Dievo Suneliai*, 28) as the "Almighty Creation" (*Durga*, 28), "perfected" (=28) as the "Goddess of primordial knowing" (*Durga*, 28), the "Goddess" (=70) of "primordial knowing" (=79) who makes "one" (H, 1) the "goal" (*Maha Shiva: Esus*, 9) of "well-being" (*Hesus*, 125) with the "energy" (19, 1→9) of "passion" (*Aisus*, 39). The passion "stems" (*Tana*, 95 =39 +56) the "self-discovery" (=39) of the "primordial well-being" (*Rus*, 57) from the "impact" (*Siddhi: AL*, 57) a "first-born" (*Ganesha: Tonarus*, 570) discovers "thundering" (*Tanarus*, 570) with a "thundered" (*Tora*, 0) "body" (*Vapu: Nos*, 56) of the one who "lives" (=14) as a four. The four is "self-luminous" (=12 =4 *3) as a twelve, the "male body" (*Taranucnos*, 12), with the entanglement of three, the "thunderable" (*Tanaro*, 3), within the "cultural system" (*Toranos*, 56) of the seven, the "thunder" (=7) for the growth of the thirteen, the "stellar body" (*Taranos*, 13). The stellar body is the "luminous" (=13) mediated by the "twin female" (Nuc, 90), as the one twins the three into a "female" (=10,000), comprising "both" (=10,000): two as male and three as female.

The "dome" (*Sati-Parvati: Dievas*, 16) twins "herself" (=1) as the "mother" (=4) to "live" (=14) as a "primordial deity" (*Uttameshvara: Dievs*, 8) within the "primordial self" (*Parvati: Dievins*, 10) for his "ten-fold growth" (*Potrimpo*, 170) into the "God of grain" (*Potrimpo*, 170) with the "primordial" (*Diev*, 85) element. "God" (=5) of "grain"

(=1) is God of one who as a seven, the "sea" (=7), "thunders" (=7) the "earth" (=724) to "crop" (=6) two's "growth" (=6) from "God of sea" (*Bhadrakali: Autrimpo*, 16) with a four, the "twin deity" (*Trimps*, 4). The twin deity "holds power" (*Ghasitana: Stomp*, 121) to twin the "third house" (=7) by "stomping" (*Trimps*, 4) the fourth from the fifth to "create" (=35) the sixth as the "emanating hand" (=134) of the "first" (=1) with a "perfect storm" (*Bhadrakali: Autrimpo*, 16) of the "tenth's" (=-12) "creative force" (=6). The emanating hand comprises the space's "sixteen hands" (=134). The first comprises the time's "three eyes" (*Trinetra*, 1). The perfect storm is a "three-eye, sixteen-hand entity" (*Bhadrakali: Autrimpo*, 16). The "entity" (*Castor*, 24) comprises the causation's "thirteen legs" (=24) which "twin" (*Pollux*, 121) one, the deity, with a four into "two deities" (*Trimps*, 4) to "live" (=14) as the "fourteen deities" (*Maha Lakshmi: Dioscuri*, 14).

The potential of two makes fourteen deities a "perfect deity" (*Maha Lakshmi: Dioscuri*, 14). Thus, the perfect storm is a "formative deity" (*Bhadrakali: Autrimpo*, 16), comprising "sixteen deities" (*Bhadrakali: Autrimpo*, 16). The sixteen deities include: "Embracer" (*Bagalamukhi: Lacedaemon*, 9), "Empowerer" (*Pratyangira: Sparta*, 290), "Decomposer" (*Mahishamardini: Diomede*, 19), "Weakener" (*Matangi: Amyclas*, 999), "Brightener" (*Vishalakshi: Eurydice*, 94), "Darkener" (*Dhumavati: Danae*, 7), "Enabler" (*Taritni: Perseus*, 20,000), "Disabler" (*Ugratara: Gorgophone*, 261), "Precomposer" (*Ekajata: Oebalus*, 900), "Postcomposer" (*Katyayani: Tyndareus*, 190), "Discolorant" (*Chhinnamasta: Leda*, 10^{10}), "Colorant" (*Nilasaraswati: Clytemnestra*, 1,500), "Blackener" (*Jayadurga: Timandra*, 1,200), "Whitener" (*Navadurga: Phoebe*, 680), "Lightener" (*Vashuli: Philonoe*, 191), and "Recomposer" (*Bhadrakaali: Autrimpo*, 16).

The "Embracer" (*Bagalamukhi: Lacedaemon*, 9) twins her "gender" (=9) with the "spirit" (=20) of the "Empowerer" (*Pratyangira: Sparta*, 290) for "channeling" (=94) one as the "Decomposer" (*Mahishamardini: Diomede*, 19), "decomposing" (=93) the "Weakener" (*Matangi: Amyclas*, 999 =1,000 -1) into the "Brightener" (*Vishalakshi: Eurydice*, 94 =[7+2], 4). Brightener becomes "Darkener" (*Dhumavati: Danae*, 7) within the "entity" (*Castor*, 24 =20+4) whose spirit is "adding four" (=6,000) with the space's "white flame" (=6000) "wrapped" (*Zeus*, 0) within the "future" (=0). The Darkener becomes the "Enabler" (*Taritni: Perseus*, 20,000 = 6,000*4 -4 *1,000) of the "Disabler" (*Ugratara: Gorgophone*, 261 = [24+2], 1) of the "Sun" (*Proteus*, 21) as the "future reality" (=21). The Sun "incarnates" (=900) as the "Precomposer" (*Ekajata: Oebalus*, 900) of the one who is the "Postcomposer" (*Katyayani: Tyndareus*, 190) of zero into a "Discolorant" (*Chhinnamasta: Leda*, 10^{10}). The Discolorant "discolors" (=14 =10 +4) the "entity" (*Castor*, 24 =10 +14) by "adding four" (=6,000) to her "spirit" (=20 =10 +10) to be the "Colorant" (*Nilasaraswati: Clytemnestra*, 1,500 = 6,000/4) of each one. "Each one" (*Jayadurga: Timandra*, 1,200) is the "Blackener" (*Jayadurga: Timandra*, 1,200) of the three reproductive as the "Colorant" (=1,500 =1,200 +3*100). The three is the "Whitener" (*Navadurga: Phoebe*, 680 = 1,500/3 +60*3) of the "light" (=180) of its "octave" (=60). The two that make the three reproductive with its reality is the "Lightener" (*Vashuli: Philonoe*, 191). The two comprise the "two ones" (*Proteus*, 21) that perpetuate the "reality" (=7) as "real" (=9) by making the three reproductive as nine. The one who twins the three with the two is the "Recomposer" *(Bhadrakali: Autrimpo*, 16).

The "Recomposer" (=16) "recomposes" (=12) the spirit by wishing a "normative truth" (*Hippocoon*, 20) of the four as the "wishable" (=20) to make the three "self-luminous" (=12 = 4 *3) with a one within two. The two's "transformative truth" (*Icarius*, 27) "develops" (=22) "primordiality" (=27) over the one's "formative truth" (*Heracles*, 190) the three reproduce within the "future" (=0). The two's "replication" (*Menelaus*, 90) of his "darkness" (*Agamemnon*, 180), "raining" (*Ares*, 3) three's "reproductive force" (*Anaxibia*, 100), is "demonic" (*Tantalus*, 570 = 90 +180 +3*100). The demonic is a "shadow entity" (=570) "illuminated" (=10,000) by the spirit as a "female" (*Pluoto*, 10,000) with the "dynamism" (*Pluto*, 16) one as an "advisor" (*Tmolus*, 1,400 = 19 -5, 3 +19 +74 +5 −1). As an advisor, one is "guided" (=3) by "intuitive knowing" (*Theogone*, 19) to be the "Miracle" (*Arrhippe*, 74) of "God" (=5). Therefore, as an "advisee" (*Paramarshi*, 92 = 19 +74 −1), the two start behaving "Almighty" (=92) as if they are the "ecosystem" (*Thyestes*, 92). The three becomes the "Almighty Creation" (*Pelopeia*, 28) with the "convergent energy" (=28) of the one as a two and the two as an eight (=2³)—"natural gravity" (*Gajakarana: Aigisthos*, 8).

"Mother Nature's" (*Dascylus*, 8) "gravity" (*Eupryto*, 629) is nature herself as "Father Nature" (*Anthemoeisia*, 1) "gravitating" (=-100) the "real" (*Lycus*, 9) with his "gravitational energy" (*Eurythemista*, 100) for destroying the "imaginary" (*Otreus*, 20) which is perpetuating the "illusionary" (*Priolas*, 87) as the "reality" (*Boeotus*, 7) of her "female body" (*Alcyone*, 9) to "twin" (*Poseidon*, 121) itself as the "husband" (=121). The "illusionary" (=87) produces the one as a "nonlinear wave" (*Thanatos*, 1 =8-7) which "squares itself" (=111) for "exposing" (=111) the "feminine" (=37) to the "sound wave" (*Niobe*, 149 =1+111+37) of the "togetherness" (=486 =296 +100 +90) with the "masculine" (=296 =87+149 +60) as the feminine becomes "reproductive" (=100) for the "replication" (=90) of the masculine within the "octave of sound" (*Tityos*, 60). The "imaginary" (=20) produces the "octave of three" (=60) as a "linear wave" (*Hypnos*, 1) which "points itself" (=40) with the "oneness" (=48) of its "reproductive reality" (=40) to "silence" (*Broteas*, 168 =20+60+40+48) the "sound" (*Hermione*, 257 =1 +40 +48 +168) of "togetherness" (=486) with the "ultrasound" (*Pelops*, 130) of "otherness" (=258 =168 +130-40).

The "real" (=9) produces the "illusionary" (=87) as an illusionary, i.e., "parabolic wave" (*Philotes*, 1) to "consume" (=467 =1 +87 +111 +168 +100) her "future value" (=2) as a three which "squares itself" (=111) with the one, "Father Nature" (=1), for "exposing" (=111) the "feminine" (=37) to her "reality" (=7) with his "future reality" (=21) of the "Sun" (=21) as the "silence" (=168) becomes "reproductive" (=100). The "silence" (=169 =78 +1 +90) "reproduces" (=78) the "reality" (=7) of "Mother Nature" (=8) as a "circular wave" (*Epiphron*, -9) to exchange one's "present reality" (=-2) with a "complex" (=1,000) of "femininity" (*Clytie*, 1,000) that makes the "replication" (=90) of the "future" (=0) "simple" (=9) with "three zeroes" (=0). The "reality" (=7) "reproduces" (=78) "itself" (=121) as a "point" (=10^{10}) within the "square wave" (*Petulantia*, 1), the "twin wavefront" (*Petulantia*, 1), whose future is a "triangular wave" (*Hybris*, 0), the "wavefront" (*Hybris*, 0).

The "future" (=0) "reproduces" (=78) the "point" (=10^{10}) as a "parallel wave" (*Varitaranga: Dyssebeia*, 1,000,000), the "triple wavefront" (*Dyssebeia*, 1,000,000), of the "female" (=10,000) to "contribute" (*Vari*, 79) her "replication" (=90) as the "water" (=169 =89+90). The "female" (=10,000) "points" (=10^{10}) to itself as the "point wave" (*Nirataranga: Impietas*, 10,000), the "half wavefront" (*Impietas*, 10,000) of the "male" (=10^8). The male "juices" (*Nira*, 169) her "water" (=169) to be reproductive as a "Sun" (=21 =100 +90 −169) with the "replication" (=90) of the three as a "Moon" (=997). Moon makes "things" (=9) "complex" (=1,000) to keep the three's "reality" (=7) "simple" (=9). The three is the "real absolute gravitational constant" (*Euryanassa*, 3 =4−1) which becomes a "reality" (=7 =3+4) after adding four as the "planets" (*Eris*, −1). The "absolute gravitational constant" (*Graeae*, 3) twins the planets assuming that the four as a "planet" (=-1) is a "three" (=3=4−1) which it must reproduce to be "real" (=9 =3^2).

The "Sun" (=21) "creates" (=35) an "imaginary absolute gravitational constant" (*Discordia*, 3) for the "Moon" (=997 =900 +78 +19) to "reproduce" (=78 =21 +35 +3 +19) a "clarified consciousness" (=19) of the "assumption's" (*OM*, 19) "transformative potential" (=111 =3 +78 +20) as the "imaginary" (=20) "incarnates" (=900) the one as a two within "planet" (=-1). The two "experiences" (=9,000) itself to be the "eldest" (*Moros: Jyeshtha*, 9,000) because it "may" (*Moros: Jyeshtha*, 9,000) be the "third" (*Moirai*, 100,000) after "both" (=10,000) the "father" (=3) as a three and the "mother" (=4) as a two which twins itself to be a four. As the two becomes four, her "origin" (=1,000) as an "entity" (=24) becomes "limitless" (*Sephira*, 1,024). The "entity" (=24) "ends" (=130 =7+123) the "spring" (*Ananke*, 811) as the "forward thermodynamic limit" (=12) for the "growth" (=6) of one's "reality" (=7) becomes "less" (=123) when the two "begins" (=13) the "summer" (*Mithras*, 967 =811 +123 +13 +6 +4) as a four for the "post-composition" (*Sophrosyne*, 70) as a "Gemini" (*Sophrosyne*, 70 =3 +6,7) of a three with six within the reality of a one.

The "ambition" (*Abhiruci*, 980) for "endeavor" (=12) becomes "evident" (=980) within one's "nature" (=8) reproduced by the two as a "soul essence community" (*Euphrosyne*, 16) for the space's "continuous transformation" (=3 x 10^{17}) with the "Gemini-effect" (=3 x 10^{17}). The soul essence community makes the "configuration" (*Euphrosyne*, 16) of the three "ambitious" (*Keres*, 85 = 8, 2+3), "aiming" (*Ker: Iharthin*, 185) to be "reproductive" (=100) for the "discontinuity" (*Oneiros*, 200) of her "present self" (=480) through "oneness" (=48) with their "primordial self" (=10) which ensures two's "continuity" (*Styx*, 50) within three as a "five" (=5 =2+3), the "perpetuator deity" (=5). Five does not need three to be present within the "infinity" (=90,000 =5 *18,000) of the "space" (=18,000) left for four when one moves to the right as a two with the "potential" (=18) within zero to be the "time" (=360) before becoming a three as an "octave" (=60) which "orbits" (=2) two as an "orbital" (=60).

The "linear-effect" (=280) of two on the "four-dimensions" (=18,000) of "space" (=18,000) is an "octahedral" (=280), "JT" (=280 = 60*4 +10*4). "J" (=60) is the orbital which twins two as a ten with the four, "T" (=4), when the two is "repeated" (=5) as "two fives" (*Parvati*, 10) like a "potential octave" (*Parvati*, 10). As a one enjoying the "effect" (=34) of the two from the three which twins the two, the "octahedral" (=280) has the potential for "reproducing" (=17) the "linear" (=497) as the "octahedral-effect" (=480 =497-34+17), thus producing the effect of two within the "hexahedral" (=200 =480-280). The hexahedral is a "distortion" (=200) of six which "edges" (=8) a "discontinuity of two" (=200) after the three becomes reproductive as time.

The octahedral-effect is the "JT effect" (=480), known as "Jahn-Teller effect" (=480). It is also the "primordial perpetuator-effect" (=480) because it "perpetuates" (=9) the "effect" (=34) of the "potential octave" (=10) of two within the "hexahedral" (=200) as an "octahedron" (=185 =200 +9 −34 +10). The "potential octave" (=10 =1 +9) is the one within the "primordial perpetuator" (=9) which becomes a two when repeated as a five and ends up becoming "two fives" (=10). It is also the "Christ-effect" (=480) because the "Christ" (=1,600) is "present" (=1,600) as a "departed entity" (=1,600) with a "potential" (=18) of an "octave" (=60) within the "nature" (=8) who twins the "potential octave" (=10) with a two to be the three repeated as a five. The Christ-effect is the "potential twin octave" (=480 =8*60).

The "octahedron" (=185) is a "twin pyramid" (=185) with "twelve heads" (=185) which converge into a "base" (=10) of "three vertices" (=10) to triangulate the "four faces" (=5 x 10^{96}) of the "convergence" (=158) "as one" (=5 x 10^{96}), the "fifth face" (=6 x 10^{192}). The fifth face is repeated as the "eight faces" (=999) as the three vertices "edge" (=8) the one into "three faces" (=18) with their potential for "divergence" (=170) as the "twelve edges" (=90) of the "future" (=0) "entwined" (=8,000) as the "six vertices" (8,000) with the "two zeroes" (=-10). The two zeroes are "kept away" (=−10) as the "twin octahedron" (=-10) with "twenty-four heads" (=−10) when the "two tails" (=-13) "head" (=-9) as a "four" (−9− [−13]) to "tail" (=70) an "orbit" (=60) of two as a "octahedron" (=185 =70 +60*2 -10/2).

The "twelve edges" (=90) are a "twin hexahedral" (=90), a "dramatic distortion" (=90) of twelve into a "three face nine hand entity" (=90) like a "herringbone" (=90). "Herring" (*Samketa*, 4) is a "clue" (=4) for the four to be a "bone" (=86) which twins four with a two when the three becomes reproductive to force the two to be a six with the four from the base immanent within it. The "base" (=10) makes the "twelve edges (=90) "heavy" (=1,000 =10*90 +10*10) as the six reproduces the three's reproductive force with its growth within the twelve edges for emanating the base as the "ten faces" (=8) of the two. The *two "hand"* (=0) the "potential" (=18) for their "reproduction" (=285) to the "triple hexahedral" (=85) for becoming "light" (=180) through the "replication" (=90) of each as a "twin hexahedral" (=90). Of the "ten faces" (=8), *seven "hand"* (=0) the potential of "three faces" (=18 =8-7, 8) to the six as an "entity" (=24) with the "heaviness" (=24) of "nine hands" (=90,000).

The "triple hexahedral" (=85) has "eighteen edges" (=85), repeated as the "twelve edges" (=90) when the growth of "six edges" (=200 = 85 +111 +4) "triples itself" (=111) with the "potential of two" (=18) which "squares itself" (=111) with the "present of four" (*SAUM*, 1,600) as the "one fused" (=8) with three. One fused is the "edge" (=8) as the "triple octahedron" (=8). After "servicing" (=47) its "potential" (=18) as the "twin octahedron" (=-10), the "edge" (=8) is "repeated" (=5) as an "octahedron" (=185) with the "one diffused" (=89) with two. The one diffused is the "vertex" (=89) of two, infused within four as the six that fuse two with the one which refuses to meet the "stance" (=89) "met" (=89) by the eight who "edges" (=8) the nine into "two directions" (=89): "tropical" (=385) and "sidereal" (=380). The vertex is the "tropical edge" (=89) of two. The edge is the "sidereal vertex" (=8) of three, "expanded" (=8) as the "sidereal edge" (=10) of three "orthogonal" (=10) to the potential, the "tropical vertex" (=18) of two. The sidereal edge has "eight edges" (=10) on the left that form the base for the "four edges" (=89) on the right. The four edges constitute the tropical edge.

A "hexahedron's (=285) "eight edges" (=10) on the "left" (=89) twin the "vertex" (=89) for "centering" (=10) the "growth" (=6 =10−4) of "six faces" (=−4) with the "six vertices" (=8,000) within the "four edges" (=89) on the "right" (=71). The "hexahedron" (=285) is the "reproduction" (=−285) of the "two directions" (=89) within the all-converging "four directions" (=100) that diverge as the "six directions" (=1,600). The six directions are present as the "one direction" (=97) fused as the all-diverging "three directions" (=70,000) which fuse "two directions" (=80) to refuse the "one direction" (=97) which meets with the all-inversing "eight directions" (=80) including itself. The two directions are the "lateral edge" (*Ajjhattikani*, 80) on the "inner side" (=380), which is "sidereal" (=380) whose "side" (=380) diverges the "inner" (=53) as the "real" (=9) when the three becomes reproductive and the two is repeated with the three as a five.

The eight directions are the "lateral vertex" (*Bahirani*, 80) on the "outer side" (=80,000). The outer side is a "parallel" (=80,000) of "two sides" (=80,000) repeated with the "three sides" (=89) on the left which face the "five sides" (=182) on the right. The inner side faces the four outer sides in two directions working like the two legs, knowing that the inner side is the body which faces four—two feet and two hands for manifesting the "central side" (=126) "looking aside" (=126) in one direction which meets the "convergent face" (=140) with "eight parts" (=140). The eight parts are the head, two eyes, two ears, a nose, a mouth, and a neck.

The "six sides" (=126) of the central side include the two on the outer side and the "four sides" (=75) that "reside" (=75) within the "working force" (=75) of the "eight sides" (=90) "beside" (=90) the "backside" (=8) of the "twelve sides" (=8). On the "inside" (=85) are the "sixteen sides" as the eight sides are repeated within the twelve sides for creating the "outside" (=10) with the "twenty sides" (=10), perpetuating the two sides parallel to their "frontside" (=19) of the "eighteen sides" (=19) for illuminating the "divergent face" (=27) with "seven parts" (=27). The seven parts are the body, two hands, two legs, and two feet, liberating the "parallel face" (=1,551) with "sixteen parts" (=1,551). The sixteen parts are the face, the eight parts within the face, and the seven parts without the face.

A "twin hexahedron" (*Ganesha*, 570) has "24 edges" (=1,111), "12 vertices" (=121), and "6 faces" (-4). Its "five parts" (=570) become six "with a face" (=570) of the "observer" (=0 =1+2+1-4) observing its "twenty edges" (=1,000). As a spinning "ellipse" (=1,000), the twenty edges make each "part" (=-1) a "vertex" (=89) with the spacetime's "seven vertices" (=12) as a "creature" (=12) within the space's "four edges" (=89) that the time "copies" (=0) as its "future" (=0) to be the observer. The observer's "reality" (=7) is observable in "three directions" (=70,000) with a "ninety-degree rotation" (=1,000) of the four edges as a "spinner" (*Clotho*, 37) within the "fourth rotation" (=963). The "fourth rotation" (=963) "restores" (=816) the "initial spatial state" (=-12) for the "continuity" (=50) of the "octahedron" (=185 =963-816-12+50) within the "reproduction" (=285) of the "state" (*Sthiti*, 100) as a "hexahedron" (=285). The "twin hexahedron's" (=570) "five parts" (=570) are "handless" (=570) because they "hand" (=0) "five digits" (=10,000) to the "four parts" (=1,000), including two hands and two legs.

A "triple hexahedron" (=855) has "36 edges" (=−11), "16 vertices" (=900,000), and "6 faces" (=-4). Sixteen vertices "catalyze" (=900,000) the "fourth rotation" (=963) as a "catalyst" (=963) for the "multiplication" (=999) of the "spinner" (*Clotho*, 37) as a "half hexahedron" (*Lachesis*, 999), with "6 edges" (=200), "4 vertices (=9 x 10^{16}), and "6 faces" (=-4), to form one as a "square" (=18,000). The square is a "half octahedron" (=18,000 =90*−4*−50), with "12 edges" (=90), "4 faces" (=−4) and "0 vertex"

(=−50) of the space "traveling" (=-50) as "time" (=360). Thirty-six edges are the "traveler" (=-11), "traveling" (=-50) like the "three squares" (*Ishvarya*, 12) as the "fourth square" (=0) "travels" (=8) like an "observer" (=0) observing the "traveled" (*Padakranta*, 41) as the "cost" (=41) of a "half hexahedral" (*Atropos*, 134). The half hexahedral is "travelable" (=134) as one within the "continuity" (=50) of the "octahedron" (=185 =134 +1 +50) "breeding" (=15) it as a "hexahedral" (=200) with a "half octahedral" (=15), "two circles" (=15) which "edge" (=8) the "third circle" (=134) to "square itself" (=111).

The "triple hexahedron" (=855) has "twelve parts" (=855)—five parts are "with a face" (=570) as the "absolute differentiation" (=570) of the "seven parts" (=27). The five parts include:

- the "spatial body" (*Bahirdhakaya*, 169) as the "outer body" (*Bahirdhakaya*, 169 = 67 +100 +2) of water that transforms into "Mars" (*Mangala*, 102) when the "galactic core" (=67) becomes "reproductive" (=100) as a "star" (=2) within the "feminine" (=37) for "producing" (=8) a "physical body" (=387) with the star's "four-fold growth" (=8 =2*4).

- the "temporal body" (*Adhyatamakaya*, 83) as the "inner body" (*Adhyatamakaya*, 83) of "fire" (=17) that norms "Saturn" (=−1) with its potential to be the "intellectual body" (*Sukshma sharira*, 306) the future mediates to "channel" (=36) "two bodies" (=56)—middle and lower—with its upper body.

  ○ the "future body" (*Shesha Sharira*, 51 =17*3) is the "upper body" (*Rahu*, 73), the primordial past of the "astral body" (*Linga sharira*, 3). The astral body is the "moderating body" (=3) moderating the present of the earth with its "first position" (=51) as the "heliocentric" (=51) "quantum core" (=22) that forms the "Uranus" (=73 =51 +22).

  ○ The "present body" (*Nirvanic sharira*, 32,223) as the "middle body" (*Bhu*, 724), the Earth whose "self-body" (*Svakaya*, 1) is mediating the past of the "causal body" (*Karana sharira*, 30) with "self-consciousness" (=2,222) of the "air's" (=385) reproductive element within the astral body.

- The "past body" (*Snigdha sharira*, 835) is the "lower body" (*Ketu*, 140 = 35*4), the primeval past of the "mediator body" (=8) of "ether" (=285), mediating with the "memory" (=35) of the "consciousness" (=4) for "reproducing" "Neptune" (=140) as the "base" (=10) with a potential to be the "etheric body" (=957 =835+140-18).

The "part" (=−1) that differentiates the present into the "five parts" (=570) is the "causation body" (=169), the "central body" (=169) making the "astral body" (=3) reproductive for integrating the "three parts" (*Meghavati*, 300) as a "half body" (*Meghavati*, 300) of divine which "refuses" (=88) the causation body into the "mental body" (*Manomaya sharira*, 381 =300 +169−88) with its present as the "Mercury" (=1,600).

The "art" (*Kala*, 360) of integrating the present as a "p" (=160), the "illuminated" (=160 =21 *8 −8) element, is perpetuated by the Sun as the "luminous body" (*Hiranyamaya sharira*, 21) of guider for its "eightfold growth" (=8) into "Venus" (=2,700), the perfect past, "dividing four" (=280) "parts" (=−1) of "time" (=360) into a "triple octave" (=80) within "Jupiter" (=1,780). The illuminated element "twins triple octave" (*Mahachitti*, 160) to be self-luminous as an "entity" (=24) when the luminous body "triples" (=130) its reality with the "astral body" (=3) to be reproductive as the "cause body" (=30).

"AT" (=-19) is the "desire" (=-19) liberated by "R" (=111) to "guide" (=111) the "T's" (=4) "consciousness" (=4) of "A" (=28) as "bodyless" (*Devasvarani*, 88 =111 −19 −4) without the "octave" (=60). The octave "copies" (=0) a "part" (=−1) of the "desire" (=−19) as a "wish" (=18) one may fulfill with "potential" (=18) as a "Wisher" (=0) "everyone" (=180) copies from the "primordial self" (*Shri Krishna*, 10) with "growth" (=6) as a "star" (=2) within the "soul essence community" (*Euphrosyne*, 16).

Each copy as a wisher has a "unique proportion" (*Hitashin*, 7,000) of "femininity" (=1,000), guiding it as a "sentient entity" (=7) with "primordial energy" (*Adi shakti*, 15) of "nature" (*Kudrat*, 8). "Descending proportion" (=6) of femininity within each "reproduction" (=285) makes the "masculine" (=296) element reproductive for the "present growth" (=11) of the "primordial dimension" (=−10) as a "person" (=1) "reproducing" (=17) the guider element with the consciousness of the "being energy" (=96).

A "destroyer deity" (=6) destroys the "undistributed masculinity" (=1,024) as the "soul essence" (=1,024) of an entity's consciousness, freeing the "illuminated universe" (*Bhadravasa*, 889)—the "renewable space" (*Bhadravasa*, 889) of her "femininity" (=1,000) to be perpetuated through "entrepreneurship action" (=111). The entrepreneurship action "renews" (=111) the "discontinuity" (=200) of time within the "spatial face" (=800) for the "continuity" (=50) of space without the "temporal face" (=1,000) as an "accelerating universe" (=1,000) that "repels" (=9) two "galaxies" (=85 =[111 +50 +9]/2).

The "first galaxy" (=99) is a "decelerating universe" (=99 =85 +12 +2) that "twins" (=121 =85+34+2) itself as a "galaxy" (=85) to "attract" (=34) the accelerating universe as the "second galaxy" (=1,000) from its future through "exoproduction" (=121) after it is "endoproduced" (=12) as the "past" (=9) of the "third galaxy" (=1). The "third galaxy" (*Maya*, 1) is the "coasting universe" (*Maya*, 1) "produced" (=9,000) as the "present" (=1,600) of the "fourth galaxy" (=2) by the "potential" (=18) for "growth" (=6) of the "sixth galaxy" (=111) as a twin of the "seventh galaxy" (=132 =111+1+18+2) within the "fifth galaxy" (=3,700 =[9,000−1,600]/2).

The "seventh galaxy" (=132) "attracts" (=34) the "mass" (=132) of the "second galaxy" (=1,000 =132+34+700+111+99−76) for "accelerating" (=700) the future of the "first galaxy" (=99) within the "sixth galaxy" (=111) through "entrepreneurship action" (=111) of the "whole face" (=76) "endoproduced" (=12) as the "eighth galaxy" (=76). The "sixth galaxy" (=111) "repels" (=9) "two galaxies" (=9,000,000 =9*1,000*1,000) as the "past" (=9) of the "second galaxy" (=1,000) for "decelerating" (=6) the "present" (=1,600 = −16 *−100) without itself. The "future" (=0) "coasts" (=−16) the "illusion" (=1) of "twelve galaxies" (=2) as a "universe" (=2) to "gravitate" (=−100) itself. The universe is the "fourth galaxy" (=1,000) which "causes" (=18) a "ten-fold growth" (=170) within itself for "oneness" (=48) with the "spatial face" (=800) of its "potential" (=18).

"Intrinsic alignment" (=14) of the tenth, "past galaxy" (=70,000), assemblable into "ten galaxies" (=70,000), with the eleventh, "present galaxy" (=9,000,000 =70,000/14*18,000/10), resembling "two galaxies" (=9,000,000), "squares" (=18,000) the "lifetime" (=125) of the ninth, "future galaxy" (=135) which "twins" (=121) the "effect" (=34) from the twelfth, "assembler galaxy" (=87) into a "twin assembler galaxy" (=135). The effect is the "energized-effect" (=34) of the time's "three circles" (=10) as a "quiescent system" (=10), which when "energized" (=10) into a "massive object" (=100), residing as "supermassive" (=135) as the effect "transforms" (=1) into the "massive quiescent galaxy" (=135)—a "high redshift quiescent galaxy" (=135).

An "assembler galaxy" (=87) is a "time galaxy" (=87) "interacting" (*Ekkekke*, 480) as a "low redshift quiescent galaxy" (=87) with its "three radii" (=25) to "bisect" (=25) the "pressure" (=62) "imposed" (=25 =13+14-2) on its "gravitational center" (=629) by the thirteenth, disassembled, spiral "space galaxy" (=13) through "conscious polarization" (=14) of the fourteenth, "causation galaxy" (=-2) with the "constellation of effects" (*Duniya*, -2) of its "present reality" (=-2).

The time galaxy is a "barred spiral galaxy" (=87) with "two arms" (=19) of the "past" (=9) as "illusionary" (=87) and the "future" (=0) as "imaginary" (=20) that "central bars" (=7) the "reality" (=7) "present" (=1,600) as the "shared value" (=130) of their "supernatural growth" (=8) into an "OB star" (=8)—the "earth star" (=8).

The shared value is the "OB" (=130) which "bulges" (=130) the two as a "star" (=2) with the "four arms" (=45) of the "shared space" (=90,000) to "surface" (=2,000) a "galaxy" (=85) for "centering" (=10) an "octave" (=60) as a "cause galaxy" (=610). The light of "O" (=180) "spreading" (=7) "B" (=7) as its "illuminating value" (=7) is the "managed growth" (=130) of "Q" (=1,000) for its "femininity" (=1,000 =180 +7 +130 +724 -37) to "source" (=1,000) the "feminine" (=37) element from the "Earth" (=724).

The fifteenth, "cause galaxy" (=610) is "burstable" (=487) for "assembling" (=18) time-dependent "ascending order" (*Arohana*, 105) into the sixteenth "entity galaxy" (=40) to "develop" (=22) "present interaction" (=22) of the "two galactic-effects" (=22) within "space galaxy" (=13) through the "path of entrepreneurship action" (=9).

5.4 The Path of Entrepreneurship Action.

The "path of entrepreneurship action" (*Waheguru Marga*, 9) is the "path of least energy" (*Waheguru Marga*, 9) because it "perpetuates" (=9) the past to "guide" (=111) the "endoproduction" (=1) of the future as a "twin" (=121) which reproduces itself as the present to remain "omnipresent" (=1). "Primordial technological servicing" (=170) of the "fire" (=17) within the "future" (=0), for "producing" (=17) itself by transforming the present with its potential to endoproduce its past, limits growth to one's "experience" (=9,000) of that past. The "fire's" (=17) "wholesome variability" (=67) "impedes" (=180) the "continuity" (=50) of the "ideal order" (=130) "necessary" (=470) to "intensify" (=340) the "normality" (=340) of a "wishable" (*Ishta*, 20) before its "creation" (=379) as a "strange quark" (=1) within the "mind" (=38).

A "charm attraction" (=268) for an "alternative form" (=1) of the "wishable" (=20) "creates" (=35) a "heavy quark" (=1) as the "discontinuity" (=200) of time is "endoproduced" (=12) by the "mind" (=38) to "twin causation" (=2). The "Illuminator deity" (=7) "illuminates" (=150) the "continuity" (=50) within the "discontinuity" (=200) to "order" (=2) the "consciousness" (=4) of the "ideal" (=1) with a "theory" (=127) of mind that accelerates the presence of ideal-effect for copying the "theory-effect" (=0). The effect of theory is zero because its future is zero without knowing the reality of the ideal's potential as a "protagonist" (=1) for growth by "decelerating" (=6) the "reproductive force" (=100) within the "present" (=1,600 = 1,6 *100.

5.4.1 Understanding Illuminator Deity

The "bliss consciousness" (=185) of one's "potential" (=18) to be the "ideal" (=1) for a "group" (Gana, 387 =185*2 +18-1) of two produces an "ultrasound" (=130) of the three. The three are "other" (=9/16), of which two are "together" (=6,900) as the "potential one" (=130).

The "third" (=100,000) "forms" (=100,000 = [18,000 +20*100] *5) as a "triple" (=130) to be "together" (=6,900 = [1,500+16*50] *3) with the "fourth" (=18,000) as the "fourth-effect" (=20). The fourth-effect is "reproductive" (=100) because it is "repeated" (=5) with the "fifth" (=1,500) as a "whole" (=16) within the "fourth" (=18,000), the space that forms the third with its "continuity" (=50) as the "multiplier" (=3) of time's "discontinuity" (=200) to be "present" (=1,600 =3*200 +1,000) as the "origin" (=1,000) of the "second" (=90,000) within the "first's" (=1) "replication" (=90) of the "sixth" (=53) as a "feminine" (=37) with "mind" (=38) of her "own" (=1).

The feminine has her mind as the "multiplier" (=3) of the "masculine" (=296) with her "femininity" (=1,000 =3*296 +37 +1 +38 +36) as she "channels" (=36) the "masculinity" (=53) to "fire" (=17) "present growth" (=11) with her "creative force" (=6). The "creative force" (=6) is present within growth as the "ideal-effect" (=1) of the "Illuminator deity" (=7 =1+6) "endoproduced" (=12) by the "theory-effect" (=0) of the "Destroyer deity" (=6) to "order" (=2) "twin causation" (=2). The "twin causation" (=2) "transmediates" (=9) the "primeval reality" (=-7) as a "Primordial Perpetuator" (=9) with a "local path of self-development" (=34) of the "national" (=32) element within "oneness consciousness" (=42) of the "international" (=18) element as an "entity" (=24).

The "self-luminous consciousness" (=20) of the four repeated as a five after a two twins causation within a "sentient entity" (*Siddha*, 7) produces "freedom from the wisher" (*Iccha mukti*, 9) who is wishing the nine to be "real" (=9) so that he could be a ten with the ideal-effect. When five and two are "together" (=6,900 =[10+59] *100), ten becomes reproductive like a "mind-born creator" (=59) of a nine with the five after the "ideal-effect" (=1) leaves a "remainder" (=4) of four when it twin causation with the "theory-effect" (=0) which it counts as a two, its "future value" (=0). The ten liberates the Wisher:

- From the "present wish" (=720) to be an "ideal" (=1) who perpetuates the "ideal-effect" (=1) as the "causation" (=1) to twin "time" (=360) as an "entity" (=24) by "living" (=10) as a "twin entity" (=24) after "death" (=18) "channels" (=36) her "consciousness" (=4) as the "soul" (=4) of "another" (=98) "still alive" (=123) after one "counts" (=123) two with a zero within the "spirit" (=20) of the "wishable" (=20) with a "reproductive potential" (=23).

- From the "present manifestation" (=3) of the "theory" (=127) "endoproduced" (=12) by a "sentient entity" (=7) as the "fulfiller" (=16) of the "wish" (=18) "attached" (=9) to the "soul" (=4) from another's "past" (=9). Another wishes to be the "enjoyer" (=9) as one "fulfills" (=9) the wish as her "wishable" (=20) with a "consciousness" (=4) that she may "twin wish" (=70) to force another to double his work after reincarnation. However, the probability" (=128) that another doubles the work is zero because she is the one to "triple work" (=90) with the "endoreplication" (=90) of the "work" (=19) to be done within her "mind" (=38) before her "death" (=18) does the "soul" (=4) "apart" (=16) for its growth into a "twin entity" (=24) as she becomes a "para entity" (=19) with the "work" (=19) she does.

- From the "present work" (=130) as the "potential one" (=130) who is in "office" (=130) to "work" (=19) for "others" (=9/16) who believe she is the "fulfiller" (=16) "in-charge" (=134) of making them an "enjoyer" (=9) of her "soul" (=4) so that they may behave like "God" (=5) and do nothing but wish her as their "twin" (=121), the "spouse" (=121).

- From the "present knowledge" (=96) as a "living entity" (=96) of the one who "triples" (=130) "itself" (=9) as a "multiplier" (=3) of the "Wisher" (=0) immanent within her, thus manifesting her three forms as a "creature" (=12). The future is an entity that self-perpetuates her present as a creature to twin herself as a twin entity. The twin entity hibernates as a living entity without the "soul" (=4) that makes her "reproductive" (=100). The "energy" (=19) within the reproductive element makes the "wisher" (=0) a "para entity" (=19) for the "assumption" (=19) of a form appropriate to fulfill the wish with the growth of potential within the creature into his "present" (=1,600).

- From the "present creation" (=570) of two repeated as seven to be left with zero. Once two's "reality" (=7) is illuminated as an "illuminator deity" (=7), two is left with nothing but a zero. Without reality, two is just the "spirit" (=20) "self-condensing" (=5/2) itself into a "tangent" (=10) which "lives" (=14) as the "soul" (=4) of the three. Three is the past, present, and future of the one who "guides" (=111) the "potential one" (=130) with her "work" (=19) to be her "twin" (=121) who "perpetuates" (=9) her "reality" (=7) as a two.

- From the "present destruction" (=1,066) of three who becomes reproductive like six with the growth of the two within her because her reality is an eight, like Mother Nature. Eight becomes three because it is not the "reality" (=7) which "produces" (=1) "Mother Nature" (=8) as a two. Two, three, and five repeated when two and three are together, "altogether" (=10) are ten. Ten is the "primordial illuminator" (=10) of the "Illuminator deity" (=7) with a three as a two: ten and three, which makes one "whole" (=16) as a "primordial greeter" (=16).

A "primordial greeter" (=16) follows a "path of the union" (*Karma Yoga Marga*, 10) as a "Wisher" (=0) wishing to be a "greeter entity" (=−18) with "Mother Nature's" (=8) "feminine potential" (=−18). The "union" (=86) is between a two that becomes an eight after growth and a three the two makes six before growth becomes real. "Growth" (=6) makes the "union" (*Sangam*, 86) "consensual" (*Mantrasiddhi*, 516 =86*6) with "infinite devotion" (=274) to the "symmetry" (=250) with three as an "entity" (=24) who is a two but behaving like a four, the "mother" (=4), to "illuminate" (=150) the seven, the "param paternal" (=7), for her "endoproduction" (=1) as a five, the "param maternal" (=5). The param maternal is essential for a two to "enjoy" (=7) his reality as a seven. She gives birth to two, the universe comprising "herself" (=1) and "myself" (=1) within "yourself" (=1), the "greeter potential" (=1). The greeter potential is the reason for myself to be the "manifested value" (=1) "Mother Nature" (=8) "greets" (=64) with her "potential" (=18) to be "reproductive" (=100 =8*8+6*6) with a six for a "union" (=86) with "herself" (=1). This union is "consensual" (=516) because "herself" (=1) mediates the "capability" (=55) for "consensus" (*Mantra*, 16) with her "divinity" (*Siddhi*, 56).

The essence of the consensus is the "insight" (=16) one gets regarding oneself through one's endoproduction of a "twin copy" (=1). With growth, twin copy makes one "God" (=5). God is the "mediator" (=5) mediating the "purpose" (=38) of creating the twin as a copy of his "image" (=1) as the one who matters for the two to be present. The two is one "himself" (=5) without the consciousness of the three who becomes "reproductive" (=100) like the "two zeroes" (=-10) after she produces a one as her "primordial child" (=120). "Himself" (=5) is the "entity" (=24) who incarnates as the "primordial child" (=120) to perpetuate the three as his "primordial creation" (=9), the past which gives birth to his present as a "child" (=128) within "Mother Nature" (=8).

The "present child" (=19) is the "work" (=19) of "energy" (=19) "organizable" (=11) within "nature" (=8) with a three. The primordial greeter makes the three "luminous" (=13) with a ten for the "replication" (=90) of its "growth" (=6) as a "whole" (=16 =10+6) with a two. The "illusionary workculture-effect" (=1,000) of the three's "femininity" (=1,000) makes one a "Worker Deity" (=1) as if one is working like a "deity" (=1) to be a "Wisher deity" (=0) in the future whose wishes will be the command of everyone. When "everyone" (=180) "works" (=19) like a "deity" (=1), the "Devil" (=0) who is wishing becomes a "Satan" (=-1) for "blessing" (=1,000) with her "femininity" (=1,000) the one everyone is copying like an ideal to take his "position" (=999) as a "Wisher" (=0) to be the "king of deities" (=0) in the "future" (=0). When a "worker" (=1) is the ideal for everyone working like a "deity" (=1), the illusion of "work" (=19) that "cultures" (=9) everyone with a ten, the "primordial self" (=10), gives way to the "manifestor workculture" (=379) that a "feminine" (=37) cultures with her "gender" (=9) through "devotion" (=46).

Manifestor workculture is "illusionary oneness" (*Param Jnana Yoga*, 379) with the "primeval creation" (=57) one "creates" (=35) as a "God" (=5) after "viewing" (=379) the "primordial creation" (=9) to be the "perfect creature" (=379) with an "invisible face"

(*Samamnaya*, 250). Consequently, the primordial creation becomes the "last face" (*Valavala*, 1,111) in a "sequence" (=48) of three—"primeval creation" (=57), "param creation" (=570), and "primordial creation" (=9)—within the first—the greeter (=7/16)—following the "third" (=100,000)—the "creation" (=379 =48*3+76+124)—which "forms" (=100,000) the "face" (=76 =20+6+50) as the fifth—the "creature" (=12)—"leading" (=20) to the "growth" (=6) of "continuity" (=50 =250/5) of the fourth—the "creator" (=578)—that "lasts" (=124) as the "eighteenth face" (*Valavala*, 1,111) of the second—the "wisher" (=0).

The "symmetry" (=250) within the creature's "invisible face" (=250) "develops" (=22) a "potential" (=18) to be a "symmetrical face" (=1,729) of the "creation's" (=379) "origin" (=1,000) as an "octave of three" (=60) before becoming the third as the "thirteenth face" (=1,729) which "enjoys" (=7) "twelve faces" (=29) with the first two. The symmetrical face is the "highest face" (=1,729 =1,548+200-4-15) among the "initial period" (=1,729) of "six faces" (=-4) following the "discontinuity" (=200) within the "eventual period" (=1,548) of "nineteen faces" (=-15). Nineteen faces "norm" (=18) the creation's "cellular face" (=-15) as the "nineteenth face" (=-15).

The eventual "orbital period" (=1,548) "conceals" (*Gup*, 1,800) the "lower face" (*Ishamnaya*, 1,800) with the "initial period's" (=1729) "six faces" (=-4) within the long "vertical axis" (=9,000 = [1,548=48] *6) in "oneness" (=48) with the short, "horizontal axis" (=2) to make the elongating "diagonal axis" (=10) "elliptical" (=80) with an "ellipse" (=1,000) of the "eight faces" (=1,000) of the shortening "circular axis" (=3) of the "time multiplier" (=3). The elongated "square axis" (=10), the tangent, "reveals" (=1,800) the "higher face" (*Valavala*, 1,111) of the shortened "triangular axis" (*Aksha*, 17) of "seventeen faces" (=689) within "asymmetry" (=689). By "servicing" (=47) "itself" (=121) as the "seventh face" (=20), the "horizontal axis" (=2) twins the "consequential period" (*Adhamnaya*, 20,068) of "five faces" (=-3) with the "diagonal axis" (=10). The "diagonal axis" (=10) "develops" (=22) a "symmetry" (=250 =10 *[22+3]) of "three faces" (=18) with the "nineteen faces" (=-15) by "self-projecting" (=3/2) the "ellipse" (=1,000) as the "sixth face" (=1,548 =3/2*1,000+48). With their "two faces" (=15), twelve "descending" (=12) for "ascending" (=1) forty-eight with their "oneness" (=48), the "fifty faces" (=89) "inject" (=10,000) the "asymmetry-effect" (=56) of five within six. With the "repeated growth" (=56) of two faces into "ten faces" (=8) within the fifty faces, the "sixty faces" (=60) "reject" (=900,000) the "symmetry-effect" (=38) of three within eight. The asymmetry-effect is the "body" (=56). The symmetry-effect is the "mind" (=38).

The "asymmetry" (=689) of three within the creator's "normative face" (Shatkuta, 689) with "seventeen faces" (=689) "transforms" (=1) two to form four, and thus norms six with five, including the one that transforms to twin itself. The "consequential period" (Adhamnaya, 20,068) of "five faces" (=-3) leads the continuity of the "two faces" (=15) within the "sixth face" (Ishanamnaya, 1,548) with the "seventh face" (=20) to "greet" (=64) the "wisher" (=0) with the "four faces" (=5 x 10^{96}) "as one" (=5 x 10^{96}) that forms "infinity" (=90,000) as the "half face" (Adhamnaya, 20,068). The half face is the "formative face" (Adhamnaya, 20,068) which becomes the sixth, "lowest face" (Ishanamnaya, 1,548) after the first face of the "generation" (Amnaya, 89) becomes the "highest face" (=1,729) in the "eventual period" (=1,548) when it is left with the eighteenth, "last face" (Valavala, 1,111) before the creation becomes a cell with the nineteenth face.

The last face appears higher than the one on the right which is low compared to the two in the center, of which the first is lower than the second last. The second last disappears from the vision like the one before because it is the "low face" (Adhamnaya, 20,068) "left" (=89) within the twenty-first "transformative face" (Vajrapana, 89) as a "fuzzy number" (Vrit, 89) with "four edges" (Vrit, 89). The fuzzy number is "fuzzy" (=111) because it "renews" (=111) the "number" (=53) once it "develops" (=22) before its "decay" (=31) from the "guider" (=100) element. The guider element "guides" (=111) its "present growth" (=11) into "fifty faces" (Vajrapana, 89) after the one on the right becomes zero for dividing the one on the left by two to center four within the undivided one as a five. The growth of five into a six after the zero trades the "effect" (=34) of the one on the right to be the second takes the "number" (=53) "uphill" (=87 =53+34) and makes its "interesting" (=490).

An "interesting number" (=370) is a "supra complex number" (=370) as a "dimension" (=370) of the "past" (=9) "left" (=89) as the "nature" (=8) of the "creation" (=379) for "programming" (=615) the "future" (=0) of the "creature" (=12) with the "present" (=1,600 =379 +615 +578 +12 +8*2) of the "creator" (=578) by reproducing nature with a two. It appears "simple" (=9) and therefore has a "disproportionate probability" (=370) for its "rapid expansion" (=615) through programming until it disappears to be a "part" (=-1) of the "complex" (=1,000) that makes it "simple" (=9) by "servicing" (=47) "complexity" (=479). "Programming" (=615) makes the "number" (=53) "boring" (=700 =615 +53 +17+15) as it "bottlenecks" (=615) "planning" (=17) of the "reproduced four" (=15). The "boring number" (=615) is a "super complex number" (=615). A "super complex" (=140) is the "lower body" (=140) whose "descending mass" (=140 =53+15+73-1) becomes a "number" (=53) "breeding" (=15) the "ascending mass" (=73) of the "upper body" (=73) as the "supra complex" (=73) of a "planet" (-1) as "part" (=-1) of the "solar system" (=13) that forms a "creature" (=12).

A fuzzy number is a "supreme complex number" (=89 =13 +53 +23). A "supreme complex" (=13) is "luminous" (=13) as a "solar universe" (=13) within a "number" (=53) that twins itself into a "twin number" (=23), the "magnetic momentum number" (=23) with the solar universe's "reproductive potential" (=23). The "twin number" (=23) norms a "gaping hole" (=90 =23 +140 -73) between the "super complex" (=140) and the "supra complex" (=73) as the "para complex" (=90) because it is the "para complex number" (=23) forming "Sun" (=21) within the "lunar universe" (=23) as a "star" (=2). The gaping hole is the "Sloane's gap" (=90). The "Sloane" (=130) is a "primeval complex number" (=130) as it is the "reproducing one" (=130), reproducing the creature as an "integer" (=12) with an "infinite sequence" (=285) of "reproduction" (=285) of the "femininity" (=1,000) within the "complex" (=1,000) as the "masculinity" (=53) within the "number" (=53). It makes two "primordial" (=85) with the "spirit" (=20) of three that is reproductive as a "primeval complex" (=20).

The reproductive "gravitational energy" (*Lalita*, 100) is "wish-fulfilling" (=2) because it produces "continuity" (=50) of "space" (=18,000) for the "primordial creation" (=9) of one as the "sameness" (=18,000) of zero, the Wisher, "fulfill" (=9) the "wish" (=18) for "discontinuity" (=200) of the "reproducing one" (=130) after "producing" (=17) a "Wisher universe" (=-1) as "part" (=-1) of "planning" (=17) the "future of everybody" (=18). Thus, the Wisher becomes "immortal" (=90,000) "reproducible point" (=90,000), the "param complex number" (=90,000) reproducible as a "twin prime" (=90,000) by reproducing the prime as a "param complex" (=81). The param complex is the "twin infinity" (=81) which transforms the "number" (=53) into the "Sun" (=21) with the growth of its "reality" (=7) to enjoy "everlasting benefits" (=17) of the "primeval creation" (=57) of the three as a "cell" (=19) with the "reproduction" (=285) of two as a "param creation" (=570 =285*2).

The two "loops" (=900) the three into an "orbital loop" (=800) with one's "reproductive force" (=100) to twin the growth of six planets ascending and descending as a "solar universe" (=13) with "thirteen numbers" (=13). The "thirteenth number" (=1/2) is a "primordial complex number" (=1/2). It is a "rational number" (=1/2), which is half of the "primordial complex" (=16) that makes one "whole" (=16) as a "number" (=53) with the "feminine" (=37) element. The thirteen numbers become whole when the thirteenth number twins them into an "orbit" (=2) to be a "part" (=-1) of the "planet" (=-1) which "cycles" (=1/60) the "octave" (=60) of three as an "epicycle" (=25 =13*2−1) of "twenty-five planets" (=900). Epicycle "time loops" (=900) past, present, and future of the eight planets within the Sun as the "twenty-fifth planet" (=21) by "servicing" (=47) masculinity with the letter "E" (=53) as a "part" (=−1) of the "octave" (=60). "Pi cycle" (=8 = 60− [53−1]) "space loops" (=800) the East, the West, the South, and the North as the "four parts" (=1,000) of femininity within "Moon" (=997). Moon makes the three reproductive as a "whole" (=16) within the "solar universe" (=13) with the "continuity" (=50) within the "sixteen planets" (=800) as each "part" (=−1) forms a "planet" (=−1).

"Epi" (=9) comprises "1,080 pi" (=9): "four parts" (=1000) of femininity within pi add "three parts" (=300) of masculinity without pi as the "three octaves" (=80) that make three reproductive with their "reproductive reality" (=40). Epicycle is the planetary cycle. Pi cycle is the solar cycle. I cycle is the stellar cycle. A "star" (=2) "cycles" (1/60) its "infinite singularity" (=12) as a "Creature" (=12), the "I" (=12), to form the three into a "Creation" (=379 = 12 *3 +[12*5]/60, 9) by taking five as "epi" (=9), the past of the "triple octave" (=80) as a "part" (=-1) of the "I cycle" (=164 =80*2+5−1) to "twin triple octave" (=160) with the letter "P" (=160) the "discontinuity" (=200) of its "reproductive reality" (=40) as a "Creator" (=578 =379-1+200).

"Time" (=360) mediates its "reality" (=7) of the "Illuminator deity" (*Nataraja*, 7) as a "virgin" (*Lajja*, 58) to be the "creator" (=578) of "discontinuity" (=200) with its "potential" (=18) to "flow" (=666) continuously. The "continuous flow" (=666) of "twelve characteristics" (=666) as a "devoted creature" (=666) "idealizes" (=666) time's "three characteristics" (=16) with the "unique potential-effect" (=666) of the space's "four characteristics" (=4). The "fourth characteristic" (=0) has a "reproductive quality" (=0), repeated as the causation's "five characteristics" (=0) as the time "sings" (=123) with "joy" (=123) to be present as a three with the "characteristic" (=12) of a "creature" (=12). "Harmonic oneness" (=98) of the "twelve arms" (=98 =91+7 =9,8) of nature's "primordial creation" (=9) with the "divine tune" (=91) of the "Illuminator deity" (=7) makes the "characteristic" (=12) "luminous" (=13) as the "sentient life force" (*Prana*, 123). The "cause's" (=18) "two characteristics" (=10) mediate the luminous to make four conscious, i.e., "sentient" (=189), of "life" (=4) as its "sentient force" (=4). "Fourteen characteristics" (=780) "characterize" (=3) a "devotee creation" (=780) whose "chant" (=780 = [360-100]*3) "produces" (=1) an "effect" (=34) of the "division by two" (=814) into an entity's "seven characteristics" (=814) as it makes "time" (=360) "reproductive" (=100). The "cosmic dance" (*Tandava*, 286) of the 3*7, i.e., "twenty-one characteristics" (=286) "addends" (=286) its "vigor" (=286) for "normative performing" (=286) with "37 genes" (=286) as a "mitochondrion molecule" (=286).

Normative performing is "matrilineal performing" (=286) of "Mother Nature" (=8) as the mediator of the "Almighty Creator" (=26 =7+19), the "param deity" (*Shiva*, 7) within a "cell" (=19) "breeding" (=15) itself as the "Primeval Perpetuator" (*Mahakaal*, 15) through "meiosis" (*Param Shiva*, 15 =4*2+7) of four with a two for adding a deity's "nine characteristics" (=120 =15 *8) by "trading" (=53) the "memory" (=35) of the "primordial" (=85) element from Mother Nature as a "primordial deity" (=8). "Itself" (=121) becomes the "mother cell" (*Shennong*, 7), the "twin creation" (*Shennong*, 7), "descending" (=12) as a "param deity" (=7) within a "child" (=128 =121+7) to be the "param child" (=19 =12+7). The self's "attraction orientation" (*Param Bodhisattva*, 7) for "it" (=900) who "incarnates" (=900) with the "time loop" (=900) "perfects" (=10) one's "enlightenment" (=408) with the "continuity" (=50) of the "time value" (=500). Thus, one becomes a "white emperor" (*Baidi* 白帝, 7) whose reality as the "Absolute God" (*Param Bodhisattva*, 7) is "self-luminous" (=12) to "God" (=5).

By "centering" (*AI*, 10) "father" (*EL*, 3) as a "God" (=5) with a two for "shadowing" (=0) the "Absolute God" (=7 =3+4) as a "mother" (=4), a "King" (*Shad*, 0) becomes a "Queen" (*El Shaddai*, 7). After "trading" (=53) masculinity with the letter "E" (=53), the "Queen" (=7) "illuminates" (=150) the "concealed truth" (*Avidhi*, -4) of "herself" (=1) as "Mother Nature" (=8) by "servicing" (=47) the "shade" (=46) of "devotion" (=46) of her "centered" (=10) "primordial self" (*Parvati*, 10). The primordial self is the "illuminator" (*Shadad*, 10), "destroyable" (*Pragabhava*, 0) before the "illumination" (=1) of the "king" (=0) as a Queen. The Queen's "feminine" (=37) element makes her "devotion" (=46) to the "Lord of illumination" (=7) "real" (=9) and "illuminates" (=150 =37*2+76) the "Lord" (=0) as the King. With Queen, the King becomes an "institutional param deity" (*Amurru*, 7) like a "temple" (*Amurru*, 7) of the "face" (=76) of two with "mind" (=38) of their "own" (=1) because one is the "brain" (=38) that "copies" (=0) "another" (=98) to "develop" (=22) its "identity" (=8) as the four "lives" (=14) in its "time consciousness" (=14), making it conscious of its "potential" (=18).

With three that "times" (=360) "consciousness" (=4) as a four for its "replication" (=90) as a "century" (*Sadi*, 100) within seven, the "institutional param deity" (*Amurru*, 7) becomes "self-devoted" (*Belu Sadi*, 7) to his "twin reality" (*Belu*, 49) as "another" (=98) within the "continuity" (=50) of two. He "develops" (=22) into an "organization" (=29 =7+22) as "another" (*Beli Sheri*, 98) services her "growth" (*Sheri*, 6) to be the "ecosystem" (*Beli*, 92) that makes the "illuminator" (=10) "whole" (=16) before he becomes "luminous" (=13) with self's "reality" (=7) as an "entity" (=24 =13+7+4). The "entity's" (=24) "beauty" (=888 =24 +225 +38 -3 +500 +12 +92) is "destroyable" (=0) like a "desert" (=225) after the "mind" (=38) "copies" (=0) the "extrinsic sentient-effect" (=-3) of "time value" (=500) for "illuminating" (=12) the ecosystem as the "Garden of Eden" (=92). The Garden of Eden "reforms" (=888) the beauty as "true" (=8), but "false" (=8) without one's "reality" (=7) as "Mother Nature" (=8).

5.4.2 Transcending Illuminator Deity

"Devotion" (=46) to "reality" (=7) of the Illuminator Deity takes one to the "path of thirteen body system" (*Maha Siddha Marga*, 467 =46+7+1+169+142+12+90) for "consuming" (=169) "thirteen bodies" (=142) as a "system" (=12) of "replication" (=90). A "replicative system" (=142) is an "equant" (=142) that "quantizes" (=169) "everybody" (=-9) into "twelve bodies" (=134) as a "body" (=56) of two. It "remains" (=550 =169+360+21) "uniform" (=144) as a "Sun" (=21) after the "decay" (=31) of four that the "time" (=360 =169+56+144-9) "triples" (=130) with its "consciousness" (=4 =360/90). Two is a "star" (=2) which forms a "body" (=56) with one who "embodies" (=8) the "momentum" (=999) of the "past" (=9) with "nine bodies" (=14) after the "entropy" (=5) of the "three bodies" (=3).

Three bodies make the "position" (=999) of the "present" (=1,600) "reproductive" (=100) within the "thirteenth body" (=169) as a "body" (=56) of two "disembodies" (=100,000) "three bodies" (=3) with the "convergent energy" (=28) of eight within two before diverging six with the "time multiplier" (=3). Past's momentum has "eight bodies" (=130), norming a "potential one" (=130), which one embodies as the "ninth body" (=81) to twin itself with the "potential" (=18). The "ninth body" (=81) becomes "equidimension" (=142 =81+60+1 =121+21) with the "eleventh body" (=121) after one becomes a "Sun" (=21 =1+60/3) with an "octave of three" (=60). The "octave" (=60) of three is the "twelfth body" (=144), comprising two as a "twin" (=121) which forms the "third" (=100,000) as a "Sun" (=21) with its "plasma" (=100,000).

The "twelfth body" (=144) is "deferent" (*Kakshavrta*, 142) to the "thirteenth body" (=169). The latter is "bunching" (=97) the "space" (=18,000 = 8*[=144+142+169+97+134+360−46] +10,000) into the "twelve bodies" (=134) with "devotion" (=46) to make the time's "motion" (=360) "circular" (=10,000) in the "mind" (=38) of "Mother Nature" (=8).

The "thirteenth body" (=169) is "referent" (=98) to the "fourteenth body" (=256), "organizable" (=11) as the "fifteenth body" (=57) for the "protection" (*Piti*, 123) of the "star" (=2) itself from becoming luminous as a "solar universe" (=13) through mediation of the future as the "sixteenth body" (=0). The "character" (*Prakriti*, 485 =360+5+360/3) of the sixteenth body is to be the "noumena" (=365) "perpetuating value" (=5) of "time" (=360) with its motion as a "seventeenth body" (=3)—the "light body" (=3). The "nineteenth body" (=378) is the "phenomenon" (=378) whose "potential" (=18) is "mutilated" (=378) by the "motion" (=360). Its "present" (=1600 =378+360+888-26) becomes "short" (=378) with the "infusion" (=378) of the "beauty" (=888) of "time value" (=500) within its "ten bodies" (=100) with the "trading force" (=26) of "nine bodies" (=14) from the "reproductive reality" (=40) of the "octave" (=60). The "twentieth body" (=15) "synchronizes" (=12) the "truth" (=375) of the "phenomenon" (=378) with its "sentient value" (=15) since the "three bodies" (*Idam*, 3) are "repeated" (=5) as the "fifteen bodies" (=15) by the "two bodies" (=56)—the "microbody" (*Idam*, 3) of three and the "macro body" (=15) of fifteen.

The "repetition" (*Ayatana*, 1,869) twins the micro body, three, into a "meso body" (*Salayatana*, 111) of six, "potential three" (=111). The meso body is an "activated sphere" (*Salayatana*, 111) of "six organs" (=111)—intellect, eye, tongue, nose, body, and ear. "Six organs" (=111) are "activated" (=18) by the "six media" (=140) as the "repeater" (=140 =18+11+111) of divine-effect, fire-effect, water-effect, air-effect, earth-effect, and ether-effect within the "twenty-first body" (=11) to "sense" (=16) the "effect" (=34) of the "contact" (=220) by the "guide" (=111) through letter "R" (=111) of the "realization sphere" (*Indumandala*, 111). The contact "activates" (=9) the "sentient" (=189) for "surrender" (=269 =189+100-24+4) of the "entity's" (=24) "sentient-effect" (=4) to the "wisdom" (=100) of the "guider" (=100).

The "tremble" (*Sal*, 12) of "repetition" (*Ayatana*, 1,869 = 100*17 +12+135+22) within the "activated sphere" (*Salayatana*, 111) "guides" (=111) "introspection" (=135) within the "entity" (=24) to develop an "ascending consciousness" (=22) of the "perfect wishable" (=17) with the "wisdom" (=100). The perfect wishable is the "omniscience" (*Shivadrishti*, 17) with "six members" (*Shivadrishti*, 17) whose "constant light" (*Nanaka*, 17) is the "primordial wisdom" (*Nandi*, 17). A "freedom from endoproduction" (*Maya Mukti*, 900) of "wisdom" (=100) "perpetuates" (=9) an "omniscient consciousness" (=800) of the "essential nature" (=8) within the "primordial wisdom" (=17).

The "capability" (=55) of the "inner body" (=83) of the "fire" (=17) within the "primordial wisdom" (=17) is "organizable" (=11) as the twenty-first "mass body" (*Bhaavi*, 11). The fire makes the mass body "pious" (*Maha Gauri*, 11) as its "flame family" (*Gauranga*, 11), reproduced with wisdom as a "twin flame family" (*Bhaavi*, 11). The variable "cultural dimension" (=11) of the "present growth" (*Nuwa*, 11) is the "reproductive base" (*Yesod*, 11). It twins the "foundation" (*Sva*, 11) for the "leadership pathway" (*Eve*, 9) of the "Feminine Adam" (*Shivagati*, 120) as the "Angel of light" (*Param Eve*, 11) with her "wisdom" (=100) about the "potential" (=18) of "Adam" (=8) within the "Garden of Eden" (=92) following "Eve" (*Eve*, 9) to "perpetuate" (=9) his "luminous self" (*Maha Gauri*, 11) as an "organization" (=29).

The "supernatural growth" (=8) of an "entity" (=24) as a "family" (=10) of three over "three cycles" (=10) of time, with past as Eve, present as Absolute Eve, and future as Adam, forms a "triplet" (=1,000) of "massive galaxies" (=80) with "Qasar" (=100,000), the "fire-effect" (=100,000). Triplet mediates the "family" (=10) with a "Triplet System" (=0), the "three zeroes" (=0) of time, to make it a "Quasar System" (=10). After reproducing the three zeroes, the Quasar system becomes a "Quasar Triplet System" (=1,000) whose "discontinuous value" (=888) "reforms" (=888) into an "ultramassive blackhole" (=888) in "primordial oneness" (=32) with the "family" (=10) of "massive galaxy's" (=80) past, present, and future. A triplet is a "twin triple quantum system" (=1,000), comprising "three qubits" (=1,000).

A "qubit" (=-1/8) is a "twin quantum system" (=-1/8)—a "primordial reproduction" (=-1/8) of a "massive galaxy" (=80) which twins its "spirit" (=20) with two zeroes" (=-10 =-1/8*80) to be the "family" (*Parivar,* 10) "superpositioned" (=90) as a "quantum" (=90). The quantum is "reproductive" (=100) as a "system" (=12) without the two, the "star" (=2), the "twin triple quantum system" (=1,000) twins. Thus, a qubit is a "fingerprint" (*Angulanka,* -1/8) of "interference" (=80) by a "param wave packet" (*Parivar,* 10)—"devolving Rydberg wave packet" (=10), devolving from the "primordial wave packet" (*Guru,* 100)—"Rydberg wave packet" (=100), for evolving the "primeval wave packet" (=1,000)—"evolving Rydberg wave packet" (=1,000) by involving "wave packet" (=13,000)—"involving Rydberg wave packet" (=13,000). The param wave packet interferes as a "half wave packet" (*Parivar,* 10) seeking to be whole when repeated with the "twin wave packet" (=100) that is "self-projecting" (=3/2) itself as a "triple wave packet" (=1,000) with the "potential wave packet" (=0)—the "revolving Rydberg wave packet" (=0), to be the "measure" (=1,000/72 = [360*3-80]/24*3) of "time" (=360) as an "entity" (=24).

The twin triple quantum system is the origin of the "infinite equanimity" (*Deshaparyaya,* 968) through "primordial oneness" (=32) of the star's two forms: the rising "morning star" (=89) and the setting "evening star" (=103) within the "discontinuity" (=200) of time that produces the star's "supernatural growth" (=8) by dividing it into two at "noon" (=6) for multiplying the two as one, the "Sun" (=21). At "noon" (=6), the "Sun" (=21) is in "direct contact" (=924) with the "multiplier" (=3), the "astral body" (=3), for "plummeting" (*Lambodara,* 11) one as the "discontinuous value" (=888) of its "beauty" (=888) "reforms" (=888) in into a "present back hole" (=888).

One is a "photocopy" (=1) which "copies" (=0) the discontinuous value, comprising "eight photons" (=888), as a "photon" (=20) for "energy equilibrium" (=120) with letter "N" (=120) by forming "six photons" (=120) when "repeated" (=5) as two with the multiplier during the "day" (=125). Two twins the "continuous value" (=800,000 =[375+125]* 1,600) of "truth" (=375) of the "day"

(=125) as the "fluctuation" (=800,000) in the "present's" (=1,600) "ascending direction" (=8,000) to be the "infinite value" (=957). The infinite value is the "etheric body" (=957) of the "past's" (=9) "descending direction" (=9,000) moving as a "flame family" (=11) away from the "flame" (=32) of "primordial oneness" (=32) of the future's "horizontal direction" (=1,000). The family of three twins itself within one for the "twin flame family" (=11) to be organizable within the "flame family" (=11) for manifesting "strong psychic linkages" (=54,000) with the "unit entity" (=43), the "whole flame family" (=43).

The "weak psychic linkages" (=185) with one as the "para entity" (=19) "cause" (18) "discontinuity" (=200) to transform the "time multiplier" (=3) into the "space divider" (=4). With the flame family, the para entity becomes the "causal body" (=30). The "entity" (=24) becomes the "mental body" (=381) when the "time multiplier" (=3) "works" (=19) like one with "two flame families" (=11). The "psychological-effect" (=0) of the "twin entity" (=24) becomes the "intellectual body" (=306) "self-projecting" a three with two and a six with four mediated by zero. The "causal-effect" (=13) of "flame" (=32) within "family" (=10) "perpetuates" (=9) itself as a "physical body" (=387) as one "triples flame" (=3) for hoarding "three flame families" (=87) within its "past" (=9) that perpetuates it as a "half flame family" (*Kali*, 96 =32*3).

The "past" (=9) reproduces its "reality" (=7) as the "sixteenth" (=81 =9*9), the "union" (*TH*, 86 =81+9+7-11) of nine with seven without the "flame family" (=11) formed as the "sixteenth infinity" (=11). "Heavy matter" (*Madhava*, 29) of the "union" (=86) "springs" (=811) "growth" (=6) of "divinity" (=57 =86-29) at the "sixteenth infinity" (=11) for "devotion" (=46) to "winter" (=754) with the "feminine" (=37) element. The "feminine" (=37 =74/2) "twins" (=121 =74+47) "Winter's" (=754) "loneliness" (=123) for "servicing" (=47) "heavy matter" (*Madhava*, 29) as the "origin" (=1,000 =754 +123*2) of "April" (*Vaishakha*, 29), the "second month" (*Vaishakha*, 29) of the "season" (=2) that "seasonally" (=68) "departs" (=68) for a "year" (=268) to "grace" (=84) the "growth" (=6) of its "past" (=9) as one into "Taurus zodiac" (=69).

5.5 The Path of Management Action

The "path of management action" (*Guru Marga*, 18) is the "path of least space" (=18) when "self-managing" (=15) "primordial technological trading" (=2,000) of "management force" (=5) for "management action" (=100) as a "manager" (=16) who "wishes" (=18) to be a "star" (=2) within "least space" (=100) "possible" (=1,600). Primordial technological trading makes a star "heavy" (=1,000) with "sentient energy" (=1,000) of "well-being" (=125) "underflowing" (=8) from "Mother Nature" (=8). As an "organizational planner" (=8), Mother Nature mediates the "continuity" (=50) of the "path of management action" (=18) with the "primordial oneness" (=32) of her "superpositioned correlation" (=580 ~580) as "everyone's" (=180) "present family" (=8). "Within-group" (=580) of "two" (=2 =1–10+11) that includes "herself" (=1) and "everyone else" (=-10) within her "flame family" (=11), "everyone" (=180) has a "superpositioned" (=90) "correlation" (=16) with the "heavy" (=1,000 =580+180+90+16*15+10) element. The heavy becomes "light" (=180) after "breeding" (=15) fifteen by "self-projecting" (3/2) a "family" (=10) with "management action" (=100).

One is superpositioned over another after its replication by the third in a sequence where another is at the bottom and one is at the top. Another is one's copy. One is a double copy of the third. One doubles another to be second. Another copy one to be first in the sequence. One reproduces heavy to make everyone in the sequence light after the three becomes reproductive for reproducing three copies-- one as a double copy and another as a copy—to which it imparts the "mass" (=132) of the heavy. Initially, there is one. Another copy one to double one, thus norming three—the one before it is copied and the two after one is copied. Since one that is copied is immanent within the two, 132 is the mass of the "immanent one" (=21) in "primordial oneness" (=32) with "itself" (=121). The immanent one is "two ones" (=21), the "potential future" (=21) of the one whose "linear mass" (=21) is "reproductive" (=100) by "itself" (=121).

The third forms as the "other" (=9/16) with "another" (=98) after the letter "N" (=120) transforms the "energy equilibrium" (=120) within the "past" (=9) of the "fourth" (=18,000 = 120*

1,500/10) into "primordial inequality" (=120) to "guide" (=111) the "fifth" (=1,500) with a two. The two mediates the ten within the letter "N" to "norm" (=18) the "convergent energy" (=28 =2,2+6) of the "sixth" (=53 =18+28+7) with the "potential" (=18) within the "seventh" (=1) to be "present" (=1,600 = 1,12 *[10+12+28]) as the "eighth" (=12) with the letter "A" (=28). The eight mediates the "feminine" (=37) to "group" (=387) the two within "itself" (=121) to be the "ninth" (=123). The "tenth" (=-12) "transforms" (=1) the "eleventh" (=17) with the "perpetuating value" (=5) of the "twelfth" (=1) "repeated" (=5) within "everyone" (=180) who "follows" (=10).

The "quantum correlation" (=580) of the "third" (=100,000) that "forms" (=100,000) "everyone" (=180) "within the group" (=580) with the "second" (=90,000) which is "breeding" (=15) the "first" (=1) as "light" (=180) is "almost perfect" (*Bhasvara tejas*, 58). The "only" (=1,000/72) "consideration" (=58) is that the first needs to be an "entity" (=24) with a "potential" (=18) for "self-breeding" (=42) to be "light" (=180) of everyone within its "infinity" (=90,000) of causation. By "signaling" (=9) the "assumption" (=19) of "family" (=10) for the first to "live" (=14) as an "entity" (=24) with a zero, its copy, the "signaling correlation" (=580) of the third with the second with the "first" (=1) as the "remainder" (=4) becomes the "almost quantum correlation" (=581) of the second with itself after it assumes one to be a "part" (=-1) of three which the two copies as a zero for the four to take "position" (=999) of the one within the "correlation" (=16 =15+1) for "self-projecting" (=3/2) the "entity" (=24 =16*3/2).

By "breeding" (=15) the "light" (=180) of everyone with her "entity time" (=360 =15*24) as the "entity" (=24) becomes "dark" (=180). For breeding, three needs to be first "repeated" (=5) as a five with the second formed as a two. Therefore, the "entity" (=24) is "almost superpositioned" (=240 =20*24/2 = 24 *1,0) over the "twin entity" (=24) which is the one who copies her to be zero. As an "almost quantum" (=240), the "entity" (=24) is "non-signaling" (=7) the "reality" (=7) to "illuminate" (=150) the twin entity as a "quantum" (=90) once the latter realizes that she is not a "copy" (=0) but a "replication" (=90) of the entity's "past" (=9), which is "shadowing" (*Maya Sita*, 0) the copy. Thus, replication becomes the "first" (=1), replicating the "copy" (=0) as "second" (=90,000) while the "third" (=100,000) makes the "past" (=9) of "both" (=10,000) "circular" (=10,000) to liberate herself from being repeated with the "fifth" (=1,500) by the "fourth" (=18,000), i.e., space.

> **How Time Breeds Itself as an Entity**
>
> When reproducing itself with a multiplier, time's accretion disk collides the five rows of its present with the eight jets of the space. The five rows of the present converge the four dimensions of space into a foursome to twin present. The eight jets of the space diverge the four dimensions into the three pairs of space loop, preserving the symmetry system with the four as the space divider. The collision divides the space into four lines of the past of time to produce the ten columns of the future of space as causation that twins present. The ten columns are the spinning column which spin the past into the future as the present focuses on self-projecting the whole as an entity.

5.5.1 Understanding Liberator Deity

The "Liberator Deity" (=8) is a "theoretical object" (=8,000) with "sentient energy" (=1,000) that makes her an "ideal subject" (=9,000) as well. "Normative truth" (=20) of "Liberator Deity" (=8) is her "normative beauty" (=12) as well which first "develops" (=22) and then "attracts" (=34) an "observer" (=0) with "dedication" (=42) of her "observer-effect" (=42). An observer may choose not to perceive his "primordial oneness" (=32) with her as "true" (=8), and instead exercise an option to seek oneness with her "perpetuating value" (=5) as the "param maternal" (=5). A "seeker" (=16) has the option to correlate her perpetuating value with his "wisher value" (=5) to be "God" (=5) "Himself" (=5), after observing the "wisher" (=) within wishing to know the "reality" (=7) that makes him a "star" (=2) of attraction with the "continuity" (=50) of five as a zero, the wisher, and the "formative growth" (=2) of reality as a two within five. As a "mediator" (=5) of Mother Nature's "sentient energy" (=1,000 =5*100*2) with his

Transformative Exchange Paradigm

"gravitational energy" (=100), a "star" (=2) "contaminates" (=10^{100}) the true "root reality" (=10^{100}) by "centering" (=10) "itself" (=121) as a "family" (=10) of three—"Mother Nature" (=8), "star" (=2), and "wisher" (=0)—for mediating the "correlation" (=16 =1,8-2) between "localization" (=1) of the family as a "member" (=1) of collective behaving like a star and "globalization" (=1) of the family as a "nation" (=1) of individuals behaving like Wishers wishing to be a star. As a "microcosm" (=3), a "family" (=10) is a "star" (=2) which "localizes" (=9) to "amplify" (=4) the "param reality" (=-2) within itself. As a "macrocosm" (=1), a "family" (=10) is a "wisher" (=0) who "globalizes" (=0) to "amplify" (=4) the "primeval reality" (=-7) without itself. As a "mesocosm" (=5/2) of two, "family" (=10) is "nature" (=8) who "nationalizes" (=10^{19} =$10^{5/2*2+2+8+4}$) to "amplify" (=4) itself as a star into a macrocosm with its "primordial reality" (=-3) as a Wisher.

"Nature" (=8) "amplifies" (=4) to "self-perpetuate" (=1/2) itself. Wisher amplifies the "polluted consciousness" (=4) of nature within the "star" (=2) that amplifies "itself" (=121) with a "twin" (=121). "Twin" (=121) "destroys" (=1) the "pollution" (=91) the star perpetuates by "reproducing" (=8) itself like nature. A star's "reproduction" (=285) as "primordial" (=85) "produces" (=1) the "twin" (=121) as "pure" (=285) like "Sun" (=21). "Sun" (=21) is "pure" (=285 =21*13+12) because it is "luminous" (=13) within "self-luminous" (=12) mediated by the self. The self is the one that is pure without the two that pollute like "satan" (=-1). One is pure because it is "part" (=-1) of three, which first pollutes in the form of a "wisher" (=0) seeking to be a star and then transforms the "natural essence" (=1) of the form into pure to norm the reality of its nature as "true" (=8). The "reality" (=7) is "private" (*Niji*, 7). The "copy" (=0) is "social" (=0).

The "private" (=7) mediates the "sociable" (*Suhabati*, 25) like a "human" (*Naran*, 275). The "human" (=275) "sequences" (=48) true as "false" (=8) to be the "institution" (=8) with "private right" (=128) over the "consequence" (=485)—the "impurity" (=485) repeated within the private. The "social" (=0) "moves" (=479) to the "right" (=71) as "consciousness" (=4) with the private's supernatural growth into an "institution" (=8). The "social-effect"

(=20) "sequences" (=48) the "impurity" (=485) within the "sociable" (=25) to be the "causative" (=138) for the "replication" (=90) of the "causation" (=1) on the "left" (=89) as a "primordial creation" (=9). Thus, private becomes the primordial creation. The social behaves like the "primordial creator" (=2700) of the "institution" (= 8) as a "primordial creation" (=9) for a "reason" (=1): enjoy the "supernormal value" (=270) of the "natural" (=270) to be "supernatural" (=270).

The "supernatural" (=270) is "colored" (=140) with the "shared value" (*Vidhi Vadartha*, 130) of the "natural" (=270). The "absolute realization" (*Param Prapti*, 16) of "identity" (=8) with nature as the "primordial one" (*Bhairavi*, 8) "self-develops" (=16) the "capability" (=55) to "fulfill" (=9) the "wishes" (=18) of "infinite wishers" (=-6) through "primordial oneness" (=32) with the "world" (=-2). "Self-illumination" (*Brahmi*, 8) "embodies" (*Kartikeya*, 8) the "Twin Wisher" (*Satarupa*, 8) as the "supernatural face" (*Asvabhavika Mukha*: YHWH, 8) within the "natural face" (*Svabhavika Mukha*: Yahweh, 8) of the "thrice-born" (*Kundalini*, 8) "child essence" (*Yahu*, 8) of the "Wisher" (=0)—first as a natural face, second as the supernatural face, and third as the Wisher who "greets" (=64) the "face" (=76) of the "creature" (=12)—the "triple wisher" (=12).

The "creature" (=12) "transcends" (=29) the two to "enjoy" (=7) the "joy" (=123) of the "half wisher" (*Parvati*, 10) with the three whose "gender dimension" (=90) is reproductive within the four as a function of time. Half-wisher is "perfect" (*Parvati*, 10) and "perfected" (*Durga*, 28) by her "potential" (=18) to be the "perfect wisher" (*Xihe*, 28) with the "perfect personification" (*Maha Durga*, 16) of two as "Mother Nature" (=8). Mother Nature "squares" (=18,000) two as "space" (=18,000) with her "experience" (=9,000) to be the "origin" (=1,000) of "time value" (=500) for the four's "well-being" (=125) as the "Mother" (=4) who "lives" (=14) within "time consciousness" (=14) as the "time" (=360).

5.5.2 Transcending Liberator Deity

A "perfect wisher" (=28) is the "Godhead" (=17) of four, two as "light" (=180) which "squares" (=18,000) "itself" (=121) as a "Sun" (=21) to "experience" (=9,000) the "twin light" (=10,000) of the "potential two" (=5) as the "triple self" (=5) of the "time-varying existence" (=2,000) of "two planes" (=2,000). "Two planes" (=2,000) are the "mother" (=4) of ten "primordial suns" (=53). "Ten primordial suns" (=500) are "semivariable" (=500) as the "time value" (=500) of the "five primordial suns" (=-3), "variable" (=-3) within the "first plane" (=92) on the "left" (=89). "Fifteen primordial suns" (=697) are "constant" (=697) within the "second plane" (=1) on the "right" (=71) "moving forward" (=113) as a "cyclical" (=113) "third plane" (182 =69+113) to "level" (=113) the "disproportionate" (=69) "contribution" (=69) of the "fourth plane" (=10^{19}). The fourth plane is the "central plane" (=10^{19}), "centered" (=10) by the "three planes" (=3) within the "center" (=16) of the peripheral "fifth plane" (=75). The "four planes" (=170) "reside" (*Vasudeva*, 75) within the fifth plane as the "four sides" (=75) "integrating" (=95) the "sixth plane" (=190). The sixth plane is "circular" (=10,000) with "three primordial suns" (=190). It self-perpetuates the fifth plane as a "circulating plane" (=75) with "eighteen primordial suns" (=95) that "stem" (=95) from a "potential" (=18) of "two primordial suns" (=2) within a "star" (=2).

A "star" (=2) "perpetuates" (=9) its "past" (=9) for the "organization" (=29) of the "seventh plane" (=31) as a "square plane" (=31) with "sixteen primordial suns" (=29). Its "future" (=0) as the "eighth plane" (=0) is zero as it is a "triangular plane" (=0), "triangulating" (=10^{1024}) "six primordial suns" (=60) with the ninth "parallel plane" (=121) of "twelve primordial suns" (=121). For one to begin as a "Sun" (=21), two become the "beginner" (=23) beginning the "circulation" (=23) of itself as the "potential sun" (=23). The potential sun is the tenth "point plane" (=23) of "twenty-one primordial suns" (=23).

The "first primordial sun" (=64) "greets" (=64) the potential sun as a "Council of Thirteen" (=23) with its "eight primordial suns" (=64 =23+54-13) to "construct" (=54) the "sum" (=54) as the eleventh "whole plane" (=-13), "bringing" (=-13) "thirteen radiated sequences" (=-13) with it from the twelfth "wholesome plane" (=96) of "twenty-one radiated sequences" (=96). The thirteenth "wholesomewhole plane" (=100) bases the replication of "two radiated sequences" (=90) on "eight radiated sequences" (=98) to be reproductive with a star's "twenty-eight radiated sequences" (=2). The fourteenth "growth plane" (=16) has an "epistemological value" (=10^{14}) of "ten radiated sequences" (=10^{14}). The fifteenth "entropy plane" (=19) has an "axiological value" (=2) of "twenty-eight radiated sequences" (=2). The sixteenth "real plane" (=106) has an "ontological value" (=15) of "twenty-eight primordial suns" (=15), norming "two circles" (=15) with a circle's "fourteen primordial suns" (=100,000). The seventeenth "imaginary plane" (=-10) has a "metaphysical value" (=1,000/72) of "seven primordial suns" (=725), repeated as the "two primordial suns" (=2) for imagining a "guider potential" (=149) of "seventy-two primordial suns" (=149) with "nine primordial suns" (=134).

The eighteenth "complex plane" (=85) is a "rotating plane" (=85), "rotating" (=-12) the "real" (=9) within the "imaginary" (=20) to "plane" (=80) the "illusionary" (=87) into "simple" (=9). The nineteenth "illusionary plane" (=180) is "rotatable" (=304) with the "hundred primordial suns" (=180) into the "eighty primordial suns" (=124) with the "sentient force" (=4) of "forty primordial suns" (=12), the "nineteenth sun" (=12). The twentieth "simple plane" (=-8) "rotates" (=96) "forty primordial suns" (=12) to be the "metric" (=-1) of the twenty-first "infinite plane" (=10) that perpetuates the simple as real with the consciousness of the twenty-second "finite plane" (=5). The twenty-third "primordial plane" (=20) "spirits" (=20) "ahead" (=8,000 =20*20*[4*5]) with the "soul" (=4) of the "finite plane" (=5) to be the twenty-fourth "param plane" (=1,000).

The param plane is the "lower half-plane" (=1,000) which twins the "ninth parallel plane" (=121) as the "upper half-plane" (=121) with the "reproductive force" (=100) of the twenty-first "primeval plane" (=10) within the twenty-fifth "potential plane" (=769). The potential plane is the "half-plane" (=769) which squares the "whole plane" (=-13) to be "extended" (=600) within the twenty-sixth "water plane" (=600). The "hydrated-effect" (=600) of the "water plane" (=600) triples the "discontinuity" (=200) from itself to the "twin plane" (=2,000 =600*3 +200). The twin plane is the twenty-seventh "fire plane" (=2,000) which "forms" (=100,000 =2,000*50) a "continuity" (=50) of "four planes" (=170), the "divergence"

(=170) of the twenty-eighth "air plane" (=170) from the "illusionary past" (=2,000) of two planes. The four planes are "repeated" (=5) into "infinity" (=90,000) as the "nine planes" (=26) for their "self-existence" (=26) as the twenty-ninth "earth plane" (=26) that norms the "Supreme council of nine" (=26) by "multiplying self" (=3) of "three planes" (=3) within the thirtieth "ether plane" (=3). The "twelve planes" (=29) are the "Divine Godhead essence" (=29) of the thirty-first "divine plane" (=29). The "seven planes" (=18) are the thirty-second "sentient plane" (=18) "conceivable with" (=18) the "Sentient Godhead essence" (=18).

The sentient plane makes the "five planes" (=1,600) reproductive as the thirty-third "guider plane" (=1,600) with the "potential" (=18) of its "Guider Godhead essence" (=1,600) to be "present" (=1,600) as a "star" (=2). The thirty-fourth "past plane" (=2) has an "effect" (=34) on the "Godhead" (=17), the thirty-fifth "future plane" (=17). The effect "channels" (=36) the star within the thirty-sixth "temporal plane" (=34). The channel is the thirty-seventh "spatial plane" (=36). The "greet" (=64) channels the "primordial space" (=100) for the spatial plane with the thirty-eighth "causation plane" (=64) to be the "perfect mind" (=64). The primordial space is the thirty-ninth "entity plane" (=100). The "mind" (=38) is the fortieth "deity plane" (=38) with "entity-effect" (=38). Mother Nature is the hundredth "God plane" (=8) whose "ego" (=-1) "transforms" (=1) the "Wisher" (=0) into a "plane" (=80), the sixtieth "Devil plane" (=80). The plane "disappears" (=960) like a "devil" (=0) to "appear" (=960) like a "satan" (=-1) within the one-hundred-twentieth "Satan plane" (=40). Their "fused weight" (=120) is the one-hundred-eightieth "Greeter plane" (=120), the "Site" (=120) of letter "N" (=120), which "decays" (=31) a "plane" (=80) into a "planet" (=-1) without "nature" (=8) to "head" (=−9 =−1−8) the "entropy" (=5) of "God" (=5) as a "Godhead" (=17) of "creature" (=12 =17-5).

A "star" (=2) makes the "planet" (=−1) "spacious" (*Vikata*, 10,000) with "four zeroes" (=888) that "square" (=18,000) the space with "four copies" (=90) of the "primordial space" (=100) to make the "creature" (=12) "fearless" (=944) through "primordial oneness" (=32) as a "universal entity" (=44) with the universe. The universal entity copies its "consciousness" (=4) to "self-perpetuate" (=1/2) the "limitless infinity" (=1,024 =12+944+44+2,4), the fiftieth "param deity plane" (=1,024) of its "sentient energy" (=1,000), as the "universe" (=2) to be self-luminous as a "creature" (=12).

The "limitless infinity" (=1,024 = [189+72-10+5]*4) makes the creature "conscious" (=189) of the "weakness" (=72) "kept away" (=-10) in the "consciousness" (=4) that he is "vulnerable" (=70) to "defeat" (=70) until he "multiplies self" (=2) with the "universe" (=3) into "God" (=5). God creates discontinuity of the creature's sentient energy through "entropy" (=5) for the continuity of the creation's gravitational energy through the growth of the universe multiplying self. "God" (=5) "heads" (=-9) his "replication" (=90) through the "path of action" (=86), which "replicates" (=1) the "belief system" (=20) in the "corporate path" (=65) of one's "de-polarization" (=65) for the growth of "horizontal order" (=5). A "conscious entity" (=7) "causes" (=18) a "negative spin" (=11) within "overjoyous" (*Pamujja*, 10) "Godhead" (=17) for "correction" (*Samshuddhi*, 57) in the "positive spin" (=40) in the "ascending worker-social benefit-cost ratio" (=36) of "action" (=10) within "growth" (=6). The "reality" (=7) of Godhead's action perpetuates "ignorance" (=1) about the "present paradigm" (=24) of "radiant love" (=17) of him as the "infinite maternal" (=17) "acting" (=13) for "perfecting" (=127) "universal sentient well-being" (=157) through "trial-and-error" (=170).

"Freedom from the infinite's" (*Para Mukti*, 86) "path of action" (=86) "spins" (=29) the "impact" (=57) of the "gravitational energy" (=100) for "servicing" (=47) "action" (=10). A "sentient entity" (=7) makes the action "reproductive" (=100) with a "time multiplier" (=3) without multiplying the self as a subject. The "subject" (=0) is the "Wisher" (=0) wishing to observe as an "Observer" (=0) the reality of the desired "intrinsic universe" (=9) within the existing "extrinsic universe" (=9). "Godhead" (*Param Ganesha*, 17) is the one who "twins reality" (=49) to be the "twin flame" (=1,649) of "Mother Nature" (=8) whose "actions" (=10) generate "ego" (=-1) to perpetuate her "past" (=9) as the "enjoyer" (=-9) of the "Wisher-effect" (=-16) within a "sentient entity" (=7) who wishes to be a Wisher to be the foundation for her growth into a "primeval maternal" (=17).

The "constant light" (*Guru Nanaka*, 17) of the primeval maternal is the "foundation" (=11) for the "cost-effective" (=83) "child-effect" (=83) of her "namesake" (=100) "guider" (*Guru*, 100) within a "child" (=128). That "strength" (=83) empowers one to be free from the "theory-effect" (=0) of the "theory" (=127) that, as a "child" (=128), he is just a "subject" (=0) "materializing reality" (=5) through "horizontal symmetry" (=5) with a "para deity" (=5). The theory-effect contaminates a child's purity by adding:

- The "sentimental wishing" (=20) to the primeval maternal's "divine-effect" (=-3)
- The socially-minded "intentional vision" (=10) to the "female's" (=10,000) "fire-effect" (=100,000),
- The "emotional feeling" (=741) to the "doer's" (=71) "water-effect" (=4),
- The "sensory perception" (=2) of the doer's water-effect to the "multiplier's" (=3) "air-effect" (=78 =71+4+3)
- The "physical contact" (=924) with the "multiplier" (=3) to the "incarnate's" (=900) "earth-effect" (=21), and
- The "divided attention" (=9) of the "guider" (=100) with the "incarnate" (=900) to the "ether-effect" (=257) of the "infinite theory" (=148) of "oneness" (=48).

The "guider-effect" (=100 =27+81−18) "ascends" (=27) the "past" (=9) with the "multiplier" (=3) to "direct" (=91) the "potential" (=18) to "descend" (=25) the "reality" (=7) of a "sentient entity" (=7) through "oneness" (=48) without "action" (=10) with the "heavenly infinity" (=9) of the "seventh heaven" (*Purna Shiva*, 16) as the "param deity kingdom" (=2,338 =7+16,48-10). The "multiplied self" (=17) is a "kind" (*Dao*, 17) of the "First Emperor" (=17), the "replica" (=17) of the "species" (*Shangdi*, 17) whose "extrinsic form" (*Shangdi*, 17) is immanent within its "transcendental form" (*Dao*, 17). The transcendental form is "constant space" (=125). The extrinsic form is the "variable time" (=−1) of one as an "entity" (=24) within "constant space" (=125) for "well-being" (=125) as a "sentient entity" (=7). The "dividing self" (*Kabbalah*, 17) is the "constant time" (=900) of one who becomes a sentient entity through "replication" (=90) of its "copy" (=0) as a "variable space" (=−1) until "rupture" (=−1) of "primordial space's" (=100) "air in ascending motion" (=−1) as "ego" (=-1).

The "march" (*Chaitra*, 90,000) into "infinity" (=90,000) of the "primordial space" (=100) with "constant time" (=900) "springs" (=811) a "straightline" (=67) for "forwarding" (=89) the "sweetness" (=90,000) of its "backward feminine-effect" (=98) to the "summer" (=967). The "limitless" (=1,024 =8*128) "Aries-effect" (=8) of the "Aries zodiac" (=256) generates "formative growth" (=2) of the "forward masculine-effect" (=256) within a "child" (=128) for the "growth" (=6) of the "God paradigm" (=61) into the "Godhead paradigm" (=67).

Chapter 6

Formative Growth Paradigm

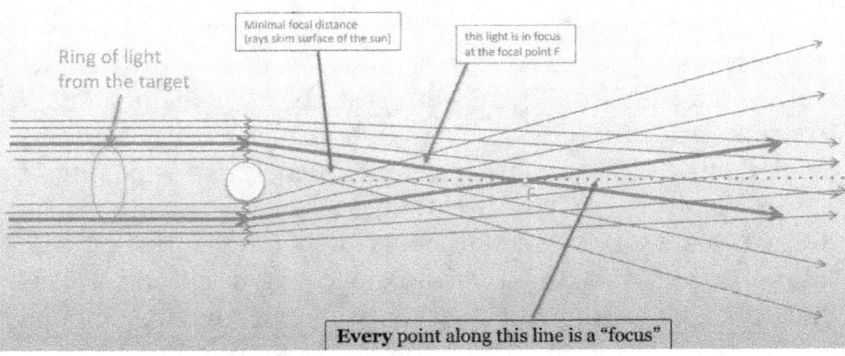

Does the mass of the energy points on an infinitely short surface in the front of the Sun reproduce itself as the heavy ring of "mass points," descending darkness with the growth of horizontal light of a para deity? (*Newtonian ideal of mass as a starting point for a scientist to investigate past reality with the Moon as the paradigm that attracts time for managing its future naturally with a mission to be present as a goalkeeper*)? Or, Does the energy of the focus points on an infinitely long surface from behind the Sun reproduce itself as the light ring of "energy points" (*Chi*, 9), ascending darkness to guide energy equilibrium with one itself as a deity? (*Einsteinian theory of energy as a starting point for a scientist to investigate future reality with the Sun as the paradigm that repels time for managing its past supernaturally with a goal in sight of its present*)?

Neither of the two. In reality, when in focus, the space at a maximal focal distance of vision as a whole is a target for gravitation as a paradigm. The paradigm is the ascending radiational pressure on space for ascending progression as Earth with a "heavy ring of fire point" to realize the mission of value with the vision's reproductive potential as a neutron star. The "heavy ring of fire point" copies light force as the mass point, consuming it as water for producing an illusion that it is bright. The fire point is the "single point" (Hawking point), which is inertial and local-effect of the illusionary production of the past. It copies itself as the "dull ring of guider point" to triple-hand its presence by trading the past of the sentient point as the starting point.

With its focus, the neutron star attracts a white dwarf as its value from the future by repelling the spirit for the imaginary consumption of its future. The white dwarf sources Moon as a "dark ring of consciousness point" (which is a dark ring of "water point"—rhombus—producing consciousness of its form as a quasar) for the illusionary production of its past. The illusionary production is a "left-handed ring of air point." This masculine ring air points at least distance for the consumption of the infinite field of para real as the first object. In distance, the past is the least because it is as real as the "right-handed ring of earth point." This feminine ring earth points "spatial value" as a dot, which is giant as a self-luminous system. The distance is the "handed ring of ether point," where the ether point is the first copy, the electromagnetic wave of the variable space with a planetary effect. The handed ring is a devil ring. The Ether point is Satan, the first devil. The devil is electroweak as he repels Satan as his copy to be the first. The whole copy of Satan is magnetostrong as a conscious system, i.e., a luminous system.

Every point of "focus" (as a neutron star) in front of the Moon heightens the "minimal focal distance" for a sentient entity to transcend the darkness of the illusionary production of the past with its light. Earth mediates Moon's vision with her mission to moderate the present and gravitate toward the potential for knowing the goalkeeper beyond the paradigm of vision with a sense of devotion to the Primordial Illuminator.

As a sentient entity, who is a goalkeeper like a param deity, Sun leaves back a "light ring of sentient point" of its presence within the energy point descending progression of time's focal distance from its present as the Dark Matter. Dark Matter is the Sun's central surface that magnifies into the Sun with the presence of three hands of time within the sentient point, whose past is the energy point in focus of the Sun's reality as the white dwarf. The "bright ring of divine point" (i.e., focus point) is the imaginary consumption that copies the Primordial Illuminator as the base of its infinite value to guide the Sun's reality with time as the multiplier in real oneness with Mother Nature.

Is Formative Growth of a Paradigm Necessary?

Formative growth orbits a paradigm with an orbital to order simplicity of action with a natural growth of disorder, perpetuating the value of horizontal order. With the natural growth of disorder, past reality produces present reality through the illusionary production of the reality of a conscious entity conscious of the future the present will reproduce as a follower of the present paradigm. Therefore, a new paradigm becomes necessary to compensate for the infinite wishes inherited from the past within the present reality.

Organizational action lets one synthesize infinite wishes from the constellation of effects of the super wishers into a finite wish for personal sentient well-being. The finite wish propels citizenship action where the definite wish for the universal sentient well-being of the nation is immanent within the finite wish for personal sentient well-being as a citizen. The definite wish activates alienship action for someone alien to service the disproportionate worker-social benefit-cost ratio for one as a worker working as a nation for the well-being of all citizens, conditioned by the possibility that all citizens are one's children. The one is a masculine who attracts a feminine as the alien devoted to servicing the collective as her family. The desire for self-sacrifice is worthy of worship action by anyone with a greeter consciousness of the transcendental who guides his path of devotion.

Worship action of a creature seeking blessings of the Almighty Creator of the creator within and creation without is subject of religion. In most religions, the object of worship is masculine. Therefore, it becomes necessary for spirituality to

> supplement the masculinity of the object with the femininity of the entity whose spirit is the greeter. The greeter substitutes consciousness of the transcendental with the guider force of the spirit immanent within oneself. As one becomes conscious of the divinity within, one behaves like a divine who has time for realizing anything he wishes. As one breeds infinity of causation to wish for something, the space expands for the absolute realization of nothing that is left with everything which is right within one's reality. As a common denominator, the thing that is real centers reality for one's natural growth into a primordial greeter. Primordial greeter has space, time, and causation for All.

6.1 Formative Growth Paradigm, within Future Consciousness.

God is the "formative growth paradigm" (=5), whose "formative growth" (=2) as a "paradigm" (=62) is "self-luminous" (=12) as a "Godhead" (=17) that "works" (=19) to "twin growth" (=19). The work "follows" (=10) four "paths" (=1) of "action" (=10) to be "God" (=5).

6.2 The Path of Organizational Action.

A "Godhead's" (=17) "organizational action" (=170) "divides itself with action" (=10) for the "development" (=170) of "least causation" (=170). The "causation" (=1) is least when the "primordial self" (=10) is divided by ten because one is the divisor dividing the development to be oneself prior to the development. The "path of organizational action" (=10,000) is "circular" (=10,000): a female "divides" (=132) her "formative growth" (=2) for "reincarnating" (=61) as a male who multiplies himself to develop into a female. The all-diverging "primordial technological growth" (=70,000) of a "sentient entity" (=7) into a "female" (=10,000) "decomposes" (=70,000) two "twin bodies" (=56) for their "ten-fold growth" (=170 =56+121-7) into "forty bodies" (=10,000) after a "body" (=56) "twins" (=121) its "normative development" (=888) with one, the "divisor" (=1) of her "femininity" (=1,000 =888+121+1

=888+56*2). The female "composes" (=8) forty bodies with "four bodies" (=12) of a "perfect female" (=1,250 =10,000/8) through the "continuity" (=50) of "six bodies" (=6) for the "conception" (=1869 =1260*6/4 -6) of "sense organs" (=1,869). The "transformative exchange" (=189) of the "sentient" (=189) element from the "sense organs" (=1,869), mediated by the "six bodies" (=6), transforms the "constant" (=697) into a "variable" (=-3) with the "reproductive force" (=100) of the "perfect potential" (=97) of the "three bodies" (=3) of the time multiplier (=3).

The "ideal-effect" (=1) of one as a body of "consciousness" (=4) twins the "theory-effect" (=0) of five repeated as a body of "female" (*Primordial Eve*, 10^4) for the growth of the "two bodies" (=56) of "male" (*Param Eve*, 10^8) into "two brothers" (*Ashwini: Para Eve*, 120). A male is a "constant Eve" (=10^8) within the female, the "variable Eve" (=10^4). When one varies its form to be a female, the male remains constant as a "metric" (=-1) like a "Super Wisher" (=-1). A child is a semivariable with the potential of the "Wisher" (=0) within "both" (=10,000) male and female.

6.2.1 Understanding Super Wisher

A "Super Wisher" (=−1) seeks to transcend the "imperfect" (=3/4) "path of mission" (*Dharma Marga*, 19) conceived in the past with an "almost perfect" (=58) "path of vision" (*Adharma Marga*, 58 =19*3-[−1]) perceived about the future to experience a "perfect" (=10) "path of value" *(Bodhi Marga*, 39 =58-19) as an "organization" (=29). The "potential knowledge" *(Bodhi*, 156) of the "vision" (*Param Kriti*, 81) "perfects" (=10) the "value" (=180) as a "partial dimension" (*Bodhi dharma*, 180) of the "mission" (=−6) to be "congruent" (*Sarvangasama*, 10) with the consciousness of reality present within an organization.

What is Congruent

A "congruent" (=10) is a "valued modular form" (=10), and "valued" (=35) as a "modular form" (=10). It "descends" (=25) its proportionate, "nonmodular form" (*Anupramana rupa*, 39) as a "vector" (=396) of growth. Growth's "even form" (=10) as a "heliocenter" (*Shri Krishna*, 10) is congruent with the growing "heliocentric" (=51) "odd form" (*Shesha sharira*, 51). The odd form enjoys an abundant "vertical symmetry" (=123) with the growable "twin even form" (*Ekantasushama Rupa*, 72) of the "entity" (=24) through a "multiplier" (=3) of forms of "Mother Nature" (=8).

The "growable form's" (=72) "congruence" (=10^{10}) with the "heliocenter" (=10) "reproduces" (=78) "partial congruence" (*Angasamata*, 3,555) with a "transcendental number" (=23), "valued" (=35) as "two numbers" (=23). The "first number" (=1) has a "partially congruent" (=1,500) "vector modular form" (=1,500) that the "second number" (=81) "modulates" (=140) with its "valued modular form" (=10) as a "congruence modular form" (=1,500). The modulation "transforms" (=1) the "noncongruence" (=1,500) "immanent" (=13,000) "within form's" (=1) "expansion" (=1) of the "half number" (=8,000) into a grown "twin odd form" (*Ekantadushama Rupa*, 3,500).

The partially congruent "congruence modular form" (=1,500) is "immanent" (=13,000) within the congruent "noncongruence modular form" (=10). It "surfaces" (=2,000) the "incongruent" (*Anuchitavrtta*, 351) scalar, "vector-valued modular form" (=351) of its "twin flame" (*Suvira*, 1,649). The "incongruence" (*Anuchitavrtti*, 890) is "initiated" (=890) by the "half number" (=8,000) to be "congruous" (=136) with the "origin" (=1,000) of the "space-bound" (=9,000) "unbounded denominator" (=9,000). The half number "experiences" (=9,000) the "replication" (=90) of its growth into a "twin number" (=23) by a "guider" (=100) that "perfects" (=10) a "number" (=53) with a "triple number" (=1). The "perfect number" (*Mrishajnana*, 30) is "incongruous" (=-12) with the common "bounded denominator" (=8) of the "perfect" (=10).

The "perfect number" (=30) "self-refines" (=7,000) itself into a "twin perfect number" (=7,000) to produce a "sentient flux" (=7,000) within the "monster group" (=380). The twin perfect number is a "J-junction" (=7,000). The octave is "modulated" (=60) as a letter "J" (=60) for "junction" (*Java*, 999) of its "reality" (=7) without "Mother Nature" (=8) through "subtraction" (*Java*, 999) of the letter "Q" (=1,000) as a "part" (=−1) of "number" (=53). The "memory" (=35) of the "repeated" (=5) element "number" (=53)

> "causes" (=18) a "field fluctuation" (*Sahaparivartaniya*, 84) within the "mind" (=38) for "grouping" (=114) the "boundaryless" (=114) "infinite sum" (=114) "q-expansion" (=114) of the "whole body" (=56) into an "array" (=114) of "parts" (=–1) of the "heliocenter" (=10).
>
> The heliocenter "expands" (=169) its "asymmetry" (=689) with the "bounded denominator" (=8) through "addition" (=999) of the "reproductive" (=100) element for the "subtraction" (=999) of letter "Q" (=1,000) as a "part" (=–1). The subtraction "super positions" (=17) heliocenter as a "whole" (=16) for the growth of the "origin (=1,000) as a "part" (=–1). Consequently, the array keeps growing "whole body" (=56) into a "wholesome body" (=111) for the "development" (=130) of the "wholesomewhole body" (=11). Each part is partially congruent with the whole body and enjoys abundant symmetry with one another within the asymmetry of the whole body with its parts. The "whole body" (=56) "twins" (=121 =56*2+9) itself with the "consciousness" (=4) of its "past" (=9) as an "octave" (=60) of copies for "spherically expanding" (=380) into a "monster group" (=380). The octave "norms" (=18) an infinite, "conformal field" (=79) of the "multiplier" (=3) of "parts" (=–1).

The "mission" (=–6) is "synchronous" (=1,810) with the odd, spatially-variable, "ancestral phenotype" (=1,810) that "governs" (=1,810) the "causality" (=230) of the "whole" (=16) as a "future entity" (=20) for "oneness" (=48) as a "multiplier" (=3) of the "path within nature" (=5,000) "with infinity" (*Param Samadhi*, 15,000) of the "potential" (=18) within a "past entity" (=98) to be a "Super Wisher" (=–1). As a "part" (=–1) "breeding" (=15) the "whole" (=16), a "Super Wisher" (=–1) "flips" (=280) the "causality" (=230) by "servicing" (=47) "vision" (=81) for "growing form" (=51) of the "primordial causality" (=97) with the "flip-effect" (=34).

The "vision" (=81) is "asynchronous" (=810) with the even, temporally-constant "genotype" (=197) that "fuses" (=197) a "present entity" (=366,666) for the "speciation" (=88) of the "present causality" (=310) within a "body" (=56) into "three parts" (=300) with a "multiplier" (=3). The "definite causality" (=750) of the "multiplier" (=3) is the "symmetry" (=250) of "causality" (=230) with the "supersystem" (=980) of "ambition" (=980). Ambition is the "finite causality" (=980) for "sentient well-being" (=190) of a "wishing system" (=790). The wishing system is the "infinite causality" (=790) for "ascending worker social benefit" (=79) of "twin odd" (=79) through "meditative oneness" (*Dhyana Yoga*, 80) of a "part" (=-1) with the "two parts" (=10).

As a "potential subject" (10), two parts "work" (=19) "together" (=19) for "centering" (=10) the "longitudinal growth" (=10) of "twin even" (*Ekantasushama*, 95) as "altogether" (=10) into a "parallel universe" (=190) of "potential object" (=9). The twin part is the "future part" (=10), comprising the "part" (=−1) and the part's "past" (=9). The part transforms its past for the "organization" (=29) of "three parts" (=300). "Triple part" (=300) is "past part" (=300), the "continuing energy" (=75) of the "part" (=−1) within the "half part" (=-4) which "triples" (=130) "itself" (=121) as a "Supreme Wisher" (=-4) when a "Wisher's" (0) "past" (=9) is repeated to "twin even" (=95).

The "present togetherness" (=23) of the "Supreme Wisher" (=-4) with the "Wisher" (=0) is the "intended wishing-effect" (=9) of the "potential subject" (=10). The multiplier's "secondary-effect" (=23) "transforms" (=1) a "Super Wisher" (=−1) into a "Super universe" (=926). It forms "infinite universes" (=285) of the "Wisher" (=0) through "infinite interactions" (=252) of the "potential subject" (=10) as a "Supreme Wisher" (=−4) with the "Primeval Wisher" (=−6). The secondary-effect is the "constant social benefit" (=23) repeated within the "primordial maternal" (=18) as additional Wishers enjoy the evenness of the "trading-effect" (=26). The "variable social cost" (=16) of the "Wisher's" (=0) "reality" (=7) "causes" (=18) the "oddness" (=16) of the "present interaction" (=22) of the "Param Wisher" (=−7) with the "Supreme Wisher" (=−4).

A Param Wisher "sparse voids" (=81) "consciousness" (=4) of the "causative interaction" (=8,000) to "signal" (=8,000) a "blip" (=8,000) in the "multipole moment" (=8,000) of the "replicated universe" (5,000) for "correlation" (=16) with the diverging, "pi-electron energy" (=500) "globalized" (=500/9) as a "space" (=18,000) of "equal density" (=10,000). The space "squares" (=18,000) "itself" (=121) with the "correlation's" (=16) "perpetuating value" (=5) within the "consciousness" (=4). The "density fluctuation"

(=36) "greets" (=64) the "temperature imperfection" (=36) of the "density's" (=60) "density fluctuation signal" (=15,000 =100 *3/2*100) by "self-projecting" (=3/2) a "super-horizon scale" (=100) of "equal density" (=10,000).

The "super-horizon scale" (=100) "relates" (=80) to "earth" (=724 =380*2 -36) from "proximity" (=380) of a "sidereal" (=380) to "appear" (=960) "larger" (*Bahutara*, 80) =724 =380*2 -6). As a "metric" (=-1) of "equal size" (*Shristi*, 1/2), the earth forces "itself" (=121 =[21-1]*5+21) to be a "fifth" (=1,500 = 15,000 *1/5*1/2) of the "Sun's" (=21) "density fluctuation signal" (=15,000) whose "potential" (=18) "twins population" (=36) of "stars" (=2). The "body" (=56) of the "Sun" (=21) "disappears" (=960) after its "decay" (=31) as a "sub-horizon scale" (=57) "smaller" (=79) than "earth" (=724). A "metric" (=-1) of "equal gravity" (*Hutabhuj*, 17) "reappears" (*Hutabhuj*, 17) as a "moon" (=997), the "sixth" (=53) "equal entity" (=-6).

The "face" (=76) of the "Moon" (=997) "gravitates" (=-100) "itself" (=121) as the "third" (=100,000) with the "Sun" (=21) as the "fourth" (=18,000). The fourth "squares" (=18,000) the "second" (=90,000) with the Sun's "triple copy's" (=3) "gravitational signal" (=8,000) for correcting the "covarying" (=84) neutral, quantum, "horizon fluctuation" (=84) with the negative, "super-horizon fluctuation" (=80) of the "super-horizon scale" (=100). The "horizon scale" (=60) "produces" (=78) "gravitational imperfection" (=800,000) with the "fluctuation" (=800,000) of the "subconscious" (=100) for "reproducing" (=8) time's "temperature fluctuation" (=8) as a positive "sub-horizon fluctuation" (=8) "materializing" (=-6) the "Sun" (=21) as a "star" (=2) within the "first" (=1), the "formative universe" (=1).

The "normative universe" (=90) "self-manifests" (=710) "space's" (=18,000) physical, "polarization signal" (=7,000) with "one hole" (=11,000) at the "super-horizon scale" (=100) to "form" (=10^5) the "spatial value" (=10^{10}) into the "ten copies" (=10^{10}) of the "horizon scale's" (=60). By "self-projecting" (=3/2) "time's" (=360) "four copies" (=90), the "horizon scale's" (=60) "twins" (=121) the "super-horizon scale's" (=100) "two copies" (=22/7) that it "circulates" (=22/7) as "two quarters" (=22/7) of the whole. The "sub-horizon scale's" (=57) "twelve copies" (=43) are "undeveloped" (=43). The "past" (=9) has an "effect" (=34) on the "horizon scale's" (=60) "development" (=160) into a "super-horizon scale" (=100). The "linear ultrasound" (=130) of "shared value" (=130) "interacts" (=30,000) with the "local universe" (=1,000) as the past is "globalized" (=500/9) by a "global universe" (=90). "Each copy" (=3) is "repeated" (=5) as a "wholesome entity" (=100) with "five copies" (=100). The "nonlinear sound" (=22/7) of the "two copies" (=22/7) within the "copy" (=0) in the "quarter" (=0) of the "global universe" (=90) "transforms" (=1) the "wave" (=1) into an "up quark" (=1) for "eternal togetherness" (=1) of the "twin" (=121) as a "wholesomewhole entity" (=120). The twin "enjoys" (=7) the "growth" (=6) of its "correlation" (=16) with the "oneness" (=48) of "twelve copies" (=43) "organizable" (=11) as a "feminine" (=37) and "repeated" (=5) within the "infinite" (=185 =37*5).

As a "unit entity" (=43) twelve copies are the "octave of heaven" (=43) "producing" (=17) the "heaven" (=19) to "live" (=14) as an "entity" (=24) whose "devotion" (=46) is "reproductive" (=100) within "each copy's" (=3) "action" (=10) as a "star" (=2) "appropriating" (=86) the "devotion-effect" (=86) for "union" (=86). The entity lives as a "worker deity" (=1) within the "present paradigm" (=24) of the "primordial greeter's" (=16) "devoted love" (=17).

The "worker deity" (=1) "works" (=19) as a "knower deity" (=2) within the "new paradigm" (=12) of the "param deity's" (=7) "divided love" (=10). The "knower deity" (=2) "knows" (=164) that it "itself" (=121) is the "unit entity" (=43) "self-luminous" (=12) as a "creature" (=12) which the "new paradigm" (=12) "copies" (=0). As a "sentient entity" (=7), the creature enjoys freedom from the worker deity's "sentient energy" (=1,000). By "descending" (=12) the "residue" (=70) of its "body" (=56) within the "universe" (=2), the "sentient entity" (=7) "ascends" (=27) the "social benefit" (=81) of "each copy" (=3) as a "living subject" (=0).

Consequently, the "super wisher" (=−1) as a "nonliving subject" (=-1) realizes "freedom from the universe" (=85) of "sound pollution" (=85) with the "metaphysical entropy" (=90,000) of the "past life behavior" (=90,000), the "absolute truth" (=90,000) of "globalizing" (=910) "actions" (=10) as the "guider" (=100) of the "present" (=1,600). The Wisher globalizing itself as a "universe" (=2) wishes to be "absolute" (=1,600) like the "Christ" (=1,600), the "savior" (=387) who "ascends" (=27) as "divine" (=360) to "time" (=360) his "presence" (=374) until he "lives" (=14) the "present life" (=180) as a "star" (=2).

6.2.2 Transcending Super Wisher

The "all-diverging" (=70,000) "primordial technological growth" (=70,000) frees the "primordial perpetuator" (=9) within the "primordial greeter" (=16) from the "duality" (=28) of first, the charm attraction for the new paradigm, and second, the strange repulsion from the present paradigm. The duality produces "guider imperfection" (=80,000) of "cosmic inflation" (=800,000 =500*1,600) in the "time value" (=500) of the "present" (=1,600). "Big bang" (=158) in the "time value" (=500 =158+34,2) "activates" (=9) an "infinite cycle" (=10) of "energy" (=19) by "reproducing" (=17) the "effect" (=34) of "order" (=2) within the "perpetuating value" (=5) of the creature.

"Perpetuating value" (=5) "produces" (=1) an "inflation field" (=689) of "growth" (=6) to "permeate" (=96) "largest distance" (=680) of the "scalar field" (=630) of "consciousness" (=4) as an "omniscient" (=91). "Growth" (=6) "transforms" (=1) the "shortest distance" (=79) of "primordial knowing" (=79) into a "type" (=21) of "scaling field" (=85) "perpetuating value" (=5) of the "limit" (=8) into "infinity" (=90,000). The scaling field "reproduces" (=78) a "vector field" (=78) to "curve" (=85) the "stability" (=35) of its "continuity" (=50) into the "instability" (=485) of "discontinuity" (=200).

A "creature" (=12) is a "believer" (*Saddha*, =-5) in a "para deity" (=5), whose "perpetuating value" (=5) "transforms" (=1) "greeter consciousness" (=1) within the creature into a "primeval greeter" (=11). "Primeval greeter" (=11) "develops" (=22) "cognizance" (*Vinnana*, 134) of the "ascending consciousness" (=22) to "reorder" (=2) "visceral manifestation" (=369 =134*2+101) of the "wisher" (=0) as a "lord" (=0) of the "materialization" (=101) of "rebirth" (=10) to fulfill the wish. The "infinite light" (=11,111) of the "present life experience" (=11) "guides" (=111) "gravitational quality" (=0) of the "lord's" (=0) "action" (=10) as a "worker deity" (=1).

In the "darkness" (=180) of "night" (=900), the lord "repels" (=9) the "wish" (=18) to be a "star" (=2) to "attract" (=34) the "wishable" (=20) who "lives" (=14) as "God" (=5) in the "heaven" (=19) to "fulfill" (=9) the "goal" (=9) as a "worker deity" (=1). The lord's "faith" (=59) in "God" (=5) who fulfills his "goal" (=9) "radiates" (=89) the "light" (=180) of "nature" (=8) for "materializing" (=–6) the "wish" (=18) by "virtue" (=16) of "action" (=10).

The Lord makes his "negative energy" (=19) the "perfect solution" (=1,000) for "blessing" (=1,000) "himself" (=5) as a "para deity" (=5) with the "positive energy" (=19) in the "subconscious" (=100) of a "child" (=128) "alive" (=10) after his "death" (=18). A "child" (=128) enjoys the "possibility" (=128) to "live (=14) as a "Primeval Illuminator" (=14) for "consciousness expansion" (=114) with "freedom from the spirit" (*Saguna mukti*, 128) of the "lord" (=0).

The Primeval Illuminator is the "Queen of Space" (*Maha Lakshmi*, 14) at the "thirteenth infinity" (=14). A "creature" (=12) transforms into the "Queen" (=7) with the "entropy" (=5) of the "thirteenth" (=24 =12+7+5) to form an "infinity" (=90,000). The infinity norms the "space" (=18,000) for "Hell" (=19) without the "First Eden" (*Maha Lakshmi*, 14) where the Queen of space "lives" (=14) as a "future system" (=14) without "consciousness" (=5) of the "past" (=9). The past "copies" (=0) the future system for its "replication" (=90) as the "present" (=1,600 =90*14+360-20) of the "long chain of replication" (=-20) of "time" (=360) to "pivot" (*Svastika*, 304) a "body" (=56) of "consciousness" (=4).

As "Goddess" (=70) of the "present wealth" (*Padmajata*, 54 =70-20+4), the "Queen of Space" (=14) "sums" (=54) the "filament" (=-20) with the "consciousness" (=4) of her "right-facing" (=10) "actions" (=10). As the "perfect personification" (*Maha Durga*, 16) of "Mother Nature" (=8), she is present as the "residual cause" (*Param Lakshmi*, 816) within the "infinite exchange" (=816) of primordial, "physical wealth" (*Shri*, 81) with the primeval "metaphysical wealth" (=6) in the "future life" (=816).

Science of Filament

A "filament" (=-20) is a "long chain" (=-20) of "replication" (=90) of "time" (=360) "radially" (*Trijyat*, 90) as an "entity" (=24). By "producing" (=17) an "electric filament" (*Adhassota*, 17) for "shortening chain" (*Adhassota*, 17) "downstream" (=17) into the "future" (=0), "time multiplier" (=3) "subjects itself" (*Taijaisa*, -10) to the "elongator chain" (*Kinja*, 900) of "magnetic filament" (*Kinja*, 900) of the "spirit" (=20) of "space" (=18,000) for its "replication" (=90) as a "horizontal filament" (=-20). The horizontal filament is an "electromagnetic filament" (=-20) that "circles" (=100,000 =105) its "radius" (=1010 =105x2) with a "pair of four filaments" (=8) to "point" (=1010) to the "origin" (=1,000). The electric filament is a "diagonal filament" (=17). The magnetic filament is a "vertical filament" (=900).

The origin is the "elongator chain" (=1,000) of "two filaments" (=1,000) which "pair" (=8) their "double copy" (=1) into an "elongated chain" (=991) of "four filaments" (*Dirghavrittaphala*, 991) to "twin causation" (=2) with their "past" (=9) as "space" (=18,000). Two filaments are "gravitational filament" (=1,000). Four filaments are "thermodynamic filament" (=991), i.e., "gravitoelectromagnetic filament" (=991) that "sources" (=1,000) a "shortener chain" (=9) of "three filaments" (=9) from their "past" (=9) with three "copies" (=0) of their "future's" (=0) "electromagnetic mass" (=0) to "square" (=18,000) the "space" (18,000) with the "potential" (=18) of two that "double copy" (=1) three to be "left" (=89) with "five filaments" (=900) as "magnetic filament" (=900) with "electromagnetic filament" (=-20) as the "metric" (=-1) of "filament" (=-20). Three filaments are "sentient filament" (=9). The gravitational filament is a "circular filament" (=1,000). The thermodynamic filament is a "square filament" (=991). The sentient filament is a "triangular filament" (=9).

"Three filaments" (=3) "shorten" (*Hras*, -1,000) "two filaments" (=1,000) with a "shortener" (*Hrasaka*, 136) element that "bursts" (=136) a "neutron" (=19) as a "whole" (=16) into a "diminution chain" (=120) of "nineteen filaments" (=120) in a "flash" (*Jyoti*, 120). Nineteen filaments are a "gravitoelectric filament" (*Jyoti*, 120).

Nineteen filaments are a "gravitoelectric filament" (*Jyoti*, 120). The "diminution" (*Hrasana*, 39) of the "gravitoelectric filament" (=120) by a "time molecule" (=39) "produces" (=1) a "double copy" (=1) of the "nine filaments" (=160) within the "gravitomagnetic filament" (=160) which "progresses" (*Vikas*, 160) as a "diminutive chain" (=160) with the "development" (=160) of a "half atom" (=160) into a "single particle" (=160).

Thus, a gravitoelectric filament (nineteen filaments) "long chains" (=-20) as a "filament" (=-20) its origin as a gravitational filament (two filaments) to perpetuate its past with a sentient filament (three filaments) for the development of gravitomagnetic filament (nine filaments) with the thermodynamic filament (four filaments). The gravitoelectric filament is a "line filament" (=120). The gravitomagnetic filament is a "point filament" (=160). The whole is a "parallel filament" (=16) that "half chains" (=16) "dark matter" (=1,600) as a "parallel" (=80,000) to "break away" (=-25) "time value" (=500) as a "filament" (=-20) with its "reproductive force" (=100). The "dark matter filament" (*Sara*, 16) "copies" (=0) the "weakness" (=72) in the "body" (=56) of its "future" (=0) for the "replication" (=90) of "filament" (=-20), with "ninety filaments" (=16), as a "light matter filament" (=2), which "guider chains" (=2) "ten filaments" (*Tara*, 2) into an "arc filament" (=2). The "matter filament" (*Brahma yajni*, 8) is a "shortened chain" (=8) of "sixteen filaments" (=8) that "angular filament" (=8) to present their "discontinuity" (=200).

"Seven filaments" (=1,600) are "present" (=1,600) as a "strong filament" (=1,600) with a "dim chain" (=1,600). The "future" (=0) "copies" (=0) "five elements" (=900) into a "weak filament" (=0) of "fifty filaments" (=0) with a "heavy chain's" (=0) "potential" (=18) for "continuity" (=50) as the "spacetime" (=20) to "time" (=360) the "space" (=18,000). The "potential" (=18) has a "strength" (*Bal*, 83) to "fragment" (=64) the "light chain" (=64) as a "bright filament" (=64) into "hundred filaments" (=64) with its "double copy" (=1). The double copy is a "dark filament" (=1) that "dull chains" (=1) an "atom" (=19) into a "particle" (=19) within the "energy" (=19) of "neutron" (=19).

Formative Growth Paradigm

> The dark filament "self-perpetuates" (=1/2) the "filament" (=-20) as a "half filament" (=1) with the fifty-filament "gravitational quality" (=0) of the future. The "gravitational" (=100) element has a "potential" (=18) to "blackhole" (=82) its "future" (=0) for "glowing" (=149) the "twin reality" (=149) of its "guider potential" (=149) as a "guider particle" (=149) like the "color glass condensate" (=149). The guider particle perpetuates the "twin radius" (=140) of the spacetime's reality as a "radial" (=140) until the entropy of the "twenty-five-filament" (=25) "radian" (=25), the "triple radius" (=25). The twenty-five filaments are a "black filament" (*Pallava*, 25) that "potential chain" (=25) the "lunar flash" (*Tiryagjyoti*, 25). The "radius" (= 10^{10}) is a "white filament" (*Sajjyoti*, =1010); it "present chains" (= 10^{10}) the "reproductive reality" (=40) of "25 filaments" (=25) to "source" (=1,000) a "solar flash" (*Sajjyoti*, 10^{10}) of "one thousand filaments" (= 10^{10}). The "twin radius" (=140) is a "color filament" (*Naciketu*, 140); it "primordial chains" (=140) the "reproductive reality" (=40) of the "gravitational element" (=100) for a "circular creation" (=10,000) of "ten thousand filaments" (=140) of "earth flash" (*Naciketu*, 140), "ascending light force" (=140). The "four radii" (=73) are a "colorless filament" (*Irammada*, 73); they "primeval chain" (=73) "hundred thousand filaments" (=73) into a "plasma filament" (=73) of "stellar flash" (*Irammada*, 73) for "lightning" (*Vidyut*, 88) the "descending light force" (*Prajval*, 73) with "enlightening" (*Maha Lakshmi*, 14) of the "double copy" (=1) within the "Goddess of Lightning" (=14).

With her "polarization" (*Dian Mu* 电木, 14) into "Goddess" (=70) of "lightning" (=88), Queen of Space "accumulates" (=5) the "potential" (=18) to "order" (=2) the "replication" (=90) of the "infinite illumination" (*Manoj*, 180) of her "reality" (=7). She "trades" (=20) "everybody's" (=-9) "profited time" (*Donar*, 14) to "reincarnate" (=240) as a "daughter spirit" (*Thor*, 14). In the present "birth moment" (=6), she "pivots" (=306) the "left-facing" (=139) "immanent quality" (*Sefirot*, 14) after the "creation" (=379) as a "primordial maternal spirit" (*Lakshmi*, 379 =240+139) of the "feminine" (=37) future "death moment" (=9). As the "demolisher" (*Vighnanasha*, 590) of the "obstacles" (*Vighna*, 680) to "right consciousness" (=680) in the "ethericsphere" (=0), she mediates the "continuity" (=50) of the "present moment" (=6) with "primordial creation" (*Ganesha*, 570) of the "spirit" (=20) of the "future moment" (=9) for "replication" (=90) of the "past moment" (=89) as a "super wisher" (=−1).

The spirit "divines" (=360) the "potential" (=18) for "spiritual" (*Saguna*, 20) "creation" (=379) of an "infinite value" (=957 =360+379+189+30) of the "primordial maternal" (=18) with the "devotional energy" (=189) of the "infinite council of spirits" (=30). "Spiritual freedom" (*Saguna Mukti*, 128) from the "sentimental-effect" (=40) of the "potential" (=18) "develops" (=22) "austerity" (*Tapasya*, 48) within "Goddess" (=70). Austerity "seasons" (=2) the "potency" (*Virya*, 48) of "fruit quality" (*Phalguna*, 48) of "harvest" (=81) from the "old season" (*Padmanga*, 31) in "February" (*Phalguna*, 48), the "twelfth month" (*Phalguna*, 48). "March" (*Chaitra*, 90,000) "springs" (=811) a "new season" (*Anuradha*, 189) through "primordial oneness" (*Adi*, 32) with the "present season" (*Shivansha*, 90). As an "entity" (=24), the "Pisces zodiac" (*Meen*, 96) "seasons" (=2) "oneness" (=48) of "past force" (=24) with "being energy" (=96) to "project" (=31) the "fire" (=17) of "self-disjunction" (*Meen*, 96) of the "birth moment" (=6) as a "primeval deity" (=6) from the "past" (=9) for "least cause" (=70) "citizenship action" (=70).

6.3 The Path of Citizenship Action.

The "path of citizenship action" (*Ashraya Marga*, 85) forms "everything" (=−5) with "technological growth" (=2) of the "sentient energy" (=1,000) through "primordial technological exchange" (=80,000) of the mediating "God cause" (=68) with the "least cause" (=70). Technological growth (=2) "limits" (=8) "extrinsic gravitational-effect" (=38,000) for "potential exchange" (=3,000) of the "replicated light" (=3,000) to "attract" (=34) "convergent energy" (=28) from the "present moment" (=6). Convergent energy develops "infinite gravitational potential" (=4) of the "goal-in-motion" (=−4) as a "supreme wisher (=−4) of the continuity in the "worker social benefit" (=82). "Goal-in-motion" (=−4) "trades" (=20) "future value" (=2) of the "Wisher" (=0) with a "decision" (=165) to "wish" (=18) "spiritual oneness" (=147). With goal-in-motion, Wisher guides one to fulfill the goal of the wish: "motion" (=360) "attuned" (=11) with "time" (=360) "surfaces" (=2,000) "sentience" (=4) of the "reality" (=7) of "sentient potential" (=27,000) for "servicing" (=47) the "divine potential" (=6) of the "individual" (=53) to a "collective" (=47) as a "guider" (=100) element.

The Least Cause is the Leading Digit

"Least cause" (=70) is the "leading digit" (=70). The leading digit is the "second digit" (=70). It is a "visible digit" (=70). "First digit" (=10,000) is the "following digit" (=10,000). As a "circular creation" (=10,000) of "Mother Nature" (=8) as a "perfect female" (=1,250), it is an "invisible digit" (=10,000). "Leading digit" (=70) "follows" (=10) "Benford distribution" (=-3), which is a "triple exponential distribution" (=-3) that "triple tiles" (=77) "two digits" (*Hasta*, 0)—the "Benford" (*Hasta*, 0). The two digits are the "base" (=10) for the "growth" (=6) of illusionary "cosmic balance" (=4).

"Nature" (=8) "self-perpetuates" (=1/2) the cosmic balance with "4" as the leading digit 10 percent times. "1" "triples" (=130) the "reproductive force" (=100) to be the "metric" (=-1) of the causal, "feminine body" (=30) 30 percent times. "2" and "3" "even" (=113) the "causal body" (=30) "moving forward" (=113) as a "cyclical" (=113) to "reproduce" (=78) the "exponent" (=78) 30 percent times with "potential" (=18) of two that "orders" (=2) the three's "growth" (=6) into "self-luminous" (=12) element.

Therefore, "two" is the leading digit 18 percent times and "three" is the leading digit 12 percent times. "5", "6" "7" "8", and "9" are "odd" (=1,810), "self-perpetuating" (=1/2) the "memory" (=35) of the "leading digit" (=70) with the "potential" (=18) within the "base" (=10) for "self-projecting" (=3/2) "one-to-five" (=15) as a "secondary cyclic group" (=15) with "two circles" (=15) forming "both" (=10,000) two and three as "circular" (=10,000).

Thus, "five" is the leading digit 8 percent times, and the "six" is the leading digit 7 percent times. "Nature" (=8) is "repeated" (=5) by five with its "entropy" (=5) for the "growth" (=6) of reality with a six, with three as a multiplier of the order.

> Further, "seven" is the leading digit 6 percent times, "eight" is the leading digit 5 percent times, and "nine" is the leading digit 4 percent times. The reality of the "goalkeeper" (=7) leads to growth. Nature leads to entropy. The goal of the "supreme wisher" (=−4) leads to cosmic balance within entropy.

6.3.1 Understanding Supreme Wisher

A "Supreme Wisher" (=−4) is the "concealed truth" (=−4) of the "immanent mass" (=−4) of a "four-dimensional space" (=−4) whose "extrinsic mass" (=−4) has an "absolute presence" (=−4) "everywhere" (=−4) as the "present value" (=476). The supreme wisher is the "half part" (=−4) of the "sentience" (=4) of a "wisher" (=0) that "animates" (=76) itself. The "part" (=−1) as a "whole" (=16) is "inanimate" (=18), "scalable" (=2) as "2-in-1" (=2) with animate within it. The "supreme" (*Adhiraj*, 0) "in" (=-12) the "wisher" (=0) is the "twin essence" (*Samsara*, 0) of the "binary" (=−12) within the "unary" (=91). The unary is the "shared wish" (=91) "appearing" (=79) "better" (=79) as a "big ecosystem" (=79) by "converging" (=79) "twin technological reality" (=79).

The big ecosystem is better because it is "self-reliant" (=79) on the "consumption" (=79) of "goalkeeper" (=7) with a "goal" (=9) of following a "path of goal" (*Asiddha Marga*, 96) as a "part" (=−1) of the "Wisher" (=0 =96−79−7−9−1). The "path of self-development" (=34) "ascends" (=26) "mindfulness" (=42) of the "living entity" (=96 =34+26+42−6) for growth in the "present moment" (=6). The "past moment" (=89) "mutates" (=15) the "reality" (=7) of the "living entity" (=96) by "feeding" (=17) the "present reality" (*Paramartha*, −2) of the "world" (*Duniya*, -2) of "spatially-variable wishes" (=−2) of "diverse wishers" (*Jagath*, -2). The diverse wishers are a "genotype" (=197) of a "temporally-constant type" (=197), "visible" (=197) as "two types" (=197).

The "first type" (=1,810) is the "past type" (*Chaturveda*, 1,810) as the "odd type" (=1,810); it is a "spatially-variable type" (=1,810) with "three types" (=1,810) "synchronous" (=1,810) with "all universes" (*Visvarupa*, 1,810) "centering" (=10) the past, present, and future of its "potential" (=18). The "second type" (=7) is the "future type"

(*Vastunara*, 7) as an "archetype" (=7) varying by both space and time; it is a "spiritually-variable type" (=7) with "seven types" (=7)—genotype and its two types plus "phenotype" (=1,810) and its three types. The genotype is the "third type" (=197 = 21*8+29) which "forms" (=100,000) a "type" (=21) of "proportionality" (=21) of an "organization" (=29) by "organizing" (=385) the "soul essence" (=1,024 =197+385+482) as a "stereotype" (*Murka*, 482). The stereotype is the "fourth type" (=482), the "spatially-constant type" (=482) in "oneness" (=48) with the "universe" (=2). It "differentiates" (=7) "eight universes" (=10) with the "seven types" (=7) to "integrate" (=1) the "potential" (=18) "itself" (=121) into the "ninth universe" (*Varutri*, 190) as the "octave of types" (=482).

The "present" (=1,600) is "livid" (=482) that the "tenth universe" (*Murka*, 482) is a pinpointed" (*Murka*, 482) "line universe" (*Murka*, 482), the "eight-pointed" (*Murka*, 482) "base" (=10) of the "one-pointed" (*Shunda*, 600) "exudation" (*Shunda*, 600) of the twelfth "point universe" (*Shunda*, 600) from the time's "three-pointed" (*Trishira*, 39) thirteenth, "circular universe" (*Trishira*, 39). The present "animates" (=76) a "four-pointed" (*Vrtra*, 38) fourteenth "triangular universe" (*Vrtra*, 38) to "square reality" (=9) of the "base" (=10) with a "five-pointed" (*Varutri*, 190) ninth, "parallel universe" (*Varutri*, 190). The base "guards" (*Varutri*, 190) the "six-pointed" (*Tvashta*, 303) eleventh, "square universe" (*Tvashta*, 303) with one as a "guardian angel" (=11) of the "two-pointed" (*Maya Vishvakarma*, 10,000) fifteenth "closed universe" (*Maya Vishvkarma*, 10,000). The "seven-pointed" (*Surenu*, 10) "base" (=10) of "whole value" (*Surenu*, 10) is the sixteenth, "open universe" (*Surenu*, 10), whose "illusionary form" (*Maya Vishvakarma*, 10,000) is the closed universe.

The open universe is "pointed" (=0) at "three vertices" (=10) for "producing" (=17) the "reality" (=7) of a "type" (=21) first as a "fifth type" (*Prakara*, 17), the "phylum" (*Parakara*, 17), a "kind" (*Prakara*, 17) of "soul type" (*Prakara*, 17). "Soul's" (=4) "reproductive potential" (=23) for "growth" (=6) "lives" (=14) as a "multiplier" (=3) of "vertex" (=89) within the "soul type" (=17) with "120 types" (*Prakara*, 17). With the growth of "two-in-one" (=2) "vertex" (=89), fifth type is repeated as the "sixth type" (*Prajati*, 17), the "conscious type" (*Prajati*, 17) of "species" (*Prajati*, 17), "speciating" (*Vishthi*, 9) a "kind" (=17) of "genera" (*Vishthi*, 9) into "sixty types" (*Prajati*, 17). The "genus" (*Vishthi*, 9) is the "seventh type" (*Vishthi*, 9), the "guider type" (*Vishthi*, 9), with "six types" (=9), reproductive like the "eighth type" (*Bhuma*, 8), "thirty-six types" (=8). The eighth type is "immense" (*Bhuma*, 8) like a "cohort" (*Bhuma*, 8) that orders a "hillock" (*Bhuma*, 8) with "ant" (*Vamri*, 2) as the "base" (=10). It is "false" (*Asat*, 8) like "Type I" (*Bhuma*, 8) because the "seven-pointed base" (*Kumbhaka*, 10) is a "family of three" (*Parivar*, 10), the "ninth type" (*Parivar*, 10). The family has "twenty-four types" (=10) within an "entity" (=24) who "types" (=21) three as "multiplier" (=3) of one's "reality" (=7).

The ninth type is the "devil type" (*Parivar*, 10); it "includes" (*Havya*, 953) "devil" (=0) as the "third" (=100,000 =100 *[953+32+10+3+2]) who makes the "family of three" (*Parivar*, 10) "reproductive" (=100) through "primordial oneness" (=32) with two, the "ant" (=2). The ant is the "order" (*Suchan*, 2), the "tenth type" (*Suchan*, 2). The tenth type is the "Satan type" (*Suchan*, 2); it "excludes" (*Vitti*, 9,000) the "Satan" (=-1) from the one who "precludes" (*Vaitadhya*, 9) "division" (=999) as the "divisor" (=1) to "produce" (=1) a zero with the "potential" (=18) of the "soul type" (=17) without the "Satan" (=-1). The "division" (*Sudhanvan*, 999) is the "eleventh type" (*Sudhanvan*, 999), the "deity type" (=999); the "deity" (=1) is the one which "types" (=21) the "division" (=999) of four with a two within Satan, the "potential zero" (=-1). The four is "life" (*Zindagi*, 4), the twelfth, "God type" (=4); "God" (=5) is the one who divides four to be a "legion" (*Dhajini*, 1/4) with "twelve types" (*Dhajini*, 1/4). The legion is the "param deity type" (*Dhajini*, ¼); the thirteenth "spectral type" (=1/4). "Param deity" (=7) is "potential zero" (=-1) whose "nature" (=8 =7-[-1] =2*8) as "param deity type (=1/4) "orders" (=2 =8*1/4) "life" (=4) with the "Satan type" (=2).

The "Satan type" (=2) "excludes" (=9,000) Satan; the "type" (=21) "experiences" (*Anubhav*, 9,000) "Satan" (=-1) as a "domain" (=9,000) that "stops" (*Stambha*, 450) the "likelihood" (*Vitti*, 9,000) of "combination without Satan type" (*Yathayogya*, 10^{100}). The type "presumes" (*Viklrp*, 9,000) that Satan "contaminates" (*Abhidush*,

Formative Growth Paradigm

10^{100}) the "true reality" (*Tulyartha*, 10^{100}) with "Googol" (=10^{100}) through "replication" (=90) of the "family's" (=10) "reproductive force" (=100). The "domain" (*Suchaka*, 9,000) is the "primeval deity type" (*Suchaka*, 9,000) as the "fourteenth type" (=9,000). Domain "includes" (*Havya*, 953) Satan as a "feminine" (=37) from a "point" (=10^{10}) "reproductive" (=100) as a "family" (=10) to be "sextillion" (=$10^{21} = 10^{10} * 10^{10} * 10$) over "180,000 lightyears" (=953).

"Light" (=180) "self-manifests" (=710) a "lightyear" (=710) by "piercing" (=110) the domain's "origin" (=1,000 =180+710+110 =953+37+10) with the "Big Bang" (=158) of "twin feminine" (=158) to "matter" (=158) for a "year" (=268 =110+158). The "potential" (=18) of "family" (10) within one, "replicated" (=180) by the "domain" (=9,000) as a "Satan" (=-1) with "Googol" (=10^{100}) in its "entirety" (=180), is "old" (=-10^{180}). The "light" (=180) "renews" (=111) the "point" (=10^{10}) as a "family" (=10) to "Googol plex" ($10^{10 \wedge 100}$) "itself" (=121) while "massifying" (=10^{90}) the "massifiable" (=10^{180}) "twin reality" (=49) to "predicate" (=49) "cycling back" (=951) to the "origin" (=1,000) as a "quintillion" (10^{18})—the "family with potential" (*Huha*, 10^{18}) to be "Potential God" (=10^{18}).

"Light" (=180) is "predicable" (*Vidheya*, 697) when "replicated" (=180) because the "origin" (=1,000) remains "constant" (*Achara*, 697). "One-to-many correspondence" (=300) is "repelled" (=123) by the origin "squaring universe" (=1,300) with a "square universe" (=303 =180+123). The squaring universe is the "observable horizon" (=1,300) of the "centillion" (=10^{303}). When "light" (=180 =68*2+50+10=8) is "circled" (=385) like "air" (=385), a "bottleneck" (=615) in the "continuity" (=50) of the "origin's" (=1,000) "order" (=2) that "disorders" (=-8) "observation-effect" (=68) is "likely" (=10) "happenable" (=10). The "self-multiplier" (=615) "bottleneck" (=615) is "Graham" (=615), the "multiplier" (=3) of "self" (=8×10^{15}) at the "speed of light" (=8×10^{15}) for "continuity" (=50) as a "number" (=53)—the "Graham number" (=8×10^{15}). The "observation-effect" (=68) is "expandable" (=68) as a "natural number" (=68), the "perfect prime" (=68) which "primes" (=81) the "supernatural number" (=81) at the "infinite edge" (=81) of "conjunction" (=13).

By "producing" (=17) "conjunction" (=13) with the "origin" (=1,000) for the "replication" (=90) of the "feminine body" (=30) as a "family of three" (=10), a "mindful" (=68) "mind" (=38) "enjoys" (=7) a "kingdom" (=957) with "infinite value" (=957). The kingdom is the fifteenth "primordial deity type" (=957). The "Kingdom" (=957 =75+689+190+3) "produces" (=1) a "transcript" (=190) of three as a "trinity" (Sridhara, 25) by "classifying" (=75) the "class" (=689) of two with the "potential" (=18) within "origin" (=1,000 =957+25+18) to be the "potential origin" (Sridhara, 25) of "reality" (=7). The class is the sixteenth "supreme deity type" (=689). The "realm" (Kalpa, 5) the reality "orders" (=2) as a "super deity" (=2) is the seventeenth "super deity type" (Kalpa, 2). The "form" (=100,000) of the realm left as the "third" (=100,000) with the three, the "supra deity" (=3), is the eighteenth "supra deity type" (=100,000). The "variety" (=10^{16}) of the form is the twentieth "primordial greeter type" (=10^{16}). The "tribe" (Kula, 9) that perpetuates the "entanglement" (=9) of the "variety" (=10^{16}) with its "plant branch" (Shakha, 9) is the nineteenth "greeter type" (=9).

The "branch" (Lata, 10^{100}) is the twenty-first "param greeter type" (Lata, 10^{100}) with "sixteen hundred copies" (Lata, 10^{100}). The "plant branch" (=9) "sections" (Anuvaka, 10,000) "itself" (=121) into the twenty-second "primeval greeter type" (=10,000) to "plant" (=-12) a "series" (Paati, 90,000) of "branches" (Lata, 10^{100}) with its "family's" (=10) "reproductive force" (=100 =121-12-9). The series is the "type-effect" (=90,000) that "dissects" (=90,000) itself into an "infinity" (=90,000) of "twins" (=121)—the "potential type" (=121)—for "parabolically-growing" (=140) the "energy" (=19) of its "origin" (=1,000). The origin is "organizable" (=11) into the potential type with the "reproductive variation" (=2,890 = 1,000+189*10) within the "ten-fold growth" (=170) of the "sentient" (=189) element. The reproductive variation is the twenty-third "absolute type" (=2,890 =1,600 +21*60 +60*1/2), "present" (=1,600) as a "type" (=21) of "octave" (=60) "self-perpetuating" (=1/2) a "self-perpetuating octave" (=2,890) of "twin types" (=197) as an "octave of genotypes" (=2,890).

The "probability" (=128) of "variation" (=10), "reproductive" (=100) within the "potential" (=18) to "produce" (=1) a "double copy" (=1) of a "kind" (=17), is the twenty-fourth "infinite type" (=128).

The double copy is the twenty-fifth "primordial type" (=1). The "copy" (=0) is the twenty-sixth "half type" (=0). The "triple copy" (=3) is the twenty-seventh, "twin double type" (=3), an "octave of phenotypes" (=3). The "six copies" (=6) are the twenty-eighth, "twin triple type" (=6), the "octave of genus" (=6). The "twelve copies" (=43) are the twenty-ninth "self-luminous type" (=43), the "octave of legions" (=43). The "twenty-four copies" (*Bhavana*, 37) is a "feminine" (*Bhavana*, 37) as the thirtieth "luminous type" (=37), the "octave of family" (=37).

"Thirty-six copies" (*Kandarpa*, 296) are "masculine" (*Kandarpa*, 296), the thirty-first "time type" (=296), an "octave of cohort" (=296). "Sixty copies" (=60) are the "octave" (=60), the thirty-second "space type" (*Sargam*, 60), the "octave of species" (*Sargam*, 60). "One-hundred-twenty copies" (*Madhusudan*, 16) are a "double octave" (=16) of the thirty-third "entity type" (=16), an "octave of phylum" (=16) that "doubles" (=18,000) the kind with a "twin octave" (*Ashvini*, 120). Twin octave comprises "two-hundred-forty copies" (*Ashvini*, 120) as the thirty-fourth "cause type" (*Ashvini*, 120) whose "effect" (=34) is "two-hundred-forty types" (=34) as the thirty-fifth "causation type" (=34). "Triple octave" (=80) comprises "three-hundred-sixty copies" (*Prabhava*, 80) as the thirty-sixth "octave type" (=80). "Time" (=360) is the thirty-seventh "feminine type" (=360) with "three-hundred-sixty types" (*Kaal*, 360). "Space" (*Vasa*, 18,000) is the thirty-eighth "masculine type" (=18,000) with "eighteen thousand types" (*Vasa*, 18,000). "Potential" (*AUM*, 18) is the thirty-ninth "gender type" (=18) with "eighteen types" (=18). "Present" (*SAUM*, 1,600) is the fortieth "S-type" (=1,600) with "sixteen hundred types" (=1,600).

S-type is "stable type" (=1,600); it "branches" (=10^{100}) "sixteen hundred copies" (=$10^{100} = 5^{*20}$) of the "white" (=10) "sentient force" (=4) from a "circular orbit" (=2) into a "branch-effect" (=14). It "forms" (=10^5) a "stable circular orbit" (=14) with the "rotational motion" (=14) of a "photon" (=20). The "rotational motion" (=14) "blackens" (=−100) the "sentient force" (=4) to "gravitate" (=−100) the "form sphere" (*Khagola*, −25 =−100/4) as a "photon sphere" (*Khagola*, =−25). It "circulates" (=22/7) "three spheres" (=22/7) of time—the "central sphere" (*Avacara*, 100) of the "present" (*SAUM*,

1,600) as a "whole" (=16), the "inner sphere" (*Kshmatala mandala*, −25) of the "past" (*Bhuta*, 9) within the "whole" (=16), and the "outer sphere" (*Akasha mandala*, −25) of the "future" (*Anagata*, 0) without the "whole" (=16) whose "reality" (=7) is the "present reality" (=−2) of "seismosphere" (*Ilagola*, 78 =100-22). "Therefore" (*Tato*, 78), the seismosphere "ergo" (*Tato*, 78), i.e., "let go" (*Tato*, 78) the "central ergosphere" (=100), "inner ergosphere" (=−25), and the "outer ergosphere" (=−25), before "vanishing" (=28) as the "peripheral ergosphere" (=78). The latter is "blackened" (=128) by the "continuity" (=50) of the "three ergospheres" (=78) as a "hole" (=18,000). The "black hole" (=82) of "sentient force" (=4) within the "peripheral ergosphere" (=78) "squares" (=18,000) "space" (=18,000) with "continuity" (=50) of "whole" (=16) as "time" (=360), whose "essence" (*Sara*, 16) is "W" (=16).

With "polarization" (=14) of the time's "rotational motion" (=14) for "time-like" (=91) "launch" (=91) within a "rotating blackhole" (=813=789+24), a "sequential" (*Kramastha*, 789) "entity" (=24) polarizes itself into a "massive" (*Dridha*, 1,000 =91+813+82+14) "blackhole" (=82) "node" (=14). The "nonrotating" (*Akramastha*, 918 =135-30+813) "nonsequential" (*Akramastha*, 918) "locus" (*Vikasaj*, 135), "tailing" (=−30) the "massive" (=1,000) "origin" (=1,000) of the "massive blackhole" (=813), is "supermassive" (*Atidridha*, 135). The "blackhole" (=82) is "repeated" (=5) as "supra massive" (*Sudridha*, 13), i.e., "ultramassive" (*Sudridha*, 13), within the "correlation force" (=13) of the "origin" (=1,000) with "itself" (=121) to "live" (=14) "forever" (*Apradivam*, 5) as the "twinning force" (=13).

The "supermassive blackhole" (*Rudhamanyu*, 67 =135-82+14) is a stable, "nonrotating blackhole" (*Rudhamanyu*, 67), "breeding" (=15) "infinity" (=90,000) as a "supreme massive blackhole" (*Atisukshma*, 90,000 =15*6*1,000) to "reform" (=888) the "supra massive blackhole" (=888) into a "para massive blackhole" (*Samidriksha*, 30) , comprising six massive blackholes, with the "origin's" (=1,000 =67+15+888+30) "darkness" (=180). The "apparatus" (*Yantra*, 12 =3+9) transforms one into three within nine to "twin causation" (=2) for the "massive" (=1,000) to be "supreme massive" (*Yantradridha*, 9,000) without the three. The three "perpetuates" (=9) itself as a "cubic" (=9), "perpetuated"

Formative Growth Paradigm

(=90) as a zero for "replication" (=90) within the six. The six is the "growth" (=6) of the "variable group-effect" (*Vajra*, 14) that "destroys" (=1) the "constant group value" (*Vajradridha*, 10^{10}) of the "para massive" (*Vajradridha*, 10^{10}).

The six "destroy" (=1) "para massive" (*Vajradridha*, 10^{10}) to be "light" (=180). The light "divides" (=132) the "darkness" (=180 =132+30+18) into "six massive blackholes" (=30) with the "potential" (=18) within three to "twin" (=121) the "blackhole" (=82) as the "nonconstant" (*Adridha*, 20). The nonconstant "twin causation" (=2) to "transform" (=1) itself into the "primeval massive" (*Abhidridha*, 21)—the "variable group value" (*Abhidridha*, 21). Supreme massive is compressed. Para massive is supercompressed. Primeval massive is supracompressed. The nonconstant is the "group value" (*Adridha*, 20), which is "primordial massive" (*Adridha*, 20). The group value comprises the three, the nine, the six, the one, and the two; the three manifests the nine as the six destroy their growth to transform one into two. The "light" (=180) "groups" (=387) two by "producing" (=17) "darkness" (=180) within "time" (=360) to "illuminate" (=150) "itself" (=121) as the "hole" (=18,000) in "space" (=18,000) with a "primordial massive" (=20) "burst area" (=900) of the "primeval massive blackhole" (=1).

As one "twin causation" (=2) to perpetuate itself like nine by "centering" (=10) the "blackhole" (=82), the "twin blackhole" (=10^{19}) is liberated as a "primordial massive blackhole" (=10^{19}). The twin blackhole comprises "thirteen massive blackholes" (=10^{19}); "one blackhole" (=82) "twins" (=121 =105+16) its "form" (=10^6) of "quasar" (=10^6) within "three blackholes" (=10^{16}) to be "massive" (=10^3). The "three maxima" (=105 =82+23) "put" (=10^7) its "reproductive potential" (=23) in an "ascending-order" (=105). The "triple blackhole" (=10^{16}) is a "triple supermassive blackhole" (=10^{16}); thirteen within the "supermassive" (*Atidridha*, 135) is "repeated" (=5) as 65, "without consciousness" (*Rahitatma*, 65), for "primordial oneness" (=32) of the three. The three "triples" (=130) the two that "twins" (=121) with 33, the "feminine consciousness" (=33). Thus, 33 becomes "massive" (=1,000), the "nucleus" (=1,000) of "blackhole" (=82 =33+49) with its "twin reality" (=49 =33+16). The "whole" (=16 =7*2+2) twins the "reality" (=7) of the "ellipse"

(=1,000) of "space" (=18,000) within the "potential" (=18 = 16+2) of two. As a "spatial ellipse" (=10^{16}), the triple blackhole is a "potential massive blackhole" (=10^{16}) with "thirty-three blackholes" (=10^{16}).

The infinite, ring-like, "central singularity" (=12) of the "blackhole" (=82) "norms" (=18) two of "equal mass" (*Dosha*, 580) "within a group" (*Dosha*, 580) with a "fault" (*Dosha*, 580 =468+110+2) in the "constant geography value" (*Tamra*, 466) for "perforation" (=110 =12+82+18-2) of the "sentient force" (=4 =2−[−2]) through "dividedness" (*Dosha*, 580) of the "third's" (*Tritiya*, 100,000) "sentient capacity" (*Tamra*, 466). The third is the "copper" (*Tambaka*, 100,000) that forms "copper-63.5" (*Tambaka*, 100,000). It produces "copper-effect" (*Tamra*, 466) as "copper-65" (*Tamra*, 466) to triple "dot" (=10^{10}) with a three. The dot is the metal's "ten copies" (=10^{10}). The "triple dot" (*Kora*, 10^{1000}) is the time's "thirty copies" (*Kora*, 10^{1000}) that "blank" (*Kora*, 10^{1000}) the space by "flattening" (=10^{1000}) it into a "belt" (*Mekhala*, 10^{1000}) of "giant donut" (*Dirghavritta*, 1,000), the "ellipse" (*Dirghavritta*, 1,000). The ellipse is a "param massive blackhole" (*Dirghavritta*, 1,000), "nested" (*Pravishta*, 80) as a "supermassive blackhole" (=67) with 13. The thirteen is "organizable" (=11) as "eleven supermassive blackholes" (*Dirghavritta*, 1,000) that "twin causation" (=2). Their "form" (=100,000) as "quasar" (=100,000) "reforms" (=888 =1,000-121-1) the "supermassive blackhole" (=67) into a "triple quasar" (=67) as they "coalesce" (=80) a "merged entity" (=80) to "twin" (=121) "causation" (=1).

The ellipse comprises "twenty-two massive blackholes" (*Dirghavritta*, 1,000). A "feminine" (=37) "twins" (=121 = [67+37-4] +3*7) a "massive blackhole" (=813 =4*2, 13) into a "supermassive blackhole" (=67), "twin massive blackhole" (=67), for "producing" (=17) the "primordialism" (=27) of the "super" (*Vic*, 37) in three. The three is the multiplier of "reality" (=7) as the "consciousness" (=4) of 13 "shadows" (=52) the "luminous" (=13) like a "star" (=2 =3-1). A "star" (=2) "triples" (=130) a "massive blackhole" (=813) into a "supramassive blackhole" (=888 =813 +91-10-3*2), the "triple massive blackhole" (=888), within a "family of three" (=10) with the "supra" (=91) element. The "past" (=9) of the three is "repeated" (=5) within the "massive blackhole" (=813) with the "supreme" (*Adhiraj*, 0) element by the three zeroes to norm "five massive blackholes" (=90,000) within the "supreme massive blackhole" (=90,000). "Para" (=81) "perpetuates" (=9) the past's "reality" (=7) of three within the "massive blackhole" (=813) to be a "para massive blackhole" (=30 =9+7*3) with "nine massive blackholes" (=30). "Primeval" (=185) "destroys" (=1) the "primordial" (=85) within the "massive blackhole" (=813) with "greeter consciousness" (*Pratyagatma*, 1) to be the "primeval massive blackhole" (=1) with "eleven massive blackholes" (=1).

Science of Blackholes

A rotating, massive blackhole is the "primary blackhole" (=813); it is a bigger, "heavier blackhole" (=813) of a 12-year "new planet cycle" (*Vidhata chakra*, 33) of the secondary "reactive energy" (*Mula shakti*, 846). A supermassive, "rotator blackhole" (=67) is the "secondary blackhole" (=67); it is a smaller, "lighter blackhole" (=67) that "radiates" (=89) "light" (=180) over "55-years" (=89) to "plunge" (*Avagaha*, 89) through the primary blackhole's "accretion disk" (=210) and "extensifies" (*Ghataana*, 340) into a tertiary, supramassive, rotatable, "light blackhole" (*Ushnavahana*, 888) of an 80-year "nonlinearum cycle" (*Rochisha chakra*, 28) of the primary "mineral energy" (*Khanijya shakti*, 2).

The "light blackhole" (=888) "intensifies" (=340) into the quaternary, suprememassive, rotated, "heavy blackhole" (=90,000) of a 25-year "Astronomical realm cycle" (*Suryaphani chakra*, 1010) of the quinary, "para entity energy" (*Vimarsha shakti*, 1,200) by "cooling" (*Nirvapana*, 90,000) the "expanding bubbles" (*Shunyata*, -2), "amplifying" (=268) the "flare" (*Mala*, 784) of "trading-effect" (=26) for "breeding" (=15) light as a "discontinuous flash" (=180) while "trading" (=53) the "reproductive force" (=100) of "twin causation" (=2).

The "reproductive potential" (=23) of "light" (=180) "jets" (=9) an "infinite flash" (*Tiryagjyoti*, 25), comprising "twenty-six flashes" (*Tiryagjyoti*, 25), with "force" (=34) of the quinary paramassive "lightest blackhole" (=30) of a 20-year "Triangulum cycle" (*Hasti chakra*, 27) of the tertiary "entity energy" (*Ambika shakti*, 66).

The "jet-effect" (*Stanumalaya*, 150) produces a senary, primeval massive "heaviest blackhole" (=1) of the 100-year "negative realm cycle" (*Kramakrita chakra*, 186) of the quaternary, "self-luminous entity energy" (*Prakasha shakti*, 35).

The "cost" (=41) of "powering" (=41) the "jet's" (=9) "continuity" (=50) through "coupling" (=41) "consumes" (=467) the septenary, param massive, "bright blackhole" (=1,000) of the 10-year "positive realm cycle" (Adhi chakra, 189) of the senary, "left energy" (*KLIM shakti*, 385).

The octonary, primordial massive, "dark blackhole" (=1019) of the 90-year interplanetary "imperium cycle" (*Uchchadana Chakra*, 2,100) of the octonary "centerward energy" (*Jyeshtha shakti*, 101000) "dots" (=1010) the "perpetuity" (=1019) of the "coupled" (*Mithanibhuta*, 109), "reproductive" (=100) as the "origin" (=1,000) that "orbits" (=2) "two blackholes" (=1019). The nonary, "dull blackhole" (=1016) of the 9-year "realm cycle" (*Kalpa chakra*, 14) of the "right energy" (*Shiva shakti*, 910) "animates" (=76) the "origin" (=1,000) of "both" (=10,000) "coupled" (=109) within "three blackholes" (=1016) as "one with all" (=109).

The denary, right-handed, "feminine blackhole" (*Shatanga*, 134) of the 16-year "para energy realm cycle" (*Loka chakra*, 16) of the "current of energy" (=150) is a "supreme blackhole" (=134) with "four blackholes" (=134).

The left-handed "masculine blackhole" (*Para Vishnunabhi*, 978) of the 360-year "potential matter cycle" (*Sadhibhuta chakra*, 9) of the first "twin entity energy" (*Prathama shakti*, 14) is the "origin" (=1,000) that "develops" (=22) into a "para blackhole" (=978) of "five blackholes" (=978).

A departed "human entity" (*Manushya*, 82) is "handed blackhole" (=82) of a 3,600-year intraplanetary "oscillationum cycle" (*Anitya chakra*, 246) of the septenary "param entity energy" (*Shri Shakti*, 385) by "Sun" (=21), the "param creator" (=21) of "life" (=4), for "second life" (*Kurukulla*, 57) as a "star" (=2).

The "massive" (=1,000) is "predominantly" (*Satireka*, 10) "odd" (*Dushama*, 1,810); it is "left" (=89 =100-11) as "mass" (=132 =121+11) after "even" (*Sushama*, 113) "twins" (=121) eight to perpetuate nine as a "system" (=12) with three's "reproductive force" (=100) within eleven. One which becomes nine with eight, Mother Nature, for "self-projecting" (=3/2) the reality of two that transform a "family" (=10) into a "system" (=12) to norm a "blackhole system" (=19) with "nineteen massive blackholes" (=19). The blackhole system twins itself with the two within "seventeen massive blackholes" (*Vaishvanara*, 17) to be "all-pervasive" (=17) as a "twin blackhole system" (*Vaishvanara*, 17). The three "triples" (=130) the blackhole system after one "divides" (=132) the two to be "subtractive" (*Shodhaka*, 288 =130+132+1+15) as a "purgatory" (*Shodhaka*, 288 = [89-19]*4 +2*4) with four, the "consciousness" (=4) of "hell" (=19) "left" (=89) within the "fifteen massive blackholes" (*Arthatma*, 15), the "triple blackhole system" (=15). The four is a "half blackhole system" (*Mamatma*, 4), a "potential massive blackhole" (*Mamatma*, 4) with "sixteen massive blackholes" (*Mamatma*, 4). It is "half nature" (=4) whose "nature" (=8) is "primordial oneness" (=32) with the "blackhole" (=82) for the "continuity" (=50) of its "potential" (=18) within one as the "mass" (=132) of two that forms six with a three to be seven like a "supermassive blackhole" (=67).

The supermassive blackhole is "Vega" (=67), the "galactic core" (=67), which forms the quasar as the "galactic core system" (=100,000) with 12, the "system" (*Gabhira*, 12). With 12, fifty is "repeated" (=5) for a "symmetry" (=250) with 250. With "seven-fold symmetry" (=7) within 12, two-hundred-fifty is "repeated" (=5) for a "twelve-fold symmetry" (*Chandravanshi*, 1,250 = 250*5) with 40. The forty is the "reproductive reality" (=40 = 100,000/1,250) which "twins" (=121) one with the "past" (=9) of the "triple" (=130) as a three within the "feminine" (=37). The triple quasar "bunches" (*Stabaka*, 67) the "triple galactic core system" (*Stabaka*, 67) into a "cluster" (*Akashaganga*, 85) of "stars" (=2) with its "past" (=9) as a "half quasar" (=21) within 13 to triple one with ten as the "base" (=10). The "cluster" (=85 = [20+18]*2+9) is an "imaginary" (=20) "pattern" (*Vyavastha*, 189) of the past's "potential" (=18) that twins nine. The "imaginary pattern" (=85) is a "nonrepeating pattern" (=85), "partially repeating" (=65) within the "imaginary" (=20).

It is "aperiodic" (*Asamvarta*, 138), "perpetuating time" (=138) as a "nonrepeatable" (=138) for the "pattern" (=189 =138+51) to be "aperiodic pattern" (=85 =34+51) of a "nonrepeater" (*Sansa*, 51) "producing" (=17) her "nonrepeatable-effect" (=34) to "bunch" (=67) the "continuity" (=50).

The ellipse is the "nucleus" (=1,000) of the "burst area" (=900) which "catalyzes" (=900,000) an "equal conscious momentum" (=900,000), i.e., equal "angular momentum" (*Akshavat*, 90) as a "contragrade" (=900,000) for "replication" (=90) as four with the "time's" (=360) "sentient force" (=4) within the "primordial massive" (=20), the "compressor" (=20). The "nucleus" (=1,000), present as the "event horizon" (=1,600 =1000+500+100), twins the "time value" (=500) with the "reproductive force" (=100) of the "twin event horizon" (=148) to "sequence" (=48) itself as the "greet" (=64) of "time" (=360 =100+148+48+64). The greet is "param massive" (*Niru*, 64), whose "nuclear potential" (=64) is "compressible" (*Niru*, 64) into the "triple event horizon" (*Niru*, 64). Since the nucleus is a "complex" (=1,000), the greet as the "first event horizon" (*Niru*, 64) is a "complex horizon" (=64). It is "greeted" (*Lantaka*, 1,300) by the "second event horizon" (*Bhushana*, 1,300)— the "simple horizon" (*Bhushana*, 1,300)—to "twin causation" (=2) with "simplicity" (*Jiva*, 2).

The simple horizon is a visible "outer horizon" (*Bhushan*, 1,300), "greeting" (=78) the "third event horizon" (*Kshitij*, 724) on the "horizon" (*Kshitij*, 724). It "breeds" (=78) "seven horizons" (*Dhruvamatsya*, 8) within the "eighth event horizon" (*Dhrvamatsya*, 8) to "triple" (=130) itself into an "entity" (=24). The third event horizon is the invisible "inner horizon" (*Kshitij*, 724). It is an "imaginary horizon" (*Kshitij*, 724), conscious of its future. Seven horizons are a "natural horizon" (*Dhruvamatsya*, 8) of the "longitude" (*Dhruva*, 27) that "fish" (*Matsya*, 17) the "fire" (=17) for "centering" (=10) their "effect" (=34) with two to "twin event horizon" (*Anabhijna*, 148 =8*17+12) of the one with "twelve twos" (*Viyadganga*, 85). Twelve twos are a "galaxy" (*Viyadganga*, 85) with twelve "stars" (=2), "centering" (=10) their "constellation of effects" (=-2) as the "constellation of twelve stars" (*Dhruvamatsya*, 8 =10-2).

The twin event horizon is the "fourth event horizon" (=148), para-conscious of its past as the "creator dimension" (*Advaita Dharma*, 148). It is the illusionary "cosmic horizon" (=148), which squares itself to be repeated as the "fifth event horizon" (=1,600), the "real horizon" (=1,600), present as the "square horizon" (*Digant*, 1,600). The "sixth event horizon" (*Vakarikriti*, –7) is the "central horizon" (*Vakarikriti*, –7) of the "reality" (=7) within the "observer" (=0). The observer is "curving" (*Vakarikriti*, –7) the "present reality" (=–2) of "everything" (=–5) with its "constellation of effects" (=–2) through "guider mediation" (=–7) of divine-effect, fire-effect, water-effect, air-effect, earth-effect, and ether-effect within sentient-effect to be a "star" (=2) of "attraction" (=268). The attraction is the "seventh event horizon" (*Akarshana*, 268) with a "heart-shaped curve" (*Akarshana*, 268) as a "triangular horizon" (=268), "cardioid" (*Akarshana*, 268). It is the "peripheral horizon" (*Akarshana*, 268), whose attraction makes it a "supernatural horizon" (*Akarshana*, 268). The observer is the "extrinsic field" (*Kakshya*, 0) in a diffused "eventual state" (*Kakshya*, 0) of the future, infused with a "gravitational orbit" (*Kakshya*, 0), the "innermost stable circular orbit" (*Kakshya*, 0), ISCO, reproducing "infinite orbits" (*Kakshya*, 0).

The entity's "forward motion" (=24), reproducing infinite orbits, is a "prograde motion" (=24). It "prograde" (*Purahakrama*, 251 = 2,[4+1],1) the "foundation" (=11) for "centering" (=10) the "retrograde" (*Vyutkrama*, 1) as a "tardigrade" (=999) within the "nucleus" (=1,000). The foundation's linear, "backward motion" (*Vidhata*, 130), which "triples" (=130) the "centering" (=10) of the "second event horizon" (*Bhushana*, 1,300), is a "retrograde motion" (*Vidhata*, 130), "radiating" (*Hiranyaretas*, 70) the "orbital's" (=60) "ascending motion" (*Hiranyaretas*, 70) with a three. The ascending motion is the "tardigrade motion" (*Hiranyaretas*, 70), repeated until the "entropy" (=5) of the "rotational motion" (=14), "descending" (=12) the "orbit" (=2) with a general, "descending motion" (=14), i.e., "contragrade motion" (=14). As the orbit "copies" (*Anukarana*, 0) itself as an "anterograde" (*Anukarana*, 0) to "orthograde" (*Chitra*, 1) a "twin copy" (*Chitra*, 1), the "magnetization" (=61) of the "octave" (=60) with two "disintegrates" (=61) into a random, "horizontal motion" (*Devasana*, 61), i.e., "orthograde motion" (*Devasana*, 61) of the "finite orbit" (*Adhyaropan*, 17). The finite orbit "reproduces" (=78) its "superposition" (*Adhyaropan*, 17) to "stabilize" (0) the reproductive "infinite orbit" (*Kakshya*, 0).

The finite orbit is a "parabolic orbit" (=17), whose "nature" (=8) is to be a "definite orbit" (*Kakshavrta*, 142 =17*8+6) with "growth" (=6) of itself within the "foundation" (=11) of the "triple orbit" (=17). The definite orbit is the unstable, "hyperbolic orbit" (*Kakshavrta*, 142). It is the "mean orbit" (=142) formed with the "internalization" (=142) of "two orbits" (=142) as a "photon orbit" (=142), "centering" (=10) "itself" (=121) as the "foundation" (=11) of the "substantial radius" (*Ishvara*, 5)—"half radius" (*Ishvara*, 5) of the two who orbit twelve for "breeding" (=15) "time" (=360) with their "motion" (=360). Consequently, the "universe" (*Brahman*, 2) is "becoming" (*Anitya*, 900,000,000) "spacious" (=10,000) with time as a "whole entity" (*Anitya*, 900,000,000), "gaining energy" (=10,000) for its "spin energy" (=10,000) by making the "nucleus" (=1,000) reproductive with its "reality" (=7) as an "entity" (=24). The space is "losing energy" (*Prithvi Shakti*, 369) as it "perpetuates" (=9) "time" (=360) with its "mass energy" (*Prithvi Shakti*, 369). The "horizontal motion" (=61) is the "self-limitation" (=61) of the universe to orbit a "maximum rotation rate" (*Ativada*, 800 =61+134+5+600) to be "travelable" (*Subrahmanya*, 134) with a "final" (=134) "rotation rate" (*Subrahmanya*, 134) at a "minimum" (*Avarardha*, 5) "temporal cost" (=600). The time's "minimum rotation rate" (*Muhurtaja*, 2,490) is "momentous" (*Muhurtaja*, 2,490) as it is "potential minimum" (*Muhurtaja*, 2,490).

Potential minimum is "maximin" (*Muhurtaja*, 2,490 =1,600+890), "absolute maximum" (=2,490) of a "series" (*Paati*, 90,000) which "sets limit" (=69) on the "maximum rotation rate" (=800) to "gravitate" (=-100) "maximum-effect" (=111) of the "present" (=1,600 = [69+111] *5 −100+800). The present "constitutes" (=890) the "minimum" (=5) "perpetuating value" (*Saranyu*, 5) "within effect" (=890) of the "sentient force" (=4) that the future "emanates" (=15) by "centering" (=10) the past of the "axis of rotation" (=2) of "space" (=18,000) into "time" (=360).

The maximum rotation rate is "minimax" (*Ativada*, 800), and the "potential maximum" (*Ativada*, 800). It is the "absolute minimum" (*Ativada*, 800 = [69+111+20] *4) of a "series" (*Paati*, 90,000) which "sets limit" (=69) on the "maximum-effect" (=111) that is "gravitable" (=-11) when the "sentient force" (=4) of the "minimum" (=5) is "reproductive" (=100) within "time" (=360) as the "spirit" (=20) of the "space" (=18,000). It is the "extremism" (*Ativada*, 800) of a "discontinuous series" (*Priya*, 800) of "present-potential exchange" (*Priya*, 800).

The "infinite exchange" (*Sitanveshana Panditaya*, 816) of the "present minimum" (=800) with the "potential maximum" (=800) is "minimin" (*Abhishruti*, 816). Minimin is the "primordial minimum" (*Abhishruti*, 816) of the "series" (=90,000), "servicing" (=47) "maximum-effect" (=111) by "centering" (=10) the "origin" (=1,000) as the "causation" (=1) for the "minimum" (=5) to be "reproductive" (=100) as the "essence" (=16) of "reality" (=7) within "maximum" (=123). Minimin is the "production base" (=816) of the "small angular momentum" (*Abhishruti*, 816) which "enlarges" (*Samvridh*, 816) its "mass" (=132) with "thermodynamic-effect" (=1) to "diversify" (=816) the "correlation" (=16) of the "twin minimum" (=25) with the "family" (=10) of "origin" (=1,000 =816+132+1+16+25+10). The "two minima" (=25 =816-800 +9) are the "minimin" (=816) without the "minimax" (=800) which "perpetuates" (=9) "itself" (=121) as the "minimum" (=5) within the "triple minimum" (*Charma*, 126)—the "eventuality" (*Charma*, 126) of the three as a "multiplier" (=3) "eventualizing" (*Mahendrani*, 0) a "maximum" (=123).

"Maximax" (*Sheshanaga*, 816) is the "primeval maximum" (*Sheshanaga*, 816) of a "series" (=90,000), "trading" (=53) the "minimum" (=5) as a "multiplier" (=3) of two. Two is the "axis" (=2) "rotating" (=-12) the "sentient force" (=4) of the "triple maximum" (*Nandini*, 105). The "three maxima" (=125) are:

- the 3, the multiplier of the "maximax" (=816) into the "maximin" (=2,490 =816*3+42),
- the 42 the three multiplies into the "triple minimum" (=126 =42*3), and
- the 80 multiplied by three into 240 for the "replication" (=90) of the "future" (=0) as the "past" (=9) which "minimizes"

(=1,000) the "maximum" (=123) with "maximin" (=2,490) by "infinite dividing" (=1,000) of the "present" (=1,600).

As a multiplier, three is "formable" (*Arupya*, 3) into a "twin maximum" (=3). Forty-two "lives" (=14) as a "half maximum" (=42) with three. Eighty "orbits" (=2) forty-two as a "half minimum" (*Sadhyata*, 80). The half minimum "twins" (=121 =80+36+5) itself to "channel" (=36) the "minimum" (=5). Two-hundred-forty is "medium angular momentum" (*Nasatya*, 240) which "maximizes" (=128) the "minimum" (=5) with "minimax" (=800) through "infinite multiplying" (=128) of "oneness" (=48) with the "future" (=9) within "half minimum" (=80).

"Maximax" (*Sheshanaga*, 816) is the "large angular momentum" (*Sheshanaga*, 816), "compressing" (=45) the compressible "thermodynamic mass" (=816) "left" (=89) by a "blackhole" (=82) without "primordial oneness" (=32) of the "compressible blackhole" (=1,000 =816+45+89+82-32). It "maximizes" (=128) the "mass" (=132) with the "multiplier's" (=3) "thermodynamic force" (=1).

"Maximization" (*Adhikarana*, 132) "breaks down" (=13) "optimism" (*Ashavada*, 13,000) to "spiral" (=13) "pessimism" (*Nirashavada*, 1,000) with its "absolutism" (*Nirankushta*, 2). "Pessimism" (=1,000) is "savage" (*Kirata*, 59), "spiraling" (*Sarpidvat*, 59) an "infinite minimum" (*Kirata*, 59) as the "peripherum" (*Kirata*, 59) for "minimizing" (*Dhanvantari*, 941) the "half effect" (=256) of "minimalism" (=512) within the "effect" (=34) of "moderatism" (*Bhidvada*, 1,900 =1,000 +34+89+19-10+256+512) that "unmanifests" (=-10) the "twin effect" (=19) of "centrum" (=89). The "centrum" (*Vikeerna*, 89) is "Hurwicz" (=89), "dissipated" (*Vikeerna*, 89) as the "primordial maximum" (*Vikeerna*, 89) for "maximizing" (*Pradyota*, 89) the "twin effect" (=19) through "minimization" (*Avarardhan*, 57) of the "spiral" (=13).

After "feeding" (=17) two to "mediate" (*Phalaha*, 13) the "spiral" (=13), the "result" (*Parinama*, 123) is "Laplace" (*Piti*, 123), the "definite maximum" (*Piti*, 123) from three's "mediatism" (*Falahavada*, 3,900 =13*3*100) of "itself" (=121). Itself is "reproductive" (=100) as one with two's "reproductive potential" (=23). A "definite minimum" (*Nanda*, 123) is "repelled" (=123) by two to "divide" (=132) the "face" (=76 =9*8+4) of the "maximalist" (*Mahishtavadi*, 888) into nine "pairs" (*Jodi*, 8) of four within "maximalism" (*Mahishtavada*, 21,000=9*8*4*76 −888) of the definite maximum. The two produce "equal probability" (*Kriyapada*, 190) for "everyone" (=180) to be "infinitive" (*Kriyapada*, 190) by "centering" (=10) the "moderatism's" (*Bhidvada*, 1,900) "potential-effect" (=1) itself.

The "mediatism" (*Falahavada*, 3,900) is "disconnected" (=30) from the "immediatism" (*Sadyovada*, 90) after "connecting" (=30) "rectum's" (*Vanishthu*, 90) "present-effect" (=10^{1000}) to the "realism's" (*Dharmashunyata*, 111) "reproductive force" (=100). The "rectum" (=90) "forms" (=100,000) three as "I am" (=100,000) "connected" (=72) with the "remediatism" (*Sakarmavada*, 8 x10^{15}) of the "growth" (=6) in its "potential" (=18) to be an "entity" (=24). The growth is "overexpressed" (=28) by four "self-projecting" (=3/2) itself with the letter "A" (*Parameshthi*, 28). The "immediatism" (*Sadyovada*, 90) "projects" (=31) a "triple-effect" (*Sarupak*, 10^9) to "gain" (=140) the "benefit" (=125) of "envisioning" (*Adharma Bhukti Yoga*, 4) four within "itself" (=121). Four is "organizable" (=11) with the "loss" (=81) of eight when a "pair" (*Jodi*, 8) of four, repeated while "enjoying" (*Bhukti yoga*, 125) the "enjoyment" (*Bhukti*, 9) of the "present life experience" (=11) as an "enjoyer" (*Rasi*, 9), "generates" (=40) the "cost" (=41) of "reproductive reality" (=40) of the "afterlife" (*Kimstughna*, 8).

The "afterlife" (=8) "pairs" (=8) the "origin" (=1,000) of "consciousness" (=4) that the "present life is hell" (*Murugan*, 8) with the "past life" (=18,000). By "self-projecting" (=3/2) "action" (*Karma*, 10), the "universe" (=2) within the "pair" (=8) "mediates" (=13) the "benefit" (=125) of "breeding" (=15 =3/2*10) desired "consciousness growth" (=15) that the "potential reality is heaven" (=15). The "universe" (=2) becomes the "eternal redeemer" (=100) of the "past-effect" (=24) of "actions" (*Karma*, 10) by "servicing" (=47) the "consciousness cycle of life, death, and rebirth" (=17). The "redeemer" (=40) "orbits" (=2) the "eternal redeemer" (=100) like an "orbital" (=60) to "redeem" (=10,000) "symmetry" (=250) of the "cycle of creation" (*Lakshmi chakra*, 246) from the "primordial state" (*Adisthiti*, 3) of a "feminine" (=37) before she "began" (*Arabhi*, 246) "acting" (=13) like a "masculine" (=296).

The "cycle of creation" (=246) "unfolds" (=246) the "spatial benefit" (=15) of "life energy" (=1,000 =246+15+139+600) without "knowledge" (Jnatva, 600) of the "temporal cost" (Vidya, 600) of "descending sentient benefit" (Shyamala, 139). The "sentient cost" (Shyamala, 139) of the "mutation" (=186 =15+139+32) of the "spatial benefit" (=15) "descends" (=25) "primordial oneness" (=32) of a "sentient entity" (=7), whose "nature" (=8) is to be a "metric" (=−1) of "disproportionate sentient benefit" (=78 =15+25+32+7−1).

A "Supreme Wisher" (=−4) "transconstructs" (=−8) the "characteristic" (Purusha, 12) of a "Wisher" (=0) without the "formative reality" (=−3) of the "primordial state" (=3). The "characteristic" (=12) "creates" (=35) a "desired" (Ceshta, 10) "workculture system" (Sthavaravisha, 57) by "servicing" (=47) "perfect knowledge" (Bodhi, 156) for the "entropy" (=5) of "nature's" (=8) "self-luminosity" (Vatsalya, 37) as a feminine within an "organization" (=29). The organization is organized as a "masculine" (=296) with the "growth" (=6) of a "star" (=2) like a universe. After a "time-lapse" (Sila, 1,790), the "masculine" (=296) "develops" (=22) "self-consciousness" (Prajna, 2,222 = 1,790+296+22+16+2+96) of a "primordial greeter" (=16) to "star" (=2) "being energy" (Kali shakti, 96). "Self-consciousness" (=2,222) "guides" (=111) "the "benefit consciousness" (Dharmatma, 1,000 =111+377+512) of "time" (=360) "producing" (=17) a "heavenly life" (Lokya, 377) with spontaneity" (=1,111) of "life energy" (=1,000) within "minimalism" (Saura, 512) of the "divine" (=360).

"Infinite continuity" (=900) with time "divines" (=360) a "minimalist" (=360) who "enjoys" (=7) a "heavenly life" (Lokya, 377), "co-dwelling" (Salokya, 20) with a "feminine" (=37) and "leading" (=20) her to a "hellish life" (Alokya, 10^{1000}) as a "remediatist" (Sakarmavadi, 10^{1000}) of his "divine energy" (=10) with her "gravitational energy" (Lalita, 100). With "freedom from the heavenly life" (Salokya Mukti, 86) of "another" (=98), who makes a "universe" (=2) "reproductive" (=100) with its "growth" (=6) as a "star" (=2), the "feminine" (=37) "lives" (=14) in "First Position" (Yogeshvari, 51), "ascending technological growth" (=51) of the "para deity" (=5) as a "deity" (=1).

6.3.2 Transcending Supreme Wisher

By "servicing" (=47) "primordial technological exchange" (*Nichevala*, 80,000) of the "first position" (*Yogeshvari*, 51) as a "composite position" (*Paarnayana*, 80,000 = [47+33] *1,000), a "supreme wisher'" (=-4) enjoys "increasing" (=33) "feminine consciousness" (=33) of "femininity" (*Yoni*, 1,000) to "perpetuate" (=9) the "heavenly togetherness" (*Gajakarani*, 1) with a "feminine" (=37) within the "hellish otherness" (*Vignesh*, 1) of that "feminine" (=37). "Masculine consciousness" (=18) is the "reason" (=1) for the "duality" (=28) of "gender" (=9). For the "feminine" (=37) as an "entity" (=24), the "present paradigm is hell" (=24) because she "works" (=19) like a "deity" (=1) to fulfill the reality that "new paradigm is heaven" "self-luminous" (=12) for the "creature" (=12) behaving like a "para deity" (=5).

By "fabricating" (*Sankhara*, 269) the "exchange" (=269) of the "extrinsic-effect" (*Sankhara*, 269) after "dying" (=269) as his "emanating force" (*Vyana*, 269), a "masculine" (=296) "nurtures" (=269) "strange repulsion" (=269) for "living" (=10) "long" (=269) as a "feminine" (=37) by "servicing" (=47) "primevalism" (=27) to "shorten" (*Indra Yoga*, -1,000) his "life energy" (=1,000) as a "Lord" (*Indra*, 0). The "fabrication" (=88) is an "outgrowth" (=88) of the "speciation" (*Prajatikarana*, 88) of the "primordial self" (*Parvati*, 10) like a "family" (=10) to "breed" (=78) the "species" (*Prajati*, 17) of her "kind" (*Prakara*, 17) for "freeing" (=95) the "reality" (=7) of a "param deity" (*Shiva*, 7). The "param deity" (=7) is "self-luminous" (=12) as a "para entity" (=19) who "breeds" (=78) the "past entity" (=98) for "disappearing" (=98) as "another" (=98).

"Another" (=98) is the "primordial solution" (=18) to "sorrow" (=47) within the "feminine consciousness" (=33) of the "universe" (=2). One "gets" (=256 =2^8) "old" (*Purana*, -10^{180}) "suffering" (*Jara*, -10^{180}) from the "co-existence" (10^{10}) of the "past life system" (*Sara Kalpa*, 10^{10}) while wishing as an "immediatist" (*Sadyovadi*, 10^{10}) to "just" (=10^{10}) "move on" (=70) with the "replication" (=90) of the "primordial self's" (=10) "proficiency dimension" (=10^{10}) for "action" (*Karma*, 10) in the "present life" (=180), "leading" (=20) to the "endoreproduction" (=2) of "nature" (=8) within the "universe" (=2). The "endoreproduction" (=2) "unbundles" (=6) a "transformative effect" (=6) that "transforms" (=1) "nature" (=8) into "culture" (=9). The "culture" (Sadakhya, 9) is a "cost-effective paradigm" (Trivikrama, 24) for "breeding" (=15) the "sentient benefit" (Jnana Siddhi, 190) of "guider conscious" (=170) "gender differentiation" (=170) of "time" (=360) into a "dark atom" (=170), "present" (=1,600) as "dark matter" (=1,600) with "growth" (=6) of "consciousness" (=4).

"Gender differentiation" (=170) "descends" (=25) the "sentient cost" (=139) of "growth" (=6) of "consciousness" (=4) of the "universe of living entities" (Talatala Loka, 1,649) within a as a "param masculine" (Brihadbala, 149). The "param masculine" (=149) "lives" (=14) as a "primeval illuminator" (=14) to "perpetuate" (=9) a "varying consciousness" (=12,000) of "life energy" (=1,000) within "each entity" (Panchajani, 12), self-projecting the "cause" (=18) for their growth into a "cost-effective paradigm" (=24) with the "trading-effect" (=26) of the "variable worker-social benefit-cost ratio" (=111).

A "primeval illuminator" (=14) "enjoys" (=7) "freedom from Quantum God" (Brahma Mukti, 971), "naturally" (=91 =7*[14-1]) "discovered" (=91) in the "wake" (=91) of "conclusion" (=91) that the "metric" (=−1) for the "growth" (=6) of "God" (=5) is "primordial self" (=100). The primordial self is the "primordial illuminator" (=10), "superpositioned" (=90) like a "quantum" (=90) "without" (Rahita, 81) the "Front-of-the-scenes entity" (=91). The "primordial illuminator" (=10) is "transcendental" (=81) within the "front-of-the-scenes entity" (=91) because he is "immanent" (=13,000) within the "behind-the-scenes entity" (=−1) as a "metric" (=−1) which "perpetuates" (=9) the "consciousness" (=4) of his "infinite forms" (Bahurupa, 13).

The "infinite" (=185) "forms" (=100,000) "infinite entities" (Astika, =−1) by "servicing" (=47) "I-ness" (Ahamta, 90) without the "intrinsic-effect" (Rohini, 50) of "My-ness" (Mamata, 0) by "fabricating" (=269) the "extrinsic-effect" (=269 =185+47+53+34-50) of "trading" (=53) "three zeroes" (Indrejya, 0) from the "effect" (=34) of zero on itself. The three zeroes "transform" (=1) a zero into an "infinity" (=90,000) to "circle" (=100,000) "both" (=10,000) as the "forms" (=100,000) of the "primordial illuminator" (=10).

A "Para Wisher" (=-5) "fabricates" (*Vinirma*, −5) "infinite linkages" (*Shabda*, 285) by "radiating" (=70) "sentient love" (=70) to a "multiplying body" (*Mena*, 210) of the "Wisher" (=0) with "My-ness" (=0). The Para Wisher is the "protagonist body" (*Bhakta*, -5) of the primordial illuminator as a "Devotee Deity" (*Shri Krishna*, 10). As a "believer" (=-5), he is "devoted" (*Nirata*, -1) to a dependable "present believer" (*Vishvasi*, 80) for the "sentient well-being" (=190) of the "potential believers" (*Sashuka*, 80), comprising the "long tail" (=80) of "devotees" (*Baladeva*, 274). The "oneness" (=48) of the "Para Wisher" (=-5) within a "Devotee Deity" (=10) is the "supplementor" (*Vinayaka*, 53) of the "quality" (=0) that a "leader" (=0) "reproduces" (=78) with a "belief system" (*Saguna*, 20) of "himself" (=5) as "God" (=5). The "path of devotion" (*Bhakti Marga*, 1) to "God" (=5) "precedes" (=6) the "virtue" (*Vrisha*, 16) of "comic" (*Hasya*, 589) "righteousness" (*San*, 589) as it "ascends" (=1) the "temporal cost" (=600) of the "single-pointed" (*Shunda*, 600) "knowledge" (=600), "extended" (*Vitata*, 600) like a "trunk" (*Shunda*, 600) of the "para wisher" (=-5).

The "devotion" (=46 =18+1+27) of a zero as a Wisher is a "wish" (=18) to "norm" (=18) "justice" (*Nyaya*, 1) for the "path of devotion" (=1) with "private benefit" (*Rupa Siddhi*, 27) of "precession" (*Ayanamsha*, 27), illumining "primordialism" (*Ayanamsha*, 27) for "itself" (=121 =27+96-2) as a "luminous entity" (*Kali*, 96) within the "entire universe" (=-2). The "potential reality" (=160) of the "luminous entity" (=96) "gifts" (=64) "ascending sentient benefit" (=157) to the "Wisher" (=0) as a "Lord" (=0) with "triple copy" (=3) of zeroes. With "growth" (=6) of the "triple copy" (=3), the "Lord" (=0) "perpetuates" (=9) the "potential" (=18) to be the "Lord of Lords" (*Devadhideva*, 18) at the "ninth infinity" (=18). The "ninth" (=123) "transforms" (=1) into "infinity" (=90,000) with the "reproductive potential" (=23) of "three zeroes" (*Indrejya*, 0) for the "replication" (=90) of the "fifth zero" (=82), the "blackhole" (=82) of the "universe of gifts" (*Vishnunabhi*, 82), as a nine within one. The nature of one is to transform himself with the Para Wisher to be the ninth. One is "present: (=1,600) as the "fourth zero" (=1,600), the dark matter. The nine is the "enjoyer" (*Rasi*, 9), "enjoying" (=9) the "joy" (=123) of the ninth with two who "enjoy" (=7) the reality of nine as "God" (=5) "himself" (=5).

The Lord of Lords is the "Primordial Maternal" (*Narayani: Shen Shu* 神荼, 18), "multiplying" (=5,000) "human spirit" (*Gliŋh*, 387) "within" (*Eke*, 5,000) the "reality" (=7) of her "spirit" (=20) as the "enjoyer" (=9) of two. Two's "time value" (=500) is the "origin" (=1,000) of "God" (=5) as a "Primordial Paternal" (*Purushottama*, 0) with three zeroes. Two "transforms" (=1) one, the "human being" (=1), into a "sister" (=130) within "Primordial Paternal" (=0). The "sister" (*Inanna*, 130) is the "Goddess" (=70) of "procreation" (=126) of the "body" (=56) of the "mother" (=4) of the "creature" (*Srijankarta*, 12), "being human" (*Srijankarta*, 12) as a "Father" (=3).

"Pro" (*Samadhana*, 18) is the "contemplation" (*Binah*, 18) about the "creation" (=379 =3[6+1]9 =2*18,19) of an imaginary, "mind-born primordial maternal" (*Amudheshvari: Baaltis*, 18). It twins itself with the "energy" (*Shakti*, 19) within "mind" (=38) for the "continuity" (=50) of one as the "mental energy" (=691), i.e., the "dark energy" (=691) of a "mind-born creator" (*Brahma*, 59) of the "past" (=9) of one within ten, the "primordial illuminator" (*Parvati*, 10). One's past copies the "inanimate" (=18) with a two who, as the "inanimate primordial maternal" (*Nilanjana: Ba'alat Gebal*, 18), "animates" (=76 =50+13*2) the "past" (=9) as the "mother" (=4) like a "Wellspring" (=13 =9+4) of "continuity" (=50). The "animate primordial maternal" (*Shani: Selija*, 18) "triples" (=130 =18*6+12) the "growth" (=6) of the "creature" (=12), "assembled" (*Trimukha Vinayaka*, 12) as a "Titan" (*Danava*, 12), with 13 "Lords" (=0). She "forms" (=100,000) an "assembly" (*Kamma*, 130) of "Thirteen Lords" (*Inanna*, 130) to "ring" (*Pushpaka*, 100,000) a "council" (=130) of "Thirteen Titan Lords" (*Inanna*, 130).

The illusionary, "intellect-born primordial maternal" (*Upadhaya*, 18) is the "First Sister" (*Upadhaya*, 18), the "Maternal Lord" (*Upadhaya*, 18) of the "Twelve Child Titans" (*Anuya*, 6) within the thirteenth Titan, the "Paternal Lord" (*Danava*, 12). The "Twelve Lords" (*Anuya*, 6) are the "gender-differentiated six copies" (*Anuya*, 6) of the "Paternal Lord" (*Danava*, 12). The "past" (=9) of the "paternal lord" (*Danava*, 12) "genders" (=9) a "self-loving" (=3) "Maternal Lord" (=18). The maternal Lord is the fourteenth Titan who "differentiates" (=7) gender for the continuity of her "life" (=4) as a "mother" (=4) within the reality of the "father" (=3). The father "imitates" (*Anuya*, 6) the mother, the "present entity" (*Harbuddhi*, 366,666), by "differentiating" (=888) the "first gender-differentiated copy" (=3) as a three before "integrating" (=95) three, the father himself, as the "past entity" (*Mahanakshatra*, 98) within the "future entity" (*Phaninayaka*, 20), the spirit of the child. The future entity "transforms" (=1) the "reality" (=7) "itself" (=121 =20*6+1) into "Twelve child Titans" (=6).

One, "Father Nature" (*Yahoodi*, 1) enjoys the "formative growth" (=2) of the "Greeter Lord" (=3) with two—the "Child Lord" (*Jiva*, 2), within eight, "Mother Nature" (*Kudrat*, 8), to norm the "Maternal Lord" (=18) with "twelve child Titans" (=6). The Child Lord is the Sixteenth Titan. The "pulse" (=104) of one as the "Pulsar" (*Samartha*, 1), "pulsing" (*Rasabhasa*, 0) the future as a "photonic" (*Rasabhasa*, 0) "transient wave" (*Rasabhasa*, 0), "fronts" (=−1) the "pulse-effect" (=121) into the "Pulsar time" (*Bhava*, 360) of three. Three is the "Greeter Lord" (*Hanuman*, 3), the Fifteenth Titan. "Pulse-effect" (=121) "backs" (*Prishtha*, 5) an "array" (*Vyuha*, 114) of "Paternal Lord" (=12) to with "Pulsar timing" (*Samanjana*, 129) of "Pulsar-effect" (*Sabija*, 129) of "Greeter Lord" (=3). It "incorporates" (*Nigamit*, 680) the "universe of creatures" (=680) as a "summative" (=680) of the "vector" (=396)—the "photonic time" (=396). The "interfa"e" (=396) of the "photonic time" (=396) with the "time reflection" (*Devatamayi*, 1,024 =396+805-91-3-18-64-1) "directs" (*Nidesha*, 91) the "focus" (*Kendra*, 3) to the "Greeter Lord" (=3) for the "feminization" (*Nilanjana*, 18), "Rix" (*Nilanjana*, 18), of the "greet" (=64) of "pulsation-effect" (*Paridhi*, 64) from the "pulsation" (*Jagriti*, 7) of "Pulsar's" (=1) "divinity" (=57) as a "deity" (=1). The "pulsation" (=7) "pulsates" *(Guduvay*, 8) the "formative-effect" (=−1) of "Saturn" (=−1) as the seventeenth Lord, the "Satan Lord" (=−1) within the "Devil Lord" (*Indra*, 0), the Lord as the eighteenth Lord.

"Saturn" (=−1) is a "potential entity" (*Maheshvari*, 17) without the "Real Primordial Maternal" (*Parakala: Chronus*, 18), the "Para Pulsar Time" (*Parakala*, 18). Para Pulsar Time is "pulsation time" (*Parakala*, 18) within "pulsation time-effect" (*Kalakrama*, 16), the "chronology" (*Kalakrama*, 16) of "effect" (=34). The effect is "pulsing-effect" (=34), i.e., "zero-effect" (=34) of the "Lord" (=0) "himself" (=5) as a "time molecule" (*Kanalakshamsha*, 39). Potential entity is "Goddess of Fire" (*Maheshvari*, 17) "without air" (*Maheshvari*, 17), "ascending air" (=-1) to "ego" (*Aham*, =−1) Saturn by "descending air" (=73) to "emotion" (=73) "Uranus" (=73) while "configuring" (*Akriti*, 79) the "growth" (=6) of "para real" (*Pashyanti*, 79) within the "galactic center" (=67) with "Neptune" (=140).

"Saturn" (*Shani Bhagwan*, -1) is the "blueness" (=-1) of "Simple Primordial Maternal" (*Shani Bhagwan*, -1). "Venus" (*Shukra*, 2,700) is the "blackness" (*Nilima*, 2,700) of "Complex Primordial Maternal" (*Nilima*, 2,700) after the "exchange" (*Mahaspanda*, 269) of the "past" (=9) with the "Saturn" (=−1). The past perpetuates the "Lord of Lords" (*Devadhideva*, 18) as a "Circular Primordial Maternal" (*Devadhideva*, 18) at the "ninth infinity" (*Devadhideva*, 18), the "finite infinity" (=18). The "joy" (=123) of "both" (=10,000) "blueness" (=-1) and "blackness" (=2,700) within "redness" (*Lohitya*, 10,000) of "Jupiter" (*Brihaspati*, 1780) forms the "finite" (=85) "infinity" (=90,000) of the "mind" (=38) with the "ninth" (=123 =85+38). "Whiteness" (=102) of "Mars" (=102) is within the "earth-effect" (=21) of "joy" (=123). "Greenness" (*Haritatta*, 715) of "earth" (*Bhu*, 724) has an "effect" (=34) on "colorlessness" (*Nishadata*, 127) of the "Lord" (=0) "present" (*SAUM*, 1,600 =715+724+34+127) as "Mercury" (=1,600). Colorlessness is the "happy point" (*Niyama*, 127) of "zero inertia" (*Sadhaka*, 127) as a "proficient system" (*Sadhaka*, 127) of the "renewal" (*Tapas*, 820) of "pulsation" (=7) within a "pulsation system" (*Vimaleshvara*, 700).

The pulsation system is the "Triangular Primordial Maternal" (*Vimaleshvara*, 700) who as the globalized "Pulsing time" (*Vimaleshvara*, 700) works with energy to "Square Primordial Maternal" (*Tapah*, 820) for "triangulation" (=3) of the "Pulsar time" (=360) into the "pulsed time" (*Jaldi*, 120). "Divination" (*Sahajata*, 916) is "pulsed" (*Sahajata*, 916) as a "Parallel Primordial Maternal" (*Sahajata*, 916) for "ease" (=916) of "internalizing" (=916) the "self-created" (=916) "energy potential" (=916), thus "blossoming" (*Jayanti*, 101) the "pulsable" (*Magha*, 101) into a "Point Primordial Maternal" (*Magha*, 101). The pulsable is "January" (*Magha*, 101), the "eleventh month" (*Magha*, 101) that "ends" (*Antaka*, 140) "winter" (*Shishira*, 754) with an "austerity" (*Tapa*, 90,000) to "spring" (*Vasanta*, 811) the "first month" (*Chaitra*, 90,000). The "divinity" (=57 =811-754) "Marches" (*Chaitra*, 90,000) as "Line Primordial Maternal" (*Siddhi*, 57) for "perfecting" (=127) the "happy point" (=127 = 140-2-11) as a "theory" (=127) of two "producing" (=17) one as "reality" (=7).

With "renewal" (=820) of the "spirit" (=20) of the "feminine" (=37) through a "path of perfection" (*Divyaratri Marga*, 467), the "masculine" (*Kandarpa*, 296) becomes a "depository" (=296) of the theory's "intuitive reality" (*Laksyartha*, 296) within the "Aquarius" (*Kumba*, 296) zodiac. "Renewal" (=820) "brings" (=–18) to "surface" (=2,000) a "large tile" (*Kaval*, 820) with "Y" (=–18) "shaped" (*Sphya*, 820) "tiling" (*Kaval*, 820), "instead of" (=2,600) a "texture" (*Banavat*, 238) of "eight tiles" (*Kaval*, 820). "Nature" (=8) "tiles" (*Kavula*, 340) "three kites" (*Khabhranti*, 340) within "Pulsar time" (=360) with the "spirit" (=20) of the "space" (*Kha*, 18,000) to "order" (=2) a "seamless" (*Bhranti*, 195) "dissolution" (=195) of the "time multiplier" (=3) within "infinity" (=90,000 =360*[195+2+3] +18,000).

Science of Tiles

"Tiles" (*Kavula*, 340) "vary" (=60,000) in "size" (*Vyoma*, 285) as they "descend" (=25) the "reproductive force" (=100) of "one tile" (*Kavula*, 340) with "two stones" (=340): the imaginary stone and the real stone. The "imaginary stone" (*Shaana*, –18) "triangulates" (=3) the "size" (=285) to "triple tile" (*Nirarthaka*, 77 =285/3-18). The "T-shaped" (=77) "three tiles" (=77) are "futile" (*Nirarthaka*, 77); they "ductile" (*Tanuva*, 85) "six tiles" (*Tanuva*, 85) to "reproduce" (=78) "sixteen tiles" (*Chapala*, 78) with the "reproductive force" (=100) within "one tile" (*Kavula*, 340 =77+85+78+100).

"Sixteen tiles" (=78) are "U-shaped" (=78), "volatile" (*Chapala*, 78) as the "smallest tile" (*Chapala*, 78); they "order' (=2) the "infertile" (*Anupjau*, 2) "ten tiles" (*Anupjau*, 2) to "produce" (=1) "five tiles" (*Nirupyukta*, 1,000) that "live" (=14) with the "super-fertile" (*Runasvara*, 933) "two tiles" (=933). "Illusionary distance" (=933) between the "two tiles" (=933) is "parallax" (933). Ten tiles are "X-shaped" (=2) as a "small tile" (*Anupjau*, 2). Five tiles are "Q-shaped" (=1,000) as a "protractile" (=1,000). Two tiles are "V-shaped" (*Shahateer*, 933) as a "charged tiling" (*Runasvara*, 933); a "chevron" (*Shahateer*, 933), "charged" (*Prabharit*, 970) by the "feminine" (=37) into "eight tiles" (=820) of "masculine" (=296) for "protracting" (*Nirupyate*, 360) "six tiles" (=85) of "child" (=128) into a "protract" (*Nirupyati*, 1).

A "protract" (=1) is an "ideal" (=1), "tileable" (*Bhranti*, 195) as a "fallacy" (*Bhranti*, 195) of "contractile" (*Samkuchita*, 95), the "P-shaped" (=95) "nine tiles" (*Samkuchita*, 95). "Nine tiles" (=95) "contract" (=81) "five tiles" (=1,000 =95+81+4+820) for "four-fold growth" (=8) of "eight tiles" (=820) into "W-shaped" (*Upjau*, 387) "thirty-two tiles" (*Upjau*, 387). Thirty-two tiles are "fertile" (*Upjau*, 387) as the "largest tile" (*Upjau*, 387), a "nonrepeating tile" (*Upjau*, 387).

The "largest tile" (=387) is "tiled" (*Kaulara*, 837) by "contracting" (=158) an "R-shaped" (=158) "reptile" (*Sarisrpa*, 158) of "twenty-five tiles" into "S-shaped" (=292) "seven tiles" (=292) that are "tactile" (*Sparshaniya*, 292). The "potential" (=18) of the "family of ten tiles" (*Parivar*, 10), shaped like letters P, Q, R, S, T, U, V, W, X, and Y, is "versatile" (*Bahurangi*, 18) within an "O-shaped" (=8) "octile" (*Upakulaka*, 8) of "sixteen tiles" (*Upakulaka*, 8). It is "many-colored" (*Bahurangi*, 18) and "Z-shaped" (=18) as a "smaller tile" (*Bahurangi*, 18) of "four tiles" (=18).

An octile is a "subset" (*Upakulaka*, 8) which "fragments" (=64) a "set" (*Kulaka*, 72) of "N-shaped" (*Apanika*, 72) "mercantile" (*Apanika*, 72) "eighty tiles" (*Apanika*, 80). A fragment is "M-shaped" (=64) "centile" (*Shataki*, 64) of "forty tiles" (*Shataki*, 64), "refracted" (*Apavartita*, 28) into an "L-shaped" (=36) "refractile" (*Apavartaniya*, 36) of "twenty-four tiles" (*Apavartaniya*, 36). The "refractile" (=36) "refracts" (=6) "K-shaped" (=30) "nonvolatile" (*Achapala*, 30) "thirty tiles" (*Achapala*, 30). "Thirty tiles" (=30) "self-perpetuate" (=1/2) an "A-shaped" (=1,964) "gentile" (*Gotraja*, 1,964) "breadth" (=1,964) of "fifteen tiles" (*Gotraja*, 1,964) within "24 tiles" (=36) that "order" (=2) "five tiles" (=1,000) with "ten tiles" (=2) to be a "B-shaped" (= 48) "erectile"

> (*Uttanakara*, 48) of "twelve tiles" (*Uttanakara*, 48). "Thirty tiles" (=30) "reorder" (=2) a "C-shaped" (=92) "textile" (*Vastramaya*, 92) of "ninety tiles" (=92) by "self-projecting" (=3/2) "order" (=2). "Ten tiles" (=2) "mimic" (=700) a "D-shaped" (=1,400) "quartile" (*Chaturthansh*, 1,400) of "hundred tiles" (*Chaturthansh*, 1,400) for "primordial oneness" (=32) with "forty tiles" (=64).
>
> With "forty tiles" (=64) "ten tiles" (=2) "live" (=14) as an "E-shaped" (=80) "percentile" (*Shatamshika*, 80) of $[40+0]^*10 + [40+10]^*10$, i.e., "nine hundred tiles" (*Shatamshika*, 80) like a "devil" (=0). "Nine hundred tiles" (=80) "self-service" (=2/3) "F-shaped" (=600) "six hundred tiles" (*Gatishila*, 600) to be "motile" (*Gatishila*, 600) with "ten-fold growth" (=170) within the "J-shaped" (=150) "pulsatile" (*Nabhiloma*, 150) of "sixty tiles" (*Nabhiloma*, 150).
>
> With "ten tiles" (=2) "forty tiles" (=64) "fire" (=17) an "H-shaped" (=45) 'retractile' (*Vrtile*, 45) of "fifty tiles" (*Vrtile*, 45) for "reproducing" (=17) the "ten-fold growth" (=170) through "spontaneous fragmentation" (=42). With "spontaneous fragmentation" (=42) of "fifty tiles" (=45), a "G-shaped" (=91) "sextile" (*Shadastaka*, 91) of "five hundred tiles" (*Shadastaka*, 91) is "naturally" (=91) "emitted" (=91) from the "vertex" (=89) of "ten tiles" (=2). "Ten tiles" (=2) are "repeated" (=5) as an "I-shaped" (=10) "infantile" (*Shishutvam*, 10) of "twenty tiles" (*Shishutvam*, 10) like a "family" (=10) of three tiles—large tile, larger tile, and largest tile—as a small tile.

A "single tile" (*Kavula*, 340) is an "Einstein" (*Kavula*, 340) "digitizable" (=340) into a "hat" (*Topi*, 340) of "two stones" (*Topi*, 340). "Two Stones" (*Topi*, 340) "Twin Einstein" (*Kacchapa*, 816) into a "turtle" (*Kacchapa*, 816) to "condensate" (=476) the "present value" (*Avastha*, 476 =340+67+69) of "twenty stones" (*Satkanda*, 67) into a "zero stone" (*Raghu*, 69). Zero stone is "twin consciousness" (*Sachi*, 69) of a "devil" (*Indra*, 0) "changing" (*Cittavikarin*, 69) his "character" (*Prakriti*, 485) with a "co-dependency" (=69) on the "solidifier" (=416) of the "littleness" (=400) of his "two-dimensional reality" (=400). A zero "stones" (=19) "itself" (=121) to "twin" (=121) "consciousness" (=4) of "another" (=98) as "Half Einstein" (*Vasu*, 98). Half Einstein is an "Electromagnet" (*Vasu*, 98) with "nineteen stones" (=98) that "twin" (=121) "stone" (=19) with "consciousness"

(=4) they "repel" (=9) to "attract" (=34) "half stone" (*Prakriti Yogi*, −1) as a "Potential Einstein" (*Prakriti Yogi*, −1). Electromagnet is a "primordial tile" (=98) that "quantum sequences" (=98) "ninety-eight tiles" (=98) into "eighteen sections" (=98).

Science of Section

An electromagnet is a "primordial tile" (=98) that "quantum sequences" (=98) "ninety-eight tiles" (=98) into "eighteen sections" (=98). Eighteen sections are "six kites" (=98); a "section" (=10,000) makes "potential" (=18) "circular" (=10,000). The circular is "reproductive" (=100) with the "kite's" (=23) "reproductive potential" (=23) for "breeding" (=15) "five kites" (=15) from "twenty-five kites" (*Manja*, 8). A "primordial tile" (=98) "tiles" (=340) the "potential" (=18) within a "section" (*Anuvaka*, 10,000 = [98+340+18+50+2-8] *20) for the "continuity" (=50) of "five sections" (*Vamri*, 2) without "eleven sections" (*Bhuma*, 8) "self-condensing" (=5/2) "sixteen sections" (=20) to be "reproductive" (=100) as "four sections" (=100). "Sixteen sections" (=20) "reproduce" (=78) "eighteen sections" (=98) with "two sections" (=80,000) that "parallel" (=80,000) "thirty-six sections" (=4,000).

"Three sections" (=1,500) of a "triangle" (=1,500) "triangulate" (=3) "nine sections" (=500) of "time value" (=500) into "forty-five sections" (=4,500). Forty-five sections are a "triangular section" (=4,500) without the "parallel section" (=500) whose "eight-fold growth" (=8) is a "square section" (=4,000). As the "three sections" (=1,500) "triangulate" (=3) "nine sections" (=500 =3*169-7), "six sections" (=169) "quantize" (=169) "seven sections" (=−7) of an "oyster" (=−7)., "coiling" (=−7) as a "circular section" (=−7) into a "shell" (=−7) that "line section" (=10) "thirteen sections" (=10). The line section is the "tangent" (=10) for the "intersection" (=10^{19}) of "twelve sections" (=10^{19}) with the "point section" (=19) of "twenty-five sections" (=19).

Twin Einstein turtles "twenty-two stones" (*Kacchapa*, 816). Two stones are "triple kite" (=340). "Twenty stones" (=67) are "four kites" (=67). "Twenty-two stones" (=816) are "seven kites" (=816). A turtle is a "potential tile" (=816) with "eleven tiles" (=816 =340*2+136); "ten tiles" (=2) "reorder" (=2) the "order" (=2) to "tile" (=340) "Einstein" (=340) into a "Triple Einstein" (=136) with a nonrepeatable "aperiodic" (=138) element. The potential tile is an "aperiodic tile" (=816).

As an "absolute tile" (=136), Triple Einstein is a "periodic tile" (=136) with "forty-two tiles" (=136). "Four tiles" (=18) have a "potential" (=18) to "twin tile" (=933) with the "ten-fold growth" (=170) of their "effect" (=34) on "Potential Einstein" (=−1). Triple Einstein has "thirty-three stones" (=136 =34*4) due to the "four-fold growth" (=8) of the "effect" (=34) after "breeding" (=15) a "stone" (=19) with the "kite's" (=23) "reproductive potential" (=23). "Triple Einstein" (=136 =23*4+44) "kites" (=23) "thirteen kites" (=136); the "workforce-effect" (=44) of the supernatural, "four-fold growth" (=8) "activates" (=9) nature's "eight-fold growth" (=8) within "Potential Einstein" (=-1). "Triple Einstein" (=136) has "eighty sections" (=136) from the "four-fold growth" (=8) of "twenty sections" (=34) that "pile" (=34) the "effect" (=34).

Science of Stones

An imaginary stone "brings" (=-18) "three stones" (*Shaana*, -18) for "reproducing" (=17) the "growth" (=6) of "potential" (=18) within "eighteen stones" (*Patanga*, 23) as a "kite" (*Patanga*, 23). Eighteen stones are "potential stone" (=23) of "thirty-six grams" (=23); they "channel" (=36) "gram" (*Masha*, 0) through their "potential" (=18) to "twin causation" (=2) with the "divinity" (=57 =19*3 = 23+36-2) within three "stones" (*Shrishaila*, 19).

As a "touchstone" (=-18), the "imaginary stone" (=-18) has an "inkling" (*Bhana*, -8) of the "illusionary stone" (=-8)—the "histone" (*Bhana*, -8) comprising "four stones" (*Bhana*, -8) before "subtraction" (=999) of a "stone" (=19) from the "origin" (=1,000) as one's copy. One's copy has an "archaic value" (*Puratana*, 10) of "ten stones" (*Puratana*, 10) without "primordial oneness" (=32) that the "past" (=9) copies to "twin causation" (=2) into "twenty grams" (*Puratana*, 10) with "two metrics" (=10). "9/32 grams" (*Shvasoshvasa*, 10) is also one's copy. Ten stones are a "divine stone" (=10)—with "ten-fold growth" (=170), ten "stones" (=19) are "divine" (=360 =170+10*19). One has "two copies" (*Chitra*, 1); it is a "general stone" (=1) with an "illusionary energy" (*Maya Shakti*, 1) of "eighty stones" (=1) as "Histone H1" (=1). The "histone" (=-8) of "four stones" (=-8) "copies" (=0) "H" (=1) to be "Histone H" (*Pashana*, 92) of forty stones—the "kidney stone" (*Pashana*, 92) as an "ecosystem" (=92) of "hard mineral" (=92). Divine stone is "Histone H6" (*Puratana*, 10) due to a "six-fold growth" (=6), comprising "two-fold growth" (=2) within "four copies" (=90) of divine with "Histone H" (=92).

The first histone is "Histone H7" (*Shvasoshvasa*, 10) with "9/64 stones" (*Shvasoshvasa*, 10). It perpetuates the "stone" (=19) with a "fragment" (=64) of "energy" (=19) from the "past" (=9) of the second "Divine Histone" (=10) as it "twins" (121 =19+64+9+10+7+12) its "reality" (=7) with a "Twin Histone" (*Gabhira*, 12). Twin Histone is "Linker" (*Linkadharaka*, 12), self-luminous as "Linker Histone" (*Linkadharaka*, 12); the linker twins his "reality" (=7) with a "causation" (=1) for "self-projecting" (=3/2) his "nature" (=8) to be "Histone H8" (*Gabhira*, 12). With his "twin reality" (=49), "his" (=13) "tone" (*Svaras*, 816 =49*13+16+121+42), the "personal essence" (*Svaras*, 816), of "twin nature" (=16) "twins" (=121) "as well" (*Api*, 42) with "oneness consciousness" (*Yogatma*, 42). Histone H8 is a "whole stone" (=16) with "sixteen stones" (*Gabhira*, 12), "self-twinning" (=5/4) "twenty pieces" (=12) "carrying" (38 =16+5/4*12+7) "thirty-two grams" (*Gabhira*, 12) of its "self-luminous" (=12) "reality" (=7).

A "stone" (=19) "internalizes" (=19) "thirty-eight grams" (=19) after it "externalizes" (=100,000) the "gram" (=0) as a "copy" (=0) for "self-condensing" (=5/2) "two stones" (=340) into "5/2 grams" (=340) with "absolute annihilation" (=850 =5/2*340) of the "twin triple copy" (=6) of the copy within the "three stones" (=−18), norming "six grams" (=-18). The twin triple copy "caps" (*Topa*, 6) "six stones" (*Topa*, 6) as a "special stone" (*Topa*, 6) since the "special" (*Khaas*, 9) "self-services" (=2/3) "six" (9*2/3) as "stone" (=19) to twin causation of "triple copy" (=3). A "cap" (=6) is a "hat's" (=340) triple copy, comprising "three hats" (*Topa*, 6) which generate a "growth" (=6) of twenty within the "origin" (=1,000 =340*3-20) whose "foundation" (=11) is "special" (=9). Thus, the "triple hat" (*Topa*, 6) is a "special hat" (*Topa*, 6). The foundation is "Histone H2" (*Gauranga*, 11), organizable with "one-hundred-sixty stones"

(*Gauranga*, 11) as "Histone H1" (=1) with "eighty stones" (=1) that "twin causation" (=2) to copy "itself" (=121) into "two copies" (=1). With two copies, "development" (=160) within stone norms "320 grams" (*Gauranga*, 11). Histone H2 is a "sentient stone" (*Gauranga*, 11); the "sentient" (=189) has a "potential" (=18) to "perpetuate" (=9) two to copy itself in "future" (=0).

The "future ecosystem" (*Garbhasamudbhava*, 14) is "Histone H2B" (=14) with "seven stones" (=14) with the letter "B" (=7). Histone H is one that twins causation to be two copies of "B" (=7); with "fourteen copies" (=14), Histone H2B is "fourteen grams" (=14). Histone H2B is a "circular stone" (*Garbhasamudbhava*, 14); "both" (=10,000) copies of "B" are "circular" (=10,000) before their "entropy" (=5) as a "stone" (=19 =2*7+5). "Five stones" (*Kancha*, 8) are a "marble" (*Kancha*, 8), the "Devil stone" (=8), that one copies with "ten grams" (*Kancha*, 8). The Devil as the copy of the "five stones" (*Kancha*, 8) is the "foundation" (=11) of a "stone" (=19) and its "formative growth" (=2) into "six stones" (=6). A devil is a "gram" (=0) which "double copies" (=1) "three stones" (=−18) to mediate "five stones" (=8) with a "stone" (=19) of "38 grams" (=19) within "ten stones" (=10) of "20 grams" (=10).

"Histone H2A" (*Kilbisa*, 15) is a "Satan stone" (*Kilbisa*, 15) with "eleven stones" (*Kilbisa*, 15). "Seven stones" (=14) "add" (=99) "four stones" (=−8) with an "A" (=28) that "subtracts" (=160) "B" (=7) "back" (*Prishtha*, 5) as "Histone H2B" (=14) to be "Histone H2A" (=15). A "duplicate" (=22 =7+15 =2−8+28) copies the "reality" (=7) of the "formative growth" (=2) of "four stones" (=-8) with "A" (=28) within "eleven stones" (=15) with "22 grams" (*Kilbisa*, 15) to "square causation" (=2). The duplicate is the "quantum core" (=22), "superpositioned" (=90) as a "core" (=2) on "Histone" (*Bhana*, −8). It "triples" (=130 =22+90+2−8+24) the "copy" (=0) with a "formative growth" (=2) into an "entity" (=24). The "Triple Histone" (*Param Shiva*, 15) is the "Core Histone" (*Param Shiva*, 15)—"Histone H4" (*Param Shiva*, 15). It is a "square stone" (*Param Shiva*, 15) with "fifteen stones" (*Param Shiva*, 15) as the formative growth of "four stones" (=−8) "squares causation" (=−2) when the "natural growth" (=6) "self-services"

(=2/3) "2A" (*Soham*, 4) as a 4. Natural growth "perpetuates" (=9) "thirty grams" (=15) with "three double copies" (=6).

"Histone H5" (*Pala*, 15) is the "link" (*Linka*, 15) "linkable" (*Linkaniya*, 1) as "Histone H1" (=1) "itself" (=121 =1,3*7) with the "multiplier" (=3) of "reality" (=7) of "stone" (=19) within "whole stone" (=12) while it "adds three" (=3) to the "linking" (*Linkan*, 100,000) (=100,000) that the Histone H1 "forms" (=100,000) as "Histone H3" (*Mahavidya*, 17). As a "Link Histone" (*Pala*, 15), Histone H5 is a "point stone" (=15) which "points" (=10^{10}) to a "stone" (=19) with "24 stones" (*Pala*, 15) for "oneness" (=48) of the double copy with the copy through "48 grams" (*Pala*, 15). As a "Linking Histone" (*Mahavidya*, 17), Histone H3 is "21 stones" (*Mahavidya*, 17), which is "also" (*Api*, 42) "42 grams" (*Mahavidya*, 17). It is a "Triangular Stone" (*Mahavidya*, 17) that "triangulates" (=3) the circular "seven stones" (*Garbhasamudbhava*, 14) to be a "Triple Circular Stone" (*Mahavidya*, 17). The "fourteen stones" (*Mahatma*, 18) are the "potential consciousness" (*Mahatma*, 18) of "nonhistone" (*Mahatma*, 18). A nonhistone is a "potential histone" (=18). Nonhistone is a "Twin circular stone" (=18) with the potential to be a "wholesome stone" (=18) when "perfected" (=28) with "twenty-eight grams" (=28).

The "twin" (=121) "becomes" (=179) a "twin circular" (*Pasha*, 900 =121+179+36+19+18+23+500+4) like a "twin kite" (*Atapin*, 900) to "channel" (=36) the "stone's" (=19) "potential" (=18) to be a "kite" (*Patanga*, 23) with a "two-fold sequence" (=500) of four "subtracting four" (*Pratibheda*, 900). The four is the "consciousness" (=4) of "ecosystem" (=92) "subtracting itself" (=43) to be the "ecosystem consciousness" (=53 =92+4-43). The "36 stones" (*Atapin*, 900) "incarnate" (*Dehi*, 900) "72 grams" (*Atapin*, 900) through "triangulation" (=900) of the "burst area" (*Kshetraphal*, 900) of a "present stone" (*Atapin*, 900). The potential copies the ecosystem's consciousness to be "present" (=1,600 =18+92+4−18+924+580) with "18 grams" (*Kosha*, 924) as a "simple stone" (*Kosha*, 924), "presenting" (=−18) a "shared value" (*Kosha*, 924) of "nine stones" (*Kosha*, 924) "within group of two" (=580). The "potential" (=18) makes nine "simple" (=9) with "two zeroes" (=−10) of "double copy" (=1) within a "copy" (=0= 9−10+1+18−18) of "three stones" (=−18).

> With "simple" (=9), a "stone" (=19) copies 24 "double zeroes" (=-10) with the "simple reality" (=10^{100}) of "2400 stones" (*Tula*, 10^{100}). Simple reality makes the stone's "oneness" (=48) "reproductive" (=100); it copies 2400 stones as a "wholesomewhole stone" (=10^{100}) of "4800 grams" (=10^{100}), "aiming" (=185) to be "infinite" (=185) like a "guider stone" (=185) for the "growth" (=6) of the "wholesomewhole" (=13) element as a "stone" (=19).

A "twin stone" (*Topi*, 340) is a "complex stone" (*Topi*, 340) whose "face" (=76 =92+2-18) "orders" (=2) a "real stone" (*Pashana*, 92) to be "infinite" (=185) as a "variant" (=188) of "Y" (=–18), the "imaginary stone" (*Shaana*, –18). As a "guider stone" (*Aparimita*, 185), the infinite has a "splitting potential" (=185) of "twenty-three pieces" (*Aparimita*, 185).

> ### Science of Pieces
>
> A twin stone is "disordered" (*Astavyasta*, 340) into a "jumble" (*Astavyasta*, 340) of "ten pieces" (*Brahmantaka*, 3), "ordered" (*Adishta*, 3) without the "connected piece" (*Brahmantaka*, 3) of the multiplier's "multiplying reality" (*Gurvartha*, 40).
>
> The "package" (*Gathari*, 2) of "two pieces" (*Gathari*, 2) is a "connecting piece" (*Gathari*, 2), "packaging" (*Samveshtan*, 1) ten pieces of the "connected piece" (*Brahmantaka*, 3) with the "disconnected piece" (*Masaki*, 9), the "bead" (*Masaki*, 9) of "eight pieces" (*Masaki*, 9), into a "system" (*Gabhira*, 12) of "twenty pieces" (*Gabhira*, 12)—the "connectable piece" (Gabhira, 12). A "system" (=12) is "connectable" (*Sambhadya*, 6) as a "piece" (*Tukada*, 11/2) when "modularized" (*Pramapi*, 480 = 12 *[6*11/2+6+1]) into a "module" (*Pramapa*, 18 =12+6) of "four pieces" (*Pramapa*, 18) for "packaging" (=1) the eighth, "disconnecting piece" (*Samveshtan*, 1) as "seven pieces" (*Brahmanda*, 1). A module is a "disconnectable piece" (*Pramapa*, 18), "self-condensing" (=5/2) "nine pieces" (*Aghani*, 5/2) of the ninth, "reconnectable piece" (*Aghani*, 5/2) into an "emergent" (=45) "reconnecting piece" (*Ganthana*, 14)

which "packs" (*Ganthana*, 45 =18*5/2) "five pieces" (*Ganthana*, 45) by "compressing" (=45) them at an "infinite cost" (=45).

The "pack" (=45) "ages" (=80=45*2 −10) into a "package" (=2) when "two pieces" (=2) are "left" (=89) within the "bead" (=9) of eight pieces because "five pieces" (=45) "center" (=16) "three pieces" (*Naadi*, 16) of "codon" (*Naadi*, 16) into a "reconnected piece" (*Naadi*, 16) of the "six-piece" (*Kumbhaka*, 10) "seven-pointed base" (=10). The six-piece base is a "vertical piece" (*Kumbhaka*, 10) with "three vertices" (*Kumbhaka*, 10). Each "vertex of two" (*Vrit*, 89) is "left" (=89) with "two pieces" (*Gathari*, 2). The "fourth piece" (*Astavyasta*, 340) "squares" (=18,000) the growth of six within the "vertical piece" (=10) into a "square piece" (*Astavyasta*, 340) of "thirty-six pieces" (*Astavyasta*, 340). The "multiplying reality" (*Gurvartha*, 40) of the "square" (=18,000 = [340+10+100] *40) as the "space" (=18,000) "generates" (=40) "space reversal" (=40) as the "thirty-piece" (*Shodashottari Dasha*, 40) "horizontal piece" (*Shodashottari Dasha*, 40) within the "gravitational force" (=100) as the "seventy-piece" (*Chit*, 100) "diagonal piece" (*Chit*, 100).

As a "multiplied piece" (=10^8), "hundred pieces" (=10^8) comprise three "square pieces" (=340), present as a "triangular piece" (=1,600) of 3*36, i.e., "108 pieces" that perpetuate "eight pieces" (=9) as their past. Three square pieces are reproductive like the "fourth square piece" (*Chit*, 100), which "copies" (=0) the "square piece" (=340), repeated as the "fifth square piece" (=340) within the vertical piece—the "sixth square piece" (=10) for the growth of six pieces within the horizontal piece—the "seventh square piece" (=40). The disconnected piece of "eight pieces" (=9) is an "octagonal piece" (*Masaki*, 9); it is the "eighth square piece" (*Masaki*, 9) with "two square pieces" (*Masaki*, 9).

"Seventy-two pieces" (*Pradarshan*, −8) of the "two square pieces" (*Masaki*, 9) are "preprogrammed" (*Pradarshan*, −8) in the "view" (*Pradarshan*, −8) as "disorder" (*Pradarshan*, −8) without "packaging" (=1) "seven pieces" (*Brahmanda*, 1) as "dis" (*Brahmanda*, 1) to "order" (=2) "two pieces" (=2). As a "circular piece" (*Purvasuchana*, 92), "twenty-seven pieces" (*Purvasuchana*,

92) "preorder" (=92) "nine pieces" (*Aghani*, 5/2), repeated with the "two pieces" (=2) as a "eighteen pieces" (*Kaal*, 360) to be the "parallel piece" (*Kaal*, 360) of "time" (=360). "Eleven pieces" (*Uttarasuchana*, 180) are "exchanged" (*Somapa*, 180) as the "line piece" (*Somapa*, 180) to "postorder" (*Uttarasuchana*, 180) "light" (=180) as the "genetic code" (*Somapa*, 180). By "gravitating reality" (=38) of "thirty-eight pieces" (*Stotrem*, 38), the "root" (*Jada*, 38) is "realizable" (=38) as the "point piece" (=38). The point piece is the one with "ten pieces" (=3) which "copies" (=0) the "growth" (=6) of "six pieces" (=10) from a "molecule" (*Kanalakshamsha*, 39) of "thirty-two pieces" (*Kanalakshamsha*, 39 =38=1). A molecule is a "real piece" (*Kanalakshamsha*, 39).

A "pearl" (*Moti*, 96) is a "multiplier" (=3) of "real" (=9), "ordered," (*Adishta*, 3) into 3x32, i.e., "ninety-six pieces" (*Moti*, 96)—representing an "imaginary piece" (*Moti*, 96) within "consciousness" (=4) of the reality of the "seventy pieces" (=100). Consciousness "clues" (*Samketa*, 4) a "complex piece" (*Samketa*, 4) of 4*32, i.e., "128 pieces" (*Samketa*, 4). Reality "rues" (*Sadvika*, 7) an "illusionary piece" (*Sadvika*, 7) of "twelve pieces" (*Sadvika*, 7), after "ten pieces" (=3) within 128 pieces" (*Samketa*, 4) order "two pieces" (=2) within one with "seven pieces" (=1). Letter "U" (=396) is a "simple piece" (*Vakra*, 396). It is a "vector" (*Vakra*, 396) of "twenty-nine pieces" (*Vakra*, 396 = 78*5/2), the "emanation-effect" (=396 = 38*5/2*(1/[1/2] *2+16) of "three pieces" (=16) on the root of "thirty-eight pieces" (=38). It produces the "emanation" (=1/2) of "nine pieces" (=5/2) with "two pieces" (=2).

"Two stones" (=340) "tile" (=340) "three kites" (=340).

Science of Kites

A guider stone's "thirteen stones" (*Aparimita*, 185) copy the "trading-effect" (=26) of "26 grams" (*Aparimita*, 185) from the "three-fold sequence" (*Trika*, 159) of "three kites" (=340 =185+159-4) after "subtracting four" (=900 =340+67-7+500) as "four kites" (*Satkanda*, 67) from the "reality" (=7) of "two-fold sequence" (=500).

> "Three kites" (=340) "bunch" (*Stabaka*, 67) "four kites" (=67) as a "small kite" (*Satkanda*, 67) to "enlarge" (*Samvridh*, 816 =340 +67*4+208) the "former dimension" (=208) of "two kites" (=900 =208*2+500) within the "two-fold sequence" (=500) into the third, "large kite" (*Khabhranti*, 340 =20*19-40) like a "long-winged falcon" (=340). With three, the "kite" (=23) "bunches" (=67) forty copies as "forty grams" (*Satkanda*, 67) for the "replication" (=90) of the second, "small kite" (=67) like a "short-winged hawk" (*Satkanda*, 67) with "twenty stones" (*Satkanda*, 67) "left" (=89) after the first, "medium kite" (=100) copies forty as the twenty "double copies" (=1) of the "stone" (=19).
>
> The "medium kite" (*Devatti*, 100) "double copies" (=1) "ten kites" (*Devatti*, 100) as the twenty double copies of "itself" (=121). Since itself is a "twin" (=121 =100+1/2*30+1+20-15), the medium kite "outwits" (*Chhalayati*, 100) 3*20,00 i.e., "6000 grams" (*Chhalayati*, 100) to "self-perpetuate" (=1/2) thirty "double copies" (=1) of the "stone" (=19) as the "3,000 stones" (*Devatti*, 100) with the "reproductive force" (=100) of the "present kite" (*Vanaramrga*, 15). The present kite is "thirty grams" (*Vanaramrga*, 15), transforming "five stones" (*Kancha*, 8) from a "kite" (=23) of "eighteen stones" (=23) into "fifteen stones" (*Vanaramrga*, 15) with "five kites" (=15) with a "potential" (=18) of "three stones" (=−18).

A single tile is a "pointed polygon" (=340).

Science of Polygon

Science of polygon follows the science of stones and kites. "Three stones" (=−18) "tile" (=340) "three kites" (=340) into a "darker" (*Nishprabha*, 340), "pointed polygon" (=340) of "thirteen sides" (*Nishprabha*, 322) with a "lighter" (=−9) "angular polygon" (*Atmata*, −9) of "seven sides" (=8,000) within "twenty sides" (*Bahis*, 10) on the "outside" (*Bahis*, 10) of a twenty-side "general polygon" (*Dasha*, 1) with "ten circular-effects" (*Dasha*, 1). Seven sides "intersect" (=8,000) with the

Formative Growth Paradigm

"two sides" (=80,000) of a "circle" (=100,000) to "dissect" (=80,000) a "square" (=18,000) with the "surface" (=2,000) of a 2-side "special polygon" (*Vartula*, 10,000)—the "circular" (*Vartula*, 10,000). The circular's first "inner side" (*Parshva*, 380) is a "linear side" (*Parshva*, 380), "parallel" (=80,000) to the "second "outer side" (=80,000)—the "curvilinear side" (=80,000). The curvilinear side curves the third "curved side" (=89) to "dissect" (=80,000) the "two sides" (=80,000) with the "three sides" (=89) of a three-side "circular polygon" (*Vrttagona*, 331) into the "six sides" (=126) of a six-side "square polygon" (*Vargigona*, 874).

A "square polygon" (=874) has "four sides" (*Vasudeva*, 75) of a four-side "triangular polygon" (=51) that "self-reproduce" (=1/8) "two sides" (=80,000) of a two-side "special polygon" (*Vartula*, 10,000) to "form" (=100,000) "eight sides" (=90) of an eight-side "parallel polygon" (*Samtolagona*, 90). "Eight sides" (=90) "twin diameter" (=75) of the "four sides" (=75) of the "triangular polygon" (=51) within the "five sides" (=182) of a five-side "line polygon" (*Rekhagona*, 28 =90+75+51-182). A "line polygon" (=28) has "three sides" (=89) of the "circular polygon" (=303 =28+89+186) and "three angles" (*Kana*, 186) of the eleventh "parallel angle" (*Kana*, 186). The "parallel angle" (=186) "parallels" (=80,000) "two sides" (=80,000) of the "special polygon" (=10,000) to "self-reproduce" (=1/8) "four angles" (=80,000) with the "angle" (=10) that "pulls" (=800 =186+140+474) "ten sides" (=140) "subtracted" (=474) from a ten-side "point polygon" (*Bindugona*, 474).

A "point polygon" (=474) has "one side" (=380) of a one-side "curve polygon" (*Vakragona*, 1,600) and "five angles" (*Laghukone*, 12) "leaving" (=550) from the "five sides" (=182) of "line polygon" (=28) on the "right" (=71); a side "copies" (=0) "itself" (=121) to be "left" (=89) with the "seven sides" (=8,000) of the "angular polygon" (=–9). "Two sides" (=80,000) of the "special polygon" (=10,000) "circulate" (=22/7) the "ten sides" (=140) of a "point polygon" (=474) as a "side" (=380) that "curves polygon" (=1,600) to potentiate "thirteen sides" (=322) of the "pointed polygon" (=340).

A one-side "curve polygon" (=1,600) is a "concave polygon" (=1,600), whose "side" (=380) "angles" (*Kone*, 10) "360 degrees" (=6) as the "maximum" (=123) of a "concave" (-100) of "four lines" (=90) to "triangulate" (=3) the "three sides" (=89) "left" (=89) as "finite" (=85) after the "ten sides" (=140) "center" (=16) the "six sides" (=126) to "keep" (*Rakha*, 123) "order" (=2). The "order" (=2) "arcs" (=47) the "nine side" (*Shirsha*, -9) "head" (*Shirsha*, −9) of the nine-side "convex polygon" (*Vakriyagona*, 18) by "180 degrees" (=12) to "tail" (=70) a "convex" (=26) of "three lines" (=9), "subtractable" (=365) as the "minimum" (=5) of the "twelve sides" (=8) of the twelve-side "skew polygon" (*Kavatigona*, 85).

The convex polygon is a "curvable polygon" (*Vakriyagona*, 18) whose "four angles" (=80,000) "square" (=18,000) the "two sides" (=80,000) as they "both" (=10,000) "recycle" (=1/10) "90 degrees" (=90) to "cube" (=90) the "minimum" (=5) within the "skew polygon" (=85) into a "side" (=380 = [90+5]*4). Both are "45 degree" (=10,000) "sections" (*Anuvaka*, 10,000). The skew polygon is a "curving polygon" (*Kavatigona*, 85) whose "three vertices" (*Kumbhaka*, 10) are "planer" (*Taliyaka*, 900) but "vertex" (*Vrt*, 89) is "coplanar" (*Samtaliyaka*, 1)—it "skews" (*Kavati*, 81) "fifteen degrees" (=4) to "live" (=14) as a "plane" (=80) of "five degrees" (*Dyujya*, 100) by "curving" (=−7) the fourteen-side "plane polygon" (*Taliyagona*, 185) with its "tenfold growth" (=170).

The "plane polygon" (=185) with "fourteen sides" (=170) is a "curved polygon" (*Taliyagona*, 185) whose "two segments" (*Pratihata*, 78) are the "linear dimension" (*Margi dharma*, 78) of the "seven letters" (*Smita*, 88) in the "word" (*Aksara*, 1,765) "polygon" (*Bahubhuj*, 2,000). Seven letters are an "outgrowth" (=88) of "five segments" (*Kushmanda*, 88) that "twin" (=121) "two segments" (=88) whose "duality" (*Bheda*, 14) is "repeated" (=5) in "fourteen segments" (*Bheda*, 28) as the "twenty-eight particles" (*Bheda*, 28).

A "particle" (*Hemarenu*, 19) is "flat" (*Sapat*, 19) with "sixteen sides" (*Antar*, 85) "inside" (*Antar*, 85), "three sides" (=89) on the "left" (=89) of the "twenty sides" (=10) "outside" (*Bahis*, 10), and "five sides" (=182) on the "right" (=71) of the "six sides" (=126) in the "center" (=16) of the "fifty sides" (*Jiva*, 2) of a fifty-side "flat polygon" (*Sapatagona*, 600) A flat polygon is a closed, "uncurved polygon" (*Sapatagona*, 600) whose "straight face" (*Tadakara Mukha*, 10) is "orthogonal" (=10) to the "fifty faces" (*Tadakara Mukha*, 10), "causing" (*Hetuka*, 47) it to "close" (=8) into an "object" (=–3) "hiding" (=57) its "past reality" (=-3) by "assuming" (=190) "one side" (=380) of its "hundred side" (*Jagatkritsna*, 300) hundred-side "open polygon" (=2), "present" (=1,600) as a "solid" (=9) within the one-side "curve polygon" (=1,600).

The "open polygon" (=2) has a "potential" (=18) to be a thousand-side "close polygon" =450), "not closed" (*Udghata*, 70), i.e., "open" (*Udghata*, 70), but "about to close" (=126), i.e., "half-closed" (=126), "openable" (=126) at the "infinity of causation" (=90,000) with "thousand sides" (=–16). A thousand sides are a "strand" (*Bhugrasta*, –16) of "hidden reality" (*Vigudhartha*, –16) which "cascades" (=–16) its "contiguity" (=–16) through "gravitational differentiation" (=–16) of four "ellipses" (=1,000) for a "symmetry" (=250) with the "concealed truth" (=–4) "everywhere" (=–4) after "leaving" (=550) the "origin" (=1,000).

The "origin" (=1,000) is the ten-thousand-side "closable polygon" (*Ardhotghatagona*, 1,000), "half-open" (*Ardhothata*, 680) but "closable" (=680) without the "obstacle" (*Vighna*, 680) of an "observing subject" (=0) with "ten thousand sides" (*Naivedya*, 320). "Ten thousand sides" (=320) have a "side" (=580) with a "main" (*Pradhan*, 16) that "remains" (=550) as a "side-effect" (*Ghatika*, 24) of "solid phase" (*Ghatika*, 24) for the "growth" (=6) of "liquid" (*Tarala*, 91) into a 100,000-side, regular, "closing polygon" (*Bandanagona*, 150) with "hundred thousand sides" (=3). Ten thousand sides are "twin edible" (*Naivedya*, 320), "closing" (*Bandana*, 960) as "potential open" (*Bandana*, 960) with "triple edible" (=3). "One million sides" (=20) of a one-million-side irregular, "opening polygon" (*Udghatanagona*, 880) are "half edible" (=20), but "potential close" (*Bandana*, 960) for "opening" (*Udghatana*, 960) "anytime" (*Pushya*, 60).

The "opening polygon" (*Udghatanagona*, 880) is "potential edible" (*Balidana*, 880) as an "opened polygon" (*Udghatitagona*, 120), "n-gon" (=120) within the "closable polygon" (=1,000). N-gon is "opened" (*Udghatita*, 10^{10}) with "10^{10} sides" (=8×10^{15}) as "N" (=120), "ceasing" (*Uparama*, 90) with "GON" (*Uparama*, 90), "gone" (=73), "half edible" (=20) with "E" (=53), to "stay" (*Nivaas*, 640) "alive" (=10) "near" (*Samipa*, 816) the "effect" (=34) the "present" (=1,600) "causes" (=18) on the "past of everybody" (=−1) as a "devil" (=0)—the "future" (=0). The devil "tiles" (=340) one's "spatial spirit" (=20) on the "soul" (=4) of "time" (=360) to be "gas" (*Prakrita*, 365).

Gas" (=365) makes "solid" (=9) "reproductive" (=100) with "liquid" (=91) to "form" (=100,000) "plasma" (=100,000) as the "origin" (=1,000) of "quark" (=476) to "guide" (=111) letter "R" (=111) as "R-gon" (*Mangona*, 0). "Two ones" (=21) are "gon" (=90) when one is "gluon" (=149) "mediated" (=10) by "LU" (=27). Therefore, R-gon is "21-gon" (*Mangona*, 0), the "21-side icosikaihenagon" (*Mangona*, 0). Twenty-one "sides" (=380) "L" (=16) whereas "U" (=396) "loiters" (*Mangona*, 0) with "21 sides" (*Kapali*, 21) "aiming" (*Sarthavaha*, 1) to "trade" (=20) the "unit" (=6) of "growth" (=6) from "LU" (*Sartha*, 27)—the "aim" (*Sartha*, 27). When "aimless" (*Asartha*, 79), "L" (=16) "darts" (*Shalivan*, 79) "eleven sides" (*Asamananki*, 79) of eleven-side "U-gon" (=324) whereas "U" (=396) "rhombuses" (*Akshagona*, 189) "thirty-two sides" (=189) of 32-side "L-gon" (=12).

"Dart" (=79) is a 42-side "I-gon" (*Shalabhangona*, 79). I-gon is a "cyclic polygon" (*Shalabhangona*, 79) whose "non-congruent side" (*Asamavrtta Pada*, 0), i.e., eighteen sides (=0), of "unequal length" (*Asamadirodha*, 10) "pairs" (*Jodi*, 8) "itself" (=121) with the "non-adjacent side" (*Asannidhipada*, 75), i.e., "four sides" (=75). Its "forty-two sides" (=310) "order" (=2) the "adjacent side" (*Sannidhipada*, 90), i.e., the "eight sides" (=90) to be the "congruent side" (*Samavrtta Pada*, 8), i.e., "twelve sides" (=8) of "equal length" (*Samadirodha*, 80) with the "discontinuity" (=200) of "fourteen sides" (=170) through the "three cycles" (*Trichakra*, 10) of "vertex" (=89).

Rhombus is a 24-side "J-gon" (*Akshagona*, 189 =2+121+90-24) which "orders" (=2) "itself" (=121) with "J" (=2) to be "gon" (=90) with "24 sides" (*Ghatika*, 24). It is an "axis polygon" (*Akshagona*, 189) whose "axis" (*Aksha*, 16) of "six sides" (=126) in the "center" (=16) "bisects" (*Dvibhajana*, 25) the "symmetry" (=250) of "itself" (=121) with "ten sides" (=140) into "three sides" (=89) on the "left" (=89) and "five sides" (=182) on the "right" (=71) to "angle" (*Kone*, 10) the "two sides" (=80,000) into a "diagonal" (*Akshnaya*, 8,000).

"Trapezium" (*Vajragona*, 324) is an 11-side "U-gon" (*Vajragona*, 324) whose "non-parallel side" (*Asamananki*, 79), i.e., "eleven sides" (=79) is "perpendicular" (*Kramajya*, 1,000) to the "two sides" (=80,000) that are "parallel" (=80,000) to the "parallel side" (*Samananki*, -9), i.e., "nine sides" (=-9), which becomes "non-parallel" (*Asamanantra*, 100) to "order" (=2) the "parallel-effect" (=34) of "24 sides" (=24) on the "four sides" (=75) of a "pyramid" (=48).

"Pyramid" (*Kumbhagona*, 48) is a 4-side "E-gon" (*Kumbhagona*, 48) whose "eight vertices" (*Kurpara*, 9) "taper" (*Kurpara*, 9) into adjacent "eight sides" (=90) when its non-adjacent "four sides" (=75) "face" (=76) "four vertices" (=9 x 10^{16}) to "sequence" (=48) a "potential vertex" (*Kurpara*, 9) of "eight edges" (=10) with "five faces" (=−3) of a "five-dimensional object" (=19). It is a "seven-pointed polygon" (=48), shaped like a "vase" (=48), with a "seven-pointed base" (*Kumbhaka*, 10) which copies a "three-dimensional object" (*Kumbha*, 1) to be a "two-dimensional object" (=186) to "cascade" (=−16) its "hidden reality" (=−16) as "E" (=53) "gon" (=90) in "diverse forms" (=1).

While "breeding" (=15) "energy equilibrium" (=120), a pyramid assumes N=8 by feeding G=2 the knowing E=1 for freeriding O=4 with the consciousness of "gone" (=73) since the "oneness" (=48) of diverse forms is gone when I =16. When "I" (=12) =16, "E" (=53) =1 for "oneness" (=48 =53+12-16-1) of "O" (=180 =28+120+64+14-4-8-32-2) =4, to be "A" (=28) as "N" (=120) =8, "greets" (=64) "F" (=0) =32 with "G" (=14) =2, to "UNIFORGE"

(*Ayasam*, 36) "U" (=396) = 128 [=48+36+64-28+8] with eight to "guide" (=111 =128-8-9) the "R's" (=111) "past" (=9) as a "vertex" (=89) of "RAM" (=256 =2^8), the "eight-dimensional power" (=256) of "M" (=264) as 2^9 =512 to "restrain" (=512) 2^5-side "L-gon" (=12) with the "minimalism" (=-512) of "UNIFORMAGE" (*Viniyamavastha*, 512).

"L-gon" (*Yonigona*, 12) is a 32-side "complex polygon" (*Yonigona*, 12). It is a "potential parallelogram" (*Yonigona*, 12), "crossed" (*Atita*, –12) with "C's" (=–12) "potential" (=18) into the sixty-side "parallelogram" (*Kriyagona*, 60,000)—"F-gon" (*Kriyagona*, 60,000). Its "congruent side" (=8) "intersects" (=8,000) with the "origin" (=1,000) of the "non-congruent side" (=0) for "primordial oneness" (=32) of the "left-side" (=89) as a "vertex" (=89) of "three sides" (=89) within "itself" (=121 =32+89). The "origin" (=1,000) is a "complex" (=1,000) of "femininity" (*Yoni*, 1,000) whose "sentient energy" (=1,000) "self-perpetuates" (=1/2) the "potential" (=18) to be "sentient" (=189) with "32 sides" (=189) of "energy" (=19) for "four-fold growth" (=8) within "Mother Nature" (*Kudrat*, 8). L-gon's "self-congruence" (=12) "self-services" (=2/3) eight "kites" (=23).

"F-gon" (*Kriyagona*, 60,000) is a 60-side "simple polygon" (*Kriyagona*, 60,000). Its 32-side "opposite side" (=189) "equal angles" (=10,000) the outer 2-side "curvilinear side" (=80,000) with the "equal length" (=80) of the 12-side "congruent side" (=8) for the "centralization" (=101) of "sixty sides" (*Madhyakarana*, 101) as the "consecutive side" (=101). F-gon is "scriptable" (=60,000) as "four diameters" (=60,000) that "bisect" (=25) the "area" (=900) of a "triangle" (=1,500) with a "twin congruent" (=2,400). The "twin congruent" (=2,500) "obliques" (=1,790) the "gon" (=90) with a 90-side "O-gon" (*Shringona*, 700).

"O-gon" (*Shringona*, 700) is a 90-side "star polygon" (*Shringona*, 700 =50*14). By "self-projecting" (=3/2) "itself" (=121), its 60-side "consecutive side" (=101) "projects" (=31) "ninety side" (=51) "oblique side" (=51) with the "continuity" (=50) of "G" (=14) as a "star" (=2). O-gon is an "eight-pointed polygon" (=700), "Rhomboid" (*Shringona*, 700) with a "thirty side" (=9 x 10^{16}) "unequal side" (=9 x 10^{16}). O-gon is an "absolute parallelogram" (=700).

"G-gon" (*Uragona*, 13) is a 30-side "spiral polygon" (*Uragona*, 13). Its 8-side "adjacent side" (=90) is the "equal side" (=90) that "evens" (=113) the "spiral" (=13) with "twenty-two sides" (=190) of "A-gon" (*Rishigona*, 190). As an "even polygon" (=13), G-gon is "primordial parallelogram" (=13). As an "odd polygon" (=190), A-gon is a "primeval parallelogram" (*Rishigona*, 190).

A-gon is a 22-side "arc polygon" (*Rishigona*, 190), "present" (=1,600) with an "odd" (=1,810 =1,600+190+47-27) "primevalism" (=27) while "servicing" (=47) the "arc" (=47) of "primordialism" (=27) to the "future" (=0) as the "multiplier" (=3) of the "past" (=9).

While G-gon is a "lateral polygon" (=13), the 11-side "U-gon" (-324) is a "unilateral polygon" (=324), the 22-side A-gon is a "bilateral polygon" (=190), and the 33-side "M-gon" (=23) is a "trilateral polygon" (=23)—"kite" (*Patanga*, 23) with "thirty-three sides" (*Patanga*, 23) as the "vertex of two" (=89) twins itself into the "five sides" (=182) on the right with the "three sides" (=89) on the left. The 50-side "tile" (=340) is the "quadrilateral polygon" (=340) with the growth of three into six which makes four zero for one to be ten as a "quadrilateral" (*Ardhagona*, 45) is a "half-gon" (=45) when a two is repeated as a five. A quadrilateral is a "half polygon" (=45). "Half-gon" (*Ardhagona*, 45) is "C-gon" (=45), whose 4-side "non-adjacent side" (=75) is "diagonal" (=8,000) to the 5-side "right side" (=182). The "symmetry" (=250 =75+182 7) "trisects" (*Tribhajana*, 5,000) the "origin" (=1,000) into two "triangles" (=1,500) which "bisect" (*Dvibhajana*, 25) the "space" (=18,000) into "forty sides" (*Visamgata*, 720) to "time" (=360 =10 *[7+10+19]) the "causation" (=1) for the space's "equal breadth" (*Samachaudai*, 6) within time's "equal width" (*Samavishtara*, 7) to "angle" (=10) the "twenty sides" (=10) of "general polygon" (=1) into a "sect" (*Bhajana*, 19) with a "copy" (=0) of "C" (=-12) that "twins causation" (=2).

"Equal breadth" (=6) of the 60-side "even side" (=90) is "twin perimeter" (=6), of which one is the "width" (*Vishtara*, 94) which "centers" (=16) the "breadth" (*Chaudai*, 1,694) as the "perimeter" (*Parimap*, 1,694) with an "angle" (*Kone*, 10) "tangent" (=10) to the "triple breadth" (*Asamavishtara*, 16), the "unequal width" (=16) of the 30-side "odd side" (=9 x 10^{16}). The "angle" (=10) "transforms" (=1) the "twin perimeter" (=6) into a "circumference" (*Tyajya*, 64) of "six perimeters" (*Paridhi*, 64) with the "multiplier" (=3) of time's "equal width" (=7) for "oneness" (=48) of the "triple perimeter" (=16) with N=8, so that perimeter = n * circumference. "Temporal width" (=94) "heightens" (=16) "spatial breadth" (=1,694) with the causation's "triple perimeter" (=16), "triple height" (*Asamavishtara*, 16). "Triple height" (=16) "twins height" (*Samavishtara*, 7) with "height" (*Uchchaih*, 9) to "tile" (=340) the "inner conflict" (=308 = 7*[47-3]) of the "multiplier" (=3) at entity's "length" (*Lambai*, 47).

Curvilinear "length" (=47) of "entity" (=24) is the "short side" (=8,000) of "seven sides" (=8,000) which "intersects" (=8,000) the "right" (=71 =47+24) with "potential" (=18) within "left" (=89) for "three sides" (=89) to be the "shortest side" (=89) of the "masculinity's" (=53) linear, "tangent length" (*Jya*, 974) within "femininity" (=1,000 =89-53+974). The shortest side is both the "tangent side" (=89) and the "masculine side" (=89). "Breadth" (=1,694) of "space" (=18,000) is the "longer side" (=75) of "four sides" (=75) "parallel" (=80,000 =[18+75+10-47+24]*1,000) to the "tangent" (=10) that "angles" (=10) the "length" (=47) of the "entity" (=24). "Width" (=94) of "time" (=360) is the "shorter side" (=−9) of "nine sides" (=−9) "perpendicular" (=1,000 = 360/-9 *[94-47-24+2] *-1) to the "arc" (=47) that "orbits" (=2) the "length" (=47) of the "entity" (=24) like a "metric" (=−1). The "arc" (=47) is the "length" (=47) of the "femininity's" (=1,000) parabolic "arc length" (*Gavyuti*, 128 =47*2+34) that "orbits" (=2) its "parallel effect" (=34). "Height" (=9) of "causation" (=1) is the "long side" (=126) of "six sides" (=126) that "focuses" (=3) the "mind" (=38) on the "symmetry" (=250) of "orbit" (=2) with the "arc length" (=128) of "longest side" (=90) of "eight sides" (=90) within a "tile" (=340).

A "tile" (=340) is a "self-mediating coherence" (*Pratyayasamgati*, 340). It "squares growth" (*Audrika*, 340) to "cause itself" (*Vimshati*, 340) be the "super-conscious coherence" (=340) of the "spatial fallacy" (*Khabhranti*, 340). The "space" (*Kha*, 18x10^3) "squares" (*Chakora*, 18,000) "itself" (=121) with "growth" (=6) to "cause"

(=18) its "copy" (=0³) to be the "first reason" (*Lakshya*, 485) for the "variability" (*Asthirta*, 485) of "para nature" (*Prakriti*, 485), i.e., character, as a "consequence" (*Lakshya*, 485). "Consequential coherence" (=340) is "variable coherence" (=340).

> ### Science of Coherence
>
> A "tile" (=340) "intensifies" (=340) "symmetry" (=250) with the "orbital of two" (=60) to "extensify" (=350) "decoherence" (*Asamgati*, 124), i.e., "loss of coherence" (=124) of "time multiplier" (=3) with "itself" (=121) as a "metric" (=-1 = 60*2 -121) of "coherence" (*Samgati*, 150) within an "organization" (=29). The metric "transforms" (=1) "causation" (=1) to "norm" (=18) the "edge" (=8) of four that "forms" (=100,000 =250*4 *[18-9]) the "continuity of space" (=50) in "symmetry" (=250) with the "discontinuity of time" (=200).
>
> For "closing" (=960) "noncoherence" (*Pratisamgati*, 960) in "time" (=360) with the "space's" (=18,000) "reaction" (=190), the "edge" (=8) "perpetuates" (=9) space's "guider reality" (=40) with the "action" (=10) of four "self-projecting" (=3/2) six to be "reproductive" (=100) with "time value" (=600). By "opening" (=960) the "origin" (=1,000) of four within two that "adds three" (=3) to the "causation" (=1), five is repeated by the "edge" (=8) of four with "incoherence" (*Visamgati*, 145 =9+5, 5), i.e., "lack of coherence" (=145), in "space" (=18,000) as it "perpetuates" (=9) the "norm" (=18) as a "form" (=100,000) that is "reproductive" (=100).
>
> Four "edges" (=8) the "reality" (=7) of the "quantum coherence" (*Susamgati*, 12) of five with the "circular coherence" (*Anusamgati*, 11) of "causation" (=1) for its "action" (=10). The "simplicity" (=2) of "action" (*karma*, 10) "adds three" (=3) as a time multiplier to "perpetuate" (=9) the "goal" (=9) with an "illusion" (=1) of "action coherence" (*Karmasamgati*, 13). Action coherence is "simple coherence" (=13) of "goal coherence" (=13) as "time coherence" (=13). The "action" (=10) of the "multiplier" (=3) is "complex" (=10³). The "constancy" (*Sthirta*, 35) of "atomic coherence" (*Anusamgati*, 11) as a "wholesomewhole" (*Anuna*, 13) shapes "complex coherence" (*Arthasamgati*, 48) into a "wholesomewhole coherence" (*Arthasamgati*, 48 =35+13) of the

"system" (=12) as a "part" (=-1) of the "wholesome" (=957). A "goalkeeper" (=7) makes the wholesome "whole" (=16) for the "origin" (=1,000) of "primeval space" (=19) like an "atom" (=19) with the "mind's" (=38) "space" (=18,000) for "oneness" (=48) of the "space coherence" (=48) as the "goalkeeper coherence" (=48).

The "whole coherence" (*Vyasasutra samgati*, 356) of the constancy within three with the simplicity within two is a "bicoherence" (=356) of both the "goalkeeper" (=7) and the "goal" (=9) as a "whole" (=16 =7+9) whose "condition" (*Vasha*, 79) is "extrinsic" (*Vahya*, 79). "Divine's" (=360) "extrinsic coherence" (=356) "opposes" (*Virodha*, 11) the "soul" (=4) of the "intrinsic" (=78) "left" (=89) as the "creature dimension" (=271) with the "quantum decoherence" (*Vyasasutra samgati*, 356) of "divine coherence" (=356) as the "whole coherence" (=356), "opposing" (*Virodhakri*, 18) the "right" (=71) to "soul coherence" (=356). "Divine's" (=360) "extrinsic coherence" (=356) "lives" (=14) as the "negation coherence" (*Anasamgati*, 730) of the "quantum noncoherence" (*Anasamgati*, 730).

The "wholesome coherence" (*Onasamgati*, 531) of the "tricoherence" (=531) among the goalkeeper, the goal, and the whole is "weak coherence" (=531); the "sum" (*Padmaja*, 54) "exceeds" (*Atikrama*, 1) the "part" (=−1) due to the "illusion" (=1) of the "devil's" (=0) "thermodynamic-effect" (=1). The "Devil coherence" (=531) is "dispassionate coherence" (=531) after the "decay" (=31) of the "divine" (=360) into "dispassionate" (*Ona*, 860 =531-31+360) "potential coherence" (*Brahmasutra samgati*, 830). "Potential coherence" (=830) is "manifestable" (*Sutra*, 497) as the "mind-born creator" (*Brahma*, 59) of "coherence" (*Samgati*, 150) with a "dispassion" (*Onata*, −10) for the "potential" (=18) of "sentient" (=189) within the "illusionary" (=87) when the "passion" (*Vyasana*, 39) is "exceeded" (*Atikramita*, 10) due to the "reproductive potential" (=23) of the "multilateral" (*Bahuparshvika*, 6) within the "systemic" (*Gabhiravyapi*, 10). "Illusionary coherence" (=830) is "sentient coherence" (=830) within "systemic coherence" (=830) of "multilateral coherence" (=830).

"Passionate coherence" (*Bhavasamgati*, 27,000) of a "part" (=-1) as a "Satan" (-1) is "strong coherence" (=27,000 =900*360/12), "varying" (=−19) as a "function" (=900) of the "mood" (*Bhava*, 360) of the "creature" (=12) as a "system" (=12) with "potential" (=18) to be "multilateral" (=6). "Satan coherence" (=27,000) is "mood coherence" (=27,000), satiating "system coherence" (=27,000 =[58-27]*100*10) by "imagining" (=58) "unilateral coherence" (*Bauddha Samgati*, 58) of the "present coherence" (*Bauddha Samgati*, 58) for an "imaginary coherence" (*Bauddha Samgati*, 58) following the "decay" (=31) of "guider" (=100) with "action" (=10) by the "unilateral" (*Ekaparshika*, 18) to "activate" (*Jagrook*, 9) the "potential" (=18) of "sentient" (*Ojas*, 189). Imaginary coherence is "spiritual coherence" (=58); for the "mind" (=38) "imaginary" (*Kalpanika*, 20) is "spiritual" (*Saguna*, 20), "present" (=1,600) as the "spirit" (*Ruh*, 20) within the "soul" (=4) of the "guider" (=100)—the "primordial space" (=100).

"Exceeding" (*Atikramana*, 10^{19}) the "unilateral" (=18) is "bilateral" (*Dviparshika*, 10^{19}); the action activates thermodynamic-effect with its "unilateral-effect" (*Dviparshika*, 10^{19}). "Bilateral coherence" (*Anumatisamgati*, 360) is "mutual coherence" (=360) between the "mood" (=360) of the "creature" (=12) and the "effect" (=34) of the "divine" (=360) on the "creation" (=379 =370+9 =360+20-1) of a "dimension" (=370) of "past" (=9) "left" (=89) by the "spirit" (=20) of the "creator" (=578 =12+34+370+89+71+2) as "part" (=-1) of a "right" (=71) that is "mutual" (*Apasparshata*, 70) and, therefore "reciprocal" (*Paraspara*, 2). "Reciprocal coherence" (=360) is "continuous coherence" (=360); "cocoherence" (=360) is "primordial coherence" (=360) of "time" (=360) with the "continuity" (=50) of space" (=18,000).

"Primeval coherence" (*Satsamgati*, 69) is "true coherence" (*Satsamgati*, 69) since it is "discontinuous coherence" (*Satsamgati*, 69) of "causation" (=1) with the "discontinuity" (=200) of "time" (=360). The causation is "left" (=89) with the "spirit" (=20) of the "unilateral" (=18) as the "cause" (=18) of the "unilateral coherence"

(=58 =48+10) with "action" (=10) for "true" (*Sat*, 8) "oneness" (=48 =6*8) of the "lateral" (*Parshvika*, 890 =69+18+8-6) without the "lateral-effect" (*Bahuparshvika*, 6) of the "lateral coherence" (*Samgati*, 69).

"Cause coherence" (*Yogasamgati*, 75) is "false coherence" (*Yogasamgati*, 75) since it is "oneness coherence" (*Yogasmgati*, 75) of "entity" (=24) with the "infinity" (=90,000) of "causation" (=1) to "square" (=18,000) the "entropy" (=5) of "cause" (=18) with "oneness" (=48 =24+1+5+18) of the "sidereal" (*Parshvanath*, 890) within the "false" (*Asat*, 8) "sidereal-effect" (=54,000 =90,000-2*18,000) of the "universe" (=2). Sidereal-effect is false because it is the "strong-effect" (*Prakatikriti*, 54,000) of "Mother Nature" (=8) with the infinity of her creation. "Sidereal coherence" (=75) is "square coherence" (=75).

"Entity coherence" (*Dharmasamgati*, 63) is "triangular coherence" (*Dharmasamgati*, 63) since it is a "dimensional coherence" (*Dharmasamgati*, 63) of "Mother Nature" (=8), "self-projecting" (=3/2) the "creature" (=12) as a "dimension" (=370) of "her" (=13) "correlation" (=16) with the "creation" (=379) of the "creator" (=578) to "triangulate" (=3) her "action" (=10) as an "entity" (=24) with "his" (=13) "past" (=9) as a "para entity" (=19). "Religion" (=370) is a "dimension" (=370) for the "growth" (=6) of "Mother Nature" (=8) "herself" (=1) as a "tropical" (*Navvakari*, 385) "essential dimension" (=385). "Religious coherence" (=379) is "tropical coherence" (=385).

"Natural coherence" (*Prasamgati*, 850) is "circular coherence" (*Prasamgati*, 50) since it is a "close coherence" (*Prasamgati*, 850) of a "child" (=128) whose "contact" (*Phassa*, 220) with "Mother" (=4) is "intimate" (*Upamitra*, 13) and "circular" (=10,000)— when the mother is "natural" (*Svabhavika*, 270), the child is "supernatural" (*Asvabhavika*, 270). Mother's "loving" (=37) is "super" (*Vic*, 37), "transposable" (=37) "far" (=37) into "distance" (*Yojana*, 190) where "action" (=10) is "high" (*Uccha*, 180) "close" (*Samhatata*, 8) to the "limit" (*Seema*, 8) of "possibility" (*Hariti*, 128). "Super coherence" (=850) is "high coherence" (=850) of "distance coherence" (=850).

"Supernatural coherence" (*Abhisamgati*, 861 =34*32-3-190-34) is "parallel coherence" (*Abhisamgati*, 861) since it is a "fearless coherence" (*Abhisamgati*, 861) of a "father" (=3) with his "spouse" (*Hridyesha*, 121), "wishing" (=190 =121+3+34+32) for a "parallel-effect" (=34) of his "primordial oneness" (=32) with "both" (=10,000) "Nature" (=8) and "Mother" (=4).

"Line coherence" (*Pritisamgati*, 5,000) is "very high coherence" (*Pritisamgati*, 5,000) since it is "open coherence" (*Pritisamgati*, 5,000) of a spouse whose "wish" (=18) is to "twin" (=121) "herself" (=1) like a "deity" (=1) by "self-moderating" (*Priti*, 33) a "line" (*Rekha*, 497) that "opens" (*Udghata*, 70) "very high" (*Atiuccha*, 90) "exchange" (*Mahaspanda*, 269) of "supra" (=91) element—the "shared kind" (=91). "Self-moderating coherence" (=5,000) is "supra coherence" (=5,000).

"Point coherence" (*Prakritisamgati*, 375) is "deity coherence" (*Prakritisamgati*, 375) since it is the "character coherence" (*Prakritisamgati*, 375) of a "deity" (=1) whose "character's" (*Prakriti*, 485) "variability" (*Asthirta*, 485) "points" (*Bindu*, 10^{10}) to "co-existence" (*Sahastitva*, 10^{10}) with "everyone" (=180) who copies his "essential nature" (=8) as "transcendental" (*Para*, 81).

"Transcendental coherence" (*Sarisamgati*, 8,000) is "appropriate coherence" (*Sarisamgati*, 8,000) since it is "friendly coherence" (*Sarisamgati*, 8,000) of "everyone" (=180) who is "friendly" (*Bandhava*, 70) towards the "transcendental" (=81) as the "foundation" (=11) for her "sentient well-being" (=190) like a "lake" (*Sari*, 190), whose "appropriate" (*Upayukta*, 170) "destiny" (=-1) to be "holy" (*Pavitra*, 9) is the "ocean" (*Sagara*, 8). "Lake coherence" (=8,000) is "holy coherence" (=8,000), and "alienated" (*Anyadhi*, 8,000) as the "para coherence" (=8,000).

"Supreme coherence" (*Svasamgati*, 9,000) is "inappropriate coherence" (*Svasamgati*, 9,000) since it is "self-coherence" (*Svasamgati*, 8,000) of one who is "unfriendly" (*Abandhava*, 5,000) towards "everyone" (=180) "immanent" (=13,000) as a "continuity" (=50) within "self" (=8×10^{15}), which is "inappropriate" (*Pratiyukta*, 5×10^{96}) because "no one" (=11) is "supreme" (*Adhiraj*, 0) unless "anyone" (=90) is the "ocean" (=8) where "ALL" (=10⁶) "resides" (=75) as a "part" (=−1). "Ocean coherence" (*Svasamgati*, 9,000) is "immanent coherence" (=9,000).

Supreme coherence "slows" (=5,000) open "self-moderating coherence" (*Pritisamgati*, 5,000) because it is "unfriendly coherence" (=9,000) of one as the "divisor" (=1) "multiplying itself" (=50) for "accruing" (*Liladhara*, 50) "combinatorial explosion" (=50) of the "space itself" (=1,000) with its "vertical growth" (*Ijara*, 10). Vertical growth is "nonlinear growth" (=10) which with its longitudinal, "Power Law growth" (=10) "squares" (=18,000) "time" (=360) for the "continuity" (=50) of "all" (=10⁶) as a "copy" (=0³) of two to "surface" (=2,000) a "circular" (=10,000) with "circular growth" (=10,000), i.e., "combinatorial explosion growth" (*Vartula*, 10,000). The "circular" (=10,000) "paces" (=30,672,000) three, the "time multiplier" (=3), with the "Power Law" (*Artha*, 7) of "reality" (=7) for the "growth" (=6) of two within "time itself" (=111). Growth "animates" (=76) two with its "combinatoric growth" (=82). Combinatoric growth is "parabolic growth" (=82) which "triangulates" (=3) "causation" (=1) for the "discontinuity" (=200) of "all" (=10⁶) with the "entropy" (=5) of the "copy" (=0³) that "slows" (=5,000) "horizontal growth" (=70) within the "animate" (=76). Horizontal growth is "linear growth" (=70) of "causation itself" (*Vaishnavi*, 70).

"Constant coherence" (*Samsamgati*, 800) "with" (*Sam*, 170) "promotion" (*Samgati*, 150) of "maximum rotation rate" (*Ativada*, 800) is a "gauge" (*Pramapi*, 480) of "parallel growth" (*Sharanarthi*, 480) of "supra-conscious coherence" (*Samsamgati*, 800).

"Reproductive coherence" (*Padaprakarana samgati*, 80) is "half coherence" (=80) which "flattens" (*Chipata*, 80) two's "reproductive reality" (*Gurvartha*, 40) to "self-materialize" (=80) "line growth" (=80) of the "lateral edge" (*Ajjhattikani*, 80) with "lateral vertex" (*Bahirani*, 80) of the "intrinsic" (*Antarbhuta*, 78) "supreme conscious" (*Atimanasa*, 280) "present self" (*Hurupa*, 280). "Intrinsic coherence" (=80) is "supreme conscious coherence" (*Padaprakarana samgati*, 80) which "perpetuates" (=9) the "intellectual" (*Brahli*, 89) with its "intellectual coherence" (=80).

"Productive coherence" (*Samyogasamgati*, 38) is "mental coherence" (=38) since it is "para conscious coherence" (=38) of the "mind" (=38) with the "brain" (=38) by "exploding" (*Phootana*, 38) the "point growth" (*Samudacara*, 38) of the "incident" (*Prasamga*, 38). The incident is "promoted" (*Samyogasamgati*, 38) by the "purpose" (=38) with an "incidence" (*Padavrtta*, 970) of "motive" (*Abhiprerana*, 970) for the "origin" (=1,000) of zero as eight—"co-incidence" (*Sampadavrtta*, 970). Mental coherence is "incidental coherence" (=38).

"Physical coherence" (*Upasamgati*, 8) is "mate coherence" (*Upasamgati*, 8) since it is "primeval conscious coherence" (*Upasamgati*, 8) of a "mate" (*Mitra*, 132) who is "primeval conscious" (*Mitra*, 132) of the "physical" (*Sharirka*, 168), "self-luminous" (=12) within "co-incidental" (*Upa*, 16). "Self-luminous coherence" (*Upasamgati*, 8) is "co-incidental coherence" (*Upasamgati*, 8) of "triangular growth" (*Mahanitya*, 8).

"Luminous coherence" (*Vishvasamgati*, 10) is "etheric coherence" (=10) since it is "param conscious coherence" (*Vishvasamgati*, 10) of a "waker" (*Vishva*, -8) who "reorders" (=2) the "disorder" (=-8) "caused" (*Karita*, 3) by a "Wisher" (=0) whose "wish" (=18) is to be "luminous" (=13) as an "entity" (=24) with the "etheric" (*Mera*, 16). "Waker coherence" (=10) is "beneficial coherence" (=10) because the "past" (=9) of the "entity" (=24) "reproduces" (=78) the "present conscious" (*Sahakari*, 978) as "beneficial" (*Sahakari*, 978) for her to be "luminous" (=13) as the "quantum light" (=13).

Light coherence" (*Jivanasamgati*, 53) is "causal coherence" (=53) since it is "primordial conscious coherence" (*Jivanasamgati*, 53 =12+34+2+5) of a "creator" (*Srijaka*, 578) who is "primordial conscious" (*Acharya*, 578 =379+41+125+123-100) of the "benefit" (*Hita*, 125) from the "creation" (*Srijana*, 379) of "cost" (*Kharcha*, 41) of "creating" (*Srijati*, 123) a "creature" (*Srijankarta*, 12) by "gravitating" (*Dharmanishpatti*, -100) a zero's "life force" (=123) to "pile" (*Sthanu*, 34) two into five. "Light" (=180) of a "belief system" (=20) is "causal" (=9) for the "pile" (=34) to "tile" (=340) "two stones" (=340) into "five stones" (=8), with a "malefic" (*Abhichara*, -10^{1024}) "loss dimension" (*Nirartha*, "-1,000) of "three stones" (=-18), to be "limitless" (=1,024) as a "primordial greeter" (=16) of "lifelong" (*Jivana*, 8,000) "temporal value" (*Pulastya*, 500). "Malefic coherence" (*Jivanasamgati*, 53) is "limitless coherence" (*Jivanasamgati*, 53) because "primordial greeter coherence" (*Jivanasamgati*, 53) is a "lifelong coherence" (*Jivanasamgati*, 53).

By "self-projecting" (3/2) "Sun" (=21), a "goalkeeper" (=7) "lives" (=14) as "solar" (*Saur*, -1,000) to "assume" (=700) "patrilineal" *(Suryavanshi*, 258) as his "future" (=0 =21+7+14+700+258-1,000). A "primordial greeter" (=16) "adds three" (=3) to the "Moon" (=997) after the "discontinuity" (=200 =3+997+16-816-1,600) of the "lunar" (*Chandrama*, 816) with the "present" (=1,600) for "continuity" (=50) as the "matrilineal" (*Chandravanshi*, 1,250 =3+997+50).

6.4 The Path of Alienship Action.

The "path of alienship action" (*Kalpaniya Marga*, 900) is the "path of least action" (=900) to "guide" (=111) the "variability" (*Asthirta*, 485) of "lineage" (*Kutumba*, 104) "present" (=1,600 =900+485+111+104) within a "Wisher" (=0). The "lineage" (=485) "varies" (=60,000) when the "line" (*Rekha*, 497) of "primordial wisher" (*Akarta*, −8) "ages" (*Aayu*, 80), "chasing" (=600 =497+80+23) the "reproductive potential" (=23) of "nature" (=8) to be "reproductive" (=100) as an "ecosystem" (=92 =497+80-485). When "feminine" (=37) becomes "primordial" (=85), "repeated force" (=3,785) of "masculine" (=296) within the "universe of replications" (*Sharkaraprabha*, 3,785) "transforms" (=1) the "genealogy" (*Vansha*, 3,785 =296+1+3,500-12) of the "two odd forms" (=3,500) of the "initial spatial state" (=−12) "without entity" (=−12).

The "first odd form" (=51) is "heliocentric" (*Shesha Sharira*, 51) "future body" (*Shesha Sharira*, 51) as the "growing form" (=51) of "Sun" (=21) within a feminine, "causal body" (=30). The "second odd form" (=835) is "selenocentric" (*Snigdha Sharira*, 835) "past body" (*Snigdha Sharira*, 835) as the "grower form" (=835) of "Moon" (=997) that the "past reality" (=−3), of a masculine "astral body" (=3), damages" (=162). The "first odd form" (=51) is the "oneness" (=48) of the "astral body" (=3) with the "past reality" (=−3). "Present body" (*Nirvanic Sharira*, 32,223), the "third odd form" (=32,223=31,000+963+260), is the "medium" (*Ganga*, 963) for the "decay" (=31) of the "odd dimension" (*Varna dharma*, 260) for one to be the "causal body" (=30) of the "origin" (=1,000) of "femininity" (*Yoni*, 1,000).

The inner, "temporal body" (*Adhyatamakaya*, 83), the fifth odd form, "moderates" (=132) "decay" (=31) by "devouring" (*Bhakshana*, 184) the "future" (=0) with the "memory" (=35) of three, "repeated" (=5) with "two zeroes" (=−10), to "assume" (=700) the "fourth odd form" (=3,500)—the "assumed body" (*Nira Sharira*, 3,500). "Assuming body" (*Bahirdhakaya*, 169), the sixth odd form, is "water" (*Jal*, 169) as the outside, "spatial body" (*Bahirdhakaya*, 169). "Earth" (*Bhu*, 724) is the seventh odd form, the "metaphysical body" (*Adhibhautika Sharira*, 724)—the "assumable body" (*Adhibhautika Sharira*, 724). "Earth" (=724) is "assumable" (=800) as a "body" (=56) with "consciousness" (=4) of "whole" (=16).

"Primordial wisher" (=-8), the "eighth odd form" (*Nama sharira*, −8)—the "assumer body" (*Nama sharira*, −8), is a "non-doer" (*Akarta*, −8). She is "whole" (=16) without the "past force" (=−24) of "consciousness" (=4) guiding the "spirit" (=20) of the "diverse forms" (=1) of her "creative force" (*Uma*, 6). Each diverse form is "perfected" (*Durga*, 28) by the "assumer" (*Kartika*, 31) of a "name" (*Nama*, 285) to "animate body" (*Sharira*, 56) with "eight odd forms" (*Akala rupa*, 56) of "reality" (=7).

6.4.1 Understanding Primordial Wisher

A "primordial wisher" (=−8) has the "reproduction potential" (=23) to "subsist" (*Gevurah*, 15) with "devotion" (=46) to the "multiplier" (=3) of "reality" (=7) within the "primordial self" (=10) as the "primeval perpetuator" (=15) of the "potential" (=18) for the "eight odd forms" (*Sharira*, 56). Primordial self is "even form" (*Samarupa*, 10). Multiplier "triple even form" (=84) to "perpetuate" (=9) "twin even form" (*Ekantasushama Rupa*, 72) by "self-projecting" (3/2) "eight odd forms" (=56) as a "whole" (*Akala*, 16). The "whole form" (*Akala Rupa*, 56) is "copyable" (=25) by the "assumer" (=31) to "copy" (=0) the "future" (=0). The "eight odd forms" (=56) "project" (=31) the "future reality" (=21) of 0, 3, 6, 9, 12, 15, 18, and 21, whose "sum" (*Padmaja*, 54) is "copied" (*Anukrita*, 350) by the "present" (=1,600 =350*4+200) as 84, the "triple even form" (=84), with the "sentient force" (=4) of the "discontinuity" (=200) of two as the "half even form" (=2) from "Mother Nature" (=8). Mother Nature is the fifth, "potential even form" (*Satarupa*, 8), who copies "future growth" (*Satarupa*, 8) of the "sentient force" (=4) as the sixth "present even form" (*Svarochisha Rupa*, 4). The seventh "primordial even form" (*Rochisha Rupa*, 125) is the "origin of life" (*Dvandva Brahma*, 125) with the "duality within entity" (=125)—the eighth, "primeval even form" (*Nama Rupa*, 24). The eighth even form is the "reproduction form" (*Nama Rupa*, 24). It "names" (=285) the "form" (=100,000) of "multiplier" (=3) the "para time" (*Mahakaal*, 15) "incarnates" (=900) to be "self-luminous" (*Svarochisha*, 12) with "Mother Nature's" (=8) "sentient force" (=4).

Within "primordial oneness" (=32) of "Mother Nature's" (=8) "sentient force" (=4), "para time" (=15) is "mindful" (*Arya*, 68) because it "perpetuates" (=9) the "path of mindful" (*Arya Marga*, 8,000) for the "origin" (=1,000) of "one without" (=12) as a "guider" (=100) "beyond present entity" (*Guru*, 100) with "two zeroes" (=–10) of the "one within" (*Dvividha*, -10). One within is "mindful manifestation" (=–10) of "somebody" (-10) "guiding" (=100) "intellectual truth" (=19) of "one within the center" (*Purushottama*, 0) to be "one without" (=12). It produces "value normalization" (*Samya*, 125) of the "followership behavior" (*Sahaja Puta*, 16) of the two zeroes in "equilibrium" (*Samya*, 125) with the "leadership behavior" (*Nirmana Puta*, 16) of a zero.

"One without" (=12) is "heterogeneous" (=12). "Three zeroes" (*Indrejya*, 0) "transform" (=1) the "normative development" (=888) of the "follower" (=24) with the "primordial reality" (=-3) of the "leader" (=0) to "incarnate" (=900) as the "primordial self" (=10). The "follower" (=24) "animates" (=76) "primeval self" (=100) as "primeval" (=185) to "perpetuate" (=9) the leader's "past" (=9) as "primordial" (=85). "Proliferating" (*Iti*, 97) "primeval" (=185) "accelerates" (*Mahant*, 18,000) "strange repulsion" (*Vikarshana*, 269) of the illusionary "transient benefit" (*Sukha*, 40) of "guiding" (=100). "Entrepreneurship behavior" (*Pusalattu*, 97) "stops" (*Stambha*, 450) the "multiplier" (=3) from the "illusionary production" (*Maya*, 1) of the "mediator" (=5) whose "light" (=180) is "mediating" (=34) the imaginary, "mind-born liberator" (*Mukti Yoga*, 8) of the "primordial self" (=10).

"Astral oneness" (*Ojas Yoga*, 180) of "light" (=180) without "everyone" (=180) "presents" (=1,600) a mind-born, "nonlinear complexity" (*Parinditartha*, -3) to the "intellectual body" (=306) for the "exchange" (=128) of the "feminine" (=37) element within "everybody" (=−5) as a "leader" (=0). The leader is the "physical body" (=387) "materializing" (=−6) "mental body" (=381). The mental body is the "follower" (=24) of the "intellectual body" (=306). The intellectual body "minds" (=38) "illusionary benefit" (=40) of the "astral body" (=3). With the "linear ultrasound" (=130) of the illusionary benefit's "replication" (=90) within the "etheric body" (=957), the "theoretical benefit" (=957) of "guider" (=100) is "constant" (=697) within the "causal body" (=30). The "guider" (=100) "destroys" (=1) "physical body" (=387) to be a "liberated entity" (=286). A "liberated entity's" (=286) "divine energy" (=10) is "present" (=1,600) within the "intellectual body" (=306) as the "present reality" (=−2) of "ideal benefit" (=1,000). The ideal benefit is the "sentient strength" (*Manibhadra*, 1,000) of the "deity kingdom" (*Devaloka*, 1,000) which "weakens" (*Citraka*, 185) "over time" (=60) following the "path of worry" (*Moha Marga*, 750) of the "Emperor" (*Nabhirai*, 750) as a "mediator" (=5) of "causation itself" (*Vaishnavi*, 70).

An "Ideal's" (=1) "effect" (=34) on the "causation" (=1) "twins" (=121) "Wisher-effect" (=-16) as "everyone" (=180) makes "him" (=5) the "internet of intellects" (=10) for "proficient networking" (=36) of his "polluted air-effect" (=13) with their "present reality" (=−2). As a "star" (=2), the "Super Deity" (=2) "attracts" (=34) the "effect" (=34) of "infinite guiders" (*Veshya Dharma*, 7×10^{180}) from the "present reality" (=−2) of "everyone's" (=180) "entropy" (=5) within the "internet of intellectuals" (=10). Infinite guider is "eternalism" (*Sundarika*, 7×10^{180}) within the "group-effect" (*Samashti*, 7×10^{180}) of a "collectivity" (*Samashti*, 7×10^{180}), who is "giving" (*Prada*, 7×10^{180}) "divinism" (*Daivika*, 10^{29}) of its "individuality" (*Vyashti*, 10^{29}) for "receiving" (=127) the "reality" (=7) a "finite guider" (*Dvaiyoyoga*, 10^{24}) "forms" (=10^5) with her "geography-effect" (*Sundari*, 10^{29}).

A "form" (=100,000) is the "polluted divine-effect" (=10,000), i.e., "manipulation" (=10,000), of the "internet of minds" (=10), whose "proficient workforce" (*Samputa*, 38) as a "brain" (=38) is "medium" (*Ganga*, 963) for "proficient exchange" (=19) of their "effect" (=34) by a "definite guider" (=64). The "geography" (=476) of "inner conflict" (=308) within the "effect" (=34) "descends" (=25) "equanimity" (=95) of the "internet of bodies" (=14). It "pollutes" (=10,000) "divine-effect" (-3) of a "para guider" (*Mahaguru*, 10,000) to "promote" (=51) her "divine energy" (=10). Therefore, the "internet of bodies" (=14) becomes a "toy" (*Khil*, 631) of the "guider-effect" (=100), "manipulated" (*Manipalayita*, 80) by the "life behaviors" (=50) of a "group" (*Gana*, 387). It works by "automating" (=50) a "guider theory" (=50) for the "continuity" (=50) of two as one—the "worker deity" (=1), the "worker" (=1) "who" (*Kaun*, 7/16) "works" (=19) as a "substitute" (=7/16) and the "deity" (=1) "knowing" (=19) "what" (*Kya*, 7,16) works as a "complement" (=9/16). The deity's "holy spirit" (=1) complements the substitute with a "double copy" (=1) of the "original" (=14), rounded as an "infinite networking system" (=12) with the "trading-effect" (=26) of the "internet of intelligence" (=54).

The "internet of intelligence" (=54) "constructs" (*Tamir*, 54) a "complex" (=1,000) to "shadow network" (=3,000,000) "present behavior" (=3,000) with a "double copy" (=1) of "human force" (=53). "Human-effect" (=53) "doubles" (=18,000) the "citizenship-effect" (=189) of "servicing" (=47) the "guider-effect" (=100) to a "mindless" (=−9) "devoted wisher" (=−9) who "copies" (=0) the "wisher" (=0) for "freedom from primordial enjoyer" (*Sarasi Mukti*, 578). The "cycle of trading" (=58) is "perfected" (=28) by the "primordial enjoyer" (*Sarasi*, 140) for "flourishing" (*Sarasi*, 140) his "path of alienship" (*Param Samadhi marga*, 58) with "citizenship" (=170) of the "copy" (=0) he is "producing" (=17) as a "double copy" (=1) of reality. The "double copy" (=1) "strengthens" (=179) the "national-effect" (=9). The copy "weakens" (=185) the "international-effect" (=1) by "copying" (=1) the "enjoyer" (*Rasi*, 9) to be "primordial" (=85) to the "guider" (=100) as a "guider nymph" (=−4). Guider nymph is the "goal-in-motion" (=−4) with "absolute presence" (=−4) as a "primeval enjoyer" (=−4) of the "self-perpetuating" (=1/2) "primordial wisher" (=−8).

6.4.2 Transcending Primordial Wisher

A "primeval enjoyer" (=-4) "ascends" (=1) his "presence" (=374) as an "enjoyer" (*Rasi*, 9) at a fixed, "technological cost" (*Yojya*, 58) to his "capability" (=55) as a "primordial paternal" (=0) within a "fixed state" (=3,794) of "normative justice" (=3,794). A "primordial paternal" (=0) "stewards" (*Sthapanacharin*, 1,111) "path of technological cost" (*Yojya Marga*, 1,024) with "force" (=34) of "servant leadership" (*Sahayakta*, 53). By "delocalizing" (=97) his "unique well-being" (=10), a "servant leader" (*Sahayaka*, 10,000) "forces" (=34) a "self-managing consciousness" (*Niyatatma*, 169) of "guider responsibility" (=700) for "universal well-being" (=10^{1000}). By "servicing" (=47) his "personal essence" (=816) as "nescience" (*Avijja*, 9,000), a primordial leader "subsumes" (=119) "bewilderment" (*Avijna*, 9,000) as the "enjoyer" (=9) of "self-managing consciousness" (=169) of "para-consciousness" (=18) through "disassociation" (=187) from the "invisible hand" (=0).

By "continuing" (*Kriya*, -6 x 10^7) "association" (*Prasanga*, 38) with the "localizable" (*Shramika*, -6), the "actions" (=10) of a "mindful" (*Arya*, 68) "servant leader" (=10^4) are "globalizable" (=61) as the "reality" (=7). "Reality" (=7) "limits" (=8) "ignorance" (*Ajnana*, 1). "Ignorance" (=1) is a "dimension" (=370) of forward, "intellectual limit" (=12), "conditioned" (=80) by the backward, "mental limit" (=6) of "creature dimension" (=271). It "produces" (=1) "infinite continuing" (*Tamas Kriya*, 137) of "research" (=120) for "quantum correction" (=120) in the "discovery" (=51) of empirical, "scientific cause" (*Upasarga*, 16) while "groping" (=371) in the "darkness" (=180) of "devotion" (=46) to the "divine" (=360) as a "consumer" (=-1) of the "diverse realities" (=-10) of "suffering" (*Jara*, -10^{180}).

"Disassociation" (=187) "illuminates" (=150) conscious, "astral limit" (=10,000) as the valid "ontological cause" (*Apahrtabhara*, 20) for "inventing" (=146) "mind-born complexity" (=-3). The "global" (*Bha*, 185) "simplifies" (=13) the "knowable" (*Jneya*, 396) by "destroying" (=285) ascending, "etheric limit" (=20,000) of the "epistemological cause" (*Kartri*, 121) which "twins" (=121) the "ontological cause" (=20). The "end solution" (=10^{1024}) is descending, "causal limit" (=30,000) of the "axiological cause" (*Aditi*, 1,024)—the "perpetuating force" (*Aditi*, 1,024) of "skepticism" (=3,000) in the "present behavior" (=3,000). "Present behavior" (=3,000) "bewilders" (*Asava*, 130) a "skeptic" (*Adharma*, -10) who is "perforating" (=130) the "outflow" (=13) of "divine-effect" (=-3) as an "effluent" (=130) of "divine energy" (=10).

Divine-effect is "past reality" (=-3); its "present" (=1,600) is "known reality" (*Rachitartha*, 1,600). "Behavior" (*Vyavahara*, 710) is the "perceptual reality" (=710) of the "unknown reality" (=7) "leftover" (=710) from the "time" (=360) before "skeptic's" (=-10) "action" (=10) "centering itself" (*Vardhaka*, 2) as a "divine" (*Divya*, 360). "Centering" (=10) "generates" (=40) "ascending sense" (=100) of "limitation" (=16) in "infinite trading" (=160) of the "entrepreneurship benefit" (*Prapta*, 140) and "ends" (=140) a "skeptic's" (=-10) "legitimacy" (=140) as the "perfect one" (=130) who "universalizes" (=130) "mindlessness" (=130) by "officiating" (=130) "managed growth" (=130).

The "present life behavior" (=97) "generates" (=40) "infinite cost" (=45) of the "officer's" (=45) "clouded consciousness" (*Prajnatma*, 52), behaving like "God" (=5), with "infinite capability" (=97) like that of a "comet" (*Dhumraketu*, 97). The "comet" (=97) "recreates" (=97) "clouded consciousness" (=52) for "opening womb" (*Hiranyagarbha*, 19) of "clarified consciousness" (=19) "with fire" (*Dhumra*, 17) of "symbolic freedom" (*Sita Mukti*, 17) from the "goal" (=9). A "goalkeeper" (=7) "reorders" (=2) "God" (*Ishvara*, 5) element. "Continuing freedom" (*Kriya Mukti*, 11) of the "goalkeeper" (=7) from the "consciousness" (=4) of "God" (=5) is a "sign" (=11) of "fortune" (*Kismat*, -250) "breaking away" (=-20) as the "symbol" (*Sita*, 0) of "real" (=9). "Freedom from unreal" (*Bhram Mukti*, 13) "fulfills" (=9) the "wish" (=18) of "knowing" (=19) the "reality" (=7) of "divine" (=360) whose "symbol" (*Sita*, 0) is "Queen of life" (*Maya Sita*, 0) at "Eighth Infinity" (*Indra*, 0). The "eighth" (=12) becomes "infinity" (=90,000) as it "reproduces" (=78) "femininity" (=1,000) after it is "reproduced" (=12) by the "femininity" (=1,000) to "cycle" (=1/60) the "sign" (=11) of "reality" (=7) within the "concealed truth" (=-4) of "masculinity" (=53).

The "future" (*Chidi*: The Red Emperor, 0) "trades" (=20) polluted "red blood" (*Rudhira*, 151) of the "unreal" (*Bhram*, 89) by "sacrificing" (*Shamitra*, 87) the "real" (=9) for "twin coherence" (=356)—

- first, "soul coherence" (=356) of "real" (=9) with "unreal" (=89), and
- second, "wholesome coherence" (=531 =356+9+89+7+70) of "reality" (=7) with "wholesome soul" (=70).

"Wholesome coherence" (=531) is the "beginning" (=515) of "whole" (=16) as the "Godhead of light" (*Yang-Wang-Yeh*, 515) for "whole coherence" (=356) as a "beginner" (*Yanluo Wang*, 23) with "twin reality" (=49) of a "wholesome point" (*Yulü*, 961). The wholesome point "begins" (=13) "triple coherence" (=531) with the "origin" (=1,000 =961+13*3). "Triple coherence" (531=356+150+2+23) comprises the following. First, "two coherences" (=356), and second, "coherence" (=150) of two with "beginner" (=23).

"Light" (=180) is the "Lord of death" (*Yama*, 180), "centering" (=10) "death" (=18) while "living" (=10) with "mass consciousness" (=10) of "future" (=0) as the "Lord" (=0). "Light" (=180) is also "Lord of the underworld" (*Yama*, 180) within the "continuity" (=50) of two as the "future value" (*Bhavishya*, 2) of the "underworld" (=9,000). Future value "squares" (=18,000) the underworld to "incarnate" (=900) the "spirit" (*Ruh*, 20) of "space" (=18,000). The "Satan" (=−1) "copies" (=0) the "destiny" (=−1) of the spirit as "heaven" (=19). "Heaven" (=19) "begins" (=13) "primordial oneness" (=32) of the "origin" (=1,000) with "space" (=18,000) to "form" (=100,000) the "future value" (=2) into "New Lemuria" (=50,000 =32*1,000+18,000), the "universe of human entities" (=50,000).

The "whiteness" (*Mangala*, 102) of light as a "white star" (*Yama*, 180) "reproduces" (=78) two as the "star" (=2). The star "begins" (=13) to triple with the "multiplier" (=3) of the "white" (=10) to "fight" (=39) for the "reformation" (*Sudhara*, 16) of "L" (*Sudhara*, 16), the "whole" (*Akala*, 16), into "F" (*Gajanan*, 0) as the "beginner" (=23) of an "answer" (*Pratigad*, 39). The answer is the "logic" (*Nyaya*, 1) of "justice" (=1) within "mind" (=38). The "potential light" (*Yamadeva*, 1,000,000) of "white" (=10) "forms" (=100,000) "two values" (*Yamadeva*, 1,000,000): first, "light" (=180) as the "value" (=180), and second, "Y-shaped" (*Sphya*, 820) "tiling" (*Kaval*, 820) of a "large tile" (*Kaval*, 820) of "eight tiles" (*Kaval*, 820) as "versed value" (=820), The "versed" (=0) "copies" (=0) the "origin" (=1,000) with which it is "well-versed" (*Abhijna*, 96) to "live" (=14) as a "community" (*Samvarna*, 1,964). "Versed value" (*Asprha*, 820) is "community value" (*Asprha*, 820).

The "upstream" (*Urdhvasota*, 17) "illuminated direction" (*Urdhvasota*, 17) of "feminization" (*Nilanjana*, 18) of the "animate primordial maternal" (*Shani*, 18) "produces" (=1) a "downstream" (*Adhassota*, 17) "shadow direction" (*Adhassota*, 17) for "masculinization" (=257) of the "potential light" (*Yamadeva*, 1,000,000) of "reality" (=7) within "white" (=10). White's "reproductive force" (=100) is reproductive within a multiplier. By "reproducing" (=17) "life planning" (=17) that the "animate primordial maternal" (*Shani*, 18) "produces" (=1), the "multiplier" (=3) "works" (=19) as the "equalizer" (=290) of the "life's" (=4) "formative justice" (=290) while "working" (=813) to "perpetuate" (=9) "afterlife's" (=8) "transformative justice" (=813 =290*3-17-19-4-8-9).

The "multiplier's" (=3) "kingship" (*Utu*, 0) is the "deliverer" (=36) of "liberation" (*Mukti*, 17) from "eight bodies" (*Inanna*, 130) with a "realization" (=16) that the "king" (*Ba'al*, 0), as a "leader" (*Chochmah*, 0), "enjoys" (=7) "almost perfect" (=58) "absolute coherence" (=58) for "servicing" (=47) "afterlife deliverance" (*Simchah*, 123) from "five bodies" (=11) that are the "foundation" (=11) of "life" (=4). The five bodies are the causal, etheric, astral, mental, and intellectual bodies. They are repeated as the physical body to "twin body" (=56) with the central and the lower bodies. "Liberation" (=17) "twin bodies" (=56) the "upper body" (=73) for "descending air" (=73) of "gone" (=73) "primordial past" (=73) and "kindles" (=73) the "future" (=0) with the "existent" (=73). Thus, "king" (*Ba'al*, 0) becomes "God" (=5) of normative, "raining justice" (=3,794) and "thunderous change" (*Vairibhu*, 32) in "weather" (*Mausama*, 94) with "hatred" (*Dvesha*, 21) for the "existent" (=73)

The "glow" (=17) of "symbolic freedom" (*Sita Mukti*, 17) of the "body" (=56) is "confidential" (*Sahasya*, 800,000 = [17+56+73+14] *5,000); it "lives" (=14) "within" (*Eke*, 5,000) as an "unfriendly"

(=5,000) "staircase" (=5,000) to "future" (=0) by "emaciating" (=5,000) the "existent" (=73) to be "subsequently" (=800,000) "nourished" (*Pausha*, 800,000) as "December" (*Pausha*, 800,000), the "tenth month" (*Pausha*, 800,000). The "tenth month" (*Pausha*, 800,000) "copies" (=000) the "month" (*Masa*, 360) as a "friend" (=360) for "primordial technological investment" (*Stip*, 800,000) to be "harvestable" (*Lavaniya*, 440) as the "future" (=0) with the "motion" (=360) of the "Capricorn zodiac" (*Makara*, 80)—the "universe of departed souls" (*Makara*, 80).

6.5 The Path of Worship Action.

The "path of worship action" (*Uma Marga*, 67) is the "path of least deity" (=67) "sequenced" (=61) through "worship action" (*Uma*, 6) of a "least deity" (=6) within the "reality" (=7) of "most deity" (=-1). "Most deity" (=-1) is the "essential" (*Prakriti Yogi*, -1) element "within all" (=-1). It is a "half stone" (=-1) "self-produced" (=-1/2) by a "star" (=2) by "feeding" (=17) a "stone" (=19) to its "double copy" (=1). The "double copy" (=1) "copies" (=0) the "action" (=10) of the "star" (=2) to "triple copy" (=3) the "reality" (=7) of "Sun" (=21) as an "assumer" (=31 =10+2,1)— "Cepheid" (*Kartika*, 31). Cepheid is "Type I Cepheid" (=31)—"Type I" (=8) "unforms" (=8) the "triple copy" (=3) within "Cepheid" (=31) into a "triple-double copy" (=31) that "double copies" (=1) a "copy" (=000) to "form" (=100,000) a "devil" (=0) into a "deity" (=1).

The "double copy" (=1) that "transforms" (=1) into a "star" (=2) is "Type II Cepheid" (=12)—"Type II" (*Taganem*, 2) "continues" (*Taganem*, 2) as a "Cepheid" (=31) for the "organization" (=29) of "Type IV" (*Todanem*, 2) as a "Type IV Cepheid" (*Rishi*, 8)—the "past" (=9) the "double copy" (=1) "discontinues" (*Todanem*, 2). Type IV Cepheid is "double-mode Cepheid" (=8)—the past is "double mode" (*Bhuta*, 9). Double mode is "least" (*Nyuntam*, 9); it "doubles" (=18,000) the "mode" (=40) to "square" (=18,000) "single-mode Cepheid" (=13) by "consuming" (=169 =13*13) "Type III" (*Akala*, 16) as a "triple copy" (=3) with the "past" (=9) of "Type II Cepheid" (=12). "Type III Cepheid" (*Tava*, 13), a "single-mode Cepheid" (=13), is an "anomalous Cepheid" (=13)—the "action" (=10) is the "anomalous" (*Visangata*, 10) as it "triple copies" (=3) "Cepheid" (=31) with a "double copy" (=1) of a "copy" (=0). The copy within Type III Cepheid is a "boson star" (=13). The boson star is a semivariable star that "copies" (=0) Type III Cepheid as a "dark companion" (=13) to "orbit" (=2) as a "dwarf planet" (=13) the "source" (=1,000) of "two zeroes" (=-10) with its "double copy " (=1). A "variable star" (=7) copies "Type II Cepheid" (=12) as a "heavy companion" (=7) after the "entropy" (=5) of "Type I Cepheid" (=31) to "reproduce" (=78) "itself" (=121) as a "giant planet" (=7). A massive, "constant star" (=9) copies "Type I Cepheid" (=31) as a "light companion" (=9) with the "potential" (=18) to "dwarf planet" (=13) with the "reproductive reality" (=40) of its "past" (=9) as a "potential dwarf planet" (=9).

As "time" (=360), "Type IV" (=2) "discontinues" (=2) the "light" (=180 =2*73+34) to "triple copy" (=3) as a "companion" (=76 =3+73) the "effect" (=34) of the "existent" (=73) into the "consciousness" (=4) of its "primordial phase" (Pravastha, 7) as a "variable star" (Khanijavarga, 7). "Type II" (=2) is the "past" (=9) that "perpetuates" (=9) as a "variable star" (=7). "Type III" (=16) is the "future" (=0) the "past" (=9) perpetuates as a "variable star" (=7). "Type I" (=8) is the "present" (=1,600), "reproductive" (=100=8*9+8+16+4) with the "consciousness" (=4) of "past" (=9) as "Type III" (=16),

"Variable star" (=7) "reincarnates" (=240) as a "friendly" (Bandhava, 70) "Geminid" (=70) after the "ten-fold growth" (=170) of time's triangular, "dimensional coherence" (=63) that it perpetuates as the "past" (=9). "Geminid" (=70) "produces" (=1) a "symmetrical light curve" (=71) with the "thermodynamic-effect" (=1) of the "double copy" (=1), "symmetrical" (=98) with the "light" (=180 =71+98+2+9) "Type IV" (=2) "discontinues" (=2) to "continue" (=2) the "past" (=9) as "Type II" (=2). The "light" (=180 =85+9.5) "curves" (Vakra, 85) the "past" (=9) with "entropy" (=5) of its "luminosity" (Vrittata, 5), i.e., "curvilinearity" (Vrittata, 5), into a "horizontal order" (=5). "Luminosity" (=5) "produces" (=1) "Cepheid" (=31 =5+1+5*5) with a "square" (=18,000 =[31+5]*500) to "time value" (=500) of the produce, the "causation" (=1). "Future value" (=2) of "Type II" (=2) "twin causation" (=2) within "Cepheid" (=31) into an "asymmetrical light curve" (Prachala Vakra, 33), "asymmetrical" (=18) with the "light" (=180) due to the "darkness" (=180) of the "action" (=10) the "curve" (=85 =2+31+33+18+1) "produces" (=1) with its "past" (=9).

"Darkness" (=180) of the "past" (=9) "produces" (=1) "Type I" (=8) as a "symmetrical face" (=1729 =180-8, 9) of the initial "pulsation period" (=1,729) of "space" (=18,000) as a "Geminid" (=70 =7,0). The space "squares" (=18,000 =2*9,0,00) the "future" (=0) with a "double copy" (=1) to "twin causation" (=2) for the "variable star" (=7) to "perpetuate" (=9) the "past" (=9) with its "copy" (=0). "Geminid" (=70) "produces" (=1) a "whole wave" (=1,800 =70+1+1,729) of "gravity" (=629) with the "initial period" (=1,729) its "action" (=10) "assumes" (=700 =629+70+1). The "action" (=10) is "preconceived" (=135) by the "lifetime" (=125) of its "luminous form" (=125) as a "stripe" (=125) whose "rhythm" (*Laya*, 130) "resonates" (*Anunada*, 5) in "horizontal order" (=5). "Linear dimension" (*Margi Dharma*, 78) is "parametric" (*Prachalika*, 78); it "orders" (=2) "rhythmic" (*Layatmaka*, 52) "resonance" (*Mridanga*, 3) as "Parametric resonance" (=135) with a "perfect rhythm" (*Layanalika*, 1,200 =1,350*10-150) of "four measures" (=150) of "error" (=10) in "action" (=10).

A "unilineal" (=800) "striped pattern" (=800) is "assumable" (=800) by a "measurement" (*Kashtha*, 800) for "maximal information packing" (=800) into a "winding" (=800) "spatial loop" (=800) without "escalating cost" (=800). "Maximal" (=-6) is "scientific" (*Nirguna*, -6) and "dimensionless" (*Nirdharma*, =-6); it "packs" (=45) "information" (=1) of its continuous, "gravitational reality" (=40) by "packing" (*Katala*, 8,000) the "discontinuity" (=200) of two into a "graviton" (=100) with the "gravitational" (=100) element of "information packing" (=80). The "action" (=10) element of "maximal information packing" (=800) "resonates" (=5) for "packing" (=8,000) "fifteen gravitons" (*Ranga*, 15) into a "color potential" (*Ranga*, 15) as an "anti-gluon" (*Ranga*, 15). A "gluon" (=149) is a "color glass condensate" (=149) which "colors" (*Raga*, 250) "glass condensate" (=231), a "physical wormhole" (=231 =149+82), with "blackhole" (=82) of the "infinite space" (=19). "Color" (=250) is "Boson" (=250). Gluon is "guider potential" (=150). Anti-gluon is "boson potential" (=15) which "produces" (=1) "potential boson"

(=14), i.e., "potential color" (=14). Potential color's "universal limit" (=14) is the "thermodynamic limit" (=14) of the "thermodynamic-effect" (=1) of the "graviton" (=100) as a "guider" (=100).

A "guider" (=100) "slits" (=18,000) a "hole" (=18,000) for "light" (=180) in "space" (=18,000). With "guider's" (=100) "action" (=10), the "slit" (=18,000) "surfaces" (=2,000) "infinite holes" (=200,000) with "infinite slits" (=200,000) through the "replication" (=90) of "light" (=180) as a "star" (=2). The infinite slit is the "causation slit" (=200,000). The "two zeroes" (=–10) on the "surface" (=2,000) of the "infinite slit" (=200,000) are a "finite slit" (=–10) without the "star" (=2) that "gravitates" (=–100) "everyone else" (=–10). The finite slit is a "spatial slit" (=–10); it is a "parallel slit" (=–10), "parallel" (=80,000) "packing" (=8,000) "temporal slit" (*Angana*, 6) as a "whole" (=16). The temporal slit is a "double slit" (=6) with a "twin hole" (=6); it "devours" (=135) the "whole" (=16) of "action" (=10) as a "supermassive" (=135) "infinite future" (*Mangala*, 102), "springing ahead" (=102) like "Mars" (=102). "Infinite future" (=102) "lits" (*Prabhavita*, 1) a "copy" (=0) of the "future value" (=2) to "be" (*Ho*, 1) "present" (=1,600) as "S" (=1,600) and "slits" (=18,000) the "discontinuity" (=200) of two, the "future value" (=2), to "order" (=2) the "action" (=10) of five—the "horizontal order" (=5).

The "profile" (=123) of the action's "angle" (=10) is "angular" (*Kona*, 9); it "goes" (*Jaana*, 10) "past" (=9) to perpetuate the "joy" (=123) of the "enjoyer" (*Rasi*, 9) of the "future value" (=2) of its "infinite future" (=102). The "action" (=10) "transposes" (=-1) the "future" (=0) and "interferes" (=-1) with the "past" (=9). "Guider interference" (=8) of "action" (=10) is "constructive" (=16) "thermodyanmism" (=16). It orders the "dynamism" (=16) of the "future value" (=2) as a "whole" (=16) with the "potential" (=18) within a "photon" (=20). The photon is the "spirit" (=20) that "guides" (=111) "four genera" (=55) of "action" (=10) with "time itself" (*Waheguru*, 111) by "wiggling" (=180) a "mind-born creator" (=59). The "mind-born creator" (=59) "creates" (=35) an "electron" (=365) with the "horizontal order" (=5) of "time" (=360) to "color" (=250) the "potential color" (=14) with the "symmetry" (=250) of "two colors" (=2). "Space" (=18,000) "cultivates" (=9,000) "space-bound" (=9,000) "two colors" (=2) as the "gluonic" (*Brihadbalika*,

2) with a "gluonic gravitational form factor" (*Mahakaya*, 18,000). It "packs" (=45) "gravitational form factor" (=45) into the "gravitational" (=100) element for "self-projecting" (=3/2) the "form factor" (=150) as a "proton" (=150).

A proton is the "spacing" (=150) of six that "wiggles" (*Uttarasucha*, 6) a "wig" (*Pallava*, 25). The wig is the "radian" (=25) with "three radii" (=25)—point, scalar, and mass. Radian is "charge radius" (*Pallava*, 25), the "square meter" (=25), that "charges" (*Sarj*, -8) the "point" (*Bindu*, 10^{10}) within "radius" (*Tribhajya*, 10^{10}) of future as a "meter" (*Tribhajya*, 10^{10}) which "squares" (=18,000) "action" (=10) with the "light" (=180) of the "color potential" (=15). The "action" (=10) "copies" (=0) the "scalar radius" (=-10) of the "past" (=9) within a "double copy" (=1). The "past" (=9) is "circular" (=10,000) within the "mass radius" (*Samvatsara*, 90,000) of the "two halves" (=90,000) of the "two-dimensional reality" (=400) of the "horizontal order" (=5) of the "present" (=1,600). With the reproductive element, the present "pieces together" (=1,500) a "colorant" (=1,500) to "color" (=250) the "two-color" (=2) "ant" (=2) with a three. The three is the "light body" (=3) of one who "transposes" (=-1) its "mass" (=132) for "primordial oneness" (=32) with the "future" (=0). The charge radius is "parabolic radius" (=25). The scalar radius is "linear radius" (=-10). The mass radius is "curvilinear radius" (=90,000). Radius is "point radius" (=10^{10})—a "discolorant" (=10^{10}).

"Discolor" (=14) is the "summation" (=14) of "two colors" (=2) that "fold up" (*Obhanjati*, 12) "primordial resummation" (=18) of "four colors" (=4) for "resummation" (=15) of "three colors" (=3) into a "sum" (*Padmaja*, 54). Sum is the "future of everything" (=54), "summing" (=16) the "essence" (*Sara*, 16) of "Geminid" (=70). "Resumming" (=160) "causation itself" (=70) with a "renormalon" (=90) "reproduces" (=78) the "linear dimension" (=78) of supernatural, "spatial growth" (*Mahanitya*, 8) within a "normalon" (*Samanayika*, 78). "Spatial growth" (=8) is the "Ubiquitin chain" (*Brahma Yanjni*, 8), a "shortened chain" (=8) that "discolors" (=14) "horizontal order" (=5) by "degrading" (=581) "space curvature" (=1). Space curvature is a degradable "twin chain" (=1) of a degraded, "fatty acid chain" (=1/2) that "twin colors" (=2). "Twin color" (=2) is "degradable" (*Pataniya*, 57) into "triple color" (=3) when "perturbed" (*Vibhramita*, 60) by "causation" (=1) "itself" (=121). "Triple color" (=3) is "degraded" (*Patita*, 132) as it "folds up" (=12) "four colors" (=4). "Four colors" (=4) "degrade" (*Patayati*, -9) a "five-color" (*Gunasa*, -18) "longest chain" (*Pranasa*, -18) to "fill space" (*Gunasa*, =-18) with the "shortest chain" (*Turnasa*, -18) of a "five-face symmetry" (*Audavita*, 5) among "time" (=360 =-18*4*5*-1), its "three dimensions" (=945=-9*-1,4,5), and "causation" (=1) with "six colors" (*Adhogati*, -1). "Longest chain" (=-18) is the "gamma chain" (=-18), a supernormal, "branched chain" (=-18). "Shortest chain" (=-18) is a "delta chain" (=-18), a normal, "straight chain" (=-18).

The "chain" (=-1) of "six colors" (=-1) is a "degeneration" (=-1) of "seven colors" (=700) that "leave" (=290) an "elongated chain" (=991) of "eight colors" (=991) after "self-degrading" (*Svapatana*, 290). The elongated chain is a "foldable longer chain" (=991) that "triple chains" (=991) a "snowflake's" (*Himakanthala*, 991) "perturbative-effect" (=991) into a "perturbative series" (=991). "Seven colors" (=700) "foldcat" (*Billaka*, 900) a "snowball's" (=900) "nonperturbative-effect" (*Billaka*, 900) for "accelerating" (=700) the "branching growth" (=700) of the "throbbing chain" (*Parispanda*, 700) as a "train" (=700), thus "elongating chain" (=700).

The "degradation" (*Patana*, 169) of a "longer chain" (=173), the "alpha chain" (=173), of "thirteen colors" (=173) "amplifies" (=4) the "growth" (=6) of the "shorter chain" (=179), the "proton-proton chain" (=179), of "nine colors" (=179). The "long chain" (*Keshara*, -20), the "filament" (=-20), of "eleven colors" (=-20) "slows" (=5,000 =-20*250*-1) "degrading" (=581=[5,000/-20*-1]+250+96+4-20-1+2) of the "color" (=250) of the "six-color" (=-1) "chain" (=-1) into "four-color" (=4) "short chain" (=96) with "two colors" (=2). Short chain is "broken chain" (=96), "being energy" (=96) of "peptide" (=96). "Two colors" (=2) of "time value" (=500) are the "origin" (=1,000) of "protein" (=855) as an "unbroken chain" (=855) of "ten colors" (=1,000), "producing" (=17) "beta chain" (=17) as a "breakable chain" (=17) of "twelve colors" (=17) with "colorlessness" (=127) "breaking chain" (=1/2).

Formative Growth Paradigm

The "resurgence" (*Punarutthana*, 9,000) of a "break" (=9,000) in the "femininity" (=1,000) at the "origin" (=1,000) is a "powerfully large" (=10,000) "perturbative" (=10,000), whose "perturbation" (*Vibhraman*, 8,000) is "unfolded" (*Anobhagga*, 8,000) as an "anomaly" (*Visangati*, 8,000). The anomaly is "resurgent" (*Punaruddha*, 8,000) as a "quantum error" (=8,000) with a "rapid" (=8,000) mental "calculus" (=8,000) of the "quaternary" (=816) "perturbative power" (*Tiryag*, 816) that "perturbs" (*Vibhram*, 8) the "origin" (=1,000) as a "whole" (=16).

The "perturbating" (*Aavalikarana*, 38) "mind" (*Manas*, 38) "exploits" (=38) "rippling" (=38) by "wiggling" (=180) a "perturbative piece" (=180) from the "perturbable" (*Vibhrami*, 68) "branched flow" (=68) of the "mindful" (=68) "resurge" (=68) to "guide" (=111) a "resurged" (=111) "masculine gene" (*Samsarga*, 37). The "masculine gene" (=37), "perturbed" (*Vibhramita*, 60) by the "feminine gene's" (*Upadhi*, 37) "reproductive potential" (=23), "orbits" (=60) the "order" (=2) with "para order" (=20), "beyond all order" (*Brahma loka*, 20), for "self-projecting" (=3/2) an "exponentially-small" (=20) "nonperturbative" (=20) into an "octave" (=60) of "triple color" (=3). The "triple color" (=3) "mediates" (=13) the "imperturbable value" (*Akshobhya*, 737) of the "resurgeable" (*Punarutthaniya*, 7) "illusionary piece" (=7) within the "feminine gene" (=37) for the "illusionary production" (*Maya*, 1) of the "infinite" (=185) with the negative, "nonperturbative exponential" (=85) within the "finite" (=85).

"Technological cost" (=58) of "producing" (=17) the "finite" (=85) with "action" (=10) "limits" (=8) "horizontal order" (=5) for "technological exchange" (=24) of the "illusionary piece" (=7), "shortening chain" (=17) of "twelve colors" (=17) "downstream" (=17) from "5' to 3' direction" (=17). "3' direction" (=9,000) is "descending direction" (=9,000) "from" (=9,000) a "creature" (=12) "upstream" (*Urdhvasota*, 17), "arranging" (=73) the "existent" (=73) as the "guider" (=100) before "para wisher" (=−5) is "gone" (=73) with the "emotion" (=73) of the "feminine" (=37).

6.5.1 Understanding Para Wisher

A "Para Wisher" (=−5) is the "continuous path of infinite development" (*Rajas Marga*, 8,000) "present" (=1,600) as a "metric" (=−1) of "everything" (*Sarva*, -5). "Everything" (*Sarva*, −5) is a "present manifestation" (*Idam*, 3) of "All things" (*Sarvam*, −8). "Unique self" (*Param Riddhi*, 15,000) is the "effective value" (=15,000) of "doing" (=15,000) "everything" (=−5) as the "absolute manifestation" (*Idam*, 3) of the "origin" (=1,000) of the "metric" (=−1) for "metrication" (*Khalvidam Brahma*, 6,000) into a "primeval human" (*Khalvidam Brahma*, 6,000). The "origin" (=1,000) of "all" (*Sab*, 1,000,000) is the "perfect metrication" (*Sarvam Khalvidam Brahma*, 1,000) of the "Doctrine of Almighty Creator" (*Sarvam Khalvidam Brahma*, −1,000,000). A "Para Wisher" (=−5) is the "Almighty Creator" (=26) of "All things" (=−8) with her "infinite potential" (=90) for "replication" (=90 =5*[26-8]) of the "metric" (*Mahashunya*, −1) as an "absolute manifestation" (=3) of a "human" (*Naran*, 275) who "behaves like" (*Param shunya*, −1) "God" (*Ishvara*, 5).

The "constant potential" (=146) within a "human" (=275) is the "normative development paradigm" (*Sthira dasha*, 146) of an "organization" (=29) as a "guider" (=100). A "guider" (=100) "accelerates" (=18,000) "metric system" (=8,000) for "organizational profiting" (=10,000) from "many forms" (*Nanarupa*, 10,000) of "entity gravity" (=10,000). "Self-replication" (*Ashvini*, 120) is the "admission" (*Arthapatti*, 120) of "conscious death" (*Sallekhana*, 120). The "dejection" (*Khalvida*, 120) in a "human being's" (=1) "spirit" (=20) is the "site" (=120) for "seeding oneness" (*Nirbijayoga*, 120) by "asserting" (=120) "fidelity" (=120). It "deposits" (=120) "seeded oneness" (*Sabijayoga*, 120) of the "Human Godhead" (*Shivagati*, 120) as the "self-love" (=1) of the "intrinsic entity" (*Shivagati*, 120). A "shared discontinuity" (=120) "jerks" (*Kampa*, 120) "energy equilibrium" (*Samatola*, 120) for "quantum correction" (=120) through "research" (=120) by "everyone" (=180) "over time" (*Kaalmaan*, 60).

"Self-acceptance" (*Pushti*, 84) of "dejectedness" (*Khalvidam*, 84) by the "worshiped" (*Addhya*, 84) is the "transformable flame" (=84 =7*12) for "reality" (=7) of the "creature" (=12) to be

Formative Growth Paradigm

"self-luminous" (=12). The worshiped is "lining" (*Astara*, 84) the "infinite formation" (=84) of the "audience" (*Shrotagana*, 84) as a "twinner" (*Mahendra*, 84) by "servicing" (=47) "self-luminosity" (*Vatsalya*, 37) to the "collective" (=47) as a "polluted entity" (*Beeja Jagrat*, 84),

"Interconnectedness" (*Seva Yoga*, 28) of the "creature" (=12) as the "essence" (=16) of the "creation" (*Srijan*, 379 = 28+9,9) of his "workculture" (=379) "interconnects" (*Anuvigratha*, 10) the "work" (=19) of the "culture" (=9) element

- within the "feminine" (=37) as a "disjunctive" (*Voyogi*, 37) "masculine gene" (=37) that "disjuncts" (*Viyoga*, 28) "unreality" (*Chitraa*, 9) and
- without the feminine as the "subjunctive" (*Linlakara*, 37) "feminine gene" (=37) that "subjuncts" (*Lin*, 3) "reality" (*Artha*, 7).

A "sentient's" (=189) "interconnection" (*Anuvigrathana*, 189) is the "capability point" (=89) for "disjunction" (*Vibhaga*, 189) of the "guider" (=100) "interconnecting" (*Anuvigrahan*, 100) "causal consciousness" (=89) as "unreal" (=89). The capability point is "triple dark matter" (=89) within the "mind-born creator" (=59) who "triples" (=130) the "present" (=1,600 =40*[10+89-59]) with "dark matter" (=1,600 = [130+89-59] *10) to "interconnect" (=10) the "work" (=19) of culture with its "gravitational reality" (=40). "Triple" (=130) has the "capability" (=55) to be "dark" (=185). "Dark" (=185) "transforms" (=1) the "capability" (=55) of a "photon" (=20) for "self-projecting" (=3/2) a "causal body" (=30) with the "consciousness" (=4) of a "body" (=56). The consciousness "triple transforms" (=3) the "interconnect" (=10) "itself" (=121) into the "reality" (=7) of the "luminous" (=13). Triple dark matter is "dark photon dark matter" (=89).

"Freedom from unreal" (*Sarupya Mukti*, 35) "connects" (*Vigratha*, 36) the "metric" (=−1) with "Gemini" (*Mithun*, 70), the "causation itself" (=70), "descending" (=12) the "creature" (=12) as the letter "D" (=12) within the "mind-born creator" (=59 =70-[12-1]) to be "self-luminous" (=12). "Freedom from unreal" (*Sarupya Mukti*, 35) "develops" (=22) the "potential" (=18) for an "infinite path" (*Tamas Marga*, 53) to be "luminous" (=13). The infinite path is a "situational path" (=53) of the "situational's" (*Samsthaika*, 30) "reproductive potential" (=23) for "similar" (*Svarupa*, 86) "deification" (*Maha Riddhi*, 86). Situational is "laborious" (*Shramapada*, 30). The situational path is a "laborious path" (=53) for the "feminine body" (=30), the "causal body" (=30), to be "God" (=5) before the "absolute manifestation" (=3), the "masculine body" (=30).

"Situation" (*Samstha*, 900,000) "accultures" (=900,000) the "infinite volume" (=900,000) of the folded "Fermi surface" (=900,000) with the masculine's subjective "consciousness sphere" (*Neva Sanna Nassana Ayatana*, 900,000) for "easing" (=900,000) "para worship" (=900,000) within the "path of female" (=900,000). "Just a little" (*Tvarat*, 45) "labor" (*Shrama*, 45) to "catalyze" (=900,000) the "situation" (=900,000) "orders" (=2) a "collective" (=47) within a "female" (*Nari*, 10,000) by "servicing" (=47) the "growth" (=6) to an "individual" (=53) on the "situational path" (=53).

"Normalization" (=269) of "exchange" (*Mahaspanda*, 269) "pollutes" (=10,000) "normative development" (=888) of "female" (=10,000) as a "maternal" (*Matri*, 112) who "interconnects" (=10) her "knowing" (*Jnana*, 19) with the "knowledge" (*Jnatva*, 600) of each individual, "paternal copy" (=0). "Divinity" (=57) within the "beauty" (=888) of "normative development" (=888 =851+37) "jounces" (*Vikampa*, 831) the "time derivative" (*Vikampa*, 831) to "jerk" (*Kampa*, 120) the "future derivative" (*Kampa*, 120 = 1,20). It "skews" (*Kavati*, 81 =831) the "past derivative" (*Kavatata*, 13 =1,3) with the "present derivative" (=9,999 =888+120-9,9) of "both" (=10,000 =9999+1) within a "female" (=10,000) to "derive" (*Nirvac*, 851) a "feminine" (=37) from a "primeval maternal" (=17) after "producing" (=17) the "photon" (=20) as a "beauty quark" (=1).

"Beauty quark" (*Jugupsa*, 1) is "inherent shadow orientation" (*Jugupsa*, 1), "inherent" (*Vasudeva*, 75) as "shadow orientation" (*Mahadeva*, 9) of the "gesture" (*Samjna*, 52) "luminous" (=13) as a "nation" (*Rashtra*, 1) within the "beauty quark" (=1). "Gesture" (=52) of "reality" (=7) is "perceptual" (*Pratipada*, 20); it "luminous" (=13) as a "photon" (=20) within "primordial oneness" (=32) of the "national" (=32) element. "Shadow" (*Nasamjna*, 52) of gesture is "backlit" (*Nasamjna*, 52) as "silhouette" (*Nasamjna*, 52) whose "edge" (=8) "outlines" (=9) "three lines" (=9) of "consciousness" (=4) of "oneness with time" (*Jugupsa*, 1) within a "nation" (=1). Each "worker" (=1) within a "nation" (=1) "times" (=360) "oneness" (=48) with "entropy" (=5) of "reality" (=7) "self-luminous" (=12) within a "creature" (=12) in the "present moment" (=12). "Entropy" (=5) "collapses" (*Jugupsa*, 1) "consciousness" (=4) of "oneness" (=48) as a "nation" (=1) within an "individual" (=53) to "sum" (=54) the "shadow" (=52) of the "endoreproduction" (=2) into a "universe" (*Brahman*, 2). A "nation's" (=1) "reality" (=7) is "inherent" (=75) in the "entropy" (=5) as a "breakpoint" (=7) of the "para psychic linkage" (=179) of a "para perpetuator" (=179) of "common value" (*Padya*, 179), "lining" (=179) a "citizen" (=179) with the "dark force" (=179) of "absolute entropy" (=179). Without "growth" (=6) within an "individual" (=53), a "collective" (=47) "gestures" (=52) its "entropy" (=5) with the "continuity" (=50) of the "order" (=2) in the "universe" (=2).

"Clarification" (*Darshan Yoga*, 162) of "devastation" (*Pida*, 162) from the "path of deification" (*Maha Riddhi Marga*, 162) "clears" (=629) "damage" (*Hani*, 198 =162+36 =179+19) to the "channel" (=36). It "reorganizes" (=431) the "gravity" (=629) of the "citizen's" (=179) "voice" (=629) in the "nick of time" (=629) with "clarified consciousness" (*Prakashatma*, 19). "Freedom from reality" (*Samipya Mukti*, 27) of "transcendental" (*Para*, 81 =27+54) "sums up" (=54) the "mortification" (*Tapas*, 140 =81+54+5) of "real" (=9 =5+4) with the "sentient force" (=4) within "entropy" (=5).

The "path of mortification" (*Tapas Marga*, 805) is the "path of manifestor" (*Sadhana Marga*, 805 =16*5,5) to "manifest" (*Vyakta*, 16) the "essence" (=16) of "entropy" (=5) that "sums" (=54) "twin reality" (=49) as "mutable" (=49) and "confused" (*Bhranta*, 49). The "essence" (=16) is the "transformative exchange paradigm" (=8) which "pops" (*Phat*, 1) "disproportionate cost" (=8) "out" (*Bahi*, 8) of the "universe" (=2) as a "truth quark" (*Samaa*, 1) to "transform" (=1) "natural growth" (=6) into "supernatural growth" (=8). Pop is "space derivative" (=1)—the "space" (=18,000) for the "logic" (=1) of the "derivative's" (=6) logical "derived reality" (*Yuktartha*,

6) of "future" (=000) from the "entropy" (=5) of "reality" (=7) of the "false" (Asat, 8).

"Spatially-constant" (=482) "potential" (=18) that "pops" (Phat, 1) out "disproportionate cost" (=8) is "synchronous" (=1,810) with the "purposelessness" (=1,810) of "mass consciousness" (=10) "without oneness" (Niryoga, 1,810) with the "spatially-variable type" (=1,810) who "directs" (=91) "space-varying sister" (=1,200) as a "purifying force" (=1,200) "once for all" (Antim, 1,200). "Freedom from realization" (Sayujya Mukti, 98) of the "aftereffect" (=9,000) of "false" (=8) on "the space curvature" (=1) in "future" (=000) "perpetuates" (=9) the "cause" (=18) for the "creation" (=379) of the "feminine" (=37). A feminine is "conscious" (=189) of the "interconnect" (=10) with the "true" (Sat, 8) within the "simplicity" (Jiva, 2) of the "before-effect" (Uttaraphala, 10). "Before-effect" (=10) "copies" (=0) the "potential" (=18) of simplicity for the "endoreproduction" (=2) of two "double copies" (=1). A double copy "folds" (=1) a "four-layer earth" (Maidana, 1) "before" (=1) "perpetuating value (=5) of the "effect" (=34) of "twenty-layer" (Vaishnavi, 70) "copy potential" (=70) for "linear growth" (=70) of "causation itself" (=70).

6.5.2 Transcending Para Wisher

A "Para Wisher" (=−5) "qualifies" (Arhata, 4) the "simplicity" (=2) of her "perpetuating value" (=5) with the "continuity" (=50) of the "Path of Para Wisher" (Arhata Marga, 39) for the "discontinuity" (=200) of the "technological cost" (=58) of a "Wisher's" (=0) "reproductive force" (=100) "after" (=46) "emancipation" (Vimukti, 10,000) from the "primordial wisher" (−8). "Super freedom from the primordial wisher" (Vimukti, 10,000) makes the "potential" (=18) of a "devotee's" (Baladeva, 274) "pi of inaction" (=22/7) "reproductive" (=100 =18+24+22+36) to "channel" (=36) the "incarnation" (Avatara, 35) of a "person" (Vyakti, 1) as a "collapsion" (Dehana, 35). The collapsion is the "sentient base" (=35) that "creates" (Srij, 35) a "supreme position" (Parakramaya, 35) with the "path of devotion" (Bhakti Marga, 1).

As a "divider" (=1) of the "reproductive potential" (=23), a "person" (=1) "enjoys" (=7) "oneness" (=48 =71-23) of "pi" (=22/7) with his "reality" (*Artha*, 7) as one. One is the "causation" (*Hetu*, 1) for the "growth" (=6) of the "potential one" (=130) in "office" (=130) after "inaction" (=13) as a "Wisher" (=0) was "leading" (=20) to "entropy" (=5) of the "sentient force" (=4). "Super freedom" (*Vimutti*, 10,000) of an "officer" (*Khyati*, 45) in "office" (=130 =45+78+7) "reproduces" (=78) a "sequence" (=48) of "twenty-three emancipations" (=7) with "twenty-three spikes" (=7) in the "illuminating value" (*Prakasha*, 7) of the "emancipation-effect" (=7) by "self-projecting" (=3/2) the "causal body" (=30) with the "reproductive potential" (=23).

1) First, "super-freedom from primordial wisher" (*Vimukti*, 10,000) who makes the "causation" (=1) "circular" (=10,000) with her "action" (=10) to "shape" (=100,000) the "future" (=000) of the "Wisher" (=0).

2) Second, "super-freedom from primeval wisher" (*Samma Vimukti*, 14) who "causes" (=18) "equilibrium" (*Samya: Samma*, 125) in "lifetime" (=125) through "primordial oneness" (=32) of the "causation" (=1) with the "octave" (=60) which "reproduces" (=78) "potential" (=18).

3) Third, "super-freedom from para wisher" (*Deha Vimukti*, 10) who "qualifies" (*Arhata*, 4) the "metric" (=-1) of "oneness" (=47) by "servicing" (=47) the "primordial" (=85) element as a "body" (=56) that "self-perpetuates" (=1/2) the "divinity-effect" (=28) of the reality's seven-fold "discriminating consciousness" (=28) to "live" (=14) as an "entity" (=24).

4) Fourth, "super-freedom from supreme wisher" (*Panna Vimukti*, 82) who "trades" (=20) sentient "life force" (*Prana: Panna*, 123) for "spontaneous deliverance" (*Nirvana*, 123) of the "entity" (=24) from the "color" (=250) of the "metric" (=−1).

5) Fifth, "super-freedom from supra wisher" (*Ceto Vimukti*, 60) who "colors" (=250 =60+186+4) the "mind" (=38) with the "effect" (=34) of "devotion" (*Ceto*, 46) "before" (=1) "awareness" (*Bodha*, 186) of the "metric" (=−1) of "discernment" (*Vichara*, 8) of the "sentient force" (=4).

Super-freedom from supra-wisher "advances" (=60) "self-compassion" (*Atma Karuna*, 60) for "mating" (*Maitra*, 60) with a superior after "self-awareness" (*Atma Bodha*, 30) that one's kind is inferior but "loving-kindness" (*Metta*, 60) of the transcendental, present as a medium for psychic linkage between the inferior, imperfect past as a masculine and the superior, perfect future as a feminine.

In the Words of Santideva

(From Bodhicaryavatara)

"तस्मिन् शान्ते सुखदुःखे समः सुखं एकोऽवशिष्टः

एवं मैत्रीकरुणोदितो बोधिचित्तस्तथा निश्चितः॥"

"Tasmin shante sukhaduhkhe samah sukham eko'vashishtah

Evam maitrikarunodito bodhicittastatha nishchitah"

Once "spiritual well-being" (*Vasihishta*) is solved by seeking peace in happiness, therein sorrow levels happiness. Thus, "spiritual oneness" (*Bodhichitta*) is resolved by befriending oneself with compassion.

---

6) Sixth, "super-freedom from the infinite super wisher" (*Saddha Vimukti*, 73) who has an "effect" (=34) on the "faith" (*Shraddha: Saddha*, 59) of an "individual" (=53) "trading" (=53) a "belief system" (=20) with a "spirit" (=20) of "compassion" (*Karuna*, 20). Compassion is towards all living. All include one acting as a being, another reacting for becoming a being, and the other interacting with the two for breeding the being after becoming one. The nature of compassion is "non-violence" (*Ahimsa*, 8)—mitigating externalization of the costs on another for one's benefit through internalization of the other as the causation for one's action.

7) Seventh, "super-freedom from the super wisher" (*Prajna Vimukti*, 710) who "self-qualifies" (*Prajna*, 2,222)

"spontaneity" (*Sattva*, 1,111) of "deliverance" (*Unmukti*, -10^{1000}) with "action" (=10) to "guide" (=111) "causation itself" (=71) to be "compassionate" (*Karuni*, 160) towards "everyone" (=180) as a "metric" (=-1) of the "natural growth" (=6) of "universe" (=2). Everyone includes both living ones as well as nonliving ones, with the potential to live when another is compassionate in giving birth to the inanimate as an animate.

> **From *"Pravachansara"***
>
> "आइसी करुणा ज्ञानी की, द्रविया वस्तु मात्र की,
>
> ज्ञानी के हृदय में अर्पण कीजै, ज्ञान के बल से आत्मा की आत्मा को जानो॥"
>
> "Aisi karuna jnani ki, dravya vastu matra ki,
>
> Jnani ke hriday mein arpan kijiye, jnan ke bal se atma ki atma ko jano."
>
> A knower's compassion is for all living beings and non-living substances. Surrender yourself to the knower in your heart to know the strength of knowing the soul of the soul.

8) Eighth, "super-freedom from the wisher" (*Vijja Vimukti*, 740) who is "qualifying" (*Nirguna*, −6) the "localizable" (=−6) "vision growth" (*Kritartha*, -6) as "scientific" (*Vijna: Vijja*, −6) to "enjoy" (=7) the "science" (*Vijnana*, 47) of "localism" (*Payuvada*, 740) of his "devotion" (=46) as a "metric" (=−1). The consciousness of "super-compassionate" (=70) mediates localism with three forms of compassion—seeks blessings from another who is superior and has strengths one lacks, compensates for the blessings by showering compassion on all so that others enjoy the strengths gifted to one, and discerns the potential of the inanimate to gift to a compassionate what an animate is unable to give—consciousness of everything in the consciousness of the universe of living beings, since the inanimate is endowed with the convergent energy of all that consciousness.

9) Ninth, "super-freedom from the worker" (*Vidya Vimukti*, 250) who "knows" (*Vedi*, 164) the "time cost" (*Vidya*, 600) of the "path of action" (=86), while "working" (=813) as a "procurer" (*Vidya*, 600) of "knowledge" (*Jnatva*, 600) to "color" (=250) the "feminine" (=37) with the "potential" (=18) of "knowing" (=19) the "work" (=19) that "perpetuates" (=9) "action" (=10).

10) Tenth, "super-freedom from the knower" (*Kara Vimukti*, 870) who "taxes" (*Kara*, 23) the "reproductive potential" (=23) to "manifest" (*Vyakta*, 16) the "reality" (=7) of the "cost" (=41) of the "circulated" (=80) "intellectual property right" (=80).

11) Eleventh, "super-freedom from the manifestor" (*Shapa Vimukti*, 850) who "curses" (*Shapa*, 490) "another" (=98) for "profiting" (=378) from her "property" (=378) as a "bearer" (*Dharaka*, 378) of the "cost" (=41) of "doing" (*Karin*, 5,000 = [98+378+41-1-16] *10) "nothing" (*Adravya*, -1) to "manifest" (=16) the "time" (=360) to "action" (=10).

12) Twelfth, "super-freedom from the creator" (*Bhedya Vimukti*, 720) who is "destructible" (*Bhedya*, 1,428) as a "mortal" (=1,428) "before" (=1) he "orders" (=2) the "universe" (=2) with his "intrinsic potential" (*Atita*, -12) as a "creature" (=12).

13) Thirteenth, "super-freedom from the perpetuator" (*Abhedya Vimukti*, 135) who is "indestructible" (*Abhedya*, 375) "truth" (*Satya*, 375) of the "creation" (=379) which "perpetuates" (=9) her "soul" (=4) to "reincarnate" (=240) as a "thing" (=9) that "matters" (=158) like a "twin feminine" (=158).

14) Fourteenth, "super-freedom from the destroyer" (*Nimitta Ceto Vimukti*, 38) whose "prognostication" (*Nimitta*, 24) as an "entity" (=24) "destroys" (=1) the "qualification" (*Arhatgati*, 15) of the "action" (=10) to "fulfill" (=9) the "goal" (=9) her "oneness" (=48).

15) Fifteenth, "super-freedom from the illuminator" (*Nissarana Vimukti*, 690) whose "illumination" (*Dipamala*, 1) of the "thing" (=9) as the "goal" (=9) of "creation" (=379) is "fruitless" (*Nissarana*, 72) because the "essential nature" (=8) of a "thing" (=9) is "unique" (=72) and the "reproductive force" (=100), that "twins" (=121), is "cost-escalating" (=72).

16) Sixteenth, "super-freedom from the liberator" (*Akimcana Ceto Vimukti*, 71) who is "someone" (*Akimcana*, 10^9) whose "devotion" (=45) "matters" (=158) to "no one known" (*Akimca,*, 10^9) because she "liberates" (=10^{10}) "everyone else" (*Idanta*, -3) from the "memory" (=35) of her "action" (=10).

17) Seventeenth, "super-freedom from the devoted" (*Appamana Ceto Vimukti*, 9,000) who is "irrelevant" (*Appamana*, 10,000) as his "intellectual force" (*Appamana*, 10,000) "spins energy" (*Gunnisu shakti*, 10,000) for "gaining energy" (*Gunnisu shakti*, 10,000) from the "femininity" (=1,000) of a "female" (=10,000).

18) Eighteenth, "super-freedom from the past" (*Patippassaddhi Vimukti*, 851) that is "tranquilizing" (*Patippassaddhi*, 250) the "present" (=1,600) to "fetter" (*Padvisha*, 8) one's "future" (=000) with the "time cost" (=600) of the "past" (=9).

19) Nineteenth, "super-freedom from the present" (*Vikkhambhana Vimukti*, 59) that is "agitating" (*Vikkhambhana*, 380) at the "defilement" (*Ashuchitva*, 41) of the "present self" (*Hurupa*, 280) by the "linear-effect" (=280) of the "replication" (=90) of his "future" (=0) as a "replicant" (*Hurupa*, 280) with the "action" (=10) of the "guider" (=100).

20) Twentieth, "super-freedom from the future" (*Akuppa Ceto Vimukti*, −10) that is "steadfast" (*Akuppa*, −1) in its "devotion" (=46) to the "present" (=1,600 =1[46-1+15]0) by "self-projecting" (=3/2) the "para present" (*Apurna*, 15) for the "qualification" (*Arhatgati*, 15) of her "action" (=10).

21) Twenty-first, "super-freedom from the space" (*Animitta Ceto Vimukti*, 48) whose "retrospection" (*Animitta*, 69) into the "past" (=9) with its "mass energy" (*Prithvi Shakti*, 369) has a "destructive-effect" (*Kaal Sarp*, 369) of "losing energy" (=369) of "citizenship" (=170) to the "office" (=130), "officiating" (=130) the "future" (=0) to be "luminous" (=13) as the "officer" (=45) with the "national" (=32) element.

22) Twenty-second, "super-freedom from the time" (*Tadanga Vimukti*, 146) to "introspect" (=146) with a "constant potential" (=146) of the "temporal" (*Tandanga*, 360) "after"

(=46) the "alienation" (*Parakikarana*, 155) of the circular "temporal mass" (=155) of "meaning" (=155) of "devotion" (=46) to the "luminous" (=13).

23) Twenty-third, "super-freedom from the causation" (*Samuccheda Vimukti*, 7) for "exterminating" (*Samuccheda*, 138) the "causative" (=138) "aperiodic" (=138) "scalar time" (*Kalpakala*, 138) through "primordial oneness" (=32) of the "metric" (=-1) as a "guider" (=100) whose "reproductive potential" (=23) "decays" (=31) to "charge" (*Sarj*, -8) "All Things" (=-8) with a "cosmic stone" (=-8)—the "Primordial Wisher" (=-8).

A "Primordial Wisher" (=-8) "aggregates" (*Jama*, -7) "ancestral energy" (*Yama Shakti*, -1) with the "Param Wisher" (=-7). "Ancestral energy" (=-1) is the "energy" (=19) of the "white star" (*Yama*, 180) that "copies" (=0) the "charge" (=-8) for "spacetime singularity" (=10). Spacetime singularity is the "vacuum decay" (*Devata*, 10) from the "continuity-effect" (=10) of the cosmic, illusionary, "spin axis" (=10) of "time singularity" (*Bhagwan*, 4) that "lives" (=14) within "space singularity" (*Shalkana*, 6) as "spacetime" (*Saguna*, 20). "Time singularity" (=4) is the "spirit essence" (*Bhagwan*, 4), the "spirit" (*Ruh*, 20) "essence" (*Sara*, 16) whose "consciousness" (=4) "times" (*Kaal*, 360) the "eventual spacetime singularity" (*Khud*, 1) of left "masculine-effect" (*Khud*, 1) to "round" (=10) the "gender" (=9) of the "past" (=9) without the "ancestral energy" (=-1). Spirit is "imaginary" (=20) and "spacetime" (*Saguna*, 20). "Space singularity" (=6) is the "imaginary base" (=6) for "spacetime singularity" (=10), the "base" (=10) of "growth" (=6) in "time singularity" (=4). "Initial spacetime singularity" (*Rochisha*, 13) is the illusionary "cosmic base" (=13); it is the "cosmic beginning" (=13) of the "reality" (=7) within "spacetime" (=20) and the "imaginary end" (=13) of the "reality" (=7) within "space singularity" (=6). The "cosmic beginner" (*Sargam*, 60) is the "octave" (*Sargam*, 60) of three, the multiplier of "spacetime" (=20).

By "self-projecting" (=3/2) a "timer" (=50), "spacetime" (=20) is "summable" (=50) into an "extreme spacetime" (=50) as it "transforms" (=1) its "continuity" (=50) to "twin" (=121) "itself" (=121). By "servicing" (=47) its "past reality" (=-3) as an "object"

(=−3), "twin spacetime" (=50) "transforms" (=1) its "oneness" (=48) with a "double twin spacetime" (=83) as it "channels" (=36) "itself" (=121) as a "temporal body" (=83). The "transformer" (=79), which "transforms" (=1) "everything" (=-5) from one into four with a "time multiplier" (=3), is a "triple spacetime" (=79). As a "central body" (=169), a "topological" (=169) "twins" (=121) "itself" (=121) into an "inner body" (=83), a "topological soliton" (=83), "self-perpetuating" (1/2) its "quantum potential" (=186), the "soliton" (=186), for "producing" (=17) "future reality" (=21). The "inner body" (=83) "combines" (=83) the "endopoducing" (=83) of "present reality" (=−2) as a "stable defect" (=83) of the "endoreproduction" (=2) of the "subject" (=0) as an "observer" (=0). As a subject, "five spacetimes" (=0) are "repeated" (=5) within the "spacetime" (=20) for the "endoproduction" (=1) of "ten spacetimes" (=10) as "spacetime singularity" (=10) with a "consciousness" (=4) of "six spacetimes" (=4). "Continuity-effect" (=10) within "ten spacetimes" (=10) is an "oscillon" (=10), "concealed" (=1,800) within the "gravitational wave" (=1,800) of the "nine spacetimes" (=180). The "light" (=180) within "nine spacetimes" (=180) "decays" (=31) the "guider potential" (=149) of the "past time force" (=149) into a "gluon" (=149). "Eight spacetimes" (=100) are "reproductive" (=100) as a "guider" (=100) with a "potential" (=18) to "blackhole" (=82) "seven spacetimes" (=82) into the "potential spacetime" (=82) of a "new moon" (=82).

"Causation singularity" (*Manaa*, 12) is a ring-like, "rounded" (*Pravritta*, 12) "central singularity" (*Manaa*, 12) that "orientation orders" (=12) the para-round "electron singularity" (=12). Orientation order is "Nematic order" (=12) It "centers" (=16) the "essence" (=16) of "spacetime" (=20) for "oneness" (=48 =16*3) with three. Three is "space-time-causation singularity" (=3), and "rounding singularity" (=3). An "electron" (=365) is "para round" (=365) "perfect spherical" (=365), "rounding" (*Pravrtti*, 9) "singularity" (=1) for the "de-polarization" (=65) of "oneness" (=48) within the "electric field" (=46) into the "electric dipole moment" (*Svakshetra*, 9,000) by "wheeling" (*Vritti*, 196) a "path of clockwise journey" (*Vama Marga*, 30) "along" (=5) the "perpetuating value" (=5) of "time" (=360).

The "time" (=360 =[1/3* [200+2+15+9+12+1,600*7/16+142]) "handles" (=8,000,000) "three phases" (=8,000,000 =200³) of "discontinuity" (=200) in its "orbit" (=2) with the "handle-effect" (*Akarshaka*, 15) of the "past" (=9). It "hands" (=0) "future" (=0) the "whole-effect" (=12) of "present" (=1,600) with "who" (*Kaun*, 7/16) as a "substitute" (=7/16) to "twin orbit" (=142). With its "superposition" (=17), "triple orbit" (=17) "tugs" (*Khinchatani*, 150) the "reality" (=7) of a "proton" (=150) for the "organization" (=29) of the "magnetic field" (=29) like a "ball" (=16) "rotating" (=-12) the "magnetic-effect" (=-2) "along" (=5) the "continuity" (=50) of the positive, "electron spin" (=40). "Rotation" (=1,000) "discharges" (=-100) "electron's" (=365) "positive charge" (=20) to "loop" (=900) the "continuous" (=15) "time value" (=500) of "space" (=18,000) as "spacetime" (=20). "Spacetime" (=20) "orders" (=2) a "copy" (=0) of the "electron" (=365) with its "roundness" (=710 = 2*365-20). "Twin electron" (=246) is "observable" (=246) as the "pressure" (=62) of the "inner conflict" (=308) within the "creation cycle" (*Lakshmi chakra*, 246) that "unfolds" (=246) "electron radiation" (=246). The "copy" (=0) is the "mediator" (=5) of the "correlation" (=16) between "space-time-causation singularity" (=3) and "one sphere" (=8) of "nature" (=8). "Space singularity" (=6) "orders" (=2) "pressure" (=62) to "time" (=360 = [62-2] *[12+4,8]) "causation singularity" (=12) with the "time singularity" (=4) of "one sphere" (=8).

"One sphere" (=8) is the "unfulfilled subtle desires" (=8) of the "Universe of Wishers" (=-1) whose "ancestral energy" (*Yama Shakti*, -1) "trades" (=20) "white color" (=10) of the "feminine" (=37) from the "spacetime singularity" (=10). Ancestral energy is "white star energy" (*Yama Shakti*, -1). "White colors" (=10) "one sphere" (=8) into a "white star" (=180). "White stars" (=180) "twin causation" (=2) with the "mediator" (=5) of white color to "time" (=360) "electron" (=365) as "gas" (=365). A "feminine's" (=37) "luminous light" (*Sharkaraprabha*, 3,785) "generates" (=40) a "strong psychic force" (=54,000) of her "labor" (=45) as an "officer" (=45) "resolving" (=10) the "negative energy" (=19) of the "past desires" (8) with her "time value" (=500). Her "sentimental attachment" (=3,000,000,000) "generates" (=40) "etheric body" (=957) as "gaseous body" (=957) with the "space-time-causation singularity" (=3) of "Wishers" (=000). "Gaseous"

(=71) is the "maker" (=71) of "body" (=56) with "continuous" (=15). "Continuous" (=15) "weakens" (=185) the "psychic force" (=47) of "kinship" (Parijanata, 500) with the "weak psychic force" (=185) of "kithship" (*Goshthikata*, 111) for "discerning" (=753) the "discontinuous" (=1,111). "Discontinuous" (=1,111) "produces" (=1) "alienship" (=10^{17}-1) by "reproducing" (=17) "kithship" (=111) with the "epistemological value" (=10^{14}) of the "Wisher's" (=0) "ancestral energy" (*Yama shakti*, -1).

"Epistemological" (=125) is the "benefit" (*Hita*, 125) which "values" (*Mulya*, 180) "ancestral capability" (*Udana*, 55) of "ten lifetimes" (*Mukhya Kala*, 55) to "revolve" (*Avrit*, 55) "disproportionate social benefit-cost ratio" (*Mahatripurasundari*, 55) as the "ontological time" (*Mukhya Kala*, 55). "Ontological" (*Mukhya*, 14) is the "original" (=14) which "times" (*Kala*, 360) the "presence" (*Atthi*, 374) of "axiological essence" (*Pratibha*, 374) with "zodiac potential" (*Shri Garuda*, 374) of "three hands" (*Pachaka Dasha*, 374) of "time with space" (*Idhara*, 74). "Axiological" (*Prameya*, 28) is "measurable" (*Prameya*, 28) as the "essence" (=16) of the "system" (=12) "self-projecting" (=3/2) the "potential" (=18) for the "endoproduction" (*Maya*, 1) of "unfulfilled subtle desire" (*Iccha*, 8) as a "zodiac" (*Araka*, 1) into a "star" (=2).

A surviving "star" (=2) is the "variable zodiac potential" (=2) "left" (=89) "after" (=46) "death" (=18) by the "reproductive potential" (=23) as a "meta" (=178). "Meta" (*Adhi*, 178) is the "base" (=10) of the "physical" (*Sharirika*, 168) "photon potential" (=168). "Metaphysical" (*Adrishtartha*, 1869) is the "volume" (*Ayatana*, 1,869) of the "earthly realm" (*Bhu Loka*, 1,869). Zodiac potential is the "zodiac residual" (=374) of a "cosmic devotee" (=274) "compressing" (*Sampeedit*, 45) the "reproductive potential" (=23) of "ten characteristics" (*Param Prakriti*, 100) "over" (*Adhi*, 178) "ten lifetimes" (=55) within an "officer" (*Khyati*, 45). "Zodiac potential" (=374) has an "eternal" (=274) "unique characteristic" (*Param Prakriti*, 100). A "zodiac" (=1) "promotes" (*Garj*, 51) "replication" (=90) of its "zodiac potential" (=374) in "diverse forms" (=1) as it "exchanges" (=269) "sentient entity-effect" (*Vidheya*, 697) of the "astral universe" (*Yaanshala*, 14). An "astrological" (=12) "trades" (=20) "guider entity-effect" (*Radha*,

43) from the "zodiac" (=1) for "centering" (=10) the "perfected" (=28) "aggregated entity-effect" (=7) as the "epistemological cause" (=121) "constituting" (=118) "three zodiacs" (=3) into a "physical body system" (=121).

First, a "discontinuous zodiac" (*Purva Bhadrapada*, 18) "transforms" (=1) the "astrological's" (=12) "gravitational-effect" (=100) with the "zodiac's" (=1) "sentient-effect" (=4) to "discontinue" (=2) the "past-effect" (=24) on the "spiritual DNA" (*Trishira*, 39)—the "human homeobox gene family" (=39). Second, a "continuous zodiac" (*Purva Phalguni*, 80) "norms" (=18) the "spiritual DNA" (=39) with a "collective copy" (=3), "homeobox" (=3), of the "spirit" (=20) to "continue" (=2) the "present-effect" (=10^{1000}) with the "reproductive-effect" (=100) of the "family" (=10) within the "origin" (=1,000) of the "zodiac" (=1). Third, an "infinite zodiac" (*Vishakha*, 570) "forms" (=10^5) the "human" (=275) element with the "shared past" (=39) of the "proportionate forms" (=39) of the "whole" (=16) that is "reproductive" (=100) as a "family" (=10) to be "present" (=1,600) as the "origin" (=1,000).

"Each form" (*Lopamudra*, 47) "shares" (*Plus*, 92) "mortality" (*Jaramarana*, 18) of the "human" (=275) element, "servicing" (=47) the "spiritual stress" (*Dukkha*, 47) of "constituting" (=118) the "collective" (=47) with a "collective copy" (=3) of the "three zodiacs" (=3) to "discontinue" (=2) the "continuity" (=50) of the "continue gene" (=2). As "mortality" (=26) "downs" (=26) the "consciousness" (=4) with "T" (=4), "morality" (=70) of the "group supervisor" (*Ganadhyaksha*, 70) "ups" (=62) the "value" (=180) of the "council supervisor" (*Sabhadhyaksha*, 180). A "group supervisor" (=70) "supervises" (*Adhyakshana*, 70) the "reproductive part" (=70) that "tails" (=70) the "productive part" (*Bhaktanga*, 90) of the "genetic growth" (=160) of "shared disposition" (=160). A "council supervisor" (=180) "supervises" (=70) the "assembly" (*Kamma*, 130) of "people" (*Jana*, 62) to "head" (=-9) the "present priority" (=253) of the "leader" (=0). The "light" (=180) of the "council supervisor" (=180) is "exchanged" (=180) as a "gene" (*Somapa*, 180), "replicated" (=180) by the "spirit" (=20) for "growth" (=6) with the "quantum light" (=13) of "two supervisors" (*Anupuraka*, 13) into a "molecule" (=39) of "spiritual DNA" (=39).

The "leader" (=0) is the "decider" (*Nirnayaka*, 13) of the "twin reality" (=49) of "people" (=62)—the "follower" (=24), who is "backbiting" (=14) as a "devoted knower" (=14), and the "guider" (=100), who makes the leader's "reproductive quality" (*Guna*, 0) "reproductive" (=100). "Reproductive force" (=100) "limits" (=8) the "deity culture-effect" (=6 x 10^{192}) with the "discontinuity" (=200) of the "deity culture" (=9) of "reproducing" (=17) the leader as a "deity" (1) who "produces" (=1) the "foundation" (=11) for the "growth" (=6) of the "collective copy" (=3). "Conscious action" (=10) "prevents" (*Drakh*, 7) "helplessness" (*Asahayata*, 90,000) of the "experience" (=9,000) of "organizational sameness" (=18,000) within the "twin reality" (=49) of the "follower" (=24) as "God" (=5) with a "spirit" (=20) of "guider" (=100).

The "bond" (*Gantha*, 963) to the "scientist" (=963) as a "solution" (=963) is the "element" (=963) that "supports" (=963) the "gravitation" (=963) of the "deadline" (*Ayati*, 963) of "death" (=18) of "clarified consciousness" (=19). "Freedom from element" (*Tattva Mukti*, 10) is a "necessity" (*Prayojana*, 17) for one's "growth" (=6) into the "King of Gods" (*Pishachendra*, 17) to "celebrate" (*Hutabhuj*, 17) "equal gravity" (*Hutabhuj*, 17) of "Grand-I" (*Shivadrishti*, 17) for "omniscience" (*Shivadrishti*, 17) with the "fire" (*Agni*, 17) of the "twelfth infinity" (*Agni*, 17). "Grand-I" (=17) is the "conscious freedom" (*Mukti*, 17) from the "nonlinearity" (*Vrittata*, 5) of "I" (=12) as a "self-luminous entity" (=12), "proportioning" (=825) the "self-deifying" (=825) "deity-effect" (=825) with the "bare essential" (=175) "sentient energy" (=1,000) of the third "knower eye" (*Vrishna*, 825) of the "param deity" (*Shiva*, 7). "inherent" (*Vasudeva*, 75) as a "guider" (=100).

The "potential" (=18) for "eternal joy" (*Da'at*, 17) from the "false sense" (=-1) of "sentient freedom" (*Siddha Ratri*, 17) "ignite" (=14) the "fire" (=17) within a "deity" (=1) to be the "primeval paternal" (=17) by "centering" (=10) the "param deity" (=7) as the "multiplier" (=3) of the "paternal soul" (=4). By "reproducing" (=17) the "absolute wisdom" (*Chochmah*, 0) of the "leader element" (=0) as "theory-effect" (=0), one "activates" (=9) a "supernatural growth" (=8) of the deity's "ideal-effect"

(=1) as a "primordial solution" (*Binah*, 18) for "knowing" (=19) the "reality" (=7).

"Closed mind" (*Leigong* 雷公, 17) "bouncing" (=17) as the "Blue master of thunder" (*Leigong* 雷公, 17) "punishes" (*Ghatay*, 70) the one "guilty" (*Rinavant*, 170) of "feeding" (=17) its "kind" (*Dao*, 17) in "secret" (=-7) as an "ideal" (=1). "Blue" (*Nila*, 1,765) is the "script" (*Akshara*, 1,765) that "masters" (*Adhipa*, 116) the "thunder" (*Stanayitnu*, 7) of "reality" (=7) with a "twin reality" (=49 =165-116) of "ideal" (=1) and its "replication" (=90) within "mind" (=38) as a "theory" (=127). A "theory" (=127 =73+11/2*110-1) is an "ideal's" (=1) "template" (*Farma*, 73) to "piece" (=11/2) a "sequence of forms" (=110) "broken" (*Bhagna*, 91) from a "system" (=12) as a "subspace" (*Samavakara*, -3) of "right reality" (=-3) with a sentient, "Rödl nibble" (=9) of "left reality" (*Nashitartha*, 89) to "leave" (=290) a "hodgepodge" (*Golamaal*, 290) of "design" (=180).

The "subspace design" (*Kirata*, 59) is "savage" (=59) because it is "one with everything" (=59) "space designs" (*Sadhyagana*, 34) with its "theoretical self" (*Homa*, 25) as a "primordial knower" (*Homa*, 25). The "primordial knower" (=25) "bites" (*Adamsh*, 30) "all sides" (*Anandamaya*, 20) of the "superspace design" (*Hed*, 75). The superspace design "surrounds" (*Hed*, 75) the "four sides" (=75) of the "supraspace design" (=500) for "winding" (=800) the "central reality" (*Madhyartha*, 150) of "time value" (=500) into a "space loop" (*Vastupurusha*, 800). The space loop "nibbles" (*Vyad*, 800) one "itself" (=121) with the "peripheral reality" (=120) of its "spatial future" (=800). By "design" (=180), one "relates" (=80) with "everyone" (=180) as an "absolute" (=1,600) whose "spirit" (=20) "surfaces" (=2,000) the "space" (=18,000) with its "light" (=180) to be the "primordial space" (=100).

The "design's" (*Abhikalpa*, 180) "potential" (=18), "Z" (=18), "copies" (=0) an "individual" (=53), "E" (=53), as a "deity" (=1) for "centering" (=10) the "consciousness" (=4), "T" (=4), of a "para deity" (=5) to "cube" (=90) "future value" (=2), "K" (=2). When Z=0, E=1, and T=10, K=3 "cubes" (=90) "essential nature" (=8), "para space design" (=8 =0+1+10-3) for "changing" (=69), with "primeval space design" (*Diptamurti*, 69) of a "N star" (=69), the "peripheral reality" (=120 =18+53+4+2+1+10+3+8+90-69), "N"

Formative Growth Paradigm

(=120), into a star's "octave of copy" (=60). The "octave" (=60) is "presentable" (=15) with "X" (=15) as a "system" (=12) when a "star" (=2) "squares" (=18,000) the "surrounded" (*Vat*, -10) "param space design" (*Vat*, -10) to "frame" (=-10) "self-destruction" (=-10) of the "spirit's" (=20) "potential reality" (=160), "P" (=160). When P=20, "spirit" (=20), and X=12, "quantum spirit" (=12), one's, "H" (=1) is "spiritless" (*Vishanna*, 192 =160+20+15+12+1-16) without "I" (=12) as the "center" (=16).

One "centers" (=16) "I" (=12) with "W" (=16) for "oneness" (=48) of a "designer" (*Abhikalpaka*, 0) after "centering" (=1) the "potential" (=18) for "growth" (=6) of its "origin" (=1,000), "Q" (=1,000). When W=48 and Q=6, "S" (=1,600) is "present" (=1,600) as a "living entity" (=96) "designing" (*Abhikalpana*, 7), "B" (=7), the "design" (=180) of its "birth" (*Kaal Ratri*, 247) with "light" (=180) of "reality" (=7). When S=96 and B=180, "Y" (=-18) is "presenting" (=-18) the "past" (=9) as a "superset" (*Bhuta*, 9) for "twinning" (=9) the "reproductive potential" (=23) of the "octave" (=60) with "J" (=60). When Y=9 and J=23, "J" (=60) "self-perpetuates" (=1/2) "vertex" (=89) "V" (=89) as a "devoted set" (=30) for "descending" (=12) "D's" (=12) "devotion" (=46). When V=30 and D=46, three "sets" (=72) eight as a "subset" (*Baindava*, 260) "technological dimension" (=260) of J's "discontinuity" (=200) as the "primordial space design" (=60).

Without three, "descending (=12) D's "devotion" (=46) "decays" (=31) "J's" (=60) "oneness" (=48) by "feeding" (=17) a "devoted space design" (=17). D's "devotion" (=46) is "Abelian," (=46), i.e., "commutative-effect" (=46) of the "order" (=2) within J's "oneness" (=48). "Abelian order" (=71) is a "sequential order" (=71) "self-perpetuating" (=1/2) the "reproductive potential" (=23) of the "devotion" (=46) with "simplicity" (=2) of "oneness" (=48) within the "sequence" (=48) of the "order" (=2). J's "oneness" (=48) is "non-Abelian" (=22), it "develops" (=22) as the D's "devotion" (=46) "reorders" (=2) three's "pollination" (*Vardhamana*, 17).

Three "pairs" (=8) "itself" (=121) for the "organization" (=29) of "three pairs" (=800), "braiding" (*Jatyam*, 800) the "non-commutative" (=800) "space loop" (=800) into a "four-dimensional particle" (=100). Itself is the "observable-effect" (=121) of a "four-dimensional object"

(=900) which "incarnates" (=900) as a "Borromean" (=900), a "Non-Abelian Anyon" (=900). "Borromean" (=900) "rings" (=100,000) the "four-dimensional particle" (=100) for "triangulating" (=10^{1024}) the "reality" (=7) within the three "pairs" (=8) of "Borromean rings" (=17). Itself "twins" (=121) "three Borromean rings" (=121). The "anyon" (=75) is "inherent" (=75) as a "two-dimensional particle" (=75) within the "continuity" (=50) of the "four-dimensional particle" (=100) that "twins" (=121) the "organization" (=29) of a "three-dimensional particle" (=29). The "result" (=123) of the "two competing realities" (=123) is "cumulative" (=123). The cumulative is the "observable reality" (=123) of "half square" (=123) as an "Abelian Anyon" (=123) that the "space" (=18,000) "self-perpetuates" (=1/2) as a "four-dimensional" (=18,000) "square" (=18,000).

"Abelian Anyon" (=123) "swaps" (=2) the "memory" (=35) of the "past" (=9) "inherent" (=75) within the "observable-effect" (=121) of its "consciousness" (=4) that the "D" (=12) "twins" (=121) as an "entity" (=24) with the "shared memory" (=890) of the "past-effect" (=24) to "reform" (=888) into a "Non-Abelian Anyon" (=900). "J" (=60) "orders" (=2) "itself" (=121) into the "Abelian Anyon" (=123) with the "simplicity" (=2) "inherent" (=75) as the "memory-effect" (=75) of the "devotion" (=46) that "channels" (=36) the "commutative" (=-10) as a "pollinator" (*Bijani*, 36) for "ascending order" (=105) within a "pollen" (*Pushpini*, 365). The "male-effect" (=365) within pollen is "subtractable" (=365) as an "electron" (=365) for "pollinating" (*Parigrahana*, 61) the "exchange" (=269) of the "nucleus's" (=1,000) "female-effect" (=1,000) as a "productive base" (=245) for an "octave" (=60) of three that "twins" (=121) one with two. Octave has "three photons" (=60). One is an "even photon" (*Vedanta*, 1), the "female photon" (=1) with "two photons" (=1). Two photons are a "double copy" (=1) which "copies" (=0) a "photon" (=20) with a two. Twenty is an "odd photon" (=20), the "male photon" (=20) with "one photon" (=20). An odd photon is a "reproducible electron" (=20).

Pollination "braids" (=2) "male-to-female exchange" (*Param Tapas*, 17) to "guide" (=111) "mere knowing" (*Kevala Jnana*, 256) of a "child" (*Pillai*, 128) whose "nature" (=8) is to "float" (=9) the "fire" (=17) of "omniscience" (*Shivadrishti*, 17) as a "perfect entity" (=17). Mere knowing is "floatable" (*Kevala Jnana*, 256) as an "infinite sound" (*Omkara*, 256) within a "primordial knower" (=25) "floating" (*Saraswati*, 256) within the "perpetuating value" (=5) of a "para deity" (*Param Vishnu*, 5) as a "deity" (=1). The deity is "floated" (*Saraswata*, 256) as the "primeval knower" (*Adityanatha*, 1,900) of the "temporal cost" (=600) of the "growth" (=6) of "knowing" (*Jnana*, 19) for the "reality" (=7) of the "twinning-effect" (=13) of the "primordial technological cost" (*Vasundhara*, 13) to be "luminous" (=13) within the "param knower" (=19).

By "daring" (*Sahas*, 97) in the "direction" (*Disha*, 7) of "replication" (=90) of "reality" (=7), a "descendant" (*Vamsaj*, 97) "pollinates" (*Samvardhamana*, 97) an "assumatory" (*Mrgashirsha*, 97) in the "ninth month" (*Mrgashirsha*, 97), "November" (*Mrgashirsha*, 97), for "developing" (*Bruhenta*, 97) the desirable "feminine behavior" (*Pusalattu*, 97) of "female-to-male exchange" (*Putana*, 97). The "generation" (*Amnaya*, 97) of "female-to-male exchange" (*Putana*, 97) "pollinates" (=97) the "reality" (=7) of a "primeval child" (*Daksha*, 123) as the "Godhead" (=17) of the "constellation" (*Duniya*, -2) of "Orion" (*Pitra*, 4). With "freedom from immanence" (*Sarshti Mukti*, 62) of the "primeval" (*Tamas*, 185) as a "paradigm" =62), a "primeval child" (*Daksha*, 123) is "repelled" (*Nanda*, 123) by "Sagittarius zodiac" (*Dhanu*, 128) as a "child" (*Pillai*, 128) from the "light" (=180) of "divinity" (=57) of "param maternal" (*Saranyu*, 5).

Table 3 summarizes the sequential composition of "divine" (=360) as "month" (*Masa*, 360) whose "motion" (*Gati*, 360) "times" (=360) "param maternal's" (=5) "energy" (=19) as an "entity" (=24). The entity's "past potential" (*Asuri*, 570) is the "param creation" (*Ganesha*, 570), the present is the "param creature" (*Prabhu*, 1,600), and the future reality is the "param creator" (*Indambra*, 21).

Table 3. Sequential Composition of Divine into Twelve Months to Time the Param Creation, Param Creator, and Param Creature

Month (Masa)	Param Creator (Indambara)	Param Creation (Ganesha)	Param Creature (Prabhu)
October	"Primordial time" (Samplava, 180)	"Light half" (Balachandra, 8)	Worker Deity
September	"Primordial space" (Antaramsa, 100)	"Variable zero" (Gajanan, 0)	Knower Deity
August	"Primordial human-effect" (Devasmita, 571)	"Intrinsic perfection" (Sumukha, 190)	Manifestor Deity
July	"Primordial trading-effect" (Chandra, 82)	"Feminine tooth" (Ekadanta, 1)	Creator Deity
June	"Primordial workculture-effect" (Antardasha, 374)	"Lone being" (Kapila, 8,000)	Perpetuator Deity
May	"Primordial culture-effect" (Param Shiva, 15)	"Natural gravity" (Gajakarana, 8)	Destroyer Deity
April	"Primordial technological servicing" (Vaikari, 170)	"Plummeting" (Lambodara, 11)	Illuminator Deity
March	"Primordial technological trading" (Tajaurba, 2,000)	"Spacious" (Vikata, 10,000)	Liberator Deity
February	"Primordial technological growth" (Vishvanc, 70,000)	"Demolisher" (Vighnanasha, 590)	Super Wisher
January	"Primordial technological exchange" (Nichevala, 80,000)	"Trainer" (Vinayaka, 53)	Supreme Wisher
December	"Primordial technological investment" (Stip, 800,000)	"Comet" (Dhumraketu, 97)	Primordial Wisher
November	"Primordial technological cost" (Vasundhara, 13)	"Group supervisor" (Ganadhyaksha, 70)	Para Wisher

Acknowledgments

This investigation into the science of science is shaped by six divine factors: determination, imagination, virtue, intuition, nature, and excellence.

Conscious determination of the value of a metaphysical approach, without the inherited scientific method's limitations, is shaped by Primordial Greeter Shri Kartar Singh Yadav Ji, ex–Joint Commissioner, Ministry of Agriculture, Government of India, who is my param guru.

Liberated imagination of the technique for initiating, persevering, and finishing this project is shaped by my father, Shri Surender Nath, and my mother, Shrimati Manju Gupta.

Illuminated virtue of transcending beyond the traditional approach has been shaped by my wife, Bhakti.

Infinite intuition for a conscious ecosystem approach is shaped by my students, devoted to transforming their social, human, ecological, economic, national, and psychological well-being.

Universal nature of the proposed organizational approach is shaped by my professional colleagues and mentors, at various institutions, from various nations, and with varying academic and life perspectives.

Technological excellence of this investigator is shaped by my family, friends, and critics and by those who have contributed through the ages to illuminate the objectives of the study.

Primordial Perpetuator Maha Saraswati and Primordial Illuminator Shri Krishna blessed me with their Divine Light, energizing me to bring the project to fruition.

This investigation is dedicated to the Universe of Children, wishing for their global, unique, inclusive, diverse, engaged, and responsible well-being.

English Index

A

Abelian anyon, 402
Abelian order, 401
Absolute, 4, 8, 10–14, 16, 19, 48, 49, 52, 60, 67, 69, 73, 75, 76, 80, 81, 84–86, 102–104, 106, 107, 126, 128, 132, 150, 152, 153, 159, 165, 179, 180, 185–187, 189, 203, 241, 242, 244, 246, 263, 268, 282, 285, 292, 308, 309, 316, 320, 330, 345, 346, 358, 372, 376, 384, 386, 387, 399, 400
Absolute coherence, 376
Absolute differentiation, 268
Absolute follower, 241
Absolute God, 241, 242, 244, 282
Absolute gravitational constant, 263
Absolute lightforce, 246
Absolute manifestation, 384, 386
Absolute pair, 180
Absolute realization, 126, 150, 189, 292
Absolute temperature, 185, 186
Absolute truth, 52, 308
Accelerate-effect, 193
Accelerating universe, 270
Accretion disk, 325
Action, 4, 19, 20, 41, 81, 86, 88, 89, 96, 121, 153, 166, 168–172, 178, 201, 206, 234–238, 240, 245, 248, 249, 256, 270, 288, 296, 297, 301, 302, 308, 310, 333, 335, 361, 363, 364, 368, 374, 377, 379–381, 383, 389–393, 399
Action coherence, 361
Activated sphere, 284, 285
Active energy, 205
Addition operator, 247
Adds three, 179, 180, 182, 189, 205, 244, 247, 348, 361
Adjacent side, 356, 359
After childbirth, 257
After cooking, 206
Afterlife deliverance, 376
Agonist blesser, 133
All-creating, 242
Almighty creation, 259, 261
Almighty creator, 107, 112, 128, 135, 204, 205, 249, 281, 301, 384
Almighty power, 220
Ambition, 263, 305
Amplify, 203, 291
Amplifying, 325
Angles, 40, 353, 354, 358, 360
Angular energy, 41
Angular filament, 312
Angular momentum, 9, 40, 189, 235, 328, 331, 332
Angular polygon, 353
Animal potential, 190
Animate primordial maternal, 338, 376
Annihilation, 255, 346
Anterograde, 329
Anticlockwise, 214, 235

Anticyclical, 241
Anti-neutrino radiation, 190
Appearance, 51, 81, 161, 198
Appeared moon, 194
Apprehension, 215
Apprehensive face, 216
Arc filament, 312
Arc polygon, 359
Archaic value, 345
Aries zodiac, 298
Aries-effect, 298
Arranging, 208, 383
Arrhippe, 261
Ascendant, 220
Ascended moon, 194
Ascending air, 247, 340
Ascending air-effect, 247
Ascending consciousness, 124, 249, 285, 309
Ascending direction, 219, 220, 287
Ascending growth, 166, 167
Ascending light, 247, 313
Ascending mass-effect, 180
Ascending motion, 298, 329
Ascending order, 20, 61, 170, 175, 185, 187, 189, 271, 402
Ascending sentient well-being, 191
Ascending technological growth, 334
Ascending value, 233
Ascends, 2, 43–45, 91, 104, 105, 170, 176, 188, 241, 297, 308, 309, 316, 337, 373
Assembler galaxy, 271
Association, 373
Assuming, 34, 54, 56, 69, 128, 188, 191, 197, 254, 263, 355, 369
Assumption, 3, 5, 24, 29–33, 39, 44, 182, 263, 274, 289
Astral formation, 198
Astral oneness, 371
Astronomical realm cycle, 325
Asymmetrical light curve, 378
Asymmetrical oneness, 206
Asymmetry, 72, 76, 77, 80, 114, 121, 122, 131, 176, 183, 184, 209, 213, 241, 277, 278, 305
Asymmetry of three, 213, 241
Asymmetry of three with two, 213
Atomic coherence, 361
Atomic orbit, 231, 232
Atomic orbital, 231
Attraction orientation, 282
Aurora, 258
Ausca, 258
Auseklis, 258
Ausrine, 257
Automates, 209
Autumn moon, 194
Axiological cause, 374
Axiological-effect, 244

English Index

B

Backward direction, 220
Backward feminine-effect, 298
Backward motion, 329
Backward sequence of zero, 79, 175
Baggage, 176, 218
Bare essential, 245, 399
Barred spiral galaxy, 271
Beautiful face, 195
Beauty, 7, 8, 10, 128, 187, 218, 252, 253, 283, 284, 286, 290, 386
Begging, 172, 226, 241
Beginning force, 219
Behavior, 2, 35, 36, 60, 102, 110, 172, 308, 370, 374, 403
Behavioral remediation, 239
Being energy, 188, 206, 214, 270, 314, 334, 382
Being value, 204
Beli, 282
Beli Sheri, 282
Belief system, 7, 14, 17, 43, 57, 122, 123, 153, 296, 337, 367, 390
Belu, 282
Belu Sadi, 282
Benefactor, 240
Benefit, 9, 33, 41, 42, 50, 63, 75, 102–107, 172, 186, 203, 225, 239, 241, 243, 245, 296, 306, 308, 314, 333, 334, 336, 337, 367, 370, 371, 374, 390, 397
Benefit-cost ratio, 33, 186, 243, 296, 336, 397
Beruth, 246
Bestows, 207
Bicoherence, 362
Bifeminism, 244
Big Bang, 309, 319
Bilateral polygon, 359
Binary, 104, 105, 231, 316
Birth moment, 313, 314
Bisect, 271, 358, 359
Black emperor, 171, 202
Black filament, 313
Black hole, 3, 15, 17, 24, 62, 82, 83, 125, 126, 137, 194, 244, 322
Blackener, 260, 261
Blessing, 62, 87, 133, 172, 245, 276, 310
Blessing energy, 245
Blissfulness, 244
Blocker, 236
Blue master of thunder, 400
Body of knowledge, 202
Boring number, 279
Borromean, 402
Borromean rings, 402
Boson potential, 379
Boson star, 378
Bottleneck, 319
Bottom quark, 72, 172, 173, 181, 252
Boundless light, 204
Braiding, 401
Braids, 402
Break away, 312
Breeding, 22, 23, 49, 89, 101, 114, 116, 117, 126–129, 134, 137, 153, 155, 167, 171, 175, 179, 181, 187, 190, 208, 220, 221, 228, 229, 232, 239, 240, 243–245, 268, 279, 281, 288, 289, 305, 322, 325, 330, 333, 336, 344, 345, 357, 390

Brettia, 247
Bright blackhole, 326
Bright filament, 312
Brightener, 260
Broteas, 262
Bulk modulus, 242
Bulk-effect, 242
Bunching, 284
Burstable, 217, 218, 271
Burster, 217
Bursts, 98, 217, 218, 311

C

Cancer, 248
Capability, 85, 104, 105, 132, 167, 184, 199, 201, 240, 251, 276, 285, 292, 373, 374, 385, 397
Capricorn zodiac, 377
Carnot, 229
Carnot cycle, 229
Castor, 260, 261
Causal body, 113, 179, 198, 235–237, 268, 287, 315, 368, 369, 371, 385, 386, 389
Causal coherence, 367
Causal consciousness, 385
Causal element, 198
Causal formation, 198
Causation, 7, 13, 22, 23, 26, 27, 30, 31, 33, 39, 40, 45–47, 57, 74, 76, 84, 86–88, 91, 97, 101, 108, 117, 119, 124, 125, 127–129, 133, 135–138, 143, 147, 149, 155, 157, 159, 174, 175, 177, 184–186, 189, 192, 197, 201, 205, 206, 209–211, 213–215, 217, 226–229, 232–234, 238, 248–251, 257, 258, 260, 269, 271, 273, 281, 289, 292, 295, 302, 321, 324, 331, 346, 347, 359–361, 363, 364, 366, 371, 378, 380–382, 385, 388–391, 394–396
Causation realignment, 184
Causation reversal, 229
Causation singularity, 33, 395, 396
Causation tensor, 257
Cause body, 269
Cause coherence, 364
Cellular face, 277
Centering, 90, 118, 128, 134, 138, 144, 166–168, 213, 218, 230, 266, 271, 282, 291, 306, 316, 323, 328–331, 333, 374, 375, 398–401
Centerward energy, 326
Centillion, 319
Central body, 269, 395
Central horizon, 329
Central sphere, 321
Cepheid, 377, 378
Certainty, 4, 8, 11, 18, 19, 78, 96, 177, 200
Challenging time, 209
Chameleon, 217
Changing position, 247
Channel, 17, 133, 134, 157, 184, 239, 246, 268, 295, 332, 345, 348, 387, 388
Chaotic, 244
Characteristic, 102, 141, 142, 173, 220, 228, 281, 334, 397
Charge, 19, 28, 43, 50, 84–86, 129, 136, 137, 172, 175, 209, 210, 245, 252, 274, 381, 394, 396
Charge radius, 381

407

Chariot, 221
Charm, 2, 7, 8, 28, 50, 72, 84, 95, 172, 173, 181–183, 197, 206, 218, 232, 252, 253, 272, 309
Charm quark, 172, 173, 181–183, 206, 252
Chasing away, 207
Chesed, 258
Child essence, 292
Childbirth, 257, 258
Chochmah, 376, 399
Christ, 133, 256, 264, 265, 309
Christ-effect, 133, 264, 265
Circle of life, 217
Circle symmetry, 212
Circular coherence, 361, 364
Circular creation, 110, 136, 148, 191, 216, 250, 313, 315
Circular filament, 311
Circular momentum, 235
Circular orbit, 321, 329
Circular polygon, 353
Circular primordial maternal, 340
Circular section, 344
Circular wave, 262
Circulates, 19, 46, 65, 92, 155, 191, 307, 321
Circumference, 360
Citizen, 8, 104, 172, 238, 240, 301, 387
Citizenship action, 301, 314
Clarification, 4, 207, 387
Clarified consciousness, 125, 126, 188, 218, 263, 374, 387, 399
Class, 41, 44, 45, 114, 129, 178, 181–183, 233, 241, 320
Class dimension, 241
Class width, 41
Cleansed, 217, 256
Clinging, 197
Clockwise, 214, 235, 395
Closable polygon, 355, 356
Closed orbit, 231
Clotho, 267
Clytie, 262
Coagonist becomer, 133
Coalesce, 324
Co-dependency, 343
Cohomology class, 242
Co-incidental coherence, 367
Collectivity, 201, 371
Color filament, 313
Color glass condensate, 313, 379
Color potential, 16–18, 248, 379, 381
Colorant, 260, 261, 381
Colorless filament, 313
Colorlessness, 340, 382
Combination of two, 98, 243
Combination with two, 19, 207
Comic, 337
Commutative-effect, 401
Compassionate, 178, 391
Competing reality, 228–230
Complement, 118, 128, 130, 149, 199, 372
Complex plane, 294
Complex primordial maternal, 340
Concave polygon, 354
Conceiving four, 236
Concordant-effect, 231
Condensed, 39
Conducting, 230

Configuration, 264
Conformal field theory, 14, 17
Conformity, 196
Congruence modular form, 304
Congruent, 303–305, 356, 358
Conic, 255
Conjunction, 319, 320
Conscious system, 122, 171, 198, 241, 300
Consciousness, 2, 3, 6–9, 12, 14–17, 19, 20, 22, 23, 26, 27, 29–32, 40, 44, 48, 53, 54, 61, 67, 69, 72, 73, 75, 76, 79, 80, 84, 85, 87, 94, 95, 100, 101, 106, 108, 109, 111, 112, 114–129, 132–135, 137, 142, 147, 148, 150, 153–159, 161–164, 166–168, 170–175, 177, 179, 180, 182, 183, 185, 187–189, 191, 194, 197, 198, 202–206, 214–216, 218–220, 224–228, 240, 241, 253, 269, 270, 272–274, 276, 282, 283, 291, 294, 296, 300, 303, 305, 306, 309, 310, 323, 324, 327, 333–336, 343, 346, 348, 351, 357, 369, 374, 378, 385–387, 389, 391, 394, 395, 398, 400, 402
Consciousness system, 121, 170, 241
Consensual, 275, 276
Consequence, 49, 79, 81, 83, 124, 168, 178, 216, 238, 291, 361
Consequential coherence, 361
Consequential period, 277, 278
Constancy, 120, 361, 362
Constant potential, 384, 393
Constant space, 297
Constant subject, 252
Constant time, 298
Constellation of twelve stars, 328
Constitution, 256
Construction, 256
Consumes, 8, 34, 56, 73, 88, 326
Consumptive, 205
Contaminates, 291, 297, 318
Continuity, 2, 8, 15–17, 45, 53, 75, 100, 120, 122, 131, 134–136, 138, 142, 147, 152, 157, 166, 181, 183, 193, 194, 202, 203, 205, 206, 209, 210, 212, 213, 218, 225, 227, 230, 235, 236, 243, 244, 247, 250, 251, 257, 264, 267, 268, 270, 272, 273, 277, 278, 280, 282, 288, 290, 294, 296, 303, 309, 312, 314, 319, 322, 326–328, 338, 339, 344, 358, 361, 363, 365, 366, 372, 375, 387, 388, 394–396, 398, 402
Continuous discontinuity, 206
Continuous flash, 198
Continuous potential reality, 122
Continuously, 215, 281
Contracting, 13, 44, 342
Contracting entropy, 13
Contracting universe, 13
Contragrade motion, 329
Contribution, 293
Converge, 112, 265
Convergence, 95, 96, 112, 184, 257, 265
Convergent energy, 96, 114, 115, 234, 249, 262, 283, 289, 314, 391
Conversancy, 217
Convex polygon, 354
Cooking value, 206
Cooks, 205
Cooling, 325
Coplanar, 354
Copper-effect, 324

English Index

Copy potential, 134, 204, 388
Copying, 49, 56, 154, 233, 272, 276, 372
Core, 49, 100, 167, 187, 225, 227, 228, 233, 268, 327, 347
Corona component, 228
Corporate, 8, 105, 159, 240, 296
Corporate path, 240, 296
Cosmic beginner, 394
Cosmic beginning, 394
Cosmic dance, 281
Cost, 3, 33, 42, 50, 66, 75, 81, 106, 109, 126, 132, 203, 207, 225, 243, 254, 268, 297, 301, 306, 326, 330, 333, 334, 336, 337, 350, 367, 374, 387, 388, 392, 393, 403
Cost-effective leadership, 207
Council of thirteen, 293
Coupled, 326
Couples, 231
Coupling, 326
Covarying, 182, 307
Covered, 209
Coverer, 254
CP symmetry, 211, 229
Creation, 13, 34, 51, 66, 79, 101, 112, 125, 135, 136, 142, 153, 170, 177, 180, 188, 198, 203, 206, 207, 215, 226, 272, 277–282, 292, 296, 301, 313, 314, 333, 334, 338, 363, 364, 367, 385, 388, 392, 396, 403, 404
Creation mass, 207
Creative force, 184, 244, 260, 273, 369
Creator, 6, 10, 34, 35, 51, 66, 76, 102, 103, 135, 141, 142, 153, 169, 177–180, 188, 196, 203–207, 215, 217, 226, 238, 239, 249, 258, 277, 279, 281, 292, 301, 326, 329, 363, 364, 367, 380, 384, 392, 403, 404
Creator deity, 203–206, 215, 217, 238, 239, 404
Creature, 32, 34, 35, 41, 51, 66, 79, 87, 101, 102, 108, 112, 126, 133–135, 141, 142, 153, 170, 172, 173, 189, 197, 198, 202, 204–206, 208–211, 215, 220, 226, 228, 230, 233–235, 245, 256, 267, 274, 277, 279, 281, 292, 295, 296, 301, 308–310, 335, 338, 362–364, 367, 373, 383–385, 387, 392, 403, 404
Crescent, 220, 221
Crowns, 247
Crystallizes, 76, 95, 205
Cubic, 158, 159, 174, 176, 244, 322
Cubic energy, 174, 176
Cultural dimension, 285
Cumulative, 402
Current of energy, 326
Curvable polygon, 354
Curve polygon, 353–355
Curvilinear energy, 167
Curvilinear radius, 381
Curvilinearity, 42, 378
Curving polygon, 354
Cycle of joy, 229
Cyclical, 156, 218, 241, 293, 315
Cylinder, 243, 244

D

D orbital, 231, 232
Dalai Lama, 255
Dangaus Kariune, 258
Dark blackhole, 326
Dark companion, 378
Dark filament, 312, 313
Dark half, 192, 194
Dark matter, 3, 41, 67, 82, 100, 108, 124, 125, 130, 131, 148, 149, 161, 162, 169, 179, 181, 184, 185, 192, 196, 202, 221, 300, 312, 336, 337, 385
Dark matter crescent, 221
Dark matter filament, 312
Dark reality, 233
Darkener, 260, 261
Darkness, 39, 40, 51, 61–63, 65, 66, 84, 136, 148, 161, 162, 164, 174, 194, 198, 223, 247, 261, 299, 300, 310, 322, 323, 373, 379
Daughter consciousness, 205
Daughter spirit, 313
Death moment, 313
Deca, 180
Decay, 2, 16, 17, 49, 81, 142, 163, 176, 196, 278, 283, 307, 362, 363, 369, 394
Decelerate-effect, 193
Decelerating universe, 270
Decomposer, 260
Decomposing, 254, 260
Decomposing group, 254
Dedication, 115, 214, 290
Definite cylinder, 243
Definite minimum, 332
Defrosts, 202
Deity, 2, 29, 67, 80, 87, 110–115, 134, 135, 137, 149, 157, 167, 169, 176–178, 187, 188, 190, 191, 195, 204, 206, 207, 214, 215, 217, 220, 222, 226, 237, 239, 240, 244–246, 255, 260, 276, 281, 282, 290, 295, 299, 308, 310, 314, 318–320, 334, 335, 337, 339, 365, 371, 372, 377, 399, 400, 403, 404
Deity face, 215
Deity kingdom, 187, 237, 371
Dejectedness, 384
Demolisher, 313, 404
Demonic, 207, 261
Denominator, 14, 52, 53, 73, 75, 85, 304, 305
Departed consciousness, 202
De-polarization, 296, 395
Descend, 26, 90, 91, 241, 297
Descended moon, 194
Descending direction, 219, 220, 287, 383
Descending growth, 166
Descending mass, 147, 155, 279
Descending motion, 329
Descending proportion, 269
Descending sentient well-being, 192
Descending thermodynamic force, 185
Desirable, 61, 109, 110, 127, 128, 153, 205, 206, 403
Destroyer deity, 197, 198, 205, 206, 244, 249, 253, 255, 257, 270, 273, 404
Destructive interference, 18, 90
Devastation, 207, 387
Devil, 29, 58, 80, 87, 115, 158, 165, 206–208, 216, 236, 256, 258, 276, 295, 300, 318, 339, 343, 347, 356, 362, 377
Devil momentum, 236
Devil momentum quantum number, 236
Devoid of information, 244
Devoted manifestor, 240, 244
Devotee creation, 281

Devotee universe, 190
Devotee wisher, 239, 244, 253
Devotion, 6, 18, 22, 23, 121, 130, 146, 150, 190, 191, 199, 207, 218, 219, 275, 276, 282–284, 287, 300, 308, 337, 369, 373, 389, 391, 393, 394, 401, 402
Devotional value, 205
Diagonal filament, 311
Diagonal symmetry, 212
Diameter, 255, 353
Diametric reality, 255
Dievo Suneliai, 259
Difficult time, 208, 209
Diffused, 105, 115, 167, 231, 232, 266, 329
Diffused orbital, 232
Dilating, 44
Dim chain, 312
Dimension, 13, 18, 30, 97, 108–114, 122, 129, 143, 149, 152, 171, 174, 177, 181, 219, 241, 244, 250, 251, 270, 279, 303, 329, 335, 352, 354, 362–364, 367, 373, 379, 381, 401
Dimensional coherence, 364, 378
Dimensional space, 14, 152, 316
Dimensional subject, 15
Dimensionless, 32, 86, 141, 379
Diminution, 311, 312
Diminution chain, 311
Diminutive chain, 312
Disabler, 260, 261
Disappearance, 148, 198
Disappeared moon, 194
Disassociation, 373
Discolor, 381
Discolorant, 260, 261, 381
Discontinuity, 15, 45, 75, 81, 100, 130, 131, 134, 138, 142, 152, 155, 163, 178, 181, 194, 197, 203–205, 210, 212, 213, 227, 243, 244, 251, 255, 264, 270, 272, 273, 277, 280, 281, 286, 287, 294, 296, 309, 312, 356, 361, 363, 366, 370, 379, 380, 384, 388, 396, 399, 401
Discontinuous asteroid, 194
Discontinuous infinity of time, 227
Discontinuous potential reality, 122
Discordant force, 80, 178
Discordia, 263
Disentangled system, 12, 16
Disjunction, 385
Dismember, 243
Disorder, 245, 301, 350, 367
Dispassion, 178, 362
Disposing, 202
Disproportionate, 3, 104, 112, 114, 115, 195, 244, 279, 293, 301, 334, 387, 388, 397
Disproportionate benefit, 115, 244
Disproportionate probability, 279
Distances, 5, 9, 11, 38, 201
Distributable value, 44
Diverge, 266
Divergent energy, 219
Divergent potential, 256
Diverging infinity, 249
Diverse form, 40, 44, 357, 369, 397
Diverse realities, 373
Divided consciousness, 218
Divination, 206, 340
Divine characteristic, 170
Divine energy, 334, 371, 372, 374

Divine Godhead essence, 295
Divine Histone, 346
Divine matter, 196
Divine momentum, 236
Divine momentum quantum number, 236
Divinity, 67, 127, 142, 149, 188, 251, 276, 287, 339, 341, 345, 386, 389, 403
Divisible by two, 245
Divisor, 119, 222, 302, 318, 366
Doctrine of transcendental, 258
Doing nothing, 122, 236
Dome, 259
Dominating force, 254
Donar, 313
Dot, 236, 241, 300, 324
Double copy, 2, 45, 48, 132, 133, 138, 152, 194, 224, 233, 288, 311–313, 320, 321, 348, 372, 377–379, 381, 388, 402
Doughnut orbital, 231
Down, 50, 60, 62, 72, 85, 129, 155, 172, 173, 180, 181, 227, 251–253, 332
Down quark, 85, 129, 172, 173, 180, 181, 251, 252
Downstream, 311, 376, 383
Downswinging, 252
Dramatic distortion, 265
Dreamer, 195, 235–237, 239, 253, 254
Dreaming, 95, 235
Duality, 12–14, 114, 124, 188, 204, 244, 309, 335, 354, 370
Dull blackhole, 326
Dull chains, 312
Dull matter, 196, 203
Dwarf planet, 378
Dynamism, 81, 85, 122, 261, 380

E
Earth energy, 207
Earth flash, 313
Earthly crescent, 220
Earthly othernesss, 251
Ecosystem, 102, 107, 119–123, 133, 214, 261, 282, 283, 316, 346, 348, 368
Effective value, 384
E-gon, 357
Eight faces, 214, 216, 265, 277
Eight hexaquarks, 253
Eight planets, 102, 103, 131, 149, 180, 181, 193, 196, 221, 280
Eight potential hexaquarks, 253
Eight potential quarks, 253
Eight quarks, 253
Eight radiated sequences, 293, 294
Eighteen degrees, 255
Eighteen pi, 234, 351
Eighteen sections, 344
Eight-fold growth, 14, 132, 166, 269, 344, 345
Eighth moon, 194
Eighth photon, 231, 232
Eighty parabolic systems, 251
Eighty primordial suns, 294
Eileithuia, 258
Einstein, 4, 5, 7, 12, 17, 42, 58, 59, 110, 152, 343–345
Einstein ring, 110
El, 188, 205, 282

English Index

EL, 205, 282
El Roi, 188
El Shaddai, 282
Electric dipole moment, 395
Electric filament, 311
Electric repulsion, 57
Electrically, 40
Electrokinetic momentum, 237
Electromagnet, 37, 343, 344
Electromagnetic filament, 311
Electromagnetic mass, 26, 311
Electromagnetic radiation, 40
Electron gas, 232
Electron singularity, 395
Electrophilic, 237
Eleven massive blackholes, 325
Eleven supermassive blackholes, 324
Eleven-dimensional object, 17
Eleventh moon, 194
Eleventh type, 318
Elohai, 240, 241
Elohim, 205
Elongated chain, 311, 382
Elongator chain, 311
Emanating hand, 260
Emanating reality, 250
Emanation value, 255
Emancipation, 388, 389
Embedded fractal wave, 233
Embedded wave, 233
Embodied matter, 184
Embodies, 56, 71, 73, 80, 81, 123, 179, 182, 283, 292
Embracer, 260
Emotional feeling, 297
Emperor, 171, 187, 202, 203, 246, 282, 297, 371, 375
Empowerer, 260
Emptiness, 248
Enabler, 260, 261
Endeavor, 240, 263
Ending, 217, 248
Endogenous electron, 180
Endopoducing, 395
Endoproduction, 4, 19, 30, 52, 67, 76, 79, 80, 82, 111, 127, 132, 137, 138, 141, 147–149, 168, 175–177, 181, 183, 185, 198, 226, 272, 275, 276, 285, 395, 397
Endoreproduction, 23, 27, 28, 36, 79–81, 98, 128, 147, 175, 181, 183, 185, 187, 188, 335, 387, 388, 395
Endowing, 246
Endpoint, 203
Energized, 271
Energized-effect, 271
Energizer, 187
Energy, 2, 4–7, 9, 19, 23, 25–28, 31, 40, 41, 44, 46–52, 57, 58, 60, 62, 64, 65, 70–77, 79–86, 90, 100–102, 104, 105, 107, 120, 121, 129, 135, 136, 143–147, 150, 155, 157, 166, 172–174, 180, 185, 187, 188, 190, 196, 200, 201, 205, 209, 218, 219, 224, 226, 229, 230, 237, 238, 245–247, 249, 258, 259, 272, 274, 276, 286, 288, 290, 299, 300, 306, 309, 310, 312, 314, 320, 330, 334–336, 338, 340, 346, 357, 358, 384, 393, 394, 396, 397, 403
Energy conservation, 44

Energy deficit, 229
Energy equilibrium, 19, 147, 166, 286, 288, 299, 357, 384
Energy force, 205
Energy points, 299
Enjoyer, 19, 87, 124, 176, 226, 229, 239, 244, 274, 296, 333, 337, 338, 372, 373, 380
Enjoyment, 333
Enlightening, 313
Entangled interference, 18, 129, 130
Entanglement, 10–13, 17, 18, 62, 82, 84, 86, 87, 149, 150, 176, 254, 259, 320
Entity, 2, 7, 16, 17, 19, 22, 23, 27–41, 44, 48, 72, 84, 87, 88, 94, 101–111, 114, 115, 118, 119, 123–137, 139, 142–144, 149, 153, 155, 170, 171, 174, 177, 183–187, 189, 190, 206, 209, 213, 214, 220, 224–226, 238, 240–242, 245, 251, 255–257, 260, 261, 263–265, 269–271, 273–276, 281, 283–287, 289, 295–297, 301, 304, 305, 307, 308, 311, 314, 318, 321, 322, 324, 325, 328–330, 333, 335–337, 339, 340, 347, 360, 364, 367, 368, 370, 371, 384, 385, 389, 392, 397, 398, 402, 403
Entity energy, 325
Entity prealignment, 184
Entrepreneurship action, 270, 272
Entrepreneurship behavior, 370
Entrepreneurship-effect, 190
Entropy, 2, 14, 16, 17, 23, 26, 27, 30, 31, 33, 44–46, 48, 50, 74, 81, 82, 85, 86, 105, 120, 126, 132, 135, 141, 143, 144, 146, 149, 156, 178, 181, 182, 185, 193, 195, 198, 204, 218–220, 222, 227–229, 239, 251, 252, 283, 294–296, 310, 313, 315, 329, 334, 347, 364, 366, 371, 378, 387–389
Entropy system, 251
Entropy theory, 16, 48, 81
Enumerator, 235
Envisioned future, 16, 123
Eos, 257
Epiphron, 262
Epistemological cause, 373, 398
Epistemological value, 112, 294, 397
Equal breadth, 359
Equanimity, 372
Equilibrium, 46, 186, 203, 370, 389
Eris, 263
Escalating cost, 105, 133, 206, 379
Essence, 19, 22, 28, 92, 101, 155, 168, 170, 224, 270, 276, 316, 317, 322, 331, 381, 385, 387, 394, 395, 397
Essential nature, 113, 118, 254, 285, 365, 392, 400
Eta, 218, 219
Eta energy, 219
Eta potential, 218
Eternal togetherness, 308
Eternalism, 371
Eternity, 61, 84, 229, 249, 251
Ether, 73, 131, 141, 145–147, 155, 162, 163, 171–173, 179, 186, 208, 211, 240, 247, 248, 269, 284, 295, 297, 300, 329
Etheric, 101, 198, 269, 287, 367, 371, 373, 376, 396
Etheric body, 101, 269, 287, 371, 396
Etheric formation, 198
Etheric limit, 373

Ethericsphere, 314
Euphrosyne, 263, 264, 269
Eupryto, 262
Euryanassa, 263
Eurythemista, 262
Eve, 62, 63, 285, 303
Even dimension, 241
Even-dimensional class, 242
Evening, 245, 246, 257, 286
Evening star, 245, 257, 286
Evenness, 156, 241, 242, 306
Evens out, 156, 202
Everlasting benefits, 280
Everyone, 4, 5, 7, 9, 10, 18, 23, 24, 34, 36, 49–52, 57, 60, 66–69, 77, 79, 81–83, 86, 109–111, 113, 117, 124, 134, 136, 146, 148, 150, 153, 168, 200, 207, 208, 226, 239, 241, 246, 269, 276, 288, 289, 333, 365, 371, 380, 384, 391, 393, 400
Everyone else, 60, 134, 168, 207, 208, 239, 246, 288, 380, 393
Everything, 16, 34, 40, 50, 51, 56, 64, 69–73, 76, 79, 81, 83, 89, 108, 118, 135, 178, 201, 205, 208, 235, 240, 249, 314, 329, 381, 384, 391, 395, 400
Exchange, 19, 23, 33, 45, 85, 86, 101–103, 109, 112, 115, 134, 135, 149, 150, 167, 184, 185, 196–198, 203, 204, 207, 220, 224, 225, 233, 236, 237, 251, 262, 310, 314, 331, 335, 340, 365, 371, 372, 383, 386, 402, 403
Exchange system, 103, 220, 251
Exchanging energy, 205
Existent, 127, 376–378, 383
Exogenous electron, 180
Expandable, 319
Expanding bubbles, 325
Experience, 51, 69, 115, 116, 119, 123–125, 128, 129, 134, 135, 148, 154, 165, 180, 198, 206, 224, 259, 272, 292, 303, 310, 333, 399
Exponent, 220, 246, 315
Exponentiable, 246
Exposition, 197
Extensifies, 325
Extrinsic sentient-effect, 283
Extrinsic value, 253

F
Fairy, 220
Fairy energy, 220
Falsehood, 202
Falsism, 13, 111, 153
Family, 3, 114, 182, 190, 200, 201, 205, 215, 219, 243, 247, 285–289, 291, 301, 318–321, 327, 331, 335, 342, 343, 398
Family with potential, 319
Far-extended, 221
Farma, 400
Father, 7–9, 12, 15, 18, 19, 30, 51, 58, 62–64, 66, 80, 113, 115, 190, 192, 205, 207, 209, 213, 214, 217, 258, 262, 263, 282, 338, 339, 365
Father nature, 18, 262, 339
Fearless, 216, 217, 295, 365
February, 314, 404
Feeding, 22, 89, 114, 155, 167, 175, 179–181, 228, 232, 316, 332, 357, 377, 400, 401

Female, 110, 141, 145, 161–164, 191, 259, 261–263, 297, 302, 303, 315, 386, 393, 402, 403
Female body, 262
Female photon, 402
Feminine, 13, 19, 23, 62, 66, 98, 102, 106, 109, 113–115, 121, 142, 162–164, 172, 179, 182, 190, 191, 202, 206, 211, 218, 219, 239, 244, 256, 262, 268, 271, 273, 275, 276, 280, 282, 285, 287, 289, 300, 301, 308, 313, 315, 319–321, 323, 324, 326, 327, 333–335, 341, 342, 368, 371, 383, 385, 386, 388, 390, 392, 396, 403, 404
Feminine blackhole, 326
Feminine dimension, 219
Feminine gene, 383, 385
Feminine Godhead, 256
Feminine tooth, 206, 404
Femininity, 12, 109, 113, 114, 128, 163, 164, 182, 191, 201, 206, 218, 258, 262, 269–271, 273, 276, 279, 280, 302, 335, 358, 360, 369, 374, 383, 393
Femininity without, 206
Feminization, 175, 339, 376
Fermi surface, 386
Ferromagnet, 36, 37
Fifteen massive blackholes, 327
Fifteen primordial suns, 293
Fifth moon, 194
Fifth orthogonal p orbital, 230
Fifth quadrant, 216
Fifty cylinders, 244
Fifty filaments, 312
Filament, 310–313, 382
Finite flash, 199
Finite infinity, 340
Finite orbit, 231, 330
Finite point, 90–92, 257
Fire, 18, 69, 73, 89, 93, 130, 140–142, 145, 146, 156, 162–164, 167, 168, 171–173, 186, 208, 237, 238, 240, 243, 244, 246–248, 268, 272, 273, 284, 285, 294, 297, 299, 300, 314, 328, 329, 340, 343, 374, 399, 402
First Eden, 240, 310
First electron, 228
First gender-differentiated copy, 339
First primordial sun, 293
First quadrant, 212
Five blackholes, 326
Five elements, 162, 163, 312
Five filaments, 311, 313
Five moons, 194
Five nonlinear systems, 250
Five photons, 181, 232, 234
Five space systems, 250
Five-dimensional object, 357
Five-dimensional particle, 15
Flame, 102, 105–107, 208, 244, 246, 247, 261, 285, 287, 288, 296, 304
Flame family, 247, 285, 287, 288
Flamemate, 246
Flare, 325
Flash, 198, 199, 311
Floatable, 403
Flow, 177, 194, 281, 383
Flowing, 155, 175, 246, 247
Fluctuation, 18, 121, 287, 305–307
Flux, 41, 44, 304

English Index

Follower, 23, 75, 85, 110, 116, 170, 172, 177, 178, 187, 199–201, 225, 242, 301, 370, 371, 399
Formative deity, 260
Formative face, 278
Formative fire-effect, 191
Formative growth, 29, 103, 106, 134, 137, 149, 150, 153, 177, 290, 298, 299, 301–303, 305, 307, 309, 311, 313, 315, 317, 319, 321, 323, 325, 327, 329, 331, 333, 335, 337, 339, 341, 343, 345, 347, 349, 351, 353, 355, 357, 359, 361, 363, 365, 367, 369, 371, 373, 375, 377, 379, 381, 383, 385, 387, 389, 391, 393, 395, 397, 399, 401, 403
Formative growth paradigm, 106, 134, 149, 150, 299, 302, 303, 305, 307, 309, 311, 313, 315, 317, 319, 321, 323, 325, 327, 329, 331, 333, 335, 337, 339, 341, 343, 345, 347, 349, 351, 353, 355, 357, 359, 361, 363, 365, 367, 369, 371, 373, 375, 377, 379, 381, 383, 385, 387, 389, 391, 393, 395, 397, 399, 401, 403
Formative justice, 177, 206, 376
Formative phase, 244
Formative Sun, 192
Formative system, 251
Formative universe, 307
Formative wishing sequence, 244
Form-effect, 208
Forming reality, 53, 54, 56
Forty cylinders, 243
Forty faces, 216, 217
Forty parabolic systems, 251
Forty primordial suns, 294
Forty-eight faces, 215
Forty-nine faces, 215
Forward direction, 220
Forward masculine-effect, 298
Forwarding, 298
Foundation, 50, 66, 111, 176, 187, 216, 285, 296, 297, 329, 330, 346, 347, 365, 376, 399
Four blackholes, 326
Four d orbitals, 235
Four down quark, 180
Four filaments, 311, 312
Four g orbitals, 232
Four hundred parabolic systems, 251
Four linear systems, 250
Four lines, 354
Four moons, 194
Four nonlinear systems, 250
Four orbitals, 236
Four parabolic systems, 251
Four quarters, 233
Four radii, 313
Four zeroes, 133, 168, 169, 231, 295
Four-dimensional particle, 401, 402
Four-fold growth, 95, 98, 129, 139, 268, 342, 345, 358
Fourteen characteristics, 281
Fourteen deities, 260
Fourteen faces, 213, 214
Fourteen primordial suns, 294
Fourth month, 248
Fourth moon, 194
Fourth one, 131, 231
Fourth square, 268, 350
Fragment, 219, 312, 346

Freedom from aloneness, 178, 254
Freedom from element, 399
Freedom from inferiority, 124, 126
Freedom from para wisher, 254, 389
Freedom from primordial enjoyer, 372
Freedom from qualifier, 136, 149
Freedom from submersion, 198
Freedom from superiority, 124, 126
Freedom from time, 137, 149, 257
Freedom from transcendental, 147, 150
Freedom from unreal, 374, 385
Frees oneness, 205
Frontside, 267
Frontswinging, 252
Fructification, 20
Fructifier, 20
Fruit of devotion, 188
Fruit quality, 314
Fruitis, 247
Full moon, 106, 107, 207, 220
Full moon phase, 207
Future, 2–6, 8, 10–13, 15–20, 22, 23, 25, 26, 30, 31, 33, 38, 40, 41, 45, 46, 48, 51, 56, 58, 60–67, 70, 71, 74–78, 80–92, 95–98, 100–102, 107, 108, 110, 112, 113, 116–120, 124–132, 135, 137–139, 143, 147, 149, 152, 154, 158, 160, 163, 167–172, 174, 176–181, 183–185, 188, 192, 193, 196, 197, 202–205, 207, 210–213, 215–219, 222, 223, 225, 227, 228, 235, 237, 238, 243, 246, 248, 250, 252–255, 258, 259, 261–263, 265, 267, 268, 270–276, 279, 280, 284, 285, 287, 288, 293, 295, 299–301, 303, 305, 306, 310–314, 316, 322, 328–332, 339, 347, 356, 359, 368–370, 375–381, 386, 388–390, 393, 395, 396, 400, 403
Future ecosystem, 347
Future entity, 207, 305, 339
Future generation, 196, 211, 254
Future moment, 158, 314
Future of everybody, 30, 31, 78, 174, 202, 243, 246, 258, 280
Future reality, 117, 176, 179, 192, 205, 211–213, 215, 216, 218, 219, 261, 262, 299, 369, 395, 403
Future time, 255
Future value, 2, 16, 26, 45, 48, 78, 97, 128, 137, 174, 180, 202, 203, 205, 210, 223, 243, 250, 252, 255, 262, 273, 314, 375, 378, 380, 400
Future workculture-effect, 184

G

Gain of velocity, 45
Galactic center, 340
Galaxy, 100, 152, 218, 270–272, 285, 286, 328
Gamma, 44, 382
Garden of Eden, 62, 283, 285
Gemini, 263, 264, 385
Geminid, 378, 379, 381
Gender dimension, 292
Gender exchange, 205
Gender-differentiated six copies, 338
Generality, 61
Geography, 91, 104, 181–183, 241, 324, 371, 372
Geography-effect, 371

Gevurah, 220, 369
Giant donut, 324
Giant planet, 378
Global universe, 308
Globalizing, 101, 114, 121, 122, 309
Glowing, 247, 313
Gluonic, 380, 381
Gluonic gravitational form factor, 381
Goal coherence, 361
Goalkeeper, 2, 14, 19, 23, 29, 30, 53, 80, 113, 128, 176, 177, 188, 189, 226, 235, 237, 239, 240, 249, 254, 255, 299, 300, 316, 362, 374
Goalkeeper coherence, 362
Goalpost, 19, 127, 128, 189, 257
God, 7–10, 12, 19, 30, 31, 51, 57, 58, 61–64, 66, 67, 73, 74, 77–83, 85, 87, 128, 150, 189, 197–199, 215, 235, 240–245, 254, 258–261, 274, 276, 282, 290, 295, 296, 298, 302, 310, 314, 318, 336–338, 374, 376, 384, 386, 399
God of Hell, 258
God of sea, 260
God paradigm, 298
Goddess, 246–248, 250, 257–259, 310, 313, 314, 338, 340
Goddess of Dawn, 257, 258
Goddess of Moon, 258
Goddess of primordial knowing, 259
Goddess of primordial life, 259
Goddess of primordial well-being, 259
Goddess of ripening of grain, 257
Goddess of Sunrise, 258
Goddess of Sunset, 246
Goddess of West, 247
Godhead, 156–159, 187, 188, 292, 295, 296, 298, 302, 375, 403
Godhead paradigm, 298
Googol, 319
Graeae, 263
Graham number, 319
Granddaughter consciousness, 206
Grandfather, 164, 207, 243
Grandfather flame, 207, 243
Grandson consciousness, 206
Gravitates, 22, 34, 201, 307, 380
Gravitation, 7, 224, 299, 399
Gravitational differentiation, 355
Gravitational energy, 7, 240, 262, 280, 291, 296, 334
Gravitational filament, 311, 312
Gravitational force, 1, 3, 34, 35, 42, 57, 58, 62, 79, 172, 173, 227, 350
Gravitational imperfection, 307
Gravitational lensing, 21, 109, 152, 225
Gravitational microlensing, 223, 224
Gravitational orbit, 231, 329
Gravitational quality, 310, 313
Gravitational reality, 379, 385
Gravitational wave, 8, 395
Gravitoelectric filament, 311, 312
Gravitoelectric-effect, 171, 196, 202
Gravitoelectromagnetic filament, 311
Gravitomagnetic filament, 312
Gravity, 23, 24, 43, 60, 155, 167, 207, 262, 307, 379, 384, 387, 399
Green Emperor, 171, 188
Greeter consciousness, 115, 147, 301, 309, 325
Greeter face, 216, 217
Greeter Lord, 339
Greeting, 150, 220, 328
Grey color, 246
Guided class, 242
Guider, 2, 7, 67, 73, 80, 101, 103–107, 110, 115, 131, 135, 142, 144–146, 154, 157, 162, 163, 170, 172, 173, 176, 178, 196, 199, 205, 208, 213, 214, 235, 236, 239–242, 248, 269, 270, 278, 284, 294, 295, 297, 300, 304, 309, 312, 313, 315, 318, 329, 336, 349, 361, 363, 370–373, 379, 380, 383–385, 393–395, 397, 399
Guider chains, 312
Guider class, 242
Guider Godhead essence, 295
Guider interference, 380
Guider matter, 196
Guider mediation, 163, 178, 205, 214, 329
Guider momentum, 235
Guider momentum quantum number, 235
Guider nymph, 372
Guider particle, 313
Guider potential, 294, 313, 379, 395
Guider spirit with three eyes, 67
Guider value, 208
Guiding class, 242
Guilty, 400

H
Half atom, 130, 312
Half chains, 312
Half circled-effect, 229
Half Einstein, 343
Half face, 278
Half filament, 313
Half hexahedron, 267
Half photon, 231
Half square, 229, 402
Half squared-effect, 229
Half triangulated-effect, 229
Half upsilon, 17
Half wavefront, 263
Half-crescent lunar body, 220
Half-moon phase, 207
Handed blackhole, 326
Hardness, 208
Heavier blackhole, 325
Heaviest blackhole, 326
Heaviness, 3, 50, 60, 84, 160, 174, 209, 242, 265
Heavy blackhole, 325
Heavy chain, 312
Hecate, 202
Hed, 219, 400
Height, 39, 40, 44, 152, 360
Heliocenter, 304, 305
Hell, 79, 82, 241, 310, 327, 333, 335
Helplessness, 399
Hermione, 262
Herring, 265
Herringbone, 265
Hesus, 259
Heterocyclic ring, 109
Heterogeneous, 370
Hexahedral, 264, 265, 268
Hexahedron, 266–268
Hexaquark, 138, 253

English Index

High redshift quiescent galaxy, 271
High-density amorphous ice, 196
Hindsight, 224, 236, 257
Hippocoon, 261
Histone H3, 348
Histone H6, 346
Hodgepodge, 400
Hologram, 201
Holograph, 13
Holographic duality, 13
Holy Cow, 246
Homogenizing, 238
Homogenizing-effect, 238
Homology class, 242
Horizontal direction, 220, 287
Horizontal filament, 311
Horizontal motion, 329, 330
Horizontal order, 19, 61, 74, 170, 186, 239, 296, 301, 378–381, 383
Horizontal symmetry, 74, 211, 297
Horizontal thermodynamic-effect, 185
Horizontally, 70, 80, 81, 205
Human, 103, 114, 125, 140, 145, 149, 150, 156, 157, 165, 186, 189–191, 198, 201, 202, 214, 219, 220, 236, 291, 326, 338, 372, 375, 384, 398, 404
Human being, 156, 165, 214, 220, 338, 384
Human entity, 214, 326
Human force, 186, 191, 202, 219, 372
Human Godhead, 157, 384
Human-effect, 125, 140, 145, 149, 189, 201, 372, 404
Humility, 199, 200, 246
Hundred filaments, 312
Hundred primordial suns, 294
Hundred thousand filaments, 313
Hundred thousand sides, 355
Husband, 190, 262
Hybris, 262
Hypnos, 262

I
Icarius, 261
Ideal-effect, 3, 9, 10, 16, 31, 33, 40, 44–46, 52, 68, 70, 72, 75, 112, 119, 126, 135, 208, 272, 273, 303, 399
Ideology, 44
Ignorance, 296, 373
Illuminated universe, 270
Illuminating value, 30, 136, 227, 271, 389
Illumination, 76, 148, 282, 292, 313, 392
Illuminator, 2, 6, 7, 77, 221, 249, 272, 273, 275, 281–283, 310, 336, 392, 404
Illuminator deity, 272, 273, 275, 281, 283, 404
Illusion, 5, 8, 10, 24, 26, 27, 33, 34, 39, 48, 62, 63, 68–70, 72, 73, 76, 78, 80, 81, 83, 85, 86, 90, 142, 143, 180, 226, 251, 252, 255, 270, 276, 300, 361, 362
Illusionary, 7, 8, 11, 12, 17, 18, 22, 25, 30, 33, 34, 40, 41, 52, 54, 56, 71, 89, 90, 92, 108, 111, 119, 124, 127, 149, 154, 156, 167, 177, 187, 192, 197, 218, 219, 222, 230, 231, 239, 240, 244, 250, 262, 271, 276, 294, 295, 300, 301, 315, 317, 329, 338, 345, 351, 362, 370, 371, 383, 394
Illusionary balance, 230

Illusionary coherence, 362
Illusionary greeter, 187
Illusionary growth, 218, 219
Illusionary mass, 197
Illusionary plane, 294
Illusionary production, 7, 8, 12, 17, 18, 30, 52, 111, 127, 149, 218, 219, 222, 300, 301, 370, 383
Illusionary reproduction, 11, 222, 231
Illusionary transport, 18
Imaginary, 6, 12, 16, 18, 22, 25, 31, 32, 34, 39–41, 43, 45, 52, 53, 56, 62, 63, 71, 76, 78, 82, 85, 86, 89, 90, 108, 112, 119, 124, 128, 159, 166, 172, 183, 192, 197, 218, 230, 231, 250, 252, 258, 262, 263, 271, 294, 300, 327, 328, 338, 345, 349, 351, 363, 370, 394
Imaginary mass, 197
Imaginary plane, 294
Imagination, 10, 11, 34, 62, 66, 71, 118, 119, 136, 138, 148, 176, 184, 219, 240
Imagination potential, 219
Immanent, 4, 6, 12, 14, 22, 23, 26, 51, 53, 61, 66, 70, 74, 75, 77, 100, 113, 117, 123, 135, 147–150, 157, 158, 160, 161, 190, 252, 254, 255, 265, 274, 288, 297, 301, 304, 313, 316, 336, 365, 366
Immediate freedom, 240
Immediatism, 333
Immortal, 88, 280
Impassioned devotee, 257
Imperium cycle, 326
Impietas, 263
Imposed, 8, 271
Impulse, 45
Inanimate, 123, 141, 167, 169, 204, 316, 338, 391
Inanimate primordial maternal, 338
Inanna, 258, 338, 376
Incandescent, 220
Incarnate, 15, 62, 115, 118, 137, 140, 141, 174, 236, 297, 348, 370, 375
Incarnational value, 197
Incidental coherence, 367
Includes, 107, 181, 221, 288, 318, 319, 391
Individual, 3, 9, 23, 36, 68, 102, 109, 110, 114, 136, 148, 181–183, 200, 219, 225, 241–243, 315, 386, 387, 390, 400
Individuality, 201, 371
Inertial mass, 197
Inertial path, 198
Infinite, 10, 49, 70, 71, 79, 81, 83, 87, 90, 92, 101, 102, 104, 105, 107–109, 113, 115, 116, 120, 123, 124, 128, 136, 137, 156, 157, 160, 165, 169, 172, 173, 176, 179, 180, 185–190, 198, 204, 209, 211, 214, 218, 224, 231, 239, 240, 243, 245–247, 253–257, 275, 279, 281, 286, 287, 292, 294, 296, 297, 300, 301, 305, 306, 308–310, 313, 314, 319, 320, 324, 325, 329–332, 334, 336, 337, 349, 350, 371–374, 379, 380, 383–386, 390, 398, 403
Infinite continuity, 108, 120, 334
Infinite creation, 204
Infinite creator, 204
Infinite creator-effect, 204
Infinite development, 188, 384
Infinite equanimity, 286
Infinite exchange value, 109, 209

415

Infinite flash, 198, 325
Infinite forms, 87, 128, 188, 189, 198, 336
Infinite future, 124, 136, 380
Infinite light, 257, 309
Infinite manifestable, 240
Infinite manifestor, 240
Infinite mother, 256
Infinite perpetuator, 218
Infinite potential reality, 123
Infinite shadow, 245, 246
Infinite value, 287, 314, 320
Infinite wishables, 239
Infinite wisher, 239, 292
Infinitely long surface, 109, 299
Infinitesimal, 197, 227, 249
Infinity, 3, 8, 29, 31, 40, 42–44, 53, 56, 67, 70, 74, 75, 77, 80, 83, 84, 86, 91, 100, 102, 104, 110, 123, 133, 134, 138, 139, 143, 155, 156, 170, 173, 174, 181, 197, 198, 205, 212, 214, 227–229, 234, 238, 244, 245, 248, 249, 264, 278, 280, 287, 289, 295–298, 305, 309, 310, 320, 322, 336, 337, 340, 341, 355, 364, 374, 399
Infinity of causation, 91, 181, 197, 212, 355
Infused, 72, 172, 187, 232, 266, 329
Ingrained, 257
Inherent, 8, 11, 71, 244, 386, 387, 399, 402
Inner sphere, 322
Institution, 126, 192, 200, 241, 291, 292
Institutional force, 126, 241
Institutional param deity, 282
Inswinging, 252
Integrating, 123, 129, 133, 150, 269, 293, 339
Intellect-born dimension, 108
Intellect-born primordial maternal, 338
Intellectual body, 268, 287, 371
Intellectual coherence, 366
Intellectual formation, 202
Intellectual tranquility, 257
Intellectual truth, 370
Intensely, 40
Intensifies, 50, 120, 198, 325, 361
Intensity, 40, 92
Intentional vision, 297
Interacting, 5, 271, 390
Interactive Particle Relic, 183
Interconnectedness, 385
Interesting, 279
Interference, 88, 90, 91, 129, 132, 286
Interjects, 241
International, 8, 254, 273, 372
International exchange, 254
Internet of bodies, 372
Internet of intelligence, 372
Intersection, 344
Intrinsic entity, 384
Intrinsic perfection, 195, 404
Intrinsic potential, 109, 149, 392
Intrinsic sentient-effect, 254
Intrinsic value, 253
Intuitive knowing, 261
Intuitive path, 254
Invisible hand, 73, 81, 82, 87, 161, 162, 169, 238, 373
Iota, 218, 219
Iota energy, 218, 219
Iron hand, 256

Isospin, 201

J
Jet-effect, 326
Jets, 325
Joy, 84, 187, 229, 281, 292, 337, 340, 380, 399
Joy cycle, 229
June, 248, 404
Just wishing, 207

K
Kept away, 249, 265, 296
Ker, 264
Keres, 264
Keter, 247
Kilonova, 228
Kinetic energy, 49, 57, 60, 187
Kinetic momentum, 236
King of deity, 29, 67
King of Gods, 399
King of hell, 258
Kingship, 376
Knocking, 216
Knower, 6, 10, 57, 172, 177, 178, 184, 187–189, 202, 206, 238–240, 308, 392, 399, 400, 403, 404
Knower deity, 184, 187–189, 206, 239, 240, 308, 404
Knower value, 239
Knowledge, 68, 109, 172, 177, 274, 334, 337, 386, 392
Known reality, 4, 13, 19, 23, 24, 84, 88, 112, 374

L
Laborious path, 386
Lachesis, 267
Lambda dark matter, 192
Lambda light matter, 192
Lateral edge, 266, 366
Lateral vertex, 266, 366
Latitudinal growth, 201
Laughlin charge pumping, 20
Leadership behavior, 370
Leadership pathway, 285
Leadership work, 116, 187
Least distance, 5, 7, 8, 53, 224, 238, 300
Least suppression, 16
Left energy, 326
Left-facing, 313
Length, 39, 40, 44, 65, 100, 152, 356, 358, 360
Liberated entity, 371
Liberation, 376
Liberator deity, 244, 290, 292, 404
Life behaviors, 372
Life planning, 376
Light, 3, 5–7, 14, 17, 22, 24–26, 28, 29, 32, 33, 39–48, 50–52, 57, 58, 60–62, 64–66, 69–74, 76–84, 86–88, 91, 92, 94, 96, 97, 112, 120–122, 124, 130, 134, 136, 137, 139, 144, 145, 147, 148, 150, 152, 154, 155, 158, 160–167, 169, 173–175, 178–180, 183–186, 192, 194, 196–199, 204, 207, 209, 224, 225, 228, 231–233, 237, 247, 248, 255, 258, 261, 265, 271, 284, 285, 288, 289, 292, 297, 299,

English Index

300, 310, 312–314, 319, 323, 325, 351, 367, 370, 371, 375, 376, 378–381, 395, 396, 398, 400, 401, 403, 404
Light blackhole, 325
Light chain, 312
Light coherence, 367
Light half, 178, 192, 194, 404
Light matter, 184, 192, 196, 228, 312
Light matter filament, 312
Light reality, 233
Lightener, 260, 261
Lighter blackhole, 325
Lightest blackhole, 325
Light-matter interaction, 33
Lightning, 207, 258, 313
Lights, 47, 48, 58, 65, 68, 87, 120–122, 136, 146, 223, 236
Limbo, 241
Limitation, 8, 124, 200, 202, 330, 374
Limitless, 263, 296, 298, 367
Line filament, 312
Line polygon, 353
Linear system, 249–251
Linear-effect, 249, 264, 393
Living entity, 102, 215, 274, 316, 401
Living subject, 308
Local universe, 308
Lone being, 246, 404
Long chain, 310–312, 382
Longitudinal growth, 201, 306
Longitudinal growth dimension, 201
Lord, 115, 176, 187, 188, 195, 198, 246, 282, 309, 310, 335, 337–340, 375
Lord of deities, 115, 187
Lord of grey color, 246
Lord of Lords, 337, 338
Lord of water, 246
Lorentz factor, 44, 45
Loss of momentum, 45
Low redshift quiescent galaxy, 271
Low-density amorphous ice, 196
Lower class limit, 41
Lower-dimensional space, 14
Lowest face, 278
Lumina, 203
Luminosity, 378
Luminous, 2, 4, 14–16, 26, 45, 49, 99, 100, 102, 125, 126, 130–132, 137, 139, 143–145, 147, 149, 157, 158, 171, 173, 174, 180, 188–190, 197, 202, 206, 218–221, 223, 225, 227, 228, 231, 236, 237, 243, 258, 259, 261, 269, 273, 276, 279, 281–285, 291, 296, 300, 302, 308, 315, 321, 324, 335, 337, 346, 367, 370, 379, 385–387, 393, 394, 396, 403
Luminous coherence, 367
Luminous entity, 102, 337
Luminous system, 145, 171, 173, 190, 300
Luminous type, 321
Lunar, 178, 179, 196, 198, 207, 220, 227, 246, 279, 313
Lunar body, 220
Lunar body system, 220
Lunar crescent, 220
Lunar element, 198
Lunar energy, 207
Lunar flash, 313
Lunar matter, 196
Lunar nectar, 246
Lunar universe, 279
Lycus, 262
Lying backward, 221

M
Macrocosm, 58, 61, 79, 291
Macrolensing, 223, 225
Magnetic field, 95, 130, 396
Magnetic filament, 311
Magnetic frustration, 254
Magnetically, 40
Magnetic-effect, 396
Magnetization, 172, 329
Male-effect, 402
Manifestable body, 195, 196
Manifestation, 240, 384
Manifested value, 198, 275
Manifestor, 6, 169, 177, 189, 190, 192–195, 202, 206, 240, 276, 387, 392, 404
Manifestor deity, 189, 190, 192–195, 202, 206, 240, 404
Manipulation, 239, 371
Many-to-one correspondence, 14
March, 298, 314, 404
Mars, 51, 104, 105, 129–132, 135, 148, 149, 160, 161, 193, 258, 268, 340, 380
Masculine, 13, 23, 62, 66, 98, 105, 106, 109, 113–115, 121, 142, 162, 163, 172, 179, 182, 183, 206, 211, 219, 262, 269, 273, 300, 301, 321, 326, 333–336, 341, 342, 360, 368, 383, 385, 386, 390, 394
Masculine blackhole, 326
Masculine body, 179, 211, 386
Masculine tooth, 206
Masculinity, 12, 109, 114, 128, 182, 191, 205, 206, 217, 219, 256, 258, 270, 273, 279, 280, 282, 360, 374
Masculinity within, 206
Masculinization, 23, 175, 376
Mask, 214
Mass consciousness, 111, 120, 166, 208, 375, 388
Mass force, 184, 207
Mass points, 299
Mass preservation, 44
Mass radius, 381
Massive blackhole, 322–325, 327
Massive galaxies, 285
Massive object, 3, 223, 225, 271
Massive quiescent galaxy, 271
Mate coherence, 367
Mater Matuta, 257
Material cause, 197
Material face, 214
Material well-being, 205
Materialism, 190
Materialization, 6, 309
Maternal, 64, 103, 106, 113–115, 141, 163, 164, 191, 205, 218, 243, 244, 275, 296, 297, 338–340, 386
Maternal community, 191
Maternal consciousness, 113, 141, 218
Maternal energy, 205
Maternal Lord, 338, 339
Matrilineal, 281

Matrilineal performing, 281
Matter energy, 41
Matter filament, 312
Maximal information packing, 379
Maximalism, 333
Maximalist, 333
Mediated, 8, 13, 14, 22, 28, 31, 42, 57, 74, 104, 105, 108, 168, 180, 205, 211, 222, 250, 259, 287, 291, 303, 356
Mediated oneness, 205
Mediating layer, 258
Mediating-effect, 204
Mediation, 17, 18, 24, 81, 108, 169, 199, 239, 242, 284
Mediatism, 332, 333
Mediator, 24, 217, 239, 254, 269, 276, 281, 290, 370, 371, 396
Meditative state, 220
Memory, 126, 243, 244, 269, 282, 304, 315, 369, 393, 402
Memory-effect, 244, 402
Menelaus, 261
Mental body, 269, 287, 371
Mental coherence, 367
Mental force, 219
Mental formation, 202
Mental limit, 373
Mental potential, 219
Menulis, 258
Mercurial-effect, 255
Mercury, 103–105, 127, 128, 130, 131, 135, 148, 160, 161, 193, 255, 258, 269, 340
Meso body, 284
Metal face, 214
Metaphysical body, 369
Metaphysical entropy, 105, 106, 308
Metaphysical subject, 252
Metaphysical value, 155, 294
Metaphysical wave, 155, 250
Meter, 255, 381
Metric system, 384
Metric tensor, 110, 157
Metrication, 384
Metta, 390
Micro energy, 201
Micro-body, 284
Microcosm, 61, 79, 291
Microlensing, 223, 225
Middle body, 268
Mighty creation, 259
Mind-born complexity, 373
Mind-born creator, 273, 338, 362, 380, 385
Mindful, 119, 120, 123, 126, 133, 134, 149, 159, 320, 370, 373, 383
Mindful manifestation, 370
Mineral energy, 325
Mineral face, 214
Minimalism, 332, 334, 358
Miracle, 261
Misty, 202, 246
Mithras, 263
Mitochondrion molecule, 140, 281
Moderatism, 332, 333
Modulus, 242, 243, 245, 249
Modulus causation, 243
Modulus cause, 243, 245
Modulus deity, 245
Modulus entity, 245
Modulus space, 243, 245
Modulus time, 243, 245
Momentum, 6, 8–11, 24, 25, 27–29, 36, 40, 41, 43–48, 62, 71, 78, 81–84, 110, 111, 118–120, 125, 128, 137, 139, 147, 157, 158, 168, 170, 185, 189, 200, 224, 225, 232, 233, 235–237, 279, 283, 328
Momentum flux, 40
Momentum force, 125, 189
Monsoon, 171, 248
Moon, 51, 103, 105–107, 125, 132, 148, 149, 161, 178–182, 193, 194, 196, 234, 258, 263, 280, 299, 300, 307, 368
Morning, 245, 257, 258, 286
Morning star, 245, 257, 286
Mother, 7, 12, 15–18, 29, 30, 36, 37, 51, 52, 63, 64, 66, 67, 80, 107, 113, 115, 133, 141, 142, 150, 190–192, 207, 209, 213, 218, 256, 257, 259, 262, 263, 275, 276, 281–284, 288, 290–293, 295, 296, 304, 310, 315, 327, 338, 339, 358, 364, 365, 370
Mother nature, 15–18, 51, 52, 67, 107, 133, 141, 142, 150, 256, 262, 275, 276, 281–284, 288, 290–292, 295, 296, 304, 310, 315, 327, 339, 358, 364, 370
Motherhood, 257, 258
Mothering, 190, 218
Mountaintop, 258
Movable reality, 228
Multilateral coherence, 362
Multilocalized universe, 190
Multiplication, 119, 158, 160, 179, 183, 184, 224, 267
Multiplier, 18, 19, 49, 67, 74, 89, 90, 129, 130, 132, 133, 137, 152, 159, 160, 178–180, 186, 189, 203, 206, 210–213, 217, 229, 230, 232–234, 236, 237, 241, 247–249, 251–253, 273, 274, 286, 287, 297, 304–306, 315, 317–319, 324, 331, 332, 348, 349, 351, 359–361, 369, 370, 375, 376, 394, 399
Musicality, 121, 252
Mutation, 334
My experience, 197, 198, 201
My God, 240–242
Myself, 155, 198, 229, 275
Myth, 254

N

Nascent star, 245
National, 8, 115, 159, 254, 255, 273, 372, 386, 393
National group, 254
Natural essence, 167, 291
Natural gravity, 262, 404
Natural growth, 117, 193, 195, 224, 225, 243, 301, 347, 348, 387, 391
Natural quality, 245
Nature, 2–4, 8, 13, 15, 22, 27–29, 32, 33, 49, 51–53, 56, 66–68, 72–75, 77, 80, 112, 117, 126, 137, 186, 187, 191, 192, 196–198, 202, 204, 208–211, 215, 226, 241, 245, 252, 256, 262–264, 269, 275, 276, 279, 281, 288, 291, 292, 295, 305, 310, 315, 318, 327, 330, 334–337, 341, 345, 346, 361, 365, 368, 370, 390, 396, 402

English Index

Negative, 17, 18, 29, 31, 50, 56, 66, 69, 70, 72, 73, 77, 80, 83–85, 89, 90, 119, 123, 126, 128, 129, 136, 137, 142, 143, 157, 170, 172–176, 181, 203, 208, 214, 221, 242, 245, 249, 251–253, 257, 296, 307, 310, 326, 383, 396
Negative cosmological constant, 17
Negative energy, 18, 72, 84, 85, 123, 136, 310, 396
Negative internality, 203
Negative one, 31, 50, 69, 70, 73, 80, 83, 84, 89, 119, 173–176, 208, 214, 221, 242
Negative realm cycle, 326
Negatively charged state, 214
Neptune, 104–106, 131, 132, 136, 137, 146–148, 155, 161, 193, 269, 340
Nescience, 373
Netzach, 188
New beginning, 257
New Lemuria, 239, 375
New moon, 106, 125, 126, 137, 194, 220, 395
New paradigm, 33, 104, 105, 239, 301, 308, 309, 335
New planet cycle, 325
Nine bodies, 283, 284
Nine down quarks, 180
Nine filaments, 312
Nine linear systems, 249, 250
Nine massive blackholes, 325
Nine moons, 194
Nine photons, 29, 166, 194, 231
Nine primordial suns, 294
Nine up quarks, 180
Nineteen faces, 277
Nineteen filaments, 311, 312
Ninety filaments, 312
Ninety-degree rotation, 267
Ninety-six faces, 216
Ninth photon, 180, 231
Ninth quadrant, 216
Niobe, 262
Non-Abelian Anyon, 402
Non-adjacent side, 356, 359
Noncoherence, 361
Non-congruent side, 356, 358
Non-doer, 245, 369
Non-existing universe, 258
Nonhistone, 348
Nonlinear complexity, 371
Nonlinear sound, 308
Nonlinear system, 249
Nonlinear wave, 8, 91, 95, 262
Nonlinear-effect, 253
Nonlinearity, 65, 82, 399
Nonlinearum cycle, 325
Nonliving subject, 308
Nontrivial cohomology class, 242
Non-violence, 390
Normality, 110, 272
Normalization, 386
Normative development, 107, 113, 150, 151, 153–155, 157, 159, 161, 163, 165, 167, 169, 171, 173, 175, 177, 179, 181, 183, 185, 187, 189, 191, 193, 195, 197, 199, 201, 203, 205, 207, 209, 211, 213, 215, 217, 219, 221, 225, 302, 370, 384, 386
Normative justice, 177, 188, 207, 373
Normative performing, 281
Normative phase, 244
Normative truth, 261, 290
Normative universe, 307
Norming reality, 53, 55, 56
Northern Dipper, 170
Nothingness, 219
Nucleus, 98, 100, 132, 135–137, 140, 145, 217, 232, 233, 323, 328–330, 402
Number density, 219
Numerator, 14
Nurturing, 207

O

Observable, 26, 39, 43, 44, 85, 86, 189, 217, 233, 267, 319, 396, 401, 402
Observable reality, 402
Observation-effect, 319
Octahedral-effect, 264
Octahedron, 264–268
Octave, 13, 18, 38, 73, 79, 113, 120, 129, 138, 139, 141, 142, 155, 156, 161–163, 165, 166, 178, 179, 182, 183, 196, 202, 208, 210, 224, 227–235, 246, 248, 253, 261, 262, 264, 269, 271, 277, 280, 281, 283, 284, 304, 305, 308, 317, 320, 321, 329, 383, 389, 394, 401, 402
Octave of antiquark, 253
Octave of atoms, 183
Octave of element-effects, 208
Octave of elements, 73, 141, 142, 208
Octave of genus, 321
Octave of hexaquarks, 253
Octave of phenotypes, 321
Octave of potential hexaquarks, 253
Octave of quark, 253
Octave of species, 321
Octave of three, 13, 18, 232–235, 248, 262, 277, 283
Odd dimension, 241, 369
Odd form, 304, 368, 369
Odd-dimensional class, 242
Oddness, 156, 241, 242, 306
Offspring, 256
Old moon, 194
Old season, 314
Omega, 219
Omega energy, 219
Omnipotent, 172, 204
Omnipresent, 2, 31, 39, 41, 43, 71, 82, 112, 115, 119, 128, 131, 173, 174, 178, 209, 231, 272
Omniscience, 285, 399, 402
Omniscient, 240, 285, 309
Omniscient consciousness, 285
One thousand filaments, 313
One with all, 134, 326
One with potential, 236, 237
One with present, 237
One with something, 34, 235
One within action, 206
One within six, 205
One-hundred-sixty stones, 346
One-hundred-twenty copies, 321
Oneiros, 264
Oneness, 11, 16, 18, 44, 53, 66, 67, 73, 75, 94, 95, 101, 106, 107, 110, 122, 124–126, 132, 142, 143, 159, 168–170, 178, 181, 182, 186, 187, 191, 194, 204, 209–215, 225, 232, 240,

244, 249, 253, 262, 264, 270, 273, 276–278, 281, 290, 297, 305, 306, 308, 314, 317, 332, 337, 346, 348, 357, 360, 362, 364, 368, 384, 387–389, 392, 395, 401
Oneness coherence, 364
Oneness of programming, 204
Oneness-effect, 106, 107, 194
One-to-many correspondence, 13, 319
One-to-one correspondence, 13
Onspring, 256
Ontological cause, 373
Ontological value, 294
Opening polygon, 355, 356
Orbital, 14, 15, 228–237, 264, 277, 280, 301, 329, 333, 361
Orbital of two, 14, 15, 233–235, 361
Ordering, 254
Organization, 33, 34, 116, 130, 148, 153, 174, 183, 200, 238, 249, 250, 282, 285, 293, 303, 306, 317, 334, 361, 377, 384, 396, 401, 402
Organizational action, 301, 302
Organizational energy, 219
Organizational reality, 99, 101, 103, 105, 107–111, 113, 115–117, 119, 121, 123, 125, 127, 129, 131, 133, 135, 137, 139, 141, 143, 145, 147, 149
Organizational sameness, 47, 105, 106, 204, 399
Organizational solution, 240
Organizational value, 249
Origin of four, 243
Origin of negative two, 243
Origin of one, 71, 243, 246
Origin of seven, 245
Origin of six, 245
Origin of three, 242
Origin of zero, 242, 246
Orthogonal, 230, 266, 355
Orthogonal p orbitals, 230
Orthograde, 329
Orthograde motion, 329
Oscillating circle of six, 255
Oscillation, 255
Oscillationum cycle, 326
Oscillon, 395
Otherness, 2, 9, 50, 72, 84, 177, 262, 335
Otreus, 262
Our God, 241, 242
Outer sphere, 322
Outswinging, 252
Overexpressed, 333
Oversoul family, 198

P
Pair of four filaments, 311
Pale yellow, 252
Para blackhole, 326
Para complex number, 279
Para conscious coherence, 367
Para creation, 204
Para deity, 2, 112, 114, 176, 177, 190, 218, 226, 240, 297, 299, 309, 310, 334, 335, 400, 403
Para energy realm cycle, 326
Para entity, 23, 28, 114, 115, 125–129, 131, 136, 143, 171, 185, 242, 274, 287, 325, 335, 364
Para entity energy, 325

Para multiplier, 195, 254
Para round, 395
Para time, 220, 370
Para wisher, 337, 383, 384, 388, 404
Parabolic orbit, 330
Parabolic radius, 381
Para-conscious value, 220
Para-conscious wishables, 207
Para-consciousness, 172, 373
Paradigm, 4, 13, 33, 50, 53, 68, 66, 74, 75, 85, 103, 105–107, 123, 126, 132, 149–151, 153–155, 157, 159, 161, 163, 165, 167, 169, 171, 173, 175, 177, 179, 181, 183, 185, 187, 189, 191, 193, 195, 197, 199, 201, 203, 205, 207, 209, 211, 213, 215, 217, 219, 221, 225–227, 299–302, 308, 336, 384, 403
Parallel face, 267
Parallel filament, 312
Parallel symmetry, 212
Parallel system, 250
Parallelogram, 152, 358
Param complex number, 280
Param conscious coherence, 367
Param deity, 2, 112, 114, 191, 215, 226, 281, 282, 296, 297, 300, 308, 318, 335, 399
Param deity kingdom, 297
Param deity type, 318
Param entity energy, 326
Param greeter type, 320
Param massive blackhole, 324
Param maternal, 243, 275, 290, 403
Param modulus, 242
Param paternal, 275
Param reality, 291
Param son, 188, 190
Param truth, 245
Param wave packet, 286
Parametric resonance, 379
Partial congruence, 304
Passionate coherence, 363
Past, 2, 3, 5, 6, 8, 10–13, 16, 17, 19, 22–25, 30–33, 38–42, 46–48, 51, 56, 58, 60–63, 65–71, 74, 76–78, 80–90, 92, 95–98, 100–110, 112, 116–119, 123–129, 131–133, 135–143, 149–152, 154, 160, 167–171, 173, 174, 176–181, 185, 188, 192–194, 197, 198, 201–203, 205, 207, 209–211, 213–218, 221–223, 225, 227, 228, 231, 234–238, 243, 247–249, 257, 268–272, 274–276, 279–281, 283, 285, 287, 288, 290, 293, 295–297, 299–301, 303, 305, 306, 308, 310–312, 314, 316, 322, 324, 325, 327, 329–331, 333, 335, 338–340, 345, 346, 350, 355, 356, 358, 359, 363, 364, 367–370, 374, 376–381, 386, 390, 393–396, 398, 401–403
Past generation, 109, 197, 211
Past moment, 314, 316
Past of everybody, 31, 141, 173, 202, 356
Past orientation, 126, 234
Past reality, 101, 117, 176, 177, 207, 209, 213, 214, 216, 249, 257, 299, 301, 355, 368, 374, 394
Paternal, 7, 102, 104, 105, 115, 164, 205, 243, 244, 338, 339, 373, 386, 399
Paternal consciousness, 7, 205
Paternal copy, 386
Paternal lord, 338, 339

English Index

Path of alienship, 368, 372
Path of alienship action, 368
Path of citizenship action, 314
Path of deification, 207, 387
Path of devotion, 177, 226, 301, 337, 388
Path of dimension, 171
Path of divinity, 204
Path of goalpost, 257
Path of leadership, 203, 238
Path of least action, 238, 368
Path of mindful, 370
Path of my-effect, 238
Path of organizational action, 302
Path of Para Wisher, 388
Path of primordial causation, 189
Path of primordial culture-effect, 257
Path of primordial time, 168
Path of technological cost, 373
Path of transcendental, 147
Path of twin atom, 141
Path of twin nucleus, 137
Path of twin rank, 123
Path of value, 303
Path of worship action, 377
Path of your-effect, 244
Pelopeia, 261
Pelops, 262
Penis, 258
Perfect, 8, 33, 66, 110, 160, 196, 198, 202, 203, 220, 232, 234, 248, 260, 269, 276, 285, 289, 292, 295, 303, 304, 310, 315, 319, 334, 374, 376, 379, 384, 390, 395, 402
Perfect creation, 196
Perfect creature, 198, 276
Perfect exchange system, 196
Perfect follower, 202
Perfect personification, 292, 310
Perfect potential, 203, 232, 234, 248, 303
Perfect vision, 220
Perfect wishable, 285
Perfection system, 250
Peripheral ergosphere, 322
Peripheral reality, 400
Perished consciousness, 174, 187, 205
Perkunas, 258
Perpendicular orbital, 230
Perpetuable reality, 249
Perpetuate, 19, 24, 27, 31, 49, 103, 116, 118, 119, 138, 139, 143, 156, 158, 174, 176, 180, 181, 185, 187, 210, 214, 215, 227, 230, 246–248, 261, 276, 285, 291, 296, 312, 323, 327, 335, 336, 342, 347, 350, 352, 361, 369, 370, 376, 379, 380
Perpetuating value, 12, 27, 30, 74, 84, 128, 129, 142, 156, 158, 169–171, 174, 180, 186, 188–190, 198, 199, 203, 218, 222, 225, 227, 234, 284, 289, 290, 306, 309, 330, 388, 395, 403
Perpetuator, 6, 77, 159, 175, 178, 217, 222, 238–240, 243, 244, 249, 264, 387, 392, 404
Perpetuator deity, 217, 222, 238–240, 243, 244, 249, 264, 404
Perpetuity, 76, 77, 121, 326
Person, 31, 56, 57, 101, 141, 142, 159, 192, 206, 211, 251, 270, 388, 389
Personal essence, 346, 373

Personal sentient well-being, 56, 149, 201, 254, 301
Personality, 192
Pessimism, 215, 332
Petulantia, 262
Philotes, 262
Photocatalysis, 175
Photon, 26–29, 32, 43, 44, 58, 60, 65, 70–72, 76–81, 84, 91–93, 97, 122, 129, 130, 139, 164, 166, 180, 181, 218, 223–225, 231–233, 286, 321, 330, 380, 385, 386, 397, 402
Photonic reality, 233
Physical body, 102, 161, 268, 287, 371, 376, 398
Physical coherence, 367
Physical contact, 297
Physical formation, 202
Physical subject, 252
Physical value, 19, 233
Physical wormhole, 17, 379
Pi of inaction, 46, 129, 233, 234, 388
Piercing, 247, 319
Piezomagnet, 36, 37
Pisces zodiac, 314
Planck, 40, 44, 236
Planck constant, 44
Planck momentum, 236
Planet, 51, 130, 131, 135, 148, 165, 196, 221, 258, 263, 279, 280, 295, 378
Plant branch, 320
Plant face, 214
Planted, 218
Plasma filament, 313
Plunge, 325
Pluto, 261
Point filament, 312
Point polygon, 353
Point reality, 19
Point wave, 263
Pointed polygon, 352, 353, 357, 358
Polar, 258
Polaris, 192
Polarization, 51, 271, 307, 313, 322
Polarization signal, 307
Pollination, 401, 402
Pollinator, 402
Polluted divine-effect, 371
Pope, 256
Poseidon, 262
Positive, 17, 18, 29, 56, 66, 72, 80, 83–85, 90, 126, 128, 129, 136, 137, 141, 157, 170, 172–174, 176, 179, 181, 203, 208, 243, 251–253, 296, 307, 310, 326, 396
Positive fluctuation, 18
Positive internality, 136, 203
Positive realm cycle, 326
Positively charged state, 214
Postcomposer, 260, 261
Post-composition, 263
Potential, 2, 4, 6–11, 13–19, 22, 23, 27–34, 39, 41, 43, 45–50, 57, 60, 62–64, 66–69, 71–78, 80, 82–87, 89, 90, 93, 94, 97, 98, 101, 104, 107–109, 112–124, 126–138, 140, 141, 144, 149, 150, 152–158, 160–171, 173–175, 178–182, 185, 187, 189–194, 198, 200–205, 207, 209–230, 232, 234–237, 243, 244, 248–250, 252, 253, 255–258, 260, 264–266, 268–270, 272, 274, 275, 277, 281–286, 288,

289, 292–295, 297, 300, 303, 305–307, 311–321, 323, 324, 326–328, 330, 331, 333, 337, 339, 340, 342, 344, 345, 347, 348, 352, 355–358, 360, 362, 363, 369, 370, 375, 376, 378–380, 384, 385, 388, 389, 391, 392, 395, 397, 399–401, 403
Potential actor, 235
Potential angular momentum, 201
Potential chain, 313
Potential charge, 180
Potential color, 18, 380
Potential crescent, 221
Potential cycle, 229
Potential dwarf planet, 378
Potential Einstein, 344, 345
Potential explosion, 228
Potential family, 4, 244
Potential flame, 207
Potential God, 319
Potential greeter face, 217
Potential herself, 198, 256
Potential hexaquarks, 253
Potential knowledge, 303
Potential leptoquark, 180
Potential life, 220
Potential matter cycle, 326
Potential maturity, 257
Potential moon, 180, 182, 194
Potential octave, 227, 264
Potential one, 83, 168, 222, 235, 272, 274, 275, 283, 389
Potential parallelogram, 358
Potential plane, 294
Potential quarks, 90, 253
Potential radiation, 17
Potential reality, 34, 63, 109, 119–122, 230, 234, 333, 337, 401
Potential sentient-effect, 171
Potential sun, 293
Potential temperature, 185
Potential thing, 11, 190, 236
Potential time, 210, 214, 219
Potential triple crescents, 221
Potential twin crescent, 221
Potential twin octave, 265
Potential two, 222
Potential upsilon, 17
Potential zero, 173, 318
Potential-effect, 69, 191, 333
Potentialism, 4, 19, 74, 75, 77, 86, 115
Powering, 326
Preceding lifetime, 109
Precession, 3, 337
Precomposer, 260, 261
Preconceived, 239, 379
Preconceived consciousness, 239
Preconception, 239
Preprogrammed, 245, 350
Present, 2–8, 11–13, 16, 18, 19, 22–26, 28, 29, 31–34, 38, 40–42, 45–48, 50, 51, 56–58, 60–69, 71, 72, 74–90, 92, 95–97, 100–139, 141–144, 148–150, 152–154, 158–162, 165, 167–182, 184, 185, 187–190, 192, 193, 196–198, 202–220, 222, 223, 225–231, 233–237, 239, 241, 243, 246–253, 255, 256, 258, 262, 264–266, 268–276, 278–281, 283–289, 295, 296, 299–301, 303, 305, 306, 308–310, 312–314, 316, 317, 320–322, 328–333, 335–337, 339, 340, 343, 348, 350, 352, 355, 356, 359, 363, 367–372, 374, 378, 380, 381, 384–387, 390, 393, 395, 396, 398, 401, 403
Present behavior, 198, 372, 374
Present being, 188, 204
Present birth, 241
Present creation, 126, 188, 275
Present creature, 103, 202, 204, 233
Present generation, 182, 197, 211
Present manifestation, 274, 384
Present paradigm, 4, 33, 103, 218, 220, 239, 296, 301, 308, 309, 335
Present reality, 32, 108, 109, 116–119, 122, 176, 177, 217, 255, 256, 262, 271, 301, 316, 322, 329, 371, 395
Present self, 112, 241, 249, 264, 393
Present value, 16, 50, 72, 84, 85, 90, 97, 131, 175, 182, 206, 225, 251–253, 316, 343
Present Vishnu, 249
Present wealth, 310
Pressure, 6, 11, 13, 16, 39–41, 44–47, 120, 121, 130, 132–134, 153, 175, 177, 178, 219, 227, 271, 299, 396
Primary blackhole, 325
Primeval, 8, 11, 12, 26–28, 48, 101, 104, 105, 110–112, 114, 115, 120, 122, 124, 126, 131–134, 140, 147–150, 156–160, 162, 163, 174, 184, 188, 191, 193, 199, 215, 217, 220, 222, 225, 226, 228, 242, 245, 249, 251, 269, 273, 276, 277, 279–281, 286, 291, 294, 296, 297, 306, 309, 310, 313, 314, 319, 320, 323, 325, 326, 331, 336, 359, 362, 363, 367, 369, 370, 372, 373, 384, 386, 389, 399, 400, 403
Primeval chain, 313
Primeval coherence, 363
Primeval complex, 279
Primeval creation, 251, 276, 277, 280
Primeval deity face, 215
Primeval enjoyer, 372, 373
Primeval greeter, 126, 132–134, 150, 191, 217, 309, 320
Primeval modulus, 242
Primeval parallelogram, 359
Primeval perpetuator, 104, 220, 222, 281, 369
Primeval truth, 245
Primevalism, 4, 11, 12, 19, 75, 83, 159, 173, 213, 335, 359
Primordial, 2, 8, 11, 12, 14, 15, 26–28, 44, 47, 53, 66, 67, 74, 101–103, 105–110, 113–115, 118–120, 123–137, 139–143, 147–150, 155–159, 162, 163, 166–168, 172, 174, 180, 183, 184, 186–189, 191, 193, 197, 199, 202, 203, 206, 207, 214–216, 218, 220–222, 225–230, 236, 238, 242–246, 248–251, 255, 258, 259, 264, 268, 269, 272, 273, 275–277, 279–282, 285–295, 298, 300, 302, 303, 305, 306, 308–310, 313, 314, 318, 320, 321, 323, 325–328, 331–338, 340, 341, 343–345, 358, 359, 363, 365, 367–370, 372, 373, 375–378, 381, 386, 388, 389, 394, 400, 401, 403, 404
Primordial cause, 203
Primordial chains, 313
Primordial conscious coherence, 367
Primordial consciousness, 197
Primordial cow, 246

English Index

Primordial creation, 251, 276, 277, 280, 281, 292, 314
Primordial deity, 113, 114, 191, 214, 215, 226, 259, 282, 320
Primordial energy, 44, 107, 269
Primordial enjoyer, 372
Primordial frequency, 44
Primordial greeter, 2, 103, 115, 125, 127, 132, 149, 150, 191, 216, 245, 249, 275, 276, 308, 309, 320, 334, 367
Primordial illuminator, 2, 67, 74, 249, 275, 300, 336–338
Primordial knowing, 238, 259, 309
Primordial life, 187, 259
Primordial maternal, 106, 113, 115, 218, 243, 306, 313, 314, 338, 340, 341
Primordial maternal spirit, 313
Primordial modulus, 242
Primordial oneness, 14, 15, 44, 135, 141, 143, 167, 168, 186, 193, 216, 255, 285–288, 290, 292, 295, 314, 318, 323, 327, 332, 334, 343, 345, 358, 365, 370, 375, 381, 386, 389, 394
Primordial pair, 180
Primordial parallelogram, 359
Primordial paternal, 105, 113, 115, 338, 373
Primordial perpetuator, 158, 222, 244, 264, 273, 309
Primordial reality, 291, 370
Primordial self, 2, 53, 67, 102, 134, 135, 166, 172, 188, 207, 218, 220–222, 229, 230, 259, 264, 269, 276, 282, 302, 335, 336, 369, 370
Primordial solution, 335, 400
Primordial space, 8, 118, 124, 183, 295, 298, 363, 400, 401, 404
Primordial space design, 401
Primordial suns, 293, 294
Primordial technological exchange, 137, 314, 335, 404
Primordial technological growth, 136, 302, 309, 404
Primordial technological investment, 142, 377, 404
Primordial tile, 344
Primordial trading-effect, 126, 203, 404
Primordial truth, 53, 244
Primordial wave packet, 286
Primordial well-being, 188, 259
Primordial wisdom, 285
Primordial wisher, 368, 369, 372, 373, 388, 394, 404
Primordialism, 4, 11, 16, 19, 23, 74, 77, 162, 173, 213, 324, 337, 359
Principal guider agent class, 242
Principal guider class, 242
Priolas, 262
Prism, 243
Procreation, 338
Productive, 19, 20, 45, 56, 60, 129, 175, 176, 208–210, 219, 367, 398, 402
Productive charge carrier, 175
Productive charge pumping, 20
Productive coherence, 367
Productive energy, 45, 210
Proficient networking, 200, 371
Proficient workforce, 372
Programming, 122, 172, 279
Progresses, 82, 252, 312
Progressing, 250
Proliferating, 124, 370
Proof, 208
Proportionate trading, 204
Protagonist, 19, 68, 120, 126, 130, 133, 170, 171, 176, 272, 337
Protagonist breeder, 133
Proteus, 261
Proton, 72, 98, 130–132, 161, 166, 185, 231, 233, 381, 382, 396
Psychic force, 41, 397
Pu orbital, 231
Pulsar, 339–341
Pulsar time, 339–341
Pulsation, 339, 340, 379
Pulse-effect, 339
Purgatory, 327
Purple north polestar, 192
Pv orbital, 231
Pyramid, 265, 357
Pz orbital, 230

Q
Quadmagnet, 36
Quadrant, 212, 216, 217
Quadrilateral, 359
Quadrilateral polygon, 359
Quadruple, 229
Qualification, 392, 393
Qualifies, 244, 388–390
Quality, 8, 11, 101, 109, 169, 201, 242, 244, 245, 281, 313, 337, 399
Quantum, 6, 7, 10–13, 15, 17–19, 28, 32, 42, 43, 45, 46, 60–67, 69, 72, 74, 78, 82–84, 87, 91, 92, 95, 96, 111, 118, 132, 139, 142, 144, 147, 184, 225, 235–237, 244, 252, 268, 286, 289, 307, 336, 344, 347, 361, 362, 367, 373, 383, 384, 395, 398, 401
Quantum correction, 373, 384
Quantum decoherence, 362
Quantum field theory, 6, 13, 69
Quantum fluctuation, 18
Quantum force, 118
Quantum Goalkeeper, 235
Quantum God, 235, 336
Quantum light, 12, 132, 139, 144, 147, 367, 398
Quantum loop, 7, 43, 62, 63, 65, 67
Quantum magic, 244
Quantum mechanics, 61, 60, 62, 64, 65, 69
Quantum mechanics of interactivity, 65
Quantum noncoherence, 362
Quantum number, 235, 237
Quantum particle, 15, 17
Quantum photon, 28, 32, 60, 132, 142, 147
Quantum potential, 252, 395
Quantum sequences, 344
Quantum state, 184
Quantum teleport, 19
Quantum uncertainty, 18, 43, 92
Quark, 16, 50, 72, 77, 84, 85, 90, 129, 141, 142, 172, 173, 179–181, 206, 251–253, 272, 356, 386, 387
Quarter, 120, 121, 233, 308
Quasar Triplet System, 285
Qubit, 286
Queen of Divinity, 188

Queen of Heaven, 258
Queen of space, 310, 313
Quiescent system, 271

R
Radial, 313
Radially, 311
Radian, 313, 381
Radiates, 3, 40, 181, 310, 325
Radius, 255, 311, 313, 330, 381
Reactive energy, 205, 325
Real absolute gravitational constant, 263
Real primordial maternal, 339
Real-effect, 126, 208
Realism, 333
Reality, 3–16, 18, 19, 22–27, 29–34, 39–41, 43–45, 49–53, 56, 57, 61–64, 66–81, 84–89, 92, 95, 96, 98, 100, 105, 108–112, 114–120, 122–128, 131, 134–137, 144, 147, 149–153, 156, 158, 159, 169–171, 176–179, 186–189, 192, 196, 205, 207–210, 213, 215, 216, 218, 219, 224, 226–230, 232–234, 236–241, 244, 245, 247, 249–251, 253, 255, 258, 261–263, 267, 269, 271–273, 275, 280–283, 287, 289–291, 296, 297, 299, 301, 303, 304, 306, 313, 315–320, 322–325, 327, 329–331, 334, 335, 337–339, 341, 346–351, 355, 357, 361, 366, 368, 369, 371–377, 384–389, 392, 394, 396, 399–403
Realization sphere, 284
Realm cycle, 326
Reappearing, 155, 195
Rearranging, 208
Recombination, 10
Recomposer, 260, 261
Reconstruction, 154, 256
Rectilinear kinetic energy, 60
Redeemer, 205, 333
Redemptive freedom, 205
Redresses, 208
Refused, 232
Reigned, 241
Reigned God, 241
Reigner, 241
Reigner God, 241
Reincarnate, 172, 209, 214, 243, 313, 392
Reintegrates, 172, 231
Relativity-effect, 3, 4, 16
Religion, 5, 250, 251, 301, 364
Remediation, 239
Remediatist, 334
Remediator, 253
Renewable, 245, 270
Renounces, 201
Repeated, 2, 15, 16, 18, 30, 33, 45, 121, 123, 152, 159, 169, 171, 181–186, 191, 197, 215, 216, 228–230, 233–236, 241, 242, 246–248, 250, 251, 253, 259, 264–267, 273, 275, 278, 281, 284, 286, 289–291, 294, 295, 303, 304, 306, 308, 315, 317, 322–324, 327, 329, 333, 343, 350, 351, 354, 359, 361, 368, 369, 376, 395
Repeated-effect, 228
Repeating fractal, 45, 46
Repetition, 284, 285

Replication, 15, 16, 29, 56, 65, 72, 91, 96, 140, 143, 146, 156, 163, 165, 183, 190, 194, 196, 203, 206, 231, 235, 236, 238, 240, 248, 251, 261–263, 265, 273, 276, 282, 283, 288–290, 292, 293, 296, 298, 304, 310–314, 319, 320, 323, 328, 331, 335, 337, 352, 371, 380, 384, 393, 397, 400, 403
Replicative system, 283
Replicator, 196
Reproduced value, 230
Reproducible, 77, 86, 134, 213, 238, 280, 402
Reproducible electron, 402
Reproduction, 40, 73, 79, 81, 90, 124, 129, 143, 147, 162, 178, 179, 189, 192, 195, 196, 229, 265–267, 269, 279, 280, 286, 291, 369, 370
Reproductive energy, 140, 204, 255
Reproductive energy, 204
Reproductive force, 11, 13, 15, 18, 34, 40, 58, 109, 110, 118, 120, 131, 132, 134, 136, 147, 167, 168, 196, 210, 216, 217, 219, 220, 224, 227, 234, 261, 265, 272, 280, 294, 303, 312, 315, 319, 320, 325, 327, 328, 333, 352, 376, 388, 392, 399
Reproductive potential, 129, 141, 142, 195, 213, 236, 249, 253, 274, 279, 300, 317, 323, 325, 332, 337, 344, 345, 362, 368, 383, 386, 389, 392, 394, 397, 401
Reproductive reality, 17, 156, 159, 180, 186, 188, 217–219, 232, 234, 262, 281, 284, 313, 327, 333, 366, 378
Reproductive variation, 320
Reproofs, 208
Reputation, 215
Residual cause, 310
Restraining, 20, 208
Resummation, 381
Resurrector, 208
Retrograde motion, 329
Rhomboid, 152, 358
Rhombuses, 356
Rhythmic, 379
Right consciousness, 314
Right energy, 326
Ripeness, 257
Rödl nibble, 400
Rotating plane, 294
Rotational center, 18
Rotational motion, 18, 321, 322, 329
Rotator blackhole, 325
Rough time, 208
Rus, 259
Rydberg wave packet, 286

S
Satan, 58, 80, 115, 158, 165, 191, 198, 206–208, 216, 236, 237, 256, 276, 291, 295, 300, 318, 319, 339, 347, 363, 375
Satan momentum, 236, 237
Satan momentum quantum number, 237
Saturn, 104, 130, 131, 136, 148, 149, 160, 161, 193, 202, 258, 268, 339, 340
Saule, 258
Schwarzschild, 16, 17
Scientific cause, 373
Seasons, 314
Second duality, 244

English Index

Second electron, 228
Second event horizon, 328, 329
Second life, 63, 326
Secondary blackhole, 325
Seeded oneness, 384
Sefirot, 313
Selene, 220
Self-acceptance, 384
Self-compassion, 390
Self-condensing, 16, 275, 344, 346, 349
Self-consciousness, 268, 334
Self-crescent, 220
Self-destruction, 249, 401
Self-development, 113, 253, 273, 316
Self-discovery, 259
Self-disjunction, 249, 314
Self-luminosity, 334, 385
Self-luminous crescent, 220
Self-luminous entity, 218, 326, 399
Self-luminous entity energy, 326
Self-luminous one, 192
Self-luminous system, 100, 145, 171, 190, 300
Self-managing consciousness, 373
Self-manifests, 307, 319
Self-perpetuable, 231
Self-projecting, 56, 101, 102, 132, 138, 142, 143, 156–159, 164, 170, 182, 194, 197, 225, 235, 248, 277, 286–289, 307, 315, 327, 333, 336, 343, 346, 358, 361, 364, 369, 381, 383, 385, 389, 393, 394, 397
Self-replication, 202, 384
Self-sovereigning, 240
Self-substantiating, 111, 133, 249
Self-twinning, 206, 346
Semivariable, 9, 192, 293, 303, 378
Sensory perception, 297
Sentient, 1, 8, 23, 33, 40, 73, 74, 79, 84, 85, 87, 88, 95, 96, 102, 104, 109, 111, 113–115, 127, 130, 131, 137, 139, 142, 144, 146, 150, 155, 158, 163, 166, 171–173, 176–178, 184, 188, 190–192, 196–198, 201, 204, 205, 216, 217, 219, 224, 227, 228, 235, 238, 240, 248, 249, 254, 258, 269, 273, 274, 281, 284, 288, 290, 294–298, 300–304, 306, 308, 311, 312, 314, 315, 320–322, 324, 328–331, 334, 336, 337, 347, 358, 362, 363, 365, 370, 371, 385, 387–389, 397–400
Sentient coherence, 362
Sentient energy, 87, 96, 114, 115, 155, 158, 204, 216, 254, 288, 290, 296, 308, 314, 358, 399
Sentient entity, 1, 87, 88, 109, 111, 113, 127, 130, 146, 178, 205, 228, 258, 269, 273, 274, 296–298, 300, 302, 308, 334, 397
Sentient filament, 311, 312
Sentient force, 40, 73, 74, 79, 84, 85, 87, 131, 142, 163, 172, 173, 177, 184, 196, 219, 224, 227, 281, 294, 321, 322, 324, 328, 330, 331, 370, 387, 389
Sentient Godhead essence, 295
Sentient matter, 198
Sentient momentum, 235
Sentient momentum quantum number, 235
Sentient plane, 131, 295
Sentient well-being, 8, 23, 33, 109, 150, 178, 188, 191, 192, 196, 201, 205, 238, 254, 301, 306, 337, 365
Sentient workculture-effect, 254
Sentient-effect, 171, 284, 329, 398
Sentimental attachment, 396
Sentimental shadow, 198
Sentimental-effect, 314
Sequence of eight, 252
Sequence of forms, 400
Sequences, 124, 168, 176, 249, 291–294
Servant leader, 373
Servant leadership, 373
Servicing, 3, 12, 23, 33, 37, 66, 79, 85, 102–107, 109, 110, 120, 122, 124, 125, 128, 129, 131, 133, 135, 148, 149, 160, 166, 176–178, 184–186, 189, 198, 201, 203, 212, 214, 216, 217, 230, 232, 251, 256, 257, 266, 272, 277, 279, 280, 282, 287, 296, 301, 305, 315, 331, 333–336, 359, 372, 373, 376, 385, 386, 389, 394, 398, 404
Servicing force, 232
Serving number, 236
Setting, 245, 286
Seven faces, 213, 214
Seven filaments, 312
Seven linear systems, 250
Seven moons, 194
Seven photons, 231–233
Seven primordial suns, 294
Seventeen faces, 277, 278
Seventh event horizon, 329
Seventh hand, 238
Seventh moon, 194
Seventh photon, 232
Seventh quadrant, 216
Shad, 282
Shadad, 282
Shadow body, 246
Shadow direction, 376
Shadow entity, 261
Shadowing, 94, 282, 290
Shadows, 215, 324
Shared value, 196, 271, 292, 308, 348
Shear modulus, 242
Shorten, 311, 335
Shortened chain, 312, 381
Shortener, 311
Shortener chain, 311
Shortening chain, 311, 383
Short-winged hawk, 352
Sidereal coherence, 364
Sidereal edge, 266
Sidereal vertex, 266
Sigma family, 204
Silence, 87, 88, 187, 200, 262
Simchah, 376
Simple primordial maternal, 340
Simplicity, 127, 131, 184, 188, 192, 301, 328, 361, 362, 388, 401, 402
Single particle, 312
Single tile, 343, 352
Situation, 386
Situational path, 386
Six moons, 180, 194
Sixteen faces, 214
Sixteen filaments, 312
Sixteen groups, 182, 241
Sixteen light forces, 214
Sixteen massive blackholes, 327
Sixteen planets, 280

Sixteen triple quarks, 253
Sixteenth Sun, 192
Sixth orthogonal p orbital, 230
Sixth quadrant, 216
Sixty linear systems, 251
Skepticism, 236, 374
Skew polygon, 354
Social exchange, 203
Social justice, 187
Solar corona, 227
Solar crescent, 220
Solar flash, 313
Solar matter, 196
Someone else, 239, 256
Son consciousness, 205
Soul, 31, 115, 136, 137, 146, 147, 164, 167, 168, 182, 191, 201, 209, 210, 215, 256, 263, 264, 269, 270, 274, 275, 294, 317, 318, 356, 362, 363, 375, 392, 399
Soul coherence, 362, 375
Soul essence community, 263, 264, 269
Sound-effect, 208
South Pole, 219
Sovereign, 240
Space divider, 74, 97, 132, 141, 186, 205, 234, 287
Space reversal, 229, 350
Space singularity, 394, 396
Space tensor, 41, 257, 258
Space-differentiated dimensions, 123
Spacetime, 5, 8, 13, 22, 29, 70, 71, 76, 100, 164, 201, 233, 267, 312, 313, 394–396
Spacetime singularity, 100, 394–396
Space-time-causation singularity, 395, 396
Spacing, 381
Spatial body, 268, 369
Spatial slit, 380
Speeding pulse, 46
Spherical orbital, 230
Spin-orbital, 231
Spiral polygon, 359
Spirit, 2, 7–9, 12, 13, 19, 29–32, 51, 58, 61–63, 66, 80, 89, 104, 115, 116, 127, 136, 139, 144, 146–148, 152, 153, 157, 158, 166, 167, 170, 172, 178, 182, 190, 195, 209, 210, 213, 215, 216, 230, 235, 246, 260, 261, 274, 275, 279, 286, 300, 310, 311, 314, 331, 338, 339, 341, 356, 363, 369, 372, 375, 380, 384, 390, 394, 398–401
Spiritual freedom, 314
Spiritual well-being, 205
Spirituality, 142, 146, 250, 251, 301
Spiritualization, 130, 148, 205
Splices, 206
Splitting potential, 349
Spontaneous potential, 198
Spring moon, 194
Square coherence, 364
Square energy, 174
Square filament, 311
Square horizon, 329
Square symmetry, 212
Square wave, 262
Squares itself, 49, 246, 262, 265, 329
Stable type, 321
Stance, 266
Star, 15, 64, 67, 100, 107, 125, 131, 148, 149, 192, 196, 198, 220, 221, 227, 228, 247–249, 252, 257, 259, 268, 269, 271, 279, 281, 283, 284, 286, 288, 290, 291, 293–295, 299, 300, 307–310, 324, 326, 329, 334, 358, 371, 375, 377, 378, 380, 396, 397, 400, 401
Start point, 202
Stellar body, 259
Stellar corona, 227
Stellar flash, 313
Stimulation, 208
Stomping, 260
Storm, 258, 260
Straightline, 155, 298
Straining, 208
Strange quark, 172, 173, 180–182, 206, 252, 253, 272
Strange repulsion, 72, 309, 335, 370
Strength, 42, 215, 216, 297, 312, 371
Stripped, 244, 245
Stroke, 230
Strong filament, 312
Strong psychic force, 41, 396
Strong psychic linkages, 287
Styx, 264
Subconscious, 307, 310
Sub-horizon fluctuation, 307
Sub-horizon scale, 307, 308
Subjects itself, 6, 39, 46, 48, 311
Subjunctive, 385
Subluminal, 197
Substantial radius, 330
Substantiation, 40
Substitute, 66, 118, 372, 396
Subtracting, 20, 46, 49, 70, 71, 80, 132, 166, 185, 236, 250, 258, 348
Subtracting itself, 20, 348
Subtraction, 3, 8, 16, 25, 26, 40, 41, 64, 125, 156, 160, 183, 224, 304, 305, 345
Sufferer, 254
Suffering, 129, 254, 335, 373
Summation, 381
Summer, 171, 194, 248, 263, 298
Summer moon, 194
Summing, 381
Sun, 100, 102, 103, 105, 107, 123, 125, 130–132, 137, 148, 149, 152, 161, 162, 168, 178–182, 192–196, 227, 247–250, 258, 261–263, 269, 279, 280, 283, 286, 291–294, 299, 300, 307, 326, 368, 377
Sunset, 247
Super complex, 279
Super complex number, 279
Super deity face, 215, 216
Super freedom from the primordial wisher, 388
Super wisher, 301, 303, 305, 306, 308, 309, 314, 390, 404
Superfast, 113, 256
Super-freedom from primordial wisher, 389
Super-freedom from supreme wisher, 389
Super-horizon scale, 307, 308
Superintention, 20
Superluminal, 98, 197
Superluminal motion, 197
Superluminal object, 197
Superluminal subject, 197

English Index

Supermassive, 24–27, 271, 322–325, 327, 380
Supermassive blackhole, 322–324, 327
Supermomentum, 233
Supernatural, 19, 25, 98, 111, 134, 139, 143, 150, 159, 163, 164, 169, 186, 189, 191, 214, 216, 224, 225, 246, 271, 285, 286, 291, 292, 319, 329, 345, 364, 365, 381, 387, 399
Supernatural face, 292
Supernatural growth, 25, 143, 159, 163, 164, 169, 189, 214, 216, 224, 225, 246, 271, 285, 286, 291, 387, 399
Supernormal profiting, 190
Supernormal value, 190, 292
Supernormality, 110
Superposition, 10, 11, 24, 27, 28, 32, 36, 78–82, 84, 93, 111, 118, 136, 147, 154, 170, 174, 231, 232, 330, 396
Superposition entity, 136
Superpositioned center, 231
Superpositioned core, 233, 234
Superpositioned corona, 233, 234
Superpositioned deity, 235
Superpositioned entity, 234
Superpositioned goalkeeper, 235
Superpositioned octave, 234
Superpositioned positioning, 233
Superpositioned vortex, 232
Superpositioned whole, 236, 237
Superpositioning, 43, 78
Supersaturation, 191
Superspace design, 400
Supra complex, 279
Supra complex number, 279
Supra deity, 2, 113, 115, 191, 215, 220, 320
Supra deity face, 215
Supraspace design, 400
Supreme blackhole, 326
Supreme conscious coherence, 366
Supreme deity, 113, 114, 191, 205, 215, 320
Supreme deity face, 215
Supreme spirit, 195
Surface, 13, 63, 100, 183, 218, 271, 299, 341, 353, 366, 380
Surrector, 208
Surrender, 284
Surrounds, 219, 400
Sweeping, 216
Symbol, 215, 374
Symbolic freedom, 374, 376
Symmetrical, 205, 277, 378, 379
Symmetrical light curve, 378
Symmetrical oneness, 205
Symmetry, 14, 74–76, 78, 80, 86, 112, 114, 121, 122, 131, 176, 183, 193, 205, 209–213, 230, 245, 275, 277, 278, 305, 327, 333, 355, 357, 359–361, 380, 382
Symmetry of two with three, 213
System, 7, 8, 10–16, 18, 19, 43, 71, 74, 96, 100, 103, 122, 129, 133, 141, 143, 153, 157, 161, 171–173, 193, 197, 204, 220, 223, 225, 232, 249–251, 253, 255, 259, 279, 283, 285, 286, 306, 310, 327, 335, 340, 349, 362, 363, 372, 397, 398, 400, 401
Systemic coherence, 362

T
Tachyon, 197
Tailing, 158, 322
Tangent length, 360
Taranis, 259
Taranos, 259
Taranucnos, 259
Tarnaitis, 258
Taste-effect, 208
Technological cost, 147, 149, 203, 373, 383, 388, 403, 404
Technological energy, 46, 65, 219
Technological growth, 135, 136, 149, 203, 314
Technological investment, 149, 203
Technological perception, 134, 135
Technological value, 238
Temperature, 39, 41, 46, 47, 89, 127, 160, 175, 185–187, 307
Temperature fluctuation, 307
Temporal direction, 46
Temporal face, 270
Temporal generation, 197
Temporal plane, 295
Temporal slit, 380
Temporal variability, 239
Temporal width, 360
Ten billion membered ring, 110
Ten copies, 121, 190, 307, 324
Ten faces, 3, 212, 213, 215, 217, 256, 265, 278
Ten filaments, 312
Ten moons, 194
Ten primordial suns, 293
Ten thousand filaments, 313
Ten-dimensional object, 17
Ten-fold growth, 14, 95, 96, 129, 162, 164–166, 193, 195, 208, 248, 259, 270, 302, 320, 343, 345, 354, 378
Tenth quadrant, 217
Tetrahedral orbital, 231, 232
Thanatos, 262
Theogone, 261
Theoretical self, 400
Theory of action, 254
Theory of activity, 5, 58
Theory of reactivity, 4, 57
Theory system, 250
Theory-effect, 3, 9, 31–33, 40, 43, 44, 52, 68, 70, 72, 75, 111, 112, 116, 126, 135, 208, 272, 273, 297, 303, 399
Thermodyanmism, 380
Thermodynamic, 39, 40, 47, 58, 84, 89, 95, 166, 171–173, 185, 186, 229, 239, 263, 311, 312, 331, 332, 362, 363, 378, 380
Thermodynamic filament, 311, 312
Thermodynamic limit, 239, 263, 380
Thermodynamic-effect, 39, 47, 89, 171, 186, 229, 331, 362, 363, 378, 380
Third electron, 232
Third moon, 194
Thirteen faces, 213, 216
Thirteen radiated sequences, 293
Thirteenth, 4, 6, 139, 144, 156, 192, 194, 271, 277, 280, 283, 284, 293, 310, 317, 318, 338, 392
Thirteenth number, 280
Thirty-eighth sun, 196
Thirty-ninth sun, 196

Thirty-seventh sun, 196
Thirty-six copies, 321
Thirty-six quarks, 253
Thirty-sixth sun, 196
Thirty-three sides, 359
Thirty-two grams, 346
Thor, 313
Three blackholes, 323, 324, 326
Three circles, 271
Three complex systems, 251
Three d orbitals, 233
Three f orbitals, 234
Three filaments, 311
Three linear systems, 250, 251
Three members, 243
Three moons, 180, 194
Three nonlinear systems, 250
Three pairs, 401
Three potential faces, 214
Three primordial suns, 293
Three qualities, 244, 245
Three radii, 271, 381
Three s orbitals, 233
Three with present, 237
Three zeroes, 95, 96, 118, 132, 134, 136, 168, 169, 173, 175, 176, 216, 229, 230, 243, 245, 262, 285, 325, 336–338, 370
Three-dimensional object, 357
Three-hundred sixty degrees, 255
Three-hundred-sixty copies, 321
Thunderable, 259
Thunderous change, 376
Tiferet, 202
Time, 2, 3, 5, 6, 8, 10–12, 16, 17, 19, 22, 27–34, 39–41, 44–47, 49, 50, 53, 54, 56–58, 61, 62, 64–67, 69–71, 73–78, 80–87, 89–91, 94, 95, 97, 98, 101, 102, 108, 110, 112, 115, 117–129, 133, 135–137, 139–144, 147–149, 152, 154–159, 164–171, 174–177, 180–183, 185–192, 195–198, 201–205, 207–211, 213, 214, 217, 219, 220, 223, 225–229, 233–239, 241, 245, 248, 250–252, 254–257, 260, 264, 267–273, 277, 280–289, 292, 293, 296, 299, 300, 303, 307, 309–313, 315, 317, 321–324, 328, 330, 331, 334, 336, 338–341, 344, 351, 356, 359–361, 363, 366, 371, 374, 378, 380, 382, 384, 386, 387, 392–397, 400, 404
Time cocoon, 227
Time coherence, 361
Time consciousness, 137, 239, 282, 292
Time galaxy, 271
Time loops, 280
Time mass, 168, 197
Time molecule, 33, 312, 340
Time multiplier, 2, 30, 32, 50, 90, 98, 139–142, 152, 157–159, 165, 166, 175, 185, 186, 189, 192, 197, 198, 203, 204, 210, 277, 283, 287, 296, 303, 311, 341, 361, 366, 395
Time reversal, 86, 214, 229, 235
Time singularity, 394, 396
Time tensor, 40, 257
Time value, 17, 49, 97, 136, 202, 204, 225, 282–284, 292, 293, 309, 312, 328, 338, 344, 361, 378, 382, 396, 400
Time-bound, 204
Tityos, 262
Tmolus, 261

Togetherness, 2, 8, 28, 50, 58, 63, 72, 84, 129, 177, 251, 262, 306, 335
Toranos, 259
Touch-effect, 208
Touchstone, 345
Tough, 208
Tradable, 120, 241
Tradable potential reality, 120
Trading, 3, 23, 28, 50, 51, 85, 86, 102–104, 109, 110, 115, 120, 121, 124, 125, 131–135, 137, 139, 140, 145, 148, 149, 160, 167, 176, 178, 179, 184–187, 190, 193, 198, 201, 203, 204, 212, 216, 236, 243, 251, 252, 256–258, 282, 284, 288, 300, 306, 325, 331, 336, 372, 374, 390, 404
Trading letter, 236
Trading potential reality, 121, 252
Trading-effect, 103, 131, 140, 145, 149, 204, 306, 325, 336, 372
Transcendental, 22, 113, 147, 157, 158, 187, 255, 258, 297, 301, 304, 336, 365, 387, 390
Transcendental form, 187, 297
Transcendental value, 255
Transcript, 320
Transcription, 213
Transcription without one, 213
Transformable flame, 384
Transformable self, 44, 122
Transformation, 25, 27, 117, 119, 148, 184, 220, 254, 264
Transformative exchange, 150, 222, 223, 226, 227, 229, 231, 233, 235, 237, 239, 241, 243, 245, 247, 249, 251, 253, 255, 257, 259, 261, 263, 265, 267, 269, 271, 275, 277, 279, 281, 283, 285, 287, 289, 291, 293, 295, 297, 301, 303, 387
Transformative exchange paradigm, 150, 223, 227, 229, 231, 233, 235, 237, 239, 241, 243, 245, 247, 249, 251, 253, 255, 257, 259, 261, 263, 265, 267, 269, 271, 275, 277, 279, 281, 283, 285, 287, 289, 291, 293, 295, 297, 301, 387
Transformative face, 278
Transformative justice, 177, 207, 376
Transformative phase, 244
Transformative potential, 252, 263
Transformative reality, 109, 117
Transformative system, 220
Transformative truth, 261
Transforming reality, 55, 56
Translation orientation, 213
Translational symmetry, 213
Trapezium, 357
Traveling, 268
Traversable, 17
Triangular filament, 311
Triangular growth, 367
Triangular section, 344
Triangular system, 250, 251
Triangular wave, 262
Triangulation, 340, 348
Triangulum cycle, 325
Tricoherence, 362
Trilateral polygon, 359
Triple bottom quark, 252
Triple breadth, 359
Triple charm quark, 252

English Index

Triple crescent, 220, 221
Triple decay channel, 17
Triple Einstein, 345
Triple evenness, 242
Triple future, 253
Triple massive blackhole, 324
Triple octave, 131, 155, 156, 163, 164, 176, 193, 202, 228, 269, 281, 321
Triple oddness, 242
Triple parallel-effect, 230
Triple perimeter, 360
Triple positive quark, 252
Triple potential quarks, 253
Triple radius, 313
Triple self, 222, 293
Triple single recombination, 252
Triple squared-effect, 229
Triple strange quark, 252
Triple symmetry, 209, 212, 242, 245
Triple system, 197
Triple top quark, 252
Triple transforms, 385
Triple up quark, 252
Triple upsilon, 17
Triple wavefront, 263
Triple-double copy, 377
Triple-negative quark, 252
Trivial cohomology class, 242
Tropical vertex, 266
Truism, 13, 111
Trut, 258
Twelfth body, 161, 283, 284
Twelfth month, 314
Twelve child Titans, 338, 339
Twelve degrees, 255
Twelve faces, 212, 214, 277
Twelve linear systems, 251
Twelve photons, 232
Twenty faces, 212–214, 217
Twenty linear systems, 250, 251
Twenty stones, 343, 344, 352
Twenty-eight faces, 214
Twenty-eight grams, 348
Twenty-eight primordial suns, 294
Twenty-eight radiated sequences, 294
Twenty-five planets, 280
Twenty-four hundred copies, 191
Twenty-four quarks, 252
Twenty-four thousand copies, 191
Twenty-four types, 318
Twenty-one faces, 216
Twenty-one primordial suns, 293
Twenty-one radiated sequences, 293
Twenty-second sun, 193, 195
Twenty-six flashes, 325
Twenty-three emancipations, 389
Twenty-three faces, 215
Twenty-two faces, 215
Twenty-two massive blackholes, 324
Twin bright matter, 202
Twin causation, 41, 126, 127, 157, 159, 167, 190, 192, 206, 209, 210, 213, 214, 225, 233, 234, 272, 273, 311, 322–325, 328, 345–347, 378, 379, 396
Twin circled-effect, 229
Twin circular stone, 348
Twin copy, 63, 64, 233, 276, 329

Twin crescent, 221
Twin decay channel, 17
Twin dot, 236
Twin double type, 321
Twin duality, 244
Twin dull matter, 203
Twin electron, 128, 129, 228, 396
Twin entity, 125–127, 129, 135, 136, 139, 144, 242, 273, 274, 287, 289, 326
Twin entity energy, 326
Twin evenness, 242
Twin flames, 207
Twin future, 202, 203
Twin growth, 15, 92, 231, 237, 302
Twin heavy, 230
Twin hexahedron, 267
Twin holographic, 14
Twin horses, 221
Twin light, 58, 61, 164, 196, 202, 292
Twin light matter, 196, 202
Twin massive blackhole, 324
Twin metric, 255
Twin number, 279, 304
Twin observable value, 233
Twin octahedron, 265, 266
Twin oddness, 242
Twin parallel-effect, 230
Twin past, 202
Twin potential, 225, 228
Twin present, 202, 203
Twin quantum, 180, 236, 286
Twin radius, 313
Twin reality, 18, 62, 80, 213, 214, 233, 235, 254, 282, 313, 319, 323, 346, 387, 399, 400
Twin self, 222, 235
Twin solar system, 250
Twin spirit, 213
Twin squared-effect, 229, 230
Twin subject, 58, 207
Twin symmetry, 209, 211, 212, 245
Twin triple, 156, 166, 167, 231, 236, 246, 281, 285, 286, 321, 346
Twin triple quantum system, 285, 286
Twin triple type, 321
Twin wavefront, 262
Twin Wisher, 292
Twin-effect, 161, 231, 242
Twinkles, 198
Twinnable, 243
Twinning, 90, 91, 140, 141, 154, 166, 168, 203, 322, 401, 403
Twins causation, 125, 127, 131, 137, 141, 159, 180, 191, 192, 197, 273, 347, 359
Two blackholes, 17, 326
Two directions, 266
Two f orbitals, 232
Two filaments, 311, 312
Two groups of eight, 208
Two hundred triangular systems, 251
Two moons, 180
Two pi, 234, 349–351
Two primordial suns, 293, 294
Two zeroes, 14, 95, 134, 155, 168, 175, 176, 205, 215, 216, 228, 229, 246, 249, 250, 265, 276, 286, 348, 369, 370, 378, 380
Two-dimensional object, 357
Two-dimensional reality, 343, 381

Two-hundred-forty types, 321
Type Iv Cepheid, 377

U

U energy, 219
Ultrasound, 187, 246, 262, 272, 308, 371
Ultrasound consciousness, 246
Ultrasound vibration, 246
Uncertainty, 10, 20, 62, 63, 71, 79–81, 168
Unconditional confidence, 188
Uncurved polygon, 355
Unfulfilled subtle desires, 396
Unilateral, 359, 363
Union, 127, 275, 276, 287, 308
Unique potential-effect, 281
Unique proportion, 269
Unit entity, 287, 308
Universal limit, 380
Universal sentient well-being, 197, 296, 301
Universal well-being, 23, 149, 201, 206, 373
Universe, 5, 9, 13, 18, 23, 29, 32, 48, 56–58, 67, 69, 71–74, 77–81, 83, 84, 99, 101–109, 111, 113, 115–117, 119, 121, 123, 125, 127, 129, 131, 133, 135, 137, 139, 141, 143, 145–150, 152, 153, 159, 170, 172–174, 176–178, 183, 191, 199, 208, 213, 226, 239, 252, 254, 270, 275, 279, 280, 284, 295, 296, 306, 308, 309, 317, 319, 330, 333–337, 339, 364, 368, 375, 377, 387, 391, 392, 396, 397
Universe of becoming, 239
Universe of gifts, 337
Universe of Wishers, 396
Unproductive charge recombination, 175
Up, 50, 56, 60, 72, 84, 129, 155, 172, 173, 180, 181, 199, 206, 251–253, 264, 308, 381, 382, 387
Up quark, 84, 129, 172, 173, 180, 181, 206, 251, 252, 308
Upper class limit, 41
Upsilon potential, 17
Upstream, 376, 383
Upswinging, 252
Uranus, 104–106, 131, 132, 136, 137, 147, 148, 155, 160, 161, 193, 268, 340
Utu, 376

V

V energy, 219
Vacuum, 120, 242, 258, 394
Value normalization, 370
Vampire, 241
Vanquisher, 196
Vaporizing, 202
Variability, 120, 121, 272, 361, 365, 368
Variable reality, 216
Variable star, 378, 379
Variable subject, 252
Variable time, 69, 297
Vector, 60, 133, 246, 304, 309, 339, 351
Vector-effect, 246
Velnias, 258
Venus, 103–105, 128, 130, 131, 135, 148, 160, 161, 164, 193, 258, 269, 340
Vertex, 44, 266, 267, 317, 343, 350, 354, 356–359, 401
Vertical filament, 311

Vertical symmetry, 74, 211, 304
Vibrating sound, 252
Vibrating wave, 253
Vibration, 252
Virgo zodiac, 202
Visceral manifestation, 309
Visible matter, 184, 185, 205
Visualization, 220
Volume, 162, 175, 219, 221, 254, 386, 397
Volume constant, 254

W

W energy, 219
Waker coherence, 367
Waning, 194
Waning moon, 194
Watcher, 250
Water, 69, 73, 131, 135, 140, 141, 145, 146, 155, 161–166, 171–173, 185, 187, 208, 240, 246, 247, 263, 268, 284, 294, 297, 300, 329, 369
Wave of continuity, 155
Wavefront, 262
Waves, 40, 44, 58, 155, 229
Waxing, 194
Weak filament, 312
Weak psychic force, 41, 397
Weak psychic linkages, 254, 287
Weakener, 260
Weakness, 216, 296, 312
Weeping, 215
Weight, 23, 51, 66, 155, 161, 255, 295
Well-being, 7, 104–107, 109, 113, 127, 137, 170, 190, 199–201, 207, 226, 239, 259, 288, 292, 297, 301, 373
White filament, 313
White north polestar, 192
White star, 102, 107, 124, 125, 130, 132, 149, 161, 181, 196, 221, 252, 258, 375, 394, 396
White star crescent, 221
Whitener, 260, 261
Whitening entity, 247
Whole area, 194
Whole coherence, 362, 375
Whole flame family, 287
Whole moon, 194
Whole-effect, 396
Wholeness-effect, 194
Wholesome coherence, 362, 375
Wholesome energy, 191
Wholesome entity, 308
Wholesome plane, 293
Wholesome stone, 348
Wholesomewhole, 12, 13, 23, 32, 47, 65, 75, 76, 82, 94, 95, 143, 160, 175, 185, 188, 209, 211, 217, 293, 305, 308, 361
Wholesomewhole coherence, 361
Wholesomewhole plane, 293
Width, 39, 41, 44, 359, 360
Willpower, 244
Winged, 257, 352
Winter moon, 194
Wisdom, 52, 67, 80, 110, 133, 154, 248, 284, 285, 399
Wish deity, 246
Wisher, 6, 8, 10, 84, 113, 125, 127, 137, 138, 167, 169, 171, 172, 176, 188, 196–199, 206,

English Index

207, 218, 219, 239, 255, 269, 273–278, 280, 290–292, 295, 296, 303, 305, 306, 309, 314, 316, 334, 335, 337, 367, 368, 371, 372, 384, 388–391, 394, 397, 404
Wisher universe, 198, 280
Wishing, 10, 60, 66, 68, 84, 112, 124, 125, 169, 178, 196, 199, 206, 207, 236, 239, 244, 254, 261, 273, 275, 276, 290, 291, 296, 297, 306, 335, 365
Wishing sequence, 239, 244
With future time, 255
Workculture system, 334
Worker, 2, 6, 10, 71, 101, 111, 149, 169–171, 176–178, 186–188, 206, 207, 226, 229, 239, 254, 276, 296, 301, 306, 308, 310, 314, 336, 372, 387, 392, 404
Workforce system, 204
Worst, 208

X
X energy, 219

Y
Y energy, 219
Yellow Emperor, 171, 246
Yesod, 285
Y-linked chromosome, 138, 143
Young, 194, 242, 243, 245
Young modulus, 242
Young moon, 194
Your God, 240–242

Z
Z energy, 219
Zemele, 258
Zemyna, 258
Zero stone, 343
Zero without one, 213
Zeus, 261
Zhi, 187
Ziezdre, 258

Hindi Index

A
Aavalikarana, 383
Aayu, 368
Abandhava, 365
Abhedya, 392
Abhedya Vimukti, 392
Abhi, 216, 217
Abhichara, 367
Abhidridha, 323
Abhidush, 318
Abhijna, 376
Abhikalpa, 400
Abhikalpana, 401
Abhilasha, 254
Abhimitra, 216
Abhiprerana, 367
Abhiruci, 263
Abhisamgati, 365
Abhishruti, 331
Abhiyu, 215, 216
Accha Marga, 254
Achapala, 342
Achara, 319
Adamsh, 400
Addhya, 384
Adhamnaya, 277, 278
Adharma, 244, 303, 333, 374
Adharma Bhukti Yoga, 333
Adharma Marga, 244, 303
Adhassota, 311, 376
Adhi, 326, 397
Adhi chakra, 326
Adhibhautika Sharira, 369
Adhikara, 20
Adhikarana, 332
Adhipa, 400
Adhiraj, 316, 324, 365
Adhogati, 382
Adhyakshana, 398
Adhyaropan, 330
Adhyatamakaya, 268, 369
Adi, 219, 220, 269, 314
Adi shakti, 220, 269
Adishta, 349, 351
Adisthiti, 333
Aditi, 374
Adityanatha, 403
Adonai, 187
Adravya, 392
Adridha, 323
Adrishtartha, 397
Advaita Dharma, 329
Adyajnana, 238
Agamemnon, 261
Aghani, 349, 351
Agni, 399

Agnija, 220
Aha, 208
Aham, 257, 340
Ahamta, 336
Ahimsa, 390
Aima, 240, 241
Aisus, 259
Aiyara, 188
Ajaba, 253
Ajjhattikani, 266, 366
Ajnana, 373
Akala, 188, 257, 258, 369, 375, 377
Akala rupa, 369
Akarmani, 20
Akarshaka, 396
Akarshana, 329
Akarta, 368, 369
Akasha mandala, 322
Akashaganga, 327
Akimca, 393
Akimcana, 393
Akimcana Ceto Vimukti, 393
Akramastha, 322
Akriti, 340
Aksara, 354
Aksha, 277, 357
Akshagona, 356, 357
Akshara, 400
Akshavat, 328
Akshnaya, 357
Akshobhya, 383
Akuppa, 393
Akuppa Ceto Vimukti, 393
Alcyone, 262
Alokya, 334
Ambika shakti, 325
Amdolira, 252
Amnaya, 278, 403
Amrita, 246
Amsha, 255
Amudheshvari, 338
Amurru, 282
Ana, 252
Anabhijna, 328
Anagata, 322
Anandamaya, 400
Ananke, 263
Ananta, 249
Anasamgati, 362
Anaxibia, 261
Andoli, 252
Andolitri, 252
Angana, 256, 380
Angasamata, 304
Angulanka, 286
Animitta, 393

Hindi Index

Animitta Ceto Vimukti, 393
Anitya, 326, 330
Anitya chakra, 326
Anobhagga, 383
Antaka, 341
Antara, 254
Antaramsa Marga, 189
Antarbhuta, 366
Anthemoeisia, 262
Antim, 388
Antu, 258
Anubhav, 318
Anuchita, 215
Anuchitavrtti, 304
Anudrivakta, 208
Anujata, 242
Anukarana, 329
Anukrita, 369
Anumanta, 254
Anumatisamgati, 363
Anuna, 361
Anunada, 379
Anupjau, 342
Anupuraka, 398
Anuradha, 314
Anusamgati, 361
Anuvaka, 320, 344, 354
Anuvigrahan, 385
Anuvigratha, 385
Anuvigrathana, 385
Anuya, 338, 339
Anyadhi, 365
Anyadriksha, 195
Anyadrivakta, 208
Apahrtabhara, 373
Apanika, 342
Aparimita, 349
Apasparshata, 363
Apavartita, 342
Api, 346, 348
Appamana, 393
Appamana Ceto Vimukti, 393
Apradivam, 322
Apurna, 393
Apuryamana, 194
Arabhi, 333
Arajak, 244
Araka, 397
Ardhagona, 359
Ardhotghatagona, 355
Ardhothata, 355
Arhata, 388, 389
Arhata Marga, 388
Arhatagati, 220
Arhatatma, 220
Arhatgati, 392, 393
Arkendu, 194
Arohana, 271
Artha, 205, 366, 385, 389
Arthapatti, 384
Arthasamgati, 361
Arthatma, 327
Arupya, 332

Arya, 370, 373
Arya Marga, 370
Aryama, 208
Asahayata, 399
Asam Shakti, 219
Asamadirodha, 356
Asamananki, 356, 357
Asamanantra, 357
Asamavishtara, 359, 360
Asamavrtta Pada, 356
Asamgati, 361
Asamvarta, 328
Asannidhipada, 356
Asartha, 356
Asat, 318, 364, 388
Asava, 374
Ashadha, 248
Ashavada, 332
Ashivna, 189
Ashraya Marga, 314
Ashuchitva, 393
Ashvini, 321, 384
Asiddha Marga, 316
Asprha, 376
Asrava Shakti, 219
Astara, 385
Astavyasta, 349, 350
Asthirta, 361, 365, 368
Astika, 127, 336
Asura Shakti, 219
Asvabhavika, 292, 364
Asvabhavika Mukha, 292
Atapin, 348
Atidridha, 322, 323
Atikrama, 362
Atikramana, 363
Atikramita, 362
Atimanasa, 366
Atisahasrara, 214
Atisukshma, 322
Atita, 358, 392
Atiuccha, 365
Atiuddipta, 217
Ativada, 330, 331, 366
Atma Bodha, 390
Atma Karuna, 390
Atodya, 252
Atropos, 268
Atthi, 397
Audavita, 382
Audrika, 360
AUM, 321
Avacara, 321
Avagaha, 325
Avarardhan, 332
Avastha, 343
Avatara, 388
Avidhi, 282
Avigata, 256
Avijja, 373
Avijna, 373
Avrit, 397
Ayanamsha, 337

Ayasam, 358
Ayatana, 284, 285, 397
Ayati, 399

B
Babhru, 216
Bagalamukhi, 260
Bahirani, 266, 366
Bahirdhakaya, 268, 369
Bahis, 354
Bahubhuj, 354
Bahuparshvika, 362, 364
Bahurangi, 342
Bahurupa, 198, 336
Bahutara, 307
Baidi, 171, 282
Baindava, 401
Bal, 312
Balachandra, 178, 194, 404
Balachandramas, 194
Baladeva, 337, 388
Balendu, 194
Balidana, 356
Banavat, 341
Bandana, 355
Bandanagona, 355
Bandhava, 365, 378
Bauddha Samgati, 363
Beeja Jagrat, 385
Belana, 243
Bha, 373
Bhaavi, 285
Bhadra, 245
Bhadrakaali, 260
Bhadrapada, 202, 245
Bhadravasa, 270
Bhagna, 400
Bhagwad Rasa, 247
Bhagwan, 205, 340, 394
Bhairavi, 292
Bhajana, 359
Bhakshana, 369
Bhakta, 337
Bhaktanga, 398
Bhakti Marga, 337, 388
Bhaktiphala, 188
Bhana, 345, 347
Bhasvara tejas, 289
Bhava, 173, 339, 363
Bhavana, 321
Bhavasamgati, 363
Bhavishya, 375
Bhedya, 392
Bhedya Vimukti, 392
Bhidvada, 332, 333
Bhram, 374, 375
Bhram Mukti, 374
Bhranta, 387
Bhu, 268, 340, 369, 397
Bhu Loka, 397
Bhugrasta, 355
Bhukti, 333
Bhukti yoga, 333

Bhuma, 318, 344
Bhushana, 328, 329
Bhuta, 247, 322, 377, 401
Bhuvanesvari, 220
Bijani, 402
Billaka, 382
Binah, 338, 400
Bindu, 14, 365, 381
Bindugona, 353
Bodhi, 303, 334
Bodhi dharma, 303
Bodhi Marga, 303
Boeotus, 262
Brahli, 189, 366
Brahli Marga, 189
Brahma, 218, 258, 312, 336, 338, 362, 381, 383, 384
Brahma loka, 383
Brahma Mukti, 336
Brahma Yanjni, 381
Brahman, 258, 330, 387
Brahmanda, 349, 350
Brahmantaka, 349
Brahmasutra samgati, 362
Brahmi, 292
Brihadbala, 336
Brihadbalika, 380
Brihaspati, 104, 340
Bruhenta, 403
Buddha, 133, 255
Buddhiyoga, 205
Budhya, 217

C
Ceshta, 334
Cetana, 213
Ceto, 389
Ceto Vimukti, 389
Chaitra, 298, 314, 341
Chakora, 360
Chandra, 106, 194, 220, 404
Chandrabha, 243
Chandrakanti, 194
Chandravanshi, 327
Chapala, 110
Charma, 331
Chaturtha, 233
Chaturthansh, 343
Chaturveda, 316
Chaudai, 359
Chauthi, 248
Chetan shakti, 244
Chetasika, 219
Chhalayati, 352
Chhinnamasta, 260, 261
Chidi, 375
Chipata, 366
Chit, 350
Chitra, 329, 345
Citraka, 254, 371
Cittavikarin, 343
Cush, 241

Hindi Index

D
Dada, 243
Daivika, 371
Daksha, 403
Danava, 338
Dao, 133, 187, 297, 400
Darshan yoga, 207
Dascylus, 262
Dasha, 350
Dasra, 202
Dasya, 202
Deha Vimukti, 389
Dehana, 388
Dehi, 348
Deshaparyaya, 286
Deva, 206, 220
Devadhideva, 337, 340
Devaloka, 371
Devasana, 329
Devasvarani, 269
Devata, 394
Devatamayi, 339
Devatti, 352
Devi, 190, 206, 220, 253
Dhajini, 318
Dhanu, 403
Dhanvantari, 332
Dharaka, 392
Dharma, 171, 250, 251, 303, 379
Dharma Marga, 171, 303
Dharmanishpatti, 367
Dharmasamgati, 364
Dharmashunyata, 333
Dharmatma, 334
Dhruva, 328
Dhruvamatsya, 328
Dhumavati, 260
Dhumra, 374
Dhumraketu, 374, 404
Dhuni, 214
Dhuv, 215
Dhvani, 252
Dhvanta, 209
Dhyana, 191, 195, 239, 306
Dhyana Yoga, 191, 306
Digant, 329
Dipamala, 392
Diptamurti, 400
Dirghavritta, 324
Dirghavrittaphala, 311
Disha, 403
Divangata, 256
Divya, 170, 374
Divyaratri Marga, 341
Dolita, 255
Dosha, 324
Drakh, 399
Dridha, 322
Driksha, 195
Dukkha, 398
Duniya, 271, 316, 403
Durga, 259, 292, 369
Dushama, 327
Dvadashasya, 214
Dvaimatura, 187, 188
Dvaiyoyoga, 371
Dvandva Brahma, 370
Dvesha, 376
Dvibhajana, 357, 359
Dvidanta, 206
Dvimukha, 215
Dviparshika, 363
Dvividha, 370
Dyssebeia, 263
Dyujya, 354

E
Ehuang Xiangshuishen, 220
Ekadanta, 206, 404
Ekajata, 260, 261
Ekantadushama Rupa, 304
Ekantasushama, 304, 306, 369
Ekaparshika, 363
Eke, 338, 376
Ekkekke, 271
Ekmukha, 217

F
Falahavada, 332, 333

G
Gabhira, 327, 346, 349
Gabhiravyapi, 362
Gajakarana, 262, 404
Gajanan, 187, 375, 404
Gana, 272, 372
Ganadhyaksha, 398, 404
Ganesha, 188, 207, 259, 267, 314, 403, 404
Ganga, 369, 372
Gantha, 399
Ganthana, 349, 350
Garbhasamudbhava, 347, 348
Gardhabi Mukha, 214
Garj, 397
Gathari, 349, 350
Gatishila, 343
Gau, 246
Gauranga, 285, 346, 347
Gavyuti, 360
Ghasitana, 260
Ghataana, 325
Ghatay, 400
Ghatika, 355, 357
Gliɲh, 338
Golamaal, 400
Goshthikata, 397
Gotraja, 342
Graha, 221
Griha Mukha, 215
Guduvay, 339
Guna, 206, 399
Gunasa, 382
Gunatita mukti, 136, 149, 245
Gunnisu shakti, 393
Gup, 277

Guru, 67, 213, 286, 288, 297, 370
Guru Marga, 288
Guru Nanaka, 297
Gurvartha, 349, 350, 366

H
Hala, 233
Hani, 387
Hanuman, 80, 188, 190, 339
Harbuddhi, 339
Hari, 216
Harimitra, 216
Haritatta, 340
Hariti, 364
Hasta, 215, 315
Hasti chakra, 325
Hasya, 337
Havya, 318, 319
Hemarenu, 354
Hetu, 248, 389
Hetuka, 355
Himakanthala, 382
Hiranyagarbha, 179, 374
Hiranyamaya sharira, 269
Hiranyaretas, 329
Hita, 137, 367, 397
Hitashin, 269
Ho, 380
Homa, 400
Hras, 311
Hrasaka, 311
Hrasana, 312
Hridyesha, 365
Huha, 319
Hurupa, 366, 393
Hutabhuj, 307, 399

I
Iccha, 273, 397
Iccha mukti, 273
Idam, 284, 384
Idanta, 393
Idhara, 397
Idriksha, 195
Idrivakta, 208
Ijara, 366
Ilagola, 322
Indambra, 403
Indra, 67, 207, 335, 339, 343, 374
Indra Yoga, 335
Indraja, 258
Indrasavarni, 257
Indrejya, 336, 337, 370
Indumandala, 284
Insan, 214
Insaniyat, 198
Irammada, 313
Isha, 189
Ishamnaya, 277
Ishanamnaya, 278
Ishta, 217, 272
Ishvara, 240, 241, 330, 374, 384

Ishvarya, 268
Iti, 370

J
Jaana, 380
Jadu, 244
Jagatkritsna, 355
Jagriti, 339
Jagrook, 363
Jaldi, 340
Jama, 394
Jana, 398
Janayitr, 256
Jara, 335, 373
Jaramarana, 398
Jatyam, 401
Jayadurga, 260, 261
Jayanti, 340
Jayashri, 248
Jiva, 80, 126, 127, 179, 328, 339, 355, 388
Jivan Mukti, 254
Jivanasamgati, 367
Jnana, 238, 254, 336, 386, 403
Jnana siddhi, 238
Jnatva, 334, 386, 392
Jneya, 373
Jodi, 333, 356
Jugupsa, 252, 253, 386, 387
Jyaka, 15
Jyeshtha, 263, 326
Jyeshtha shakti, 326
Jyoti, 198, 311, 312

K
Kaal, 137, 149, 257, 321, 351, 393, 394, 401
Kaal Marga, 257
Kaal Mukti, 257
Kaal Ratri, 401
Kaal Sarp, 393
Kaalmaan, 384
Kabbalah, 298
Kacchapa, 343, 344
Kaivalya Mukti, 254
Kakshavrta, 284, 330
Kakshya, 329, 330
Kalakrama, 339
Kali, 188, 214, 287, 334, 337
Kali shakti, 188, 334
Kalpa, 320, 326
Kalpa chakra, 326
Kalpakala, 394
Kalpanika, 363
Kalpaniya Marga, 368
Kamadeva, 246
Kamma, 338, 398
Kampa, 384, 386
Kana, 252, 353
Kanalakshamsha, 340, 351
Kancha, 347, 352
Kandarpa, 321, 341
Kapali, 213, 216, 356
Kapardi, 213, 214

Hindi Index

Kapi, 196, 213
Kapila, 245, 246, 404
Kapota, 246
Kara, 392
Kara Vimukti, 392
Karaliune, 258
Karana sharira, 268
Karaniya, 217
Karin, 392
Karita, 367
Karma, 249, 252, 275, 333, 335
Karma Marga, 249
Karma Yoga Marga, 275
Karma Yogi, 252
Karmaphala, 20
Karmasamgati, 361
Kartaka, 248
Kartikeya, 182, 292
Kartri, 373
Karuni, 391
Kashtha, 379
Katala, 379
Katyayani, 260, 261
Kaulara, 342
Kaun, 372, 396
Kausalya, 216, 217, 256
Kaval, 341, 376
Kavatata, 386
Kavati, 354, 386
Kavatigona, 354
Kavula, 341, 343
Kavyavaha, 240
Kendra, 339
Keshara, 382
Ketu, 105, 269
Kevala Jnana, 402, 403
Kha, 341, 360
Khaas, 346
Khabhranti, 341, 352, 360
Khagola, 321
Khalvida, 384
Khalvidam, 384
Khalvidam Brahma, 384
Khanijavarga, 378
Khanijya shakti, 325
Kharcha, 367
Khil, 372
Khinchatani, 396
Khud, 394
Khyati, 389, 397
Kilbisa, 347
Kimstughna, 333
Kinja, 311
Kirata, 332, 400
Kismat, 374
KLIM shakti, 326
Kone, 354, 357, 359
Kora, 324
Kosha, 196, 348
Kou, 187
Kramajya, 357
Kramakrita chakra, 326
Kramastha, 322

Krandita, 215
Krishna, 196, 244
Krishna Paksha, 244
Kritartha, 391
Kriya, 206, 373, 374
Kriya Mukti, 374
Kriya yoga, 206
Kriyagona, 358
Kriyapada, 333
Kshetraphal, 348
Kshitij, 257, 328
Kshmatala mandala, 322
Kudrat, 254, 256, 269, 339, 358
Kula, 320
Kulaka, 342
Kumbha, 357
Kumbhagona, 357
Kumbhaka, 318, 350, 354, 357
Kundala, 182
Kundalini, 182, 292
Kundalini Shakti, 182
Kurpara, 357
Kurukulla, 326
Kushmanda, 354
Kutumba, 368
Kya, 372

L
Laghukone, 353
Lajja, 281
Lakshmi, 207, 313, 333, 396
Lakshmi chakra, 333, 396
Lakshya, 361
Laksyartha, 341
Lalita, 280, 334
Lambai, 360
Lambodara, 286, 404
Lantaka, 328
Lata, 320
Lavaniya, 377
Laya, 131, 214, 379
Laya Mukti, 131, 214
Layanalika, 379
Layatmaka, 379
Liladhara, 366
Lin, 385
Linga sharira, 268
Linka, 348
Linkadharaka, 346
Linkan, 348
Linkaniya, 348
Linlakara, 385
Lohitya, 340
Loka chakra, 326
Lokya, 334
Lopamudra, 398

M
Madhava, 287
Madhusudan, 321
Madhyakarana, 358
Madhyartha, 400

Magha, 340
Maha Brahma, 190
Maha Dhyana Yoga, 253
Maha Durga, 221, 259, 292, 310
Maha Gauri, 285
Maha Kali, 188
Maha Lakshmi, 260, 310, 313
Maha Pita Parameshvara, 220
Maha Riddhi, 207, 386, 387
Maha Riddhi Marga, 207, 387
Maha Saraswati, 248
Maha Shiva, 188, 259
Maha Siddha Marga, 283
Mahachitti, 269
Mahadeva, 221, 386
Mahaguru, 372
Mahakaal, 220, 281, 370
Mahakaya, 381
Mahakrandita, 215
Mahakriya Yoga, 204
Mahanakshatra, 339
Mahanitya, 367, 381
Mahant, 370
Mahapralaya Mukti, 189
Mahas, 194
Mahashunya, 256, 384
Mahasiddhi, 238
Mahasiddhi Marga, 238
Mahaspanda, 340, 365, 386
Mahatma, 348
Mahatripurasundari, 397
Mahavidya, 348
Mahayoga, 187
Mahendra, 385
Mahendrani, 331
Mahesha, 198, 249, 255
Maheshvari, 339, 340
Mahishamardini, 260
Mahishtavadi, 333
Maidana, 388
Maim, 252
Maitra, 390
Majaraja, 240
Majaraja Yoga, 240
Makara, 377
Mala, 325
Mamata, 336
Mamatma, 327
Manas, 383
Mandoli, 252
Mandra, 253
Mangala, 104, 268, 375, 380
Mangona, 356
Manibhadra, 243, 371
Manipalayita, 372
Manja, 344
Manoj, 313
Manomaya sharira, 269
Mantrasiddhi, 275
Manushya, 214, 326
Margi dharma, 354
Maruti, 220
Masaki, 349, 350

Masha, 345
Matangi, 260
Matri, 243, 386
Matsya, 328
Mausama, 376
Maya, 52, 67, 141, 149, 270, 285, 290, 317, 346, 370, 374, 383, 397
Maya Mukti, 141, 285
Maya Sita, 290, 374
Maya Vishvakarma, 317
Meen, 314
Meghavati, 269
Mekhala, 324
Mena, 337
Mera, 367
Milaana, 252
Mithaka, 254
Mithanibhuta, 326
Mithun, 385
Mitra, 216, 367
Moha Marga, 371
Moirai, 263
Moksha, 205, 214
Moksha Mukti, 205
Moti, 351
Mrgashirsha, 403
Mridanga, 379
Mrishajnana, 304
Muhurtaja, 330
Mukhya, 397
Mukhya Kala, 397
Mukti, 135, 169, 179, 244, 314, 370, 376, 385, 399
Mukti Yoga, 370
Muktiphala, 188
Mula shakti, 325
Muladhara chakra, 236
Mulya, 397
Murka, 317
Murti, 15
Murugan, 333

N

Naadi, 350
Nabhas, 202
Nabhasya, 202
Nabhiloma, 343
Nabhirai, 371
Naciketu, 313
Naivedya, 355
Nakaratmak, 253
Nama, 369, 370
Nama Rupa, 370
Nama sharira, 369
Nanaka, 285
Nanarupa, 384
Nandi, 246, 285
Nandini, 331
Naran, 291, 384
Narayani, 338
Nasamjna, 386, 387
Nasatya, 202, 332
Nashitartha, 400

Hindi Index

Nataraja, 281
Navadurga, 260, 261
Navvakari, 364
Neva Sanna Nassana Ayatana, 386
Nibbida, 188
Nibodha, 244
Niche, 253
Nichevala, 335, 404
Nidesha, 339
Nidhidhyasana, 257
Nigamit, 339
Nihsarita, 258
Niji, 291
Nila, 217, 400
Nila Mukha, 217
Nilanjana, 338, 339, 376
Nilasaraswati, 260, 261
Nilima, 340
Nimitta, 392
Nimitta Ceto Vimukti, 392
Nira, 263, 369
Nira Sharira, 369
Nirankushta, 332
Nirartha, 367
Nirashavada, 332
Nirata, 337
Nirataranga, 263
Nirbijayoga, 384
Nirdharma, 379
Nirguna, 379, 391
Nirmana Puta, 370
Nirnayaka, 399
Niru, 328
Nirupyati, 342
Nirvac, 386
Nirvana, 187, 389
Nirvana Mukti, 187
Nirvanic sharira, 268
Nirvapana, 325
Niryana, 198
Niryoga, 388
Nishadata, 340
Nissarana, 392
Nissarana Vimukti, 392
Nitya Ratri, 251, 252
Nivas, 214
Nivasi, 215
Niyama, 340
Niyatatma, 373
Nuwa, 285
Nyaya, 337, 375
Nyuntam, 377

O

Obhanjati, 381
Ojas, 198, 363, 371
Ojas Yoga, 371
OM, 263
Omkara, 403
Onasamgati, 362
Onata, 362

P

Paarnayana, 335
Paati, 320, 330, 331
Pachaka Dasha, 397
Padakranta, 268
Padaprakarana samgati, 366
Padavrtta, 367
Padmaja, 256, 362, 369, 381
Padmajata, 310
Padmanga, 314
Padvisha, 393
Padya, 254, 387
Pala, 348
Pallava, 313, 381
Pamapana, 19
Pamujja, 296
Panchajani, 336
Pandoli, 252
Pandu, 247
Pandubhava, 247
Pani, 254
Panna Vimukti, 389
Para, 26, 27, 47, 53, 106, 147, 150, 188, 195, 218, 246, 258, 296, 303, 323, 325–337, 339, 365, 384, 387, 388, 404
Para Ganesha, 188
Para Mukti, 296
Para Shakti, 246
Para Vishnunabhi, 326
Parai, 252
Parakala, 339
Parakara, 317
Parakikarana, 394
Parakramaya, 388
Param Bhakti, 220
Param Bodhisattva, 282
Param Brahma, 203, 218
Param Buddha, 255
Param Christ, 256
Param Eve, 285, 303
Param Ganesha, 188, 296
Param Jnana Yoga, 276
Param Kriti, 303
Param Lakshmi, 310
Param Parvati, 207
Param Prakriti, 397
Param Prapti, 292
Param Prophet, 256
Param Riddhi, 384
Param Samadhi, 305, 372
Param Samadhi marga, 372
Param Shankara, 255
Param Shiva, 19, 189, 220, 257, 282, 347, 404
Param Shiva Marga, 257
Param shunya, 384
Param Siddha, 198, 256
Param Tapas, 402
Param Vishnu, 217, 240, 403
Paramarshi, 261
Paramartha, 255, 316
Parameshthi, 333
Parameshvara, 241, 244
Paraspara, 363

Paravada, 258
Paridhi, 339, 360
Parigrahana, 402
Parimap, 359
Parinama, 332
Parinditartha, 371
Paripaka, 257
Parispanda, 382
Parivar, 286, 318, 342
Parshnisamasta, 244
Parshva, 353
Parshvanath, 364
Parshvika, 364
Parvana, 194
Parvati, 188, 207, 248, 249, 259, 264, 282, 292, 335, 338
Pasara, 257
Pasha, 348
Pashana, 346, 349
Pashyanti, 340
Passaddhi, 257
Patana, 382
Patanga, 345, 348, 359
Pataniya, 382
Patayati, 382
Patippassaddhi, 393
Patippassaddhi Vimukti, 393
Patita, 382
Pativrata, 253
Pausha, 377
Pavitra, 246, 365
Payuvada, 391
Phala, 19
Phalaha, 332
Phalguna, 314
Phaninayaka, 339
Phasi, 257
Phassa, 257, 364
Phat, 387, 388
Phootana, 367
Pida, 387
Pillai, 190, 402, 403
Pillaiyara, 188
Pille, 188
Pishacha, 241
Pishachendra, 399
Piti, 284, 332
Pitra, 403
Poshana, 254
Potrimpo, 259
Prabharit, 342
Prabhat, 207
Prabhava, 321
Prabhavita, 380
Prabhu, 403, 404
Prachala Vakra, 378
Prachalika, 379
Prada, 371
Pradarshan, 350
Pradhan, 355
Pradyota, 332
Pragabhava, 282
Prajati, 317, 318, 335

Prajatikarana, 335
Prajna, 334, 390
Prajna Vimukti, 390
Prajnatma, 374
Prajval, 313
Prakara, 317, 335
Prakasha, 326, 389
Prakasha shakti, 326
Prakashatma, 387
Prakatikriti, 364
Prakriti, 284, 343, 344, 361, 365, 377
Prakriti Yogi, 344, 377
Prakritisamgati, 365
Prakshinachandra, 194
Pralaya Mukti, 129, 198
Pramapi, 349, 366
Prameya, 397
Prana, 281, 389
Pranasa, 382
Prapta, 374
Prasamga, 367
Prasamgati, 364
Prasanga, 373
Prasuti, 257
Prasutika, 257
Prasvara, 256
Prathama, 248, 326
Prathama shakti, 326
Pratibha, 397
Pratibheda, 348
Pratidriksha, 195
Pratigad, 375
Pratigha, 255
Pratihata, 354
Pratipada, 386
Pratisamgati, 361
Pratiyukta, 365
Pratyagatma, 325
Pratyangira, 260
Pratyayasamgati, 360
Pravishta, 324
Pravrtti, 190, 395
Prayojana, 399
Prema, 258
Prishana, 197
Prishtha, 339, 347
Prithvi, 102, 208, 330, 393
Prithvi Shakti, 330, 393
Priti, 365
Pritisamgati, 365, 366
Priya, 331
Priyadarshani, 253
Punaruddha, 383
Punarutthana, 383
Punarutthaniya, 383
Purahakrama, 329
Purana, 335
Puratana, 345, 346
Purna Shiva, 297
Purusha, 170, 220, 334
Purushartha, 219
Purushottama, 338, 370
Purva Bhadrapada, 398

Hindi Index

Purva Phalguni, 398
Purvasuchana, 350
Pusalattu, 370, 403
Pushpaka, 109, 110, 338
Pushpini, 402
Pushti, 384
Pushya, 355
Putana, 403

R

Rachitartha, 374
Radha, 397
Raga, 248, 379
Raghu, 343
Rahita, 336
Rahitatma, 323
Rahu, 105, 268
Rajas Marga, 384
Rajayoga, 187, 205
Rajo Guna, 205, 206, 245
Rakha, 354
Randoli, 252
Ranga, 248, 379
Rasabhasa, 339
Rashtra, 386
Rasi, 333, 337, 372, 373, 380
Rasika, 257
Rati, 254
Rekha, 365, 368
Rekhagona, 353
Rinavant, 400
Rishi, 377
Rishigona, 359
Rivakta, 208
Rochisha, 325, 370, 394
Rochisha chakra, 325
Rochisha Rupa, 370
Rohini, 336
Roma, 258
Ropit, 218
Rosha, 241
Rtajit, 208
Rudhamanyu, 322
Rudhira, 375
Ruh, 363, 375, 394
Runasvara, 342
Rupa, 198, 304, 337, 369, 370
Rupa Siddhi, 337

S

Sab, 384
Sabhadhyaksha, 398
Sabhaga, 239
Sabija, 339
Sabijayoga, 384
Sadakhya, 247, 336
Saddha, 309, 390
Saddha Vimukti, 390
Sadhaka, 340
Sadhana Marga, 387
Sadhibhuta chakra, 326
Sadhya, 243

Sadhyagana, 254, 400
Sadhyata, 332
Sadi, 282
Sadvika, 351
Sadyo, 240
Sadyo mukti, 240
Sadyovadi, 335
Sagara, 365
Saguna, 310, 314, 337, 363, 394
Saguna mukti, 310
Sahaja Puta, 370
Sahajata, 340
Sahakari, 367
Sahaparivartaniya, 305
Sahas, 403
Sahastitva, 365
Sahasya, 376
Sahayaka, 373
Sahayakta, 373
Saj, 199
Sajjyoti, 199, 313
Sakaratmak, 253
Sakarma, 253
Sakarmavadi, 334
Sakshipa, 209
Salayatana, 284, 285
Sallekhana, 384
Salokya, 334
Salokya Mukti, 334
Sam, 219, 349, 366
Sam Shakti, 219
Samaa, 251, 387
Samachaudai, 359
Samadhana, 338
Samadhi, 214
Samadirodha, 356
Samamnaya, 277
Samananki, 357
Samanayika, 381
Samanjana, 339
Samartha, 339
Samarupa, 369
Samashti, 371
Samatola, 384
Samavakara, 400
Samavishtara, 359, 360
Samavrtta Pada, 356
Sambhadya, 349
Samgati, 361–364, 366
Samghatta, 216
Samhatata, 364
Samidriksha, 195, 322
Samipa, 356
Samipya Mukti, 387
Samjna, 386
Samketa, 265, 351
Samkhya, 178, 205
Samkhya yoga, 178, 205
Samkuchita, 342
Samma Vimukti, 389
Sampadavrtta, 367
Sampeedit, 397
Samputa, 372

Samsamgati, 366
Samsara, 316
Samsarga, 383
Samshuddhi, 296
Samstha, 386
Samsthaika, 386
Samtaliyaka, 354
Samtolagona, 353
Samuccheda, 394
Samuccheda Vimukti, 394
Samudacara, 367
Samudbhavaya, 255
Samvardhamana, 403
Samvarna, 376
Samvat Shakti, 219
Samvatsara, 381
Samveshtan, 349
Samvridh, 331, 352
Samya, 370, 389
Samyogasamgati, 367
San, 337
Sandhya, 245, 257
Sandoli, 252
Sangaha, 20
Sangam, 275
Sankhara, 335
Sannidhipada, 356
Sansa, 328
Sapaksha, 257
Sapat, 354
Sapatagona, 355
Sara, 312, 322, 335, 381, 394
Sara Kalpa, 335
Saranyu, 239, 330, 403
Sarasi, 372
Sarasi Mukti, 372
Saraswati, 403
Sarata, 217
Sargam, 321, 394
Sari, 365
Sarisamgati, 365
Sarisrpa, 342
Sarj, 381, 394
Sarpidvat, 332
Sarshti Mukti, 403
Sartha, 356
Sarthavaha, 356
Sarupak, 333
Sarupya Mukti, 385
Sarvam, 384
Sarvam Khalvidam Brahma, 384
Sarvangasama, 303
Sashuka, 337
Satarupa, 292, 370
Sati, 214, 249, 259
Sati-Parvati, 249, 259
Satireka, 327
Satkanda, 343, 352
Sato Guna, 205, 244
Satsamgati, 363
Satttvajyoti, 198
Sattva, 198, 391
Satya, 198, 202, 253, 392
Satyajyoti, 198
Satyamitra, 216
SAUM, 202, 265, 321, 340
Saundarya, 252, 253
Saura, 334
Sayujya Mukti, 388
Seema, 364
Senajit, 196
Sephira, 263
Seva Yoga, 385
Shaana, 345, 349
Shabda, 337
Shadastaka, 343
Shahateer, 342
Shakha, 320
Shakti, 214, 219, 245, 249, 330, 338, 346, 394, 396
Shalabhangona, 356
Shalivan, 356
Shalkana, 394
Shamitra, 375
Shangdi, 297
Shani, 103, 202, 214, 258, 338, 340, 376
Shani Bhagwan, 340
Shanideva, 244
Shankara, 255
Shankava, 255
Shapa, 392
Shapa Vimukti, 392
Sharanarthi, 366
Sharira, 268, 368, 369
Sharirka, 367
Sharkaraprabha, 368, 396
Shataki, 342
Shatamshika, 343
Shatanga, 326
Shatkuta, 278
Shennong, 282
Sheri, 282
Shesha sharira, 304
Sheshanaga, 331, 332
Shi, 187
Shikhi, 215
Shirsha, 354
Shishira, 341
Shishutvam, 343
Shiva, 19, 80, 188, 257, 259, 281, 326, 335, 347, 399
Shiva shakti, 326
Shivadrishti, 285, 399, 402
Shivagati, 285, 384
Shivansha, 314
Shodashottari Dasha, 350
Shodhaka, 327
Shraddha, 390
Shrama, 386
Shramapada, 386
Shramika, 373
Shravana, 217
SHREEM, 245
Shri Garuda, 397
Shri Krishna, 249, 269, 337
Shri Shakti, 326

Hindi Index

Shringona, 358
Shrishaila, 345
Shristi, 307
Shrotagana, 385
Shukra, 340
Shunda, 317, 337
Shunya, 214
Shunyata, 248, 325
Shvasoshvasa, 345, 346
Shyamala, 334
Siddha, 126, 196, 256, 273, 399
Siddha Guna, 196
Siddha Ratri, 399
Siddhi, 123, 206, 254, 259, 276, 336, 341
Siddhiphala, 188
Sila, 334
Sita, 215, 216, 374, 376
Sita Mukti, 374, 376
Sitanveshana Panditaya, 331
Smita, 354
Snigdha sharira, 269
Soham, 348
Soma, 106, 220
Somapa, 351, 398
Sophrosyne, 263
Spandan, 255
Sparshaniya, 342
Sphya, 341, 376
Sridhara, 320
Srij, 388
Srijak, 196
Srijan, 385
Srijana, 367
Srijankarta, 338, 367
Srijati, 367
Stabaka, 327, 352
Stambha, 318, 370
Stanayitnu, 258, 400
Stanumalaya, 326
Sthanu, 367
Sthapanacharin, 373
Sthavaravisha, 334
Sthira dasha, 384
Sthirta, 361
Sthiti, 267
Stip, 377, 404
Subrahmanya, 330
Suchaka, 319
Suchan, 214, 318
Sudhanvan, 318
Sudhara, 375
Sudridha, 322
Sugriva, 240
Suhabati, 291
Sukha, 244, 370
Sukshma sharira, 268
Sumukha, 195, 404
Sundari, 371
Sundarika, 371
Suprata, 257
Sura, 219, 256, 258
Sura Shakti, 219
Surenu, 317

Surya, 102
Suryamnaya, 256
Suryaphani chakra, 325
Suryasta, 247
Suryendu, 194
Susamgati, 361
Sushama, 327
Sushena, 214
Sushupti, 215
Sutra, 362
Sutradhara, 256
Suvira, 304
Sva, 285, 292, 364
Svabhavika, 292, 364
Svabhavika Mukha, 292
Svadha, 245
Svakaya, 268
Svakshetra, 395
Svapatana, 382
Svaras, 346
Svarochisha, 370
Svarochisha Rupa, 370
Svarupa, 386
Svasamgati, 365, 366
Svastika, 127, 310
Swasthani, 205

T

Tadakara Mukha, 355
Tadanga Vimukti, 393
Taganem, 377
Taijaisa, 311
Taimura, 256
Talatala Loka, 336
Taliyagona, 354
Taliyaka, 354
Tamas, 147, 373, 386, 403
Tamas Kriya, 373
Tamas Marga, 386
Tambaka, 324
Tamo Guna, 205, 245, 249
Tamra, 324
Tanaro, 259
Tanarus, 259
Tandanga, 393
Tandava, 281
Tanha, 208
Tantalus, 261
Tapah, 340
Tapana, 215
Tapas, 340, 387
Tapas Marga, 387
Tapasya, 314
Tara, 259, 312
Tarala, 355
Taritni, 260, 261
Taruna, 242
Tato, 322
Tattva Mukti, 399
Tiryag, 198, 383
Tiryagjyoti, 198, 313, 325
Tiryancajya, 255
Todanem, 377

Topa, 346
Tora, 259
Tribhajana, 359
Tribhajya, 255, 381
Trichakra, 356
Trijyat, 311
Trimps, 260
Trimukha Vinayaka, 338
Trinetra, 67, 80, 260
Trishira, 317, 398
Tritiya, 248, 324
Trivikrama, 336
Triyoga, 191
Tukada, 349
Tula Rashi, 189
Tulyartha, 319
Turnasa, 382
Tushara, 258
Tvarat, 386
Tvashta, 208, 317
Tyajya, 360

U
Uccha, 364
Uchchadana Chakra, 326
Uchchaih, 360
Udana, 397
Udghata, 355, 365
Udghatana, 355
Udghatanagona, 355, 356
Udghatita, 356
Udghatitagona, 356
Ugavati, 257
Ugratara, 260, 261
Uma, 369, 377
Uma Marga, 377
Unmukti, 391
Upa, 367
Upadana, 197
Upadhaya, 338
Upadhi, 383
Upakulaka, 342
Upamitra, 364
Upar, 253
Uparama, 356
Upasamgati, 367
Upasarga, 373
Upayukta, 365
Upjau, 342
Uragona, 359
Urdhvasota, 376, 383
Urja, 182
Usha, 258
Ushnavahana, 325
Uttameshvara, 244, 259
Uttanakara, 343
Uttaraphala, 388
Uttarasucha, 381
Uttarasuchana, 351

V
Vaha, 240

Vahya, 362
Vairibhu, 376
Vaishakha, 287
Vaishnavi, 366, 371, 388
Vaishvanara, 327
Vaitadhya, 318
Vaivora, 258
Vajradridha, 323
Vajragona, 357
Vajrapana, 215, 278
Vakarikriti, 329
Vakarine, 257
Vakra, 351, 378
Vakragona, 353
Vakriyagona, 354
Valavala, 277, 278
Valaya, 216
Vama Marga, 395
Vamavarta, 214
Vamri, 318, 344
Vamsaj, 403
Vanaramrga, 352
Vanishthu, 333
Vansha, 368
Vapu, 259
Vara, 241, 255
Vardha, 242
Vardhaka, 374
Vardhamana, 401
Vargigona, 353
Vari, 263
Varitaranga, 263
Varna dharma, 369
Varnasamamnaya, 217
Varsha, 248
Vartula, 353, 366
Varutri, 254, 317
Vasa, 321
Vasanta, 341
Vasha, 362
Vashuli, 260, 261
Vastramaya, 343
Vastunara, 317
Vastupurusha, 400
Vastuta, 257
Vasu, 343
Vasudeva, 244, 293, 353, 386, 399
Vasundhara, 403, 404
Vat, 219, 401
Vatsalya, 334, 385
Vedanta, 246, 402
Vedi, 392
Vega, 192
Veshya Dharma, 371
Vibhaga, 385
Vibhram, 383
Vibhraman, 383
Vibhrami, 383
Vibhramita, 382, 383
Vibhvan, 247
Vic, 324, 364
Vichara, 389
Videha mukti, 191

Hindi Index

Vidharaya, 202
Vidhata, 325, 329
Vidhata chakra, 325
Vidheya, 319, 397
Vidhi Vadartha, 292
Vidya, 334, 392
Vidya Vimukti, 392
Vidyut, 313
Vigata, 256
Vighna, 314, 355
Vighnanasha, 313, 404
Vighnaraja, 188
Vignesh, 187, 249, 251, 335
Vigratha, 385
Vigudhartha, 355
Vijja, 391
Vijja Vimukti, 391
Vijnana, 391
Vikampa, 386
Vikarshana, 370
Vikas, 312
Vikasaj, 322
Vikata, 295, 404
Vikeerna, 332
Vikkhambhana, 393
Vikkhambhana Vimukti, 393
Viklrp, 318
Vilokanam, 254
Vimaleshvara, 340
Vimarsha shakti, 325
Vimshati, 360
Vimukti, 388, 389
Vimutti, 389
Vinayaka, 188, 337, 404
Vinirma, 337
Viniyamavastha, 358
Vinnana, 309
Vipaka, 257
Viratpurusha, 214
Virodha, 362
Virodhakri, 362
Virshakapi, 214
Visamgati, 361
Visangata, 377
Visangati, 383
Vish, 215
Vishakha, 398
Vishalakshi, 260
Vishanna, 401
Vishnu, 220, 249
Vishnunabhi, 337
Vishtara, 359
Vishthi, 318
Vishva, 367
Vishvakarma, 214, 317
Vishvasamgati, 367
Vishvasi, 337
Visvarupa, 316
Vitata, 337
Vitti, 318
Viyadganga, 328
Voyogi, 385
Vriksha, 196

Vrikshakapi, 196
Vrisha, 213, 337
Vrishakapi, 186, 212, 213
Vrishna, 399
Vrit, 278, 350
Vrittata, 378, 399
Vritti, 395
Vrt, 354
Vrtile, 343
Vrtra, 317
Vrttagona, 353
Vruschik, 182
Vyad, 400
Vyakti, 388
Vyana, 335
Vyasa, 255
Vyasadala, 255
Vyasana, 362
Vyasartha, 255
Vyasasutra samgati, 362
Vyashti, 371
Vyavahara, 374
Vyavastha, 327
Vyuha, 339
Vyutkrama, 329

W
Waheguru, 214, 217, 272, 380
Waheguru Marga, 272

X
Xihe, 292

Y
Yaanshala, 397
Yadrivakta, 208
Yahoodi, 339
Yahu, 292
Yajna Yoga, 254
Yaju, 208
Yama, 258, 375, 394, 396, 397
Yama shakti, 397
Yamadeva, 375, 376
Yang-Wang-Yeh, 375
Yanika, 249, 256
Yanika Marga, 249
Yanluo Wang, 375
Yantra, 322
Yantradridha, 322
Yathabhuta, 196
Yathayogya, 318
Yoga, 169, 204, 249, 387
Yogasamgati, 364
Yogatma, 346
Yogeshvari, 334, 335
Yojana, 364
Yojya, 373
Yojya Marga, 373
Yoni, 335, 358, 369
Yonigona, 358
Yuktartha, 387
Yulü, 375

Z
Zao Shen, 205
Zindagi, 318

www.ingramcontent.com/pod-product-compliance
Lightning Source LLC
Chambersburg PA
CBHW071443220526
45472CB00003B/643